AF358843

Seismic Impact on Building Structures: Assessment, Design, and Strengthening

Seismic Impact on Building Structures: Assessment, Design, and Strengthening

Editors

Rajesh Rupakhety
Dipendra Gautam

Basel • Beijing • Wuhan • Barcelona • Belgrade • Novi Sad • Cluj • Manchester

Editors
Rajesh Rupakhety
University of Iceland
Selfoss
Iceland

Dipendra Gautam
University of Iceland
Selfoss
Iceland

Editorial Office
MDPI
St. Alban-Anlage 66
4052 Basel, Switzerland

This is a reprint of articles from the Special Issue published online in the open access journal *Buildings* (ISSN 2075-5309) (available at: https://www.mdpi.com/journal/buildings/special_issues/seismic_impact_building_structures).

For citation purposes, cite each article independently as indicated on the article page online and as indicated below:

Lastname, A.A.; Lastname, B.B. Article Title. *Journal Name* **Year**, *Volume Number*, Page Range.

ISBN 978-3-7258-1361-2 (Hbk)
ISBN 978-3-7258-1362-9 (PDF)
doi.org/10.3390/books978-3-7258-1362-9

Cover image courtesy of Dipendra Gautam

Contents

 *buildings*

 MDPI

Editorial

Seismic Impact on Building Structures: Assessment, Design, and Strengthening

Rajesh Rupakhety * and Dipendra Gautam

Earthquake Engineering Research Center, Faculty of Civil and Environmental Engineering, University of Iceland, Austurvegur 2a, 800 Selfoss, Iceland; dig17@hi.is
* Correspondence: rajesh@hi.is

1. Genesis and Motivation

The changing landscape of building technology, seismic engineering understanding, data, innovative rehabilitation strategies, and computing efficiency have morphed the field of structural earthquake engineering and closely allied fields into one of the most dynamic and vibrant fields of civil engineering, both in research and practice. From seismic code compliance in design to high-fidelity predictions through human–artificial intelligence hybrid applications, rapid changes have occurred over the past decade. Recent earthquakes, on the other hand, have provided a momentous understanding of the seismic behavior of various structural forms, which should serve as the basis for understanding and innovation. This Special Issue aims at gather cutting edge innovations and state-of-the-art knowledge to encourage a more robust understanding for beaconing the future on the topic of seismic impact on building structures. The scope of the Special Issue is seismic vulnerability to loss assessment, damage quantification to rehabilitation, and forensics to intelligent predictions and designs. This gives rise to a one-stop solution for researchers, practitioners, and general practitioners in terms of understanding and advancements.

2. Seismic Impact on Buildings: A Bird's Eye View

This Special Issue encompasses 14 cutting-edge research papers from different parts of the world. Xu et al. [1] propose an ontology-based seismic risk assessment approach for building structures. Through the hybridization of ontology and semantic web rule language, the proposed method is justified to be capable of predicting consequences. The implications are generalized for decision making and risk assessment applications. Bessason et al. [2] create vulnerability models using loss data recorded after the notable earthquakes of 2000 and 2008 in Iceland. They conclude that the near-field mean loss will be constrained to 5% of the replacement value for timber and reinforced concrete buildings. Considering notable earthquake scenarios, Ademovic et al. [3] shed light on historical constructions during the Zagreb earthquake. Using visual inspection, in situ testing, and numerical modeling, they report grave improvements in the ultimate load capacity after FRCM interventions as opposed to non-strengthened structures. Biglari et al. [4] adopt two common index-based formulations to assess the damage index spectra of a case study school building. Their findings highlight the importance and acceptance of damage index spectra in swift vulnerability assessment campaigns of masonry buildings in active seismic regions. Papazafeiropoulos and Plevris [5] used strong motion and structural damage paradigms to underline the devastating impact of the 2023 M_w 7.8 Kahramanmaraş—Gaziantep earthquake in Turkiye. They infer that the affected structures had inferior capacity compared to the seismic excitation they observed during the earthquake. They highlight the occurrence of strong vertical shaking in damage aggravation. Based on the observations, construction improvements and seismic code updating are recommended as the way forward.

Citation: Rupakhety, R.; Gautam, D. Seismic Impact on Building Structures: Assessment, Design, and Strengthening. *Buildings* **2024**, *14*, 1545. https://doi.org/10.3390/buildings14061545

Received: 23 May 2024
Accepted: 24 May 2024
Published: 27 May 2024

Biglari et al. [6] assemble empirical and numerical modeling outcomes to compare fragility functions for reinforced concrete (RC) and steel buildings. Based on the hybrid approach, they conclude that the Iranian RC and steel building stocks are seismically more vulnerable than the prevalent code assumes. On the other hand, Isik et al. [7] address a very pertinent issue of the short column effect in RC buildings through numerical modeling. Using both force- and deformation-based approaches, they conclude that the short column formation notably enhances plastic rotation demand and shear force in the columns. In the meantime, the first story drift was found to be decreased because of the band type window and slope effect. Tresnjo et al. [8] conducted an experimental dynamic campaign and integrated that with a seismic assessment for a historical minaret. The comparison between experimental and eigen analyses shows a good agreement in terms of dynamic characteristics. Beside this, the applied element method was deployed in nonlinear time history analysis (NLTHA) for seismic assessment. Their results note that the transition regions observed stress concentration. Beam–column joints were noted as one of the most critical components of RC buildings. Owing to the severity and widespread concern, Ahmad et al. [9] proposed a computationally efficient approach to model the nonlinear behavior of beam–column joints in a substandard RC construction. They further reinforce the numerical modeling proposal with a full-scale quasistatic cyclic test. Congregating numerical and experimental results, they concluded that the collapse likelihood of a substandard RC under design earthquake increases from 4.20% to 29.20%. Phan et al. [10] integrated regression and classification schemes to predict shear strength and failure modes in rectangular RC columns. Using 541 experimental datasets, they concluded that the K-nearest neighbor (KNN) outperformed any other classifier or regressor. They highlighted the importance of machine learning applicability in addressing structural engineering, and also developed a graphical user interface (GUI) for shear strength prediction and failure characterization.

Despite addressing damage, vulnerability, and risk, strengthening and seismic improvements are also incorporated into the themes of this Special Issue. Shreekeshava et al. [11] used half-scaled samples to assess the efficacy of strengthened and non-strengthened masonry infilled RC frames under cyclic lateral in-plane loading. They concluded that the maximum drift can be lowered by 24% under in-plane actions if geo-fabric is used to strengthen the infill–frame interfaces. Sharifi Ghalehnoei et al. [12] reported the outcomes of numerical modeling for RC beams strengthened by carbon-fiber-reinforced polymer grid-strengthened engineering cementitious composites. They compared analytical and numerical results and concluded that the difference between these approaches is not significant. Moreover, their findings highlight that the shear capacity of strengthened beams is around 35–50% more than that of common RC beams. Wu et al. [13] assessed the performance of active and passive control techniques and concluded that the multiple active tuned mass damper with an inerter system surpasses its peers in controlling structural response and stroke of devices. They also highlighted the capacity of the multiple active tuned mass damper system with an inerter system in checking structural and environmental fluctuations, which leads to superior robustness and stability. Finally, Avinash et al. [14] provide a comprehensive review of sliding isolation systems considering several dimensions of base isolation systems. They incorporated hybrid and active isolation systems, shedding light on passive sliding isolation systems.

3. Précis and a Way Forward

Recalling the famous quote by Prof. Nick Ambraseys, "earthquakes do not kill people-building do", the evolution of structural earthquake engineering focusing on the seismic impact on buildings has remained spurious throughout, particularly due to the contributions and lessons from a vast swath of resources pivoted to damage, loss, fatalities, innovations, etc. The progress made to achieve a particular milestone is usually stirred by occasional devastating earthquakes, so the field has remained greatly dynamic. The congregation of empirical, numerical, experimental, analytical, heuristic, and hybrid approaches in the era of high-performance computing can be streamlined to check the devastation during one of the most destructive phenomena of nature. The collective conclusion of the published papers highlights the need for cross-border and cross-field collaboration in knowledge exchange and rational solutions. Although much has already been done, Barry Schwartz's famous insight, "more is less", holds true in structural earthquake engineering under the unprecedentedly large dimensions of the propagating issues. The structural earthquake engineering field, as addressed in this Special Issue, awaits more rational, innovative, economical, versatile, and globally localized solutions to protect humankind from earthquake impacts.

Conflicts of Interest: The authors declare no conflict of interest.

References

1. Xu, M.; Zhang, P.; Cui, C.; Zhao, J. An Ontology-Based Holistic and Probabilistic Framework for Seismic Risk Assessment of Buildings. *Buildings* **2022**, *12*, 1391. [CrossRef]
2. Bessason, B.; Rupakhety, R.; Bjarnason, J.Ö. Scenario-Based Seismic Risk Assessment for the Reykjavik Capital Area. *Buildings* **2023**, *13*, 2919. [CrossRef]
3. Ademović, N.; Toholj, M.; Radonić, D.; Casarin, F.; Komesar, S.; Ugarković, K. Post-Earthquake Assessment and Strengthening of a Cultural-Heritage Residential Masonry Building after the 2020 Zagreb Earthquake. *Buildings* **2022**, *12*, 2024. [CrossRef]
4. Biglari, M.; Hadzima-Nyarko, M.; Formisano, A. Seismic Damage Index Spectra Considering Site Acceleration Records: The Case Study of a Historical School in Kermanshah. *Buildings* **2022**, *12*, 1736. [CrossRef]
5. Papazafeiropoulos, G.; Plevris, V. Kahramanmaraş—Gaziantep, Türkiye Mw 7.8 Earthquake on 6 February 2023: Strong Ground Motion and Building Response Estimations. *Buildings* **2023**, *13*, 1194. [CrossRef]
6. Biglari, M.; Hosseini Hashemi, B.; Formisano, A. The Comparison of Code-Based and Empirical Seismic Fragility Curves of Steel and RC Buildings. *Buildings* **2023**, *13*, 1361. [CrossRef]
7. Işık, E.; Ulutaş, H.; Harirchian, E.; Avcil, F.; Aksoylu, C.; Arslan, M.H. Performance-Based Assessment of RC Building with Short Columns Due to the Different Design Principles. *Buildings* **2023**, *13*, 750. [CrossRef]
8. Trešnjo, F.; Humo, M.; Casarin, F.; Ademović, N. Experimental Investigations and Seismic Assessment of a Historical Stone Minaret in Mostar. *Buildings* **2023**, *13*, 536. [CrossRef]
9. Ahmad, N.; Rizwan, M.; Ilyas, B.; Hussain, S.; Khan, M.U.; Shakeel, H.; Ahmad, M.E. Nonlinear Modeling of RC Substandard Beam–Column Joints for Building Response Analysis in Support of Seismic Risk Assessment and Loss Estimation. *Buildings* **2022**, *12*, 1758. [CrossRef]
10. Phan, V.T.; Tran, V.L.; Nguyen, V.Q.; Nguyen, D.D. Machine Learning Models for Predicting Shear Strength and Identifying Failure Modes of Rectangular RC Columns. *Buildings* **2022**, *12*, 1493. [CrossRef]
11. Sreekeshava, K.S.; Rodrigues, H.; Arunkumar, A.S. Response of Masonry-Infilled Reinforced Concrete Frames Strengthened at Interfaces with Geo-Fabric under In-Plane Loads. *Buildings* **2023**, *13*, 1495. [CrossRef]
12. Ghalehnoei, M.S.; Javanmardi, A.; Izadifar, M.; Ukrainczyk, N.; Koenders, E. Finite Element Analysis of Shear Reinforcing of Reinforced Concrete Beams with Carbon Fiber Reinforced Polymer Grid-Strengthened Engineering Cementitious Composite. *Buildings* **2023**, *13*, 1034. [CrossRef]
13. Wu, X.; Liu, X.; Chen, J.; Liu, K.; Pang, C. Parameter Optimization and Application for the Inerter-Based Tuned Type Dynamic Vibration Absorbers. *Buildings* **2022**, *12*, 703. [CrossRef]
14. Avinash, A.R.; Krishnamoorthy, A.; Kamath, K.; Chaithra, M. Sliding Isolation Systems: Historical Review, Modeling Techniques, and the Contemporary Trends. *Buildings* **2022**, *12*, 1997. [CrossRef]

Article

Scenario-Based Seismic Risk Assessment for the Reykjavik Capital Area

Bjarni Bessason [1,*], Rajesh Rupakhety [1] and Jón Örvar Bjarnason [2]

[1] Faculty of Civil & Environmental Engineering, University of Iceland, 107 Reykjavik, Iceland; rajesh@hi.is
[2] Natural Catastrophe Insurance of Iceland, 201 Kópavogur, Iceland; jon@nti.is
* Correspondence: bb@hi.is

Abstract: About two-thirds of the population in Iceland lives in the Reykjavik capital area (RCA), which is close to active volcanoes and seismic zones. In the period 1900–2019, a total of 53 earthquakes of $M_w \geq 5.0$ struck in these zones. The two largest events in the Reykjanes Peninsula, $M_w6.36$ and $M_w6.12$, occurred in 1929 and 1968, respectively. Both events were less than 20 km from the outskirts of the RCA. Late in the year 2020, the seismicity on the peninsula greatly increased due to magma intrusion and volcanic activity, which has so far resulted in three eruptions, in 2021, 2022, and 2023, and six earthquakes of $M_w \geq 5.0$. Based on historical and geological data, the ongoing activity is probably the initial phase of an active period ahead that could continue for many decades, and has the potential to trigger larger earthquakes like those in 1929 and 1968. Further east, in the South Iceland Seismic Zone, two earthquakes of $M_w6.52$ and 6.44 struck in June 2000, and in May 2008, a $M_w6.31$ earthquake occurred. In both cases, around 5000 buildings were affected. Insurance loss data from these events have been used to develop empirical vulnerability models for low-rise buildings. In this study, the loss data are used to calibrate seismic vulnerability models in terms of the source-site distance. For a given magnitude scenario, this provides a simpler representation of seismic vulnerability and is useful for emergency planning and disaster management. These models are also used to compute different types of scenario risk maps for the RCA for a repeat of the 1929 earthquake.

Keywords: disaster preparedness; emergency planning; seismic vulnerability; seismic fragility; risk maps

Citation: Bessason, B.; Rupakhety, R.; Bjarnason, J.Ö. Scenario-Based Seismic Risk Assessment for the Reykjavik Capital Area. *Buildings* **2023**, *13*, 2919. https://doi.org/10.3390/buildings13122919

Academic Editors: Xavier Romão and Annalisa Greco

Received: 29 September 2023
Revised: 14 November 2023
Accepted: 20 November 2023
Published: 23 November 2023

1. Introduction

Seismic resilience is crucial for communities and infrastructure exposed to seismic hazards. It is defined as the ability of a system to withstand, adapt to, and recover from destructive earthquakes. Strengthening infrastructure, improving building codes, enhancing emergency preparedness, and increasing educational and knowledge level can collectively reduce the potential consequences of seismic events. To formulate strategies and policies aimed at increasing seismic resilience, it is vital to have knowledge of the seismic risk for the community in question [1–3]. This study focuses on a vulnerability assessment and a scenario-based seismic risk assessment for the Reykjavik capital area (RCA) in Iceland. A known past $M_w6.36$ earthquake is used as a scenario event, representing one of the most damaging scenarios for the RCA among all known earthquakes in its vicinity.

In addition to seismic hazard and exposure models, the seismic vulnerability of structures and infrastructures is a key factor in seismic risk assessment. Vulnerability models can be based on (1) judgement-based methods [4,5]; (2) analytical simulations and experiments [6,7]; (3) empirical methods using loss data from post-earthquake surveys [8–10]; and (4) hybrid methods with combinations of two or more of these methods. All these methods have their advantages and drawbacks. When local loss datasets exist, they always offer valuable information to learn from, and to use to predict losses for similar future events.

In this century, three destructive, shallow (5–8 km), strike-slip earthquakes have occurred in the South Iceland Seismic Zone (SISZ). On 17 and 21 of June 2000, M_w6.52 and M_w6.44 earthquakes hit in the eastern part of the zone [11,12]. Both were on north–south-oriented faults approximately 16 km apart (Figure 1). The fault rupture lengths shown in the figure are based on the post-earthquake micro seismicity. These earthquakes affected nearly 5000 low-rise residential buildings [13]. Then, the western part of the zone was hit on 29 May 2008 by a M_w6.31 earthquake [14,15]. This event also affected nearly 5000 buildings.

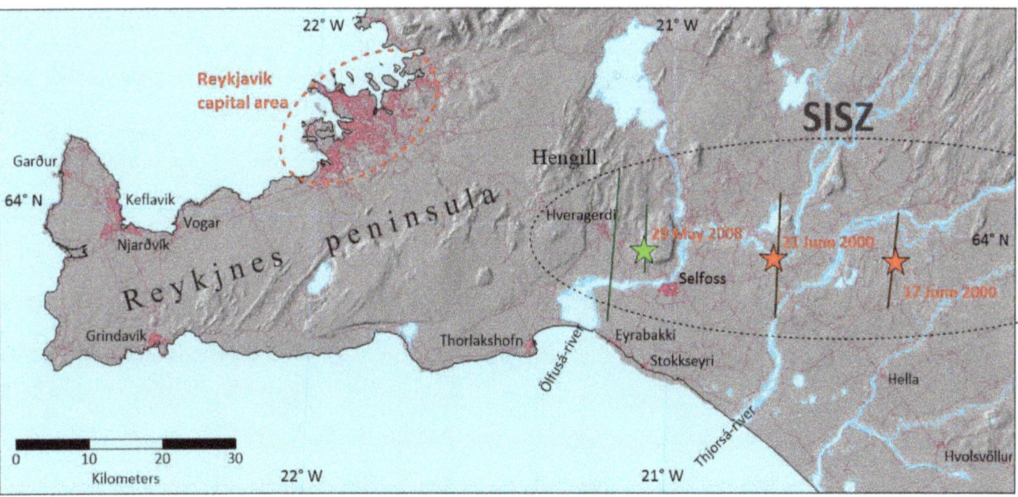

Figure 1. South Iceland Seismic Zone (SISZ). The map shows the epicentre and fault rupture of the two South Iceland earthquakes of 17 June 2000 ($M_w = 6.52$) and 21 June 2000 ($M_w = 6.44$), and the Ölfus earthquake of 29 May 2008 ($M_w = 6.31$).

The monetary losses of all damaged buildings caused by the 2000 and the 2008 earthquakes were collected and compiled for insurance claim settlements. Two loss datasets were created and named the 2000 and 2008 datasets. These datasets have, to some extent, been analysed, and advanced empirical vulnerability models have been developed independently for each dataset [13,16,17]. In all these studies, the peak ground acceleration (PGA) has been used as an intensity measure (IM) using a local ground motion prediction model (GMPM) reported in [18]. Bessason et al. [17] report differences for the same building typologies in the vulnerability models derived from the two datasets and suggest that the model calibrated from the 2000 dataset is used for events in the M_w range 6.4–6.6, while the one based on the 2008 dataset is used for quakes in the M_w range 6.2–6.4. The scenario event of M_w6.36 is just below the magnitude range covered by the 2000 dataset, but within the range covered by the 2008 dataset.

The applied local GMPM used in recent local vulnerability assessments [13,16,17] predicts the PGA using three variables, as follows:

$$PGA = f\left(M_w, R_{JB}, S\right) \tag{1}$$

where M_w is the moment magnitude. R_{JB} is the Joyner–Boore distance [19], which is the distance from the site to the surface projection of the rupture plane. For a vertical fault plane like in the June 2000 and 2008 strike-slip earthquakes, this is the distance to the surface trace of the causative fault (see Figure 1). Finally, S is a site factor which is taken as 0 for rock sites and 1 for stiff soil conditions. Based on geological maps from the Icelandic Institute of Natural History [20], all the main settlements and almost all the built areas affected by the May 2008 earthquake are rock or lava sites (see also [21]). For the affected areas in June 2000, most of the building sites are also rock or lava sites, although some areas along the coast, mainly east of the Thjórsá River and south of the town of Hella (see Figure 1), contain sediments and can be classified as stiff soil sites [20]. For simplicity, rock site conditions are considered for all the building sites in this study. This means that the PGA estimated at each site is only a function of M_w and R_{JB} (see Equation (1)). Therefore, for a fixed M_w and S = 0 (rock site), the PGA and R_{JB} are linked and fully correlated using the GMPM.

While vulnerability models based on ground shaking intensity such as the PGA are more commonly used in the literature because of their application in probabilistic risk studies, scenario-based studies can benefit from alternative models. For example, for a given magnitude scenario, vulnerability models based on the source-to-site distance can be useful. They are easier to comprehend as they relate directly to the expected damage zone. In areas which are characterised by earthquakes of a similar characteristic size, vulnerability models based on the source-site distance are suitable for scenario-based risk assessment. Although this parameter is not, in the true sense, a ground motion intensity measure, it can be thought of as a proxy for it for a given magnitude.

The first main objective and novelty of the present work is to independently calibrate vulnerability models for the 2000 dataset ($\sim M_w6.5$) and the 2008 dataset ($\sim M_w6.3$) as a function of R_{JB}. The benefits of using R_{JB} instead of the PGA is twofold. Firstly, it is easier in emergency planning for stakeholders, civil defense officials, decision-makers, and politicians with a limited technical and/or earthquake engineering background to understand and use vulnerability and fragility curves as functions of distance rather than functions of the PGA or some other complicated intensity measure. Secondly, GMPMs that estimate ground motion intensity measures tend to evolve over time with new data and are associated with significant uncertainties. In this aspect, the use of a median intensity measure predicated using a GMPM in calibrating vulnerability models fails to capture this inherent uncertainty. On the other hand, for a well-understood tectonic environment, with reliable seismic scenarios, the source-site distance is relatively well defined. It is, nevertheless, associated with certain uncertainties regarding the length of rupture and location of the epicentre on the fault, but the same uncertainties are also associated with the PGA. Such uncertainties in scenario modelling can be considered by defining different potential scenarios. This approach also makes the vulnerability model independent of a GMPM. Despite the suggested simple presentation of the explanatory variable, calibration of the model is nevertheless based on advanced statistical modelling, such as that reported by [13,17].

The second novel objective of this study is to prepare different types of scenario risk maps for the RCA. The considered scenario is a repeat of the 1929 earthquake in the Reykjanes Peninsula. This quake was probably the most damaging scenario for buildings in the RCA of all known earthquakes in its vicinity. Given the heightened and ongoing volcanic and earthquake activity in the Reykjanes Peninsula, as described in the next section, these risk maps can be employed to enhance the seismic resilience of the Reykjavik capital area (RCA).

The upcoming section discusses the seismic hazards for the RCA and outlines the exposure data for residential buildings. This is followed by a section describing the 2000 and the 2008 loss datasets, along with the theoretical background for the empirical vulnerability model. The concluding section encompasses the results and discussion, divided into two parts. The first part presents calibrated vulnerability and fragility curves as functions of the Joyner–Boore distance. The second part focuses on presenting scenario-based risk maps for the RCA.

2. Seismic Hazard and Exposure Data

2.1. Seismic Hazards in the Reykjanes Peninsula

About two-thirds of the population of Iceland lives in the Reykjavik capital area (RCA). The capital area is in the north-east part of the Reykjanes Peninsula, which is an active seismic and volcanic zone (Figure 2) [22]. In a new harmonised earthquake catalogue for Iceland covering the instrumental period from 1900–2019, a total of 53 earthquakes of magnitude $M_w \geq 5.0$ are reported in the Reykjanes Peninsula and Ölfus, the western part of the South Iceland Seismic Zone (SISZ) (Figure 2) [12]. Of these, three earthquakes had a magnitude greater than six: M_w6.36 in 1929, 6.12 in 1968, and 6.31 in 2008. The first two had an epicentral distance of less than 20 km from the outskirts of the RCA. In addition, three historical earthquakes in the period 1700–1900, all with an estimated M_w6.1, struck in Ölfus in 1706, 1766, and 1896, respectively (Figure 2) [12].

The main characteristic of the seismicity on both the Reykjanes Peninsula as well as in the SISZ are shallow (5–10 km), strike-slip earthquakes with N–S faulting and vertical fault planes (dip angle ~90°) [23]. The estimated upper magnitude bound for earthquakes in the Reykjanes Peninsula is M_w6.5. This is supported by historical and instrumentally recorded earthquakes as well as geological evidence. Further east, in the SISZ, bigger events are known to occur [12]. In that area, the crust is thicker and capable of storing larger seismic strains. The fault lengths shown in Figure 2 are based on models by Wells and Coppersmith [24], and the epicentre is placed in the middle of the fault.

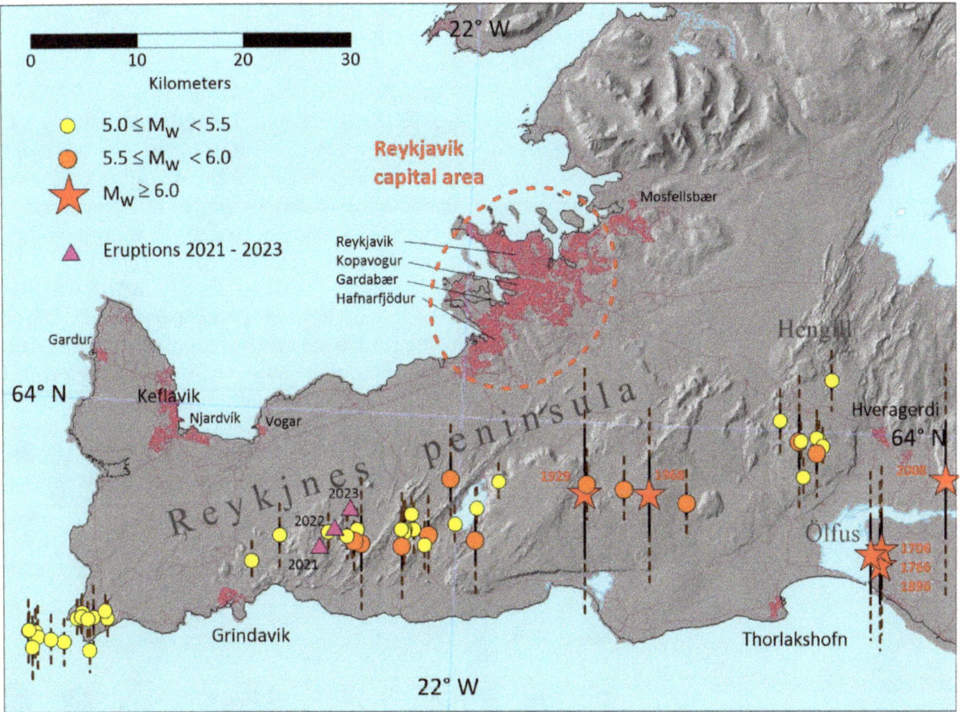

Figure 2. Reykjanes peninsula. Red thin lines show the road system and thicker red lines can be used to identify settlements and towns. Instrumental recorded earthquakes in the period from 1900–2019 with $M_w \geq 5.0$ and three historical earthquakes from 1700 to 1900 with $M_w \geq 6.0$ in the Ölfus region are shown [12]. The largest events (red stars) are marked with the year they struck. The black lines indicate computed surface fault rupture and the dotted brown lines the subsurface rupture based on the rupture model by [24]. Eruptions in the period 2021 to 2023 are marked with violet triangles and year.

Late in the year 2020, the seismic activity greatly increased in the western part of the Reykjanes Peninsula due to magma intrusion and volcanic activity, which has so far resulted in three eruptions. The first eruption started on 19 March 2021 and lasted for a few months. The second eruption, which began on 3 August 2022, was relatively brief, lasting only two weeks. The third eruption initiated on 10 July 2023 and endured for few weeks (Figure 2). Intense seismicity preceded all these eruptions. In the period from October 2020 to October 2023, six earthquakes of $M_w \geq 5.0$ occurred in the area close to the eruptions. The largest event was $M_w = 5.6$ on 24 February 2021 [25]. Historical and geological information indicate that the incipient volcanic and associated seismic activity in the region might last for many decades to come [26]. The seismicity in the period prior to the eruptions caused discomfort among the residents in the small towns closest to eruption, mainly in Grindavik, which was the closest, but also in Vogar and Njardvik. The ground shaking was also clearly felt in the RCA. Nevertheless, these events only caused minor damage related to cosmetic and non-structural losses (objects falling on flooring, etc.). In total, 25 insurance claim payouts were made by Natural Catastrophe Insurance of Iceland [27], which is a public institution that oversees the insurance of buildings and other properties in Iceland against natural disasters like earthquakes. In Iceland, all real estate is mandated by law to be insured against natural disasters. At present, the deductible is EUR ~2600 for each property for a given damaging event. This means that although the losses in the above 25 cases were minor, they were nevertheless higher than the deductible. The ongoing activity in the area has the potential to trigger larger tectonic earthquakes in a magnitude range like those in 1929 and 1968 (see Figure 2).

2.2. Exposure Data

The building stock in the Reykjavik capital area (Reykjavík, Kópavogur, Gardabær, Hafnafjördur and Seltjarnarnes) has quite a different composition than the building mass in the SISZ contained in the two datasets (see Section 3.2). In Figure 3, the dwellings in RCA are classified according to the building material, number of storeys, and code level. Around 95% of the dwellings are in buildings made of concrete or concrete+other material, whilst only 1% of the dwellings are in masonry buildings. Around 26% of dwellings are in 1–2-storey buildings and 56% are in 3–5-storey buildings. Finally, 62% of the dwellings are buildings designed in the moderate-code and high-code period (see Table 1). The lateral load resisting system for most of the building stock is structural walls like in the SISZ. The RCA has expanded in recent decades. Therefore, most of the moderate-code and high-code buildings are in the suburbs of the municipalities closest to the fault rupture of the scenario event of 1929 (Figure 2). Apartment buildings are common in the new districts and all the storey classes in Figure 3b can on the other hand be found almost everywhere within the RCA.

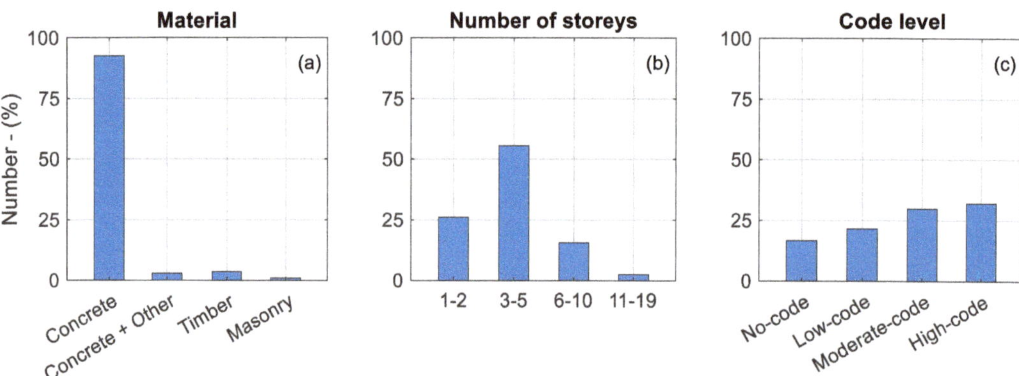

Figure 3. Classification of dwellings in the Reykjavik capital area, according to; (**a**) building material, (**b**) number of storeys, and (**c**) code level. The data behind "material" and "number of storeys" are from the official property database in Iceland accessed in August 2023 [28].

Table 1. Status of seismic design codes in Iceland for different construction periods.

Status	Description	Comment	Period
CDN	No-code	No seismic design code	<1958
CDL	Low-code	First generation of seismic codes	1958–1975
CDM	Moderate-code	Second generation of seismic codes	1976–2001
CDH	High-code	Latest generation of seismic codes	≥2002

3. Loss Data and Empirical Vulnerability Model

3.1. Loss Data

It is mandated by law in Iceland to have insurance against destructive natural events. This includes events like earthquakes, landslides, avalanches, floods, etc. The insurance is a part of the fire insurance terms. If a property has fire insurance, it also has natural disaster insurance. The insurance is managed by Natural Catastrophe Insurance of Iceland (NCI) [27]. After the June 2000 earthquakes, the repair cost of damaged properties was estimated by trained assessors to fulfil insurance claims. The affected area is an agricultural region with farms, settlements, small towns, and all the infrastructure assets of a modern society. The two events in June 2000 occurred within only four days, and therefore the observed damage may include accumulate damage from both events. The distance between the fault ruptures of the 17 June and 21 June events is approximately 16 km (Figure 1). The attenuation of wave propagation is high in Iceland due to a geologically young and cracked volcanic bedrock and layered lava fields. In [13], it was argued that almost all the damaged buildings in June 2000 were only affected by the earthquake closest to the building in question. Only very few buildings were in the middle area at a similar distance from each fault, and thereby affected "equally" by both events (Figure 1). In the 2000 dataset, each loss unit refers to a "building" where a building is defined from a street address (see [16] for more details).

After the May 2008 earthquake, similar methods as for the 2000 earthquakes were used to assess repair cost to establish the 2008 loss dataset. In the 2008 dataset, the loss unit is a "dwelling". In 2000 and 2008, the insurance deductible was low (EUR 650 per dwelling). This low value encouraged all owners to report damage to their properties to receive insurance payback. This is the main reason for the assumption that both the 2000 and the 2008 datasets are complete and include all affected buildings in the region (PGA > 0.05 g) which is rare to find in other studies [29]. The deductible has now been raised to EUR ~2600 per dwelling.

The two datasets cover the loss estimates for both structural and non-structural elements. The term "non-structural elements" includes cladding, flooring, fixtures, and technical systems (electrical installations, plumbing, etc.). Damage to the building content, that is, loose household items like furniture, TVs, computers, etc., is not included. In this study, the presented vulnerability models are assessed for the combined losses of the structural and non-structural damage given in the datasets. A damage factor, DF, is computed for each building unit, which is defined as:

$$DF = \frac{\text{Estimated repair cost}}{\text{Fire Insureance Value}} = \frac{ERC}{FIV} \tag{2}$$

where ERC is the combined estimated repair cost. The fire insurance value (FIV) is obtained from an official property database. The FIV is estimated as the depreciated replacement value (DRV) plus the cost of dismantling and transporting the debris. The dismantling and transport cost is taken as 12% of the DRV. The DRV is based on age, construction material, and general condition. The FIV is used by NCI to define the maximum payback that a building owner can receive for a dwelling/building with "total damage" or total "loss" (DF = 1). In practice, damage equivalent to "total loss" was assigned to most of the buildings that suffered an estimated repair cost of more than 70% of their FIV value in the 2000 dataset. In the 2008 dataset, total loss was estimated on a case-by-case basis for dwellings with an estimated loss in the range 50–70% of their FIV.

3.2. Building Taxonomy of the Loss Data

The Property Registry in Iceland [28] maintains a detailed property database for all building units in Iceland. It contains information on the construction material, year of construction, number of storeys, usage (residential, service, industry, etc.), floor area, street address, GPS coordinates, fire insurance value, etc. It does not include information on the lateral load resting system or main structural system, nor about the site conditions (bedrock, soil, etc.). In this study, the GEM taxonomy was used to classify all the building units in the datasets [30]. Only residential buildings were considered. The 2000 dataset is a building-by-building set, while in the 2008 dataset is a dwelling-by-dwelling set. Most of the buildings in the datasets are one or two storeys, and in such cases, building units and dwelling units are the same.

In the 2000 dataset, 54% of the residential buildings were built of concrete (C), almost all cast-in-place (CIP); 37% were timber buildings (W) made of light wood (WLI); 9.3% unreinforced masonry buildings (MUR + CBH + MOC) built before 1976 when seismic codes were first implemented in Iceland; and the rest, 0.3%, used other construction material. Only 23 buildings, mostly made of timber, were built before 1900 and the oldest building was built in 1875. Regarding height, 68% of the buildings were one-storey, 23% two-storey, 7.9% three-storey, and 0.3% four-storey. No buildings were higher. In the 2008 dataset, where all losses refer to dwellings, 45% of them were in buildings built of concrete, 48% in timber buildings, and 7.6% in masonry buildings. Furthermore, 74% of the dwellings were in one-storey buildings, 19% in two-storey, 5.9% in three-storey, and 0.5% in four-storey buildings. In the 2008 dataset, 19 dwellings were in buildings built before 1900, and the oldest one was constructed in 1875.

Since only a low fraction of the affected buildings are three-storey or higher in both datasets, they were excluded in the modelling and the evaluated models therefore only consider one- to two-storey buildings. These buildings were combined and the class marked with HBET:1,2 according to the [30] taxonomy.

The lateral load resisting system for almost all the buildings in the two datasets are structural walls. Based on the GEM taxonomy, LWAL is therefore used to identify the structural system. In both datasets, the masonry buildings are built of unreinforced, hollow concrete blocks using light-weight pumice as the main aggregate (high-porosity volcanic rock). The foundations and the bottom slabs in both timber and masonry buildings are usually made of reinforced concrete. Concrete frame buildings with stone or brick infills, which are common in South Europe, do not exist in Iceland. For more details of the building characteristic, see [17].

Crowley et al. [31] classified the status of seismic design codes in different European countries, including Iceland, into four categories based on the construction period (Table 1). In [17], it was concluded that no-code and the low-code (CDN + CDL) buildings could be combined into one class, and the moderate-code and the high-code (CDM + CDH) buildings into one class. However, in the 2000 dataset, no building belongs to the high-code period, which started in 2002. Only one class, CDN + CDL, is available for masonry buildings, as most of them, 98%, were built before 1976. The 2% remainder were not used in the model calibration. Table 2 sums up how the building units in the two loss datasets are distributed into the five building typologies used in this study.

Table 2. Classification of residential buildings affected by the June 2000 and May 2008 earthquakes.

Short Name	GEM Building Taxonomy	2000 Dataset		2008 Dataset	
		Number	(%)	Number	(%)
C-NL	CR + CIP/LWAL/HBET:2,1/CDN + CDL	1665	35	1112	23
C-M or C-MH [1]	CR + CIP/LWAL/HBET:2,1/CDM + CDH	907	19	1003	21
W-NL	W + WLI/LWAL/HBET:2,1/CDN + CDL	692	15	649	14
W-M or W-MH [1]	W + WLI/LWAL/HBET:2,1/CDM + CDH	1047	22	1623	34
M-NL	MUR + CBH + MOC/LWAL/HBET:2,1/CDN + CDL	443	9.3	359	7.6
	Total sum	4754	100	4746	100

[1] C-M and W-M refer to the 2000 dataset since that dataset includes no high-code buildings, whilst C-MH and W-MH refer to the 2008 dataset.

3.3. Vulnerability Model

In the June 2000 and May 2008 earthquakes, a high proportion of the buildings/dwellings in the affected area (PGA > 0.05 g) had no losses, DF = 0, whilst buildings with total losses, DF = 1, were rare in both datasets. In between these extremes are a number of building units with some loss, that is, a DF in the range 0 to 1. Since the loss dataset includes "zero" and "one" values and then many values in between, it is preferable to use a mixed continuous–discrete regression to model the data, that is, discrete models to cover the DF = 0 and DF = 1 cases, and then continuous regression for loss data in the range (0, 1). The beta probability distribution is flexible and can take a wide range of shapes. It is also bounded in the interval (0, 1) and is therefore suitable to cover the continuous regression in the model. Beta distribution was used in ACT-13 [4,5] to model losses, where judgement-based methods were applied. In the GEM guidelines for empirical vulnerability assessment [29], beta regression is mentioned as potential future method. In this study, where the data include a high fraction of zero-loss incidents but only small number of total losses, a zero-inflated beta regression model is well suited [32]. This is an improved beta regression model to take care of zeros. The discrete modelling of the total loss buildings (DF = 1) is excluded, but instead the DF for these buildings is assigned a value less than 1. This model has been used earlier to model both the 2000 dataset and the 2008 dataset using the PGA as the intensity measure in a five-parameter model for each building typology [13,17].

A two-step regression process is used to construct the vulnerability model. This approach is explained schematically in Figure 4 using the PGA as intensity measure and the loss data from the 2000 dataset for the C-NL class (1665 datapoints, see Table 2). The PGA can be replaced with any other desired intensity measure or its proxy, like for instance, the Joyner–Boore distance, R_{JB}, as in Section 4. In Figure 4a, the loss data (raw data) are shown. Each point indicates the estimated DF based on Equation (2) for the given building unit and the computed PGA at the property site in question. In Figure 4b, the loss data are transformed into binomial variables, i.e., Y = 0 if DF = 0 and Y = 1 if DF > 0. The data points are jittered in the range [−0.05, 0.05] for Y = 0 and in the range [0.95, 1.05] for Y = 1 to better show the density of the data points. A logistical regression model (LM), which is a type of generalised linear model, is then used to analyse the relationship between Y and the PGA, that is, to model the probability of obtaining loss as a function of the PGA. The logistical model for each building typology is given as:

$$\log_e\left(\frac{p}{1-p}\right) = \beta_0 + \beta_1 \cdot PGA \tag{3}$$

where β_0 and β_1 are regression parameters and p is the probability of sustaining losses (DF > 0) (Figure 4b). The parameters of the model are computed using the general linear model package, *glm*, in R [33].

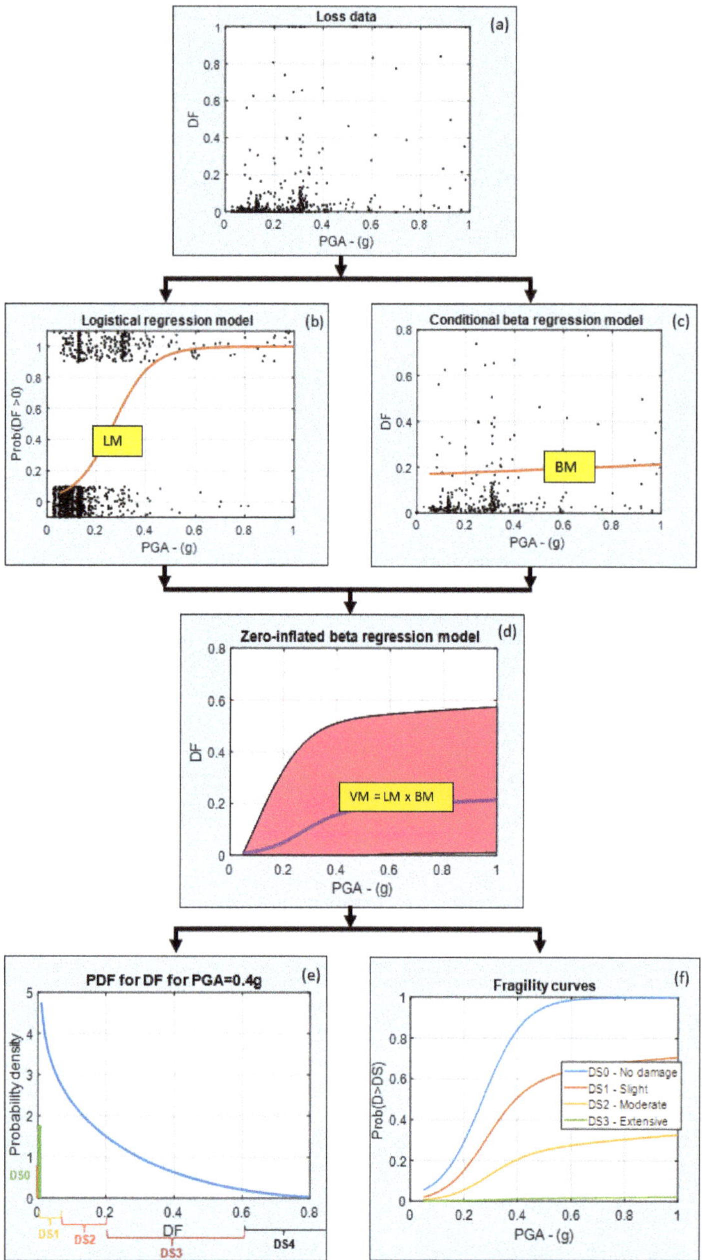

Figure 4. Flowchart that explains the main steps in the zero-inflated beta regression model: (**a**) loss data form the 2000 dataset for C-NL (black dots); (**b**) logistical regression model (LM); (**c**) conditional beta regression model (BM); (**d**) vulnerability model (VM) obtained by combining the LM and BM; (**e**) probability density function for the DF for PGA = 0.4 g; (**f**) fragility functions based on predefined DF bins (see Table 3).

Table 3. Definitions of damage states in this study.

Damage State	Description	DF Bins
DS0	No damage	DF = 0
DS1	Slight damage	$0.00 < \text{DF} \leq 0.05$
DS2	Moderate damage	$0.05 < \text{DF} \leq 0.20$
DS3	Extensive damage	$0.20 < \text{DF} \leq 0.50$
DS4	Complete damage	DF > 0.5

A beta regression [34] is carried out for the data points with a DF > 0 to model the continuous loss distribution (Figure 4c). The beta probability density function (PDF) is given as:

$$f(x; \mu, \varphi) = \frac{\Gamma(\varphi)}{\Gamma(\mu\varphi)\Gamma(1 - \mu)\varphi} x^{\mu\varphi - 1}(1 - x)^{(1-\mu)\varphi - 1} \tag{4}$$

where x is the random variable in the range (0, 1); μ is the mean; φ the precision; and $\Gamma(\cdot)$ is the gamma function. The mean and precision of the beta PDF are related to the linear predictors η_1 and η_2 using link functions, $g_1(\cdot)$ and $g_2(\cdot)$. The linear predictors are a function of the ground shaking intensity measure or its proxy. The link functions must be strictly monotonic and differentiable twice. The logit link function was adopted for μ, and the log link function for φ:

$$g_1(\mu) = \text{logit}(\mu) = \log\left(\frac{\mu}{1 - \mu}\right) = \eta_1 \tag{5}$$

$$g_2(\varphi) = \log(\varphi) = \eta_2 \tag{6}$$

The first predictor is a function of the intensity measure while the second is a constant:

$$\eta_1 = \theta_0 + \theta_1 \cdot \log_e(PGA) \tag{7}$$

$$\eta_2 = \theta_0' \tag{8}$$

where θ_0, θ_1, and θ_0' are regression coefficients. The beta regression is carried out in R [33] using the *betareg* package. Finally, the logistical model and the conditioned beta model are combined to obtain the vulnerability model (Figure 4d). Hence, for a given building typology, five parameters $\{\beta_0, \beta_1, \theta_0, \theta_1, \theta_0'\}$ define the vulnerability model. The expected value and the variance of the DF for the combined model are given as:

$$E[DF] = p \cdot \mu \tag{9}$$

$$Var[DF] = p \cdot \frac{\mu \cdot (1 - \mu)}{\varphi + 1} + (1 - p) \cdot p \cdot (\mu)^2 \tag{10}$$

The total probability theorem can then be used to compute the desired prediction interval:

$$P[X < x] = 1 + p \cdot (F_X(x, \mu, \varphi) - 1) \tag{11}$$

where $F_X(x, \mu, \varphi)$ is the beta cumulative distribution function (CDF) for a given building typology, which is a function of the PGA. Hence, from the vulnerability model, it is possible to directly compute the PDF for the random variable DF for any given PGA. An example of this is shown in Figure 4e for a PGA = 0.4 g. Finally, by defining bins for different damage stages, fragility curves can be constructed using Equation (11). The loss bins and the verbal description of every damage state here are based on [35] (see Table 3). Fragility curves based on the procedure are shown in Figure 4f. For more details, see Bessason et al. [17].

4. Results and Discussion

4.1. Vulnerability Curves and Fragility Curves

By replacing the PGA in Equations (3) and (7) with the Joyner–Boore distance, R_{JB}, a new set of model parameters were estimated for the ZIBR model. To account for outliers in the two datasets, as applied in previous studies [13,17], all data points with a DF > 0.85 were replaced with a max value of $DF_{max} = 0.85$. This was carried out for 15 concrete, 5 timber, and 13 masonry buildings in the 2000 dataset, and for 4 concrete, 7 timber, and 12 masonry dwellings in the 2008 dataset. Furthermore, the same type of data weighting was performed as in [13,17]. The model parameters are given in Table 4 for the building typologies defined in Table 2. The vulnerability curves using R_{JB} as an intensity parameter are shown in Figure 5 for five building typologies for each dataset (Table 2). For the 2000 dataset, corresponding to a $M_w6.5$ event, the mean DF is down to 0.01 (1% loss) at a less than 20 km distance for all the four concrete and timber building classes (Figure 5a,c,e,g) and at 23 km for the masonry buildings. For the 2008 dataset, corresponding to a $M_w6.3$ event, the mean DF is at 0.01 at less than 15 km for all the five building typologies (Figure 5b,d,f,h,j). In Figure 6, the fragility curves computed directly from the vulnerability model using the damage state definitions given in Table 3 and Equation (11) are shown. As an example, at a 20 km fault distance (R_{JB}), the probability of exceeding DS1 (slight damage) is less than 5% for all the 1–2-storey concrete and timber buildings based on the model from the 2000 dataset. For the 2008 dataset, the corresponding distance is 15 km. Similarly, the probability of exceeding DS2 (moderate damage), that is, when the DF exceeds 0.20 (20% loss), is approximately zero at a Joyner–Boore distance of 20 km or more for no-code and low-code concrete buildings based on the 2000 dataset (Figure 6a). For the 2008 dataset and the same building typology, the probability of exceeding DS2 at a 10 km distance or more (Figure 6b) is approximately zero. In all cases, the probability of exceeding DS4 is very low.

Table 4. Estimated model parameters for the ZIBR model using the Joyner–Boore distance as intensity measure. See Table 2 for definition of building typologies.

Dataset	Building Typology	β_0	β_1	θ_0	θ_1	θ_0'
2000	C-NL	1.748	−0.202	−1.798	−0.148	1.592
2000	C-M	0.800	−0.167	−2.505	−0.155	2.648
2000	W-NL	1.147	−0.192	−1.490	−0.215	1.480
2000	W-M	1.098	−0.268	−2.765	−0.029	2.371
2000	M-NL	1.823	−0.191	0.0075	−0.616	0.964
2008	C-NL	2.551	−0.388	−2.327	−0.201	2.851
2008	C-MH	2.018	−0.386	−2.928	−0.204	3.756
2008	W-NL	0.764	−0.185	−2.389	−0.020	2.395
2008	W-MH	0.748	−0.175	−2.997	−0.160	3.635
2008	M-NL	2.094	−0.302	−1.307	−0.247	1.185

An indicator of the reliability of the vulnerability models in Figure 5 can be measured by using them to simulate both the mean DF, and the accumulated loss in the 2000 and 2008 earthquake and compare the results to actual observations. The result of this simulation for each building typology and each dataset is shown in Table 5. The ratios are within reasonable limits: the lowest ratio for R_{DF} is 0.81 and the highest is 1.18. For R_{Loss}, the lowest ratio is 0.98 and the highest is 1.17. This indicates that the ZIBR model calibrated for the Joyner–Boore distance provides acceptable results which are comparable to the results obtained using the PGA as an intensity measure (see [17]).

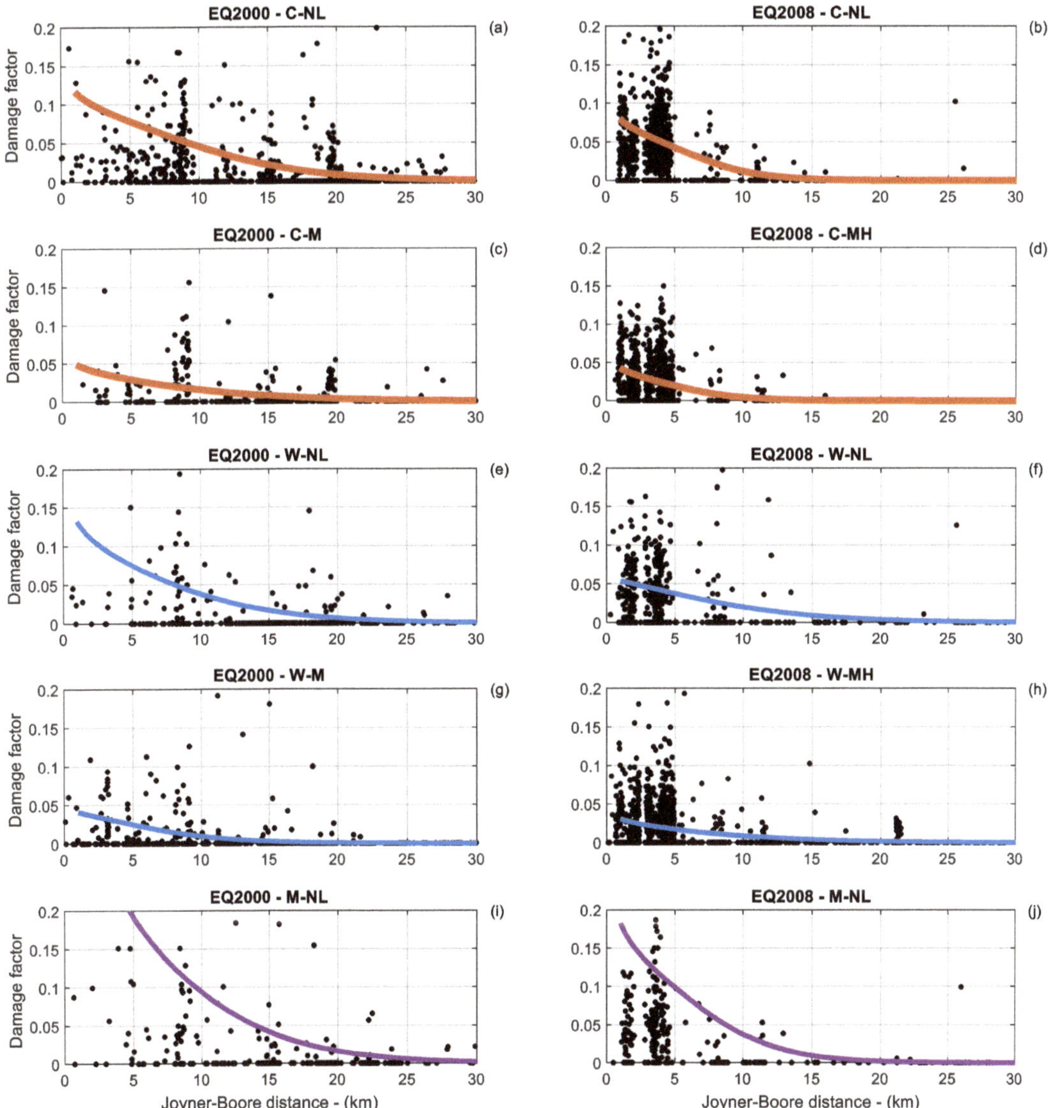

Figure 5. Raw data (black dots) and fitted vulnerability functions for 1–2-storey buildings/dwellings based on the ZIBR model with the Joyner–Boore distance as intensity measure for no- and low-code concrete buildings/dwellings in (**a**) the 2000 dataset and (**b**) the 2008 dataset; (**c**) moderate-code concrete buildings in the 2000 dataset; (**d**) moderate- and high-code concrete dwellings in the 2008 dataset; no- and low-code timber buildings/dwellings in (**e**) the 2000 dataset and (**f**) the 2008 dataset; (**g**) moderate-code timber buildings in the 2000 dataset; (**h**) moderate- and high-code timber dwellings in the 2008 dataset; and no- and low-code masonry buildings/dwellings in the (**i**) 2000 dataset and (**j**) in the 2008 dataset.

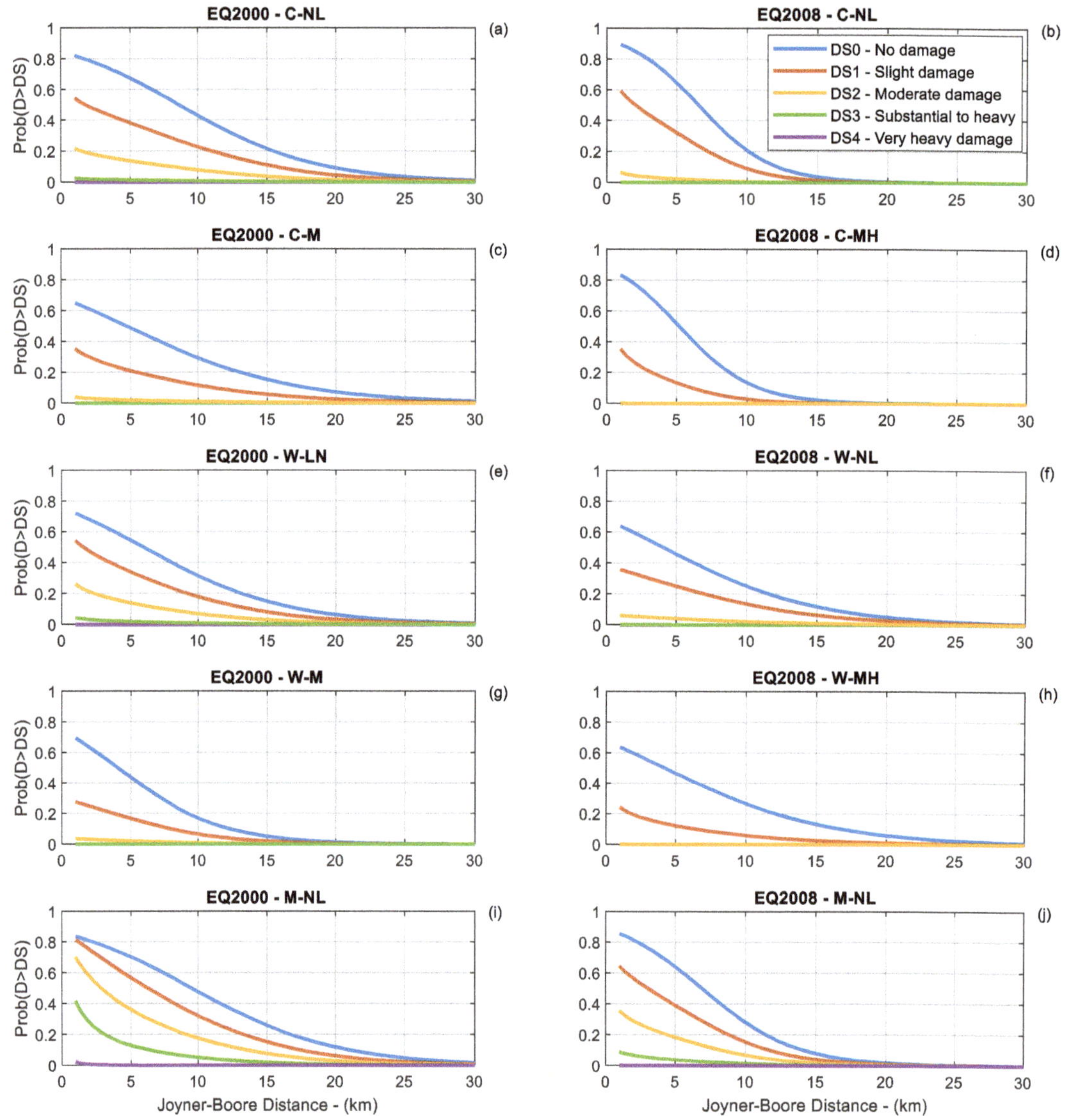

Figure 6. Fragility functions for one- to two-storey buildings/dwellings based on the ZIBR model with the Joyner–Boore distance as intensity measure for no- and low-code concrete buildings/dwellings (**a**) in the 2000 dataset and (**b**) in the 2008 dataset; (**c**) moderate-code concrete buildings in the 2000 dataset; (**d**) moderate- and high-code concrete dwellings in the 2008 dataset; no- and low-code timber buildings/dwellings in (**e**) the 2000 dataset and (**f**) the 2008 dataset; (**g**) moderate-code timber buildings in the 2000 dataset; (**h**) moderate- and high-code timber dwellings in the 2008 dataset; and no- and low-code masonry buildings/dwellings in the (**i**) 2000 dataset and (**j**) in the 2008 dataset.

Table 5. Ratio of simulated mean DF to actual mean DF from loss data (R_{DF}), and ratio of simulated accumulated loss to actual accumulated loss (R_{Loss}) for the five building typologies.

	Dataset	C-LN	C-M or C-MH	W-NL	W-M or W-MH	M-NL	Mean
R_{DF}	2000	0.94	0.99	0.81	0.90	0.87	0.90
R_{DF}	2008	1.07	0.98	1.09	0.94	1.18	1.05
R_{Loss}	2000	1.17	1.06	1.12	1.11	1.11	1.11
R_{Loss}	2008	1.06	0.99	1.00	1.03	1.13	1.04

4.2. Scenario-Based Risk Maps

By using the vulnerability functions in Figure 5, scenario risk maps can be constructed. The M_w6.36 earthquake on 23 July 1929 with an epicentre 21.75° W and 63.95° N [12] is used as the scenario (see Figure 1). This event can be considered to have the most effect in the RCA of all known historical and instrumentally recorded events in the vicinity of Reykjavik. The subsurface rupture model by Wells and Coppersmith [24] is used to estimate the projected fault line on the surface (brown dotted line in Figure 1), which is then used to compute the Joyner–Boore distance. It is assumed that the epicentre lies at the centre of the fault. Since the residential building stock in the RCA mainly (>92%) consists of concrete buildings (Section 2.2), only such buildings are considered. Figure 7 shows risk maps in the form of the mean damage factor for one- to two-storey residential concrete buildings. Contours for the mean loss are shown for no-code and low-code buildings in Figure 7a using the vulnerability functions from the 2000 dataset and in Figure 7b for the vulnerability curves based on the 2008 dataset. Figure 7c shows the contours for mean damage for moderate-code concrete buildings based on the 2000 dataset and Figure 7d for moderate- and high-code buildings based on the 2008 dataset. If the dataset from 2000 is considered as representative for an M_w6.5 event and the dataset from 2008 as representative for an M_w6.3 event, it is clear how the magnitude affects the extent of the damage. The effect of the seismic code level is also clearly distinguished.

The maps in Figure 7 give useful information on the mean losses, which is important for disaster insurance purposes, but they provide no information on the distribution of the losses at different sites, which is important for emergency management. For instance, information on how many buildings (in percentage) are undamaged and how many are extensively damaged is not available in this representation. This information can be displayed by constructing scenario risk maps for damage states based on the fragility curves in Figure 6. In Figure 8, risk maps for one- to two-storey concrete residential buildings designed in the no-code and low-code period, i.e., before 1976, are shown for damage states DS0, DS1, DS2, and DS3 based on the 2000 dataset corresponding to an event of M_w6.5. As an example, Figure 8a shows the probability of exceeding DS0 (no damage). For the greatest part of the RCA, the probability of sustaining losses is 40–60%. For central Reykjavik, the probability is between 20–40%. On the other hand, the probability of exceeding DS3 and obtaining more than a 50% loss is in all cases less than 2% (Figure 8d). In Figure 9, corresponding risk maps are shown for the same building typology, except now the results are based on fragility curves calibrated from the 2008 dataset, corresponding to an event of M_w6.3. From Figure 9c, the probability of exceeding DS2 with more than a 20% loss is less than 2% in all cases.

In Figure 10, scenario risk maps are shown for one- to two-storey concrete residential buildings designed in the moderate-code period after 1976, for damage states DS0 and DS1, based on the 2000 dataset (M_w6.5). As an example, the probability of exceeding DS1 (minor damage) is close to 20% for the most exposed building sites (Figure 10b). Finally, in Figure 11, risk maps for one- to two-storey concrete residential buildings designed in the moderate-code and new-code period, i.e., after 1976, are shown for damage states DS0 and DS1, based on the 2008 dataset (M_w6.3). The probability of exceeding DS1 (minor damage) is close to 10% for the most exposed building at the outskirts of the RCA.

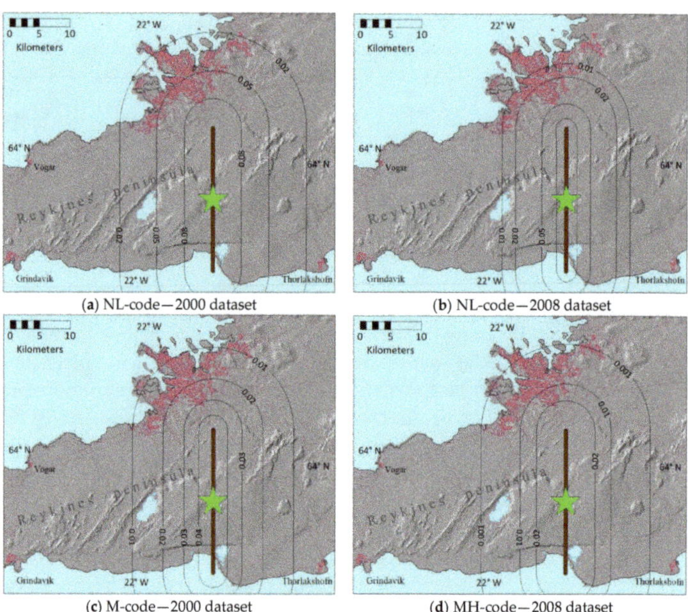

(**a**) NL-code—2000 dataset

(**b**) NL-code—2008 dataset

(**c**) M-code—2000 dataset

(**d**) MH-code—2008 dataset

Figure 7. Scenario risk map showing predicted mean damage factor for one- to two-storey concrete buildings based on the vulnerability functions in Figure 5: (**a**) no- and low-code buildings based on the 2000 dataset (M_w6.5); (**b**) no- and low-code buildings based on the 2008 dataset (M_w6.3); (**c**) moderate-code buildings based on the 2000 dataset (M_w6.5); and (**d**) moderate- and high-code buildings based on the 2008 dataset (M_w6.3).

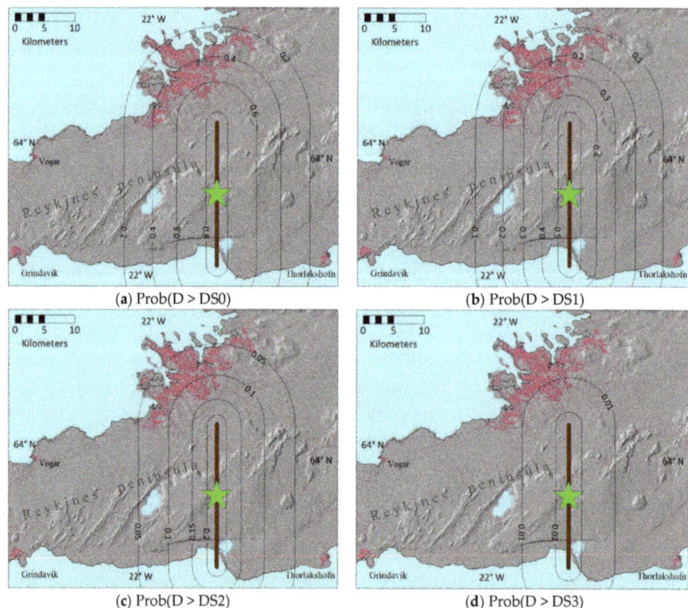

(**a**) Prob(D > DS0)

(**b**) Prob(D > DS1)

(**c**) Prob(D > DS2)

(**d**) Prob(D > DS3)

Figure 8. Scenario risk map for one- to two-storey, no-code and low-code, reinforced concrete buildings, C-NL (built before 1976) based on a vulnerability model from the 2000 dataset (M_w6.5). The maps show the probability that the damage state will (**a**) exceed DS0; (**b**) exceed DS1; (**c**) exceed DS2; and (**d**) exceed DS3.

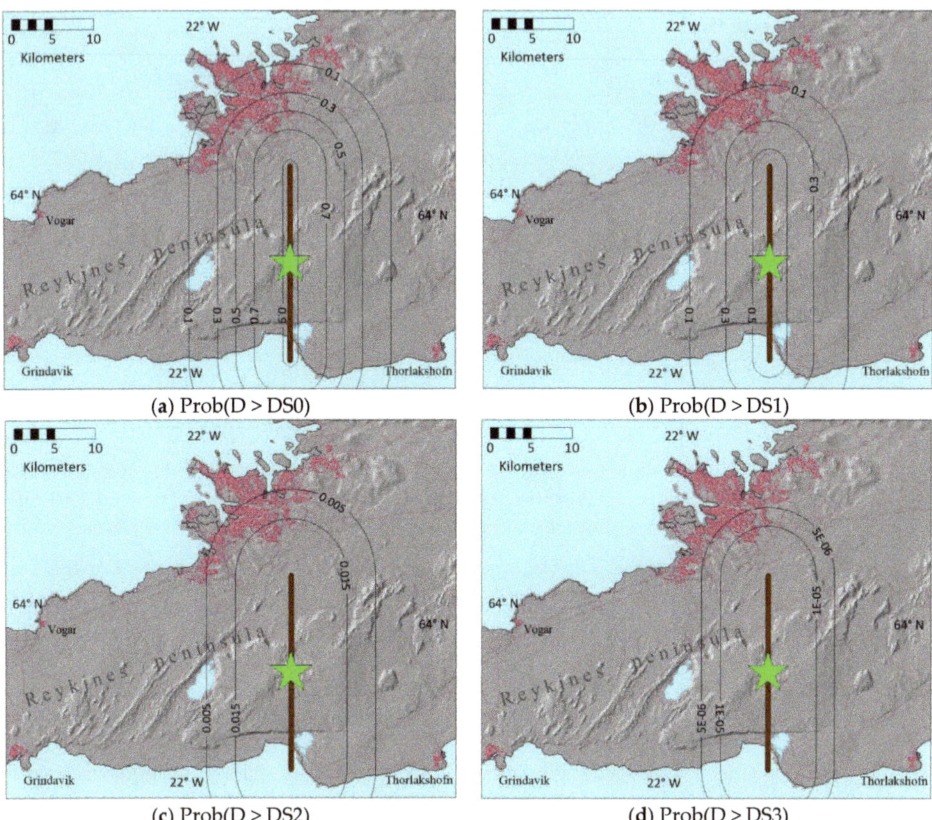

(a) Prob(D > DS0)

(b) Prob(D > DS1)

(c) Prob(D > DS2)

(d) Prob(D > DS3)

Figure 9. Scenario risk map for one- to two-storey, no-code and low-code, reinforced concrete buildings, C-NL (built before 1976) based on the vulnerability model from the 2008 dataset (M_w6.3). The maps show the probability that the damage state will (**a**) exceed DS0; (**b**) exceed DS1; (**c**) exceed DS2; and (**d**) exceed DS3.

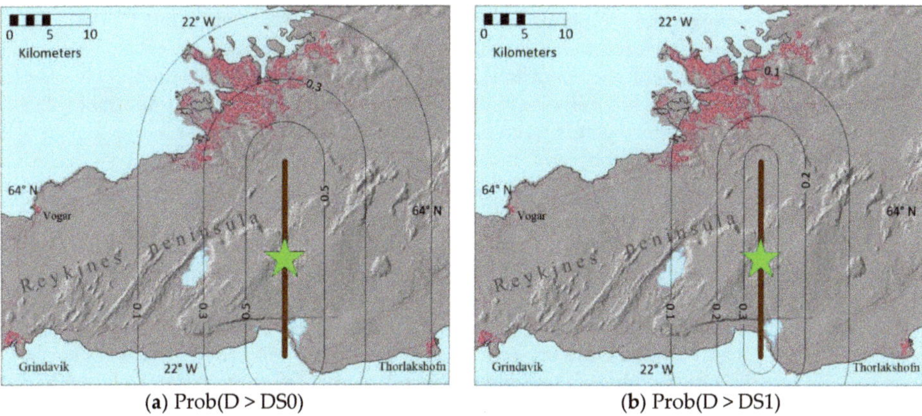

(a) Prob(D > DS0)

(b) Prob(D > DS1)

Figure 10. Scenario risk map for one- to two-storey, moderate-code, reinforced concrete buildings, C-M (built after 1976) based on the vulnerability model from the 2000 dataset (M_w6.5). The maps show the probability that the damage state will (**a**) exceed DS0 and (**b**) exceed DS1.

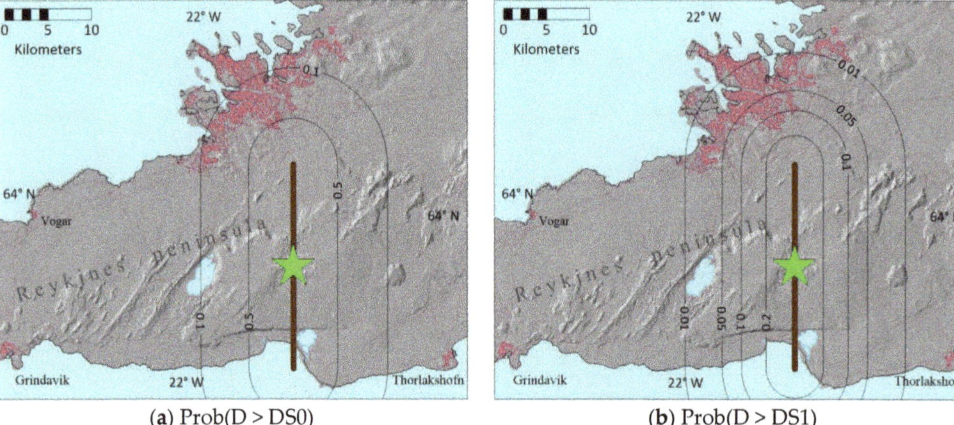

(**a**) Prob(D > DS0)   (**b**) Prob(D > DS1)

Figure 11. Scenario risk map for one- to two-storey, moderate-code and high-code, reinforced concrete buildings, C-MH (built after 1976) based on the vulnerability model from the 2008 dataset (M_w6.3). The maps show the probability that the damage state will (**a**) exceed DS0 and (**b**) exceed DS1.

5. Conclusions

In this study, advanced empirical vulnerability models and fragility functions for low-rise, one- to two-storey residential buildings are presented. The building typology is defined in terms of the main building material (concrete, timber, and masonry) as well as the level of seismic codes, where no-code and low-code buildings are grouped together, and moderate-code and high-code buildings are grouped together. The vulnerability model based on zero-inflated beta regression was calibrated for two loss datasets, independently. One of the datasets is from two ~M_w6.5 strike-slip shallow earthquakes in 2000 and other one is from an M_w6.3 strike-slip shallow event in 2008. Both datasets are from South Iceland, where the earthquakes affected buildings with similar characteristics. In this study, the Joyner–Boore distance, R_{JB}, is used as a proxy for an intensity measure, instead of the commonly used measures like the PGA, Sa, AvgSa, etc. This approach makes the vulnerability model independent of a particular ground motion prediction model (GMPM) and is well suited for scenario-based applications where the earthquake size is fixed, and the intensity is mainly a function of the source-site distance. The application is valid for uniform site conditions, and local amplifications are not accounted for directly. However, if data allows, local effects could be incorporated by creating separate vulnerability models for different site conditions, which is not the case in this study. This representation of seismic vulnerability is simpler and more comprehensible to important stakeholders such as emergency planners, civil protection authorities, and decision-makers in risk reduction and management operations. It is also well suited for demarking spatial zones of different risk levels for a given earthquake scenario, and is therefore valuable for emergency response planning.

The presented vulnerability models show that at a distance of 10 km or more (R_{JB}) from the fault rupture, the mean loss is less than 5% of the replacement value for concrete and timber buildings, but higher for masonry buildings. The fragility curves show that the probability of exceeding DS2, i.e., moderate damage and exceeding losses greater than 20% of the fire insurance value, is very low (<4%) at all distances for moderate-code and high-code concrete and timber buildings. These results are valid for both M_w6.3 and M_w6.5 earthquakes.

Scenario seismic risk maps, based on the most damaging earthquake scenario in the vicinity of the RCA, that is the July 1929 M_w6.36 earthquake, were computed for low-rise concrete buildings, showing both the predicted mean loss and probability of exceeding different damage states. Although there is a lot of uncertainty about fault lengths, magnitude size, exact epicentre, as well as wave propagation and attention, the maps give a useful indication of what damage can be expected for earthquakes in the range 6.3 to 6.5 for one- to two-storey buildings. The study shows that there is considerable difference in the expected damages for 6.3 and 6.5 events. Furthermore, the code levels used in the design are easily visible on the maps. Here, it should be kept in mind that the newer buildings, which are moderate-code or high-code buildings, are generally located in the outskirts of the RCA (Reykjavik, Kopavogur, Gardabær, and Hafnafjördur), and therefore closest to the active seismic zones on the Reykjanes Peninsula.

The two complete datasets from the 2000 and the 2008 South Iceland earthquakes mainly contain one- to two-storey residential buildings (>90%) and therefore the available local empirical vulnerability models only cover these types of buildings. Nearly 60% of dwellings in the RCA are in three- to five-storey concrete apartment buildings, where the lateral load resisting system is structural walls. It is therefore a great challenge and a subject for further research to develop reliable vulnerability models for such buildings. An analytical approach based on non-linear numerical modelling of structures is a viable alternative for such research.

Author Contributions: Conceptualization, B.B., R.R. and J.Ö.B.; Methodology, B.B.; Validation, B.B.; Formal analysis, B.B.; Resources, J.Ö.B.; Writing—original draft, B.B.; Writing—review & editing, R.R. and J.Ö.B.; Visualization, B.B. All authors have read and agreed to the published version of the manuscript.

Funding: The authors thank Natural Catastrophe Insurance of Iceland for putting the earthquake damage database and other relevant information at their disposal. This research was supported by the Icelandic Research Fund (grant no. 218149-051) and the University of Iceland Research Fund.

Data Availability Statement: Data are contained within the article.

Conflicts of Interest: The authors declare no conflict of interest.

References

1. Bruneau, M.; Reinhorn, A. Exploring the concept of seismic resilience for acute care facilities. *Earthq. Spectra* **2007**, *23*, 41–62. [CrossRef]
2. Sangaki, A.H.; Rofooei, F.R.; Vafai, H. Probabilistic integrated framework and models compatible with the reliability methods for seismic resilience assessment of structures. *Structures* **2021**, *34*, 4086–4099. [CrossRef]
3. Forcellini, D. An expeditious framework for assessing the seismic resilience (SR) of structural configurations. *Structures* **2023**, *56*, 105015. [CrossRef]
4. *ATC-13*; Earthquake Damage Evaluation Data for California. Applied Technology Council: Redwood City, CA, USA, 1985; pp. 1–492.
5. *ATC-13-1*; Commentary on the Use of ATC-13 Earthquake Damage Evaluation Data for Probable Maximum Loss Studies of California Buildings. Applied Technology Council: Redwood City, CA, USA, 2002; pp. 1–66.
6. Rota, M.; Penna, A.; Magenes, G. A methodology for deriving analytical fragility curves for masonry buildings based on stochastic nonlinear analyses. *Eng. Struct.* **2010**, *32*, 1312–1323. [CrossRef]
7. Ruggieri, S.; Calò, M.; Cardellicchio, A.; Uva, G. Analytical-mechanical based framework for seismic overall fragility analysis of existing RC buildings in town compartments. *Bull. Earthq. Eng.* **2022**, *20*, 8179–8216. [CrossRef]
8. Rossetto, T.; Elnashai, A. Derivation of vulnerability functions for European-type RC structures based on observational data. *Eng. Struct.* **2003**, *25*, 1241–1263. [CrossRef]
9. Rota, M.; Penna, A.; Strobbia, C.L. Processing Italian damage data to derive typological fragility curves. *Soil Dyn. Earthq. Eng.* **2008**, *28*, 933–947. [CrossRef]
10. Colombi, M.; Borzi, B.; Crowley, H.; Onida, M.; Meroni, F.; Pinho, R. Deriving vulnerability curves using Italian earthquake damage data. *Bull. Earthq. Eng.* **2008**, *6*, 485–504. [CrossRef]
11. Pedersen, R.; Jónsson, S.; Árnadóttir, T.; Sigmundsson, F.; Feigl, K.L. Fault slip distribution of two June 2000 MW6. 5 earthquakes in South Iceland estimated from joint inversion of InSAR and GPS measurements. *Earth Planet. Sci. Lett.* **2003**, *213*, 487–502. [CrossRef]
12. Jónasson, K.; Bessason, B.; Helgadóttir, Á.; Einarsson, P.; Gudmundsson, G.B.; Brandsdóttir, B.; Vogfjörd, K.S.; Jónsdóttir, K. A harmonised instrumental earthquake catalogue for Iceland and the northern Mid-Atlantic Ridge. *Nat. Hazards Earth Syst. Sci.* **2021**, *21*, 2197–2214. [CrossRef]
13. Bessason, B.; Bjarnason, J.Ö.; Rupakhety, R. Statistical modelling of seismic vulnerability of RC, timber and masonry buildings from complete empirical loss data. *Eng. Struct.* **2020**, *209*, 109969. [CrossRef]

14. Sigbjörnsson, R.; Snæbjörnsson, J.T.; Higgins, S.M.; Halldórsson, B.; Ólafsson, S. A note on the M_w6.3 earthquake in Iceland on 29 May 2008 at 15:45 UTC. *Bull. Earthq. Eng.* **2009**, *7*, 113–126. [CrossRef]
15. Halldórsson, B.; Sigbjörnsson, R. The M_w6.3 Ölfus earthquake at 15:45 UTC on 29 May 2008 in South Iceland: ICEARRAY strong-motion recordings. *Soil Dyn. Earthq. Eng.* **2009**, *29*, 1073–1083. [CrossRef]
16. Ioannou, I.; Bessason, B.; Kosmidis, I.; Bjarnason, J.Ö.; Rossetto, T. Empirical seismic vulnerability assessment of Icelandic buildings affected by the 2000 sequence of earthquakes. *Bull. Earthq. Eng.* **2018**, *16*, 5875–5903. [CrossRef]
17. Bessason, B.; Bjarnason, J.Ö.; Rupakhety, R. Comparison and modelling of building losses in South Iceland caused by different size earthquakes. *J. Build. Eng.* **2022**, *46*, 103806. [CrossRef]
18. Rupakhety, R.; Sigbjörnsson, R. Ground-motion prediction equations (GMPEs) for inelastic response and structural behaviour factors. *Bull. Earthq. Eng.* **2009**, *7*, 637–659. [CrossRef]
19. Joyner, W.B.; Boore, D.M. Peak horizontal acceleration and velocity from strong-motion records including records from the 1979 Imperial Valley, California, earthquake. *Bull. Seismol. Soc. Am.* **1981**, *71*, 2011–2038. [CrossRef]
20. Icelandic Institute of Natural History. 2023. Available online: https://jardfraedikort.ni.is/ (accessed on 15 October 2023).
21. Atakan, K.; Brandsdóttir, B.; Halldorsson, P.; Fridleifsson, G.O. Site response as a function of near-surface geology in the South Iceland seismic zone. *Nat. Hazards* **1997**, *15*, 139–164. [CrossRef]
22. Einarsson, P. Plate boundaries, rifts and transforms in Iceland. *Jökull* **2008**, *58*, 35–58. [CrossRef]
23. Einarsson, P.; Hjartardóttir, Á.R.; Hreinsdóttir, S.; Imsland, P. The structure of seismogenic strike-slip faults in the eastern part of the Reykjanes Peninsula Oblique Rift, SW Iceland. *J. Volcanol. Geotherm. Res.* **2020**, *391*, 106372. [CrossRef]
24. Wells, D.L.; Coppersmith, K.J. New empirical relationships among magnitude, rupture length, rupture width, rupture area, and surface displacement. *Bull. Seismol. Soc. Am.* **1994**, *84*, 974–1002. [CrossRef]
25. Hernández-Aguirre, V.M.; Rupakhety, R.; Ólafsson, S.; Bessason, B.; Erlingsson, S.; Paolucci, R.; Smerzini, C. Strong ground motion from the seismic swarms preceding the 2021 and 2022 volcanic eruptions at Fagradalsfjall, Iceland. *Bull. Earthq. Eng.* **2023**, *21*, 4707–4730. [CrossRef]
26. Sæmundsson, K.; Sigurgeirsson, M.Á. Reykjanes peninsula. In *Natural Hazard in Iceland, Volcanic Eruptions and Earthquakes*; Sólnes, J., Sigmundsson, F., Bessason, B., Eds.; University of Iceland Press: Reykjavik, Iceland; Natural Catastrophe Insurance of Iceland: Kópavogur, Iceland, 2013; pp. 379–401. (In Icelandic)
27. Natural Catastrophe Insurance of Iceland. 2023. Available online: https://island.is/en/o/nti (accessed on 15 October 2023).
28. Property Register. 2023. Available online: https://www.fasteignaskra.is/english (accessed on 15 October 2023).
29. Rossetto, T.; Ioannou, I.; Grant, D.N.; Maqsood, T. *Guidelines for the Empirical Vulnerability Assessment*; GEM Foundation: Pavia, Italy, 2014.
30. Brzev, S.; Scawthorn, C.; Charleson, A.W.; Allen, L.; Greene, M.; Jaiswal, K.; Silva, V. *GEM Building Taxonomy*; Version 2.0; GEM Technical Report No. 2013-02; GEM Foundation: Pavia, Italy, 2013.
31. Crowley, H.; Despotaki, V.; Silva, V.; Dabbeek, J.; Romão, X.; Pereira, N.; Castro, J.M.; Daniell, J.; Veliu, E.; Bilgin, H.; et al. Model of seismic design lateral force levels for the existing reinforced concrete European building stock. *Bull. Earthq. Eng.* **2021**, *19*, 2839–2865. [CrossRef]
32. Ospina, R.; Ferrari, S.L.P. A general class of zero-or-one inflated beta regression models. *Comput. Stat. Data Anal.* **2012**, *56*, 1609–1623. [CrossRef]
33. R Core Team. *R: A Language and Environment for Statistical Computing*; R Foundation for Statistical Computing: Vienna, Austria, 2023. Available online: https://www.R-project.org/ (accessed on 15 October 2023).
34. Ferrari, S.L.P.; Cribari-Neto, F. Beta Regression for Modelling Rates and Proportions. *J. Appl. Stat.* **2004**, *31*, 799–815. [CrossRef]
35. Dolce, M.; Kappos, A.; Masi, A.; Penelis, G.; Vona, M. Vulnerability assessment and earthquake damage scenarios of the building stock of Potenza (Southern Italy) using Italian and Greek methodologies. *Eng. Struct.* **2006**, *28*, 357–371. [CrossRef]

Article

Response of Masonry-Infilled Reinforced Concrete Frames Strengthened at Interfaces with Geo-Fabric under In-Plane Loads

K. S. Sreekeshava [1], Hugo Rodrigues [2,*] and A. S. Arunkumar [3]

1 Jyothy Institute of Technology, Affiliated to Visvesvaraya Technological University, Belagavi 590018, India; sreekeshava.ks@jyothyit.ac.in
2 RISCO, Civil Engineering Department, University of Aveiro, Campus of Santiago, 3810-193 Aveiro, Portugal
3 BMS College of Engineering, Affiliated to Visvesvaraya Technological University, Belagavi 590018, India; arunkumar.civ@bmsce.ac.in
* Correspondence: hrodrigues@ua.pt

Abstract: The interfaces between masonry infill and reinforced concrete (MI-RC) frames are identified as the weakest regions under lateral loads. Hence, the behavior of such frames under lateral loads can be understood mainly through experimental investigations. The deformation demands induced by horizontal loads on RC frames with infill masonry walls change due to contact losses between the infill masonry and the RC frames. This can be controlled by providing proper reinforcements at the interfaces. In the present experimental investigation, three half-scaled models subjected to reversed cyclic lateral in-plane loads were tested. In detail, the specimens considered are the MI-RC frame model, an MI-RC frame with geo-fabric reinforcement at the interface and an MI-RC frame with geo-fabric reinforcement at interfaces with an open ground story. The models were subjected to reversed cyclic lateral in-plane loads, and the post-yield responses of the models with respect to stiffness degradation, drift, energy dissipation, ductility and failure mode have been discussed.

Keywords: geo-fabrics; stiffness; ductility; energy dissipation; lateral drift; strengthening

Citation: Sreekeshava, K.S.; Rodrigues, H.; Arunkumar, A.S. Response of Masonry-Infilled Reinforced Concrete Frames Strengthened at Interfaces with Geo-Fabric under In-Plane Loads. *Buildings* **2023**, *13*, 1495. https://doi.org/10.3390/buildings13061495

Academic Editors: Andreas Lampropoulos and Ciro Del Vecchio

Received: 7 April 2023
Revised: 30 May 2023
Accepted: 6 June 2023
Published: 9 June 2023

1. Introduction

The masonry infill (MI) plays a significant role under lateral loads on reinforced concrete (RC) frames. All major codes have ignored, in the past, the contribution of MI to strength and stiffness by treating it as non-structural elements [1]. Different types of MI used worldwide. The brick masonry has been widely used in construction of highrise buildings because of huge benefits it has when compared with other types of MI. Masonry behavior under the action of lateral loads showed poor performance because of its insufficient shear and flexure strengths. The brittle nature of masonry creates a lot of damage to the structure and life of humans under heavy lateral loads [2].

The damage to buildings during past earthquakes reveals that unreinforced infill masonry may exhibit a poor performance under the action of lateral loads, and it also depends on several factors, such as workmanship and design factors adopted in construction. In recent studies, several researchers focused on strengthening the masonry in seismic areas to increase the strength and stiffness of the structure to resist lateral loads [3,4].

The strengthening of infill masonry elements may play a crucial role in avoiding in-plane and out-of-plane failure in masonry elements. Several researchers reported the merits and demerits observed under different reinforcing techniques. The use of fiber-reinforced polymer with organic (epoxy-resin-based) composites for applications in masonry structures is discouraged because of the drawbacks pertaining to diminished performance at elevated temperatures, requirements of protective coatings, degradation of mechanical properties after continuous exposure to certain environmental surroundings [5], lack of compatibility between the resin and masonry surface, production of toxic gases associated with the use of FRP resin in epoxy application and relatively higher level of care and supervision required in application [6,7]. To overcome these drawbacks, environmentally friendly, thermal-resistant, non-corrosive composite material is used to reinforce walls by replacing FRPs with geo-fabric materials [8–10], and results showed that these geo-fabrics help to increase the shear and flexural capacity [11,12].

The behavior of MI-RC frames under critical combinations of lateral loads is very complex. Analytical studies conducted on MI-RC frames by modelling masonry using different methods have illustrated the importance of MI in enhancing the lateral stiffness of the frame [13]. The change in the lateral stiffness and mass will change the dynamic characteristics of the frame. Hence, to account for the composite action between the RC frame and MI, proper analytical models need to be developed to design MI-RC structures [14,15]. Recent studies conducted on MI-RC frames being subjected to pseudo-dynamic tests have indicated an increased structural response with respect to load-carrying capacity and stiffness when the connectivity of MI-RC frames is improved [16,17].

The analytical models developed to predict the structural response of infilled reinforced RC frames must consider the geometric and material non-linearity. The developed analytical models can only be validated through experimental studies. Interfacial behavior between MI and RC also plays a major role in the case of MI-RC frames. In the Indian context, there are scant studies, especially related to the interfacial strengthening of MI-RC frames subjected to in-plane lateral loads. Further, the increased strength and lateral stiffness contributed by MI influence the post-yield response significantly. These structural responses can be well understood through experimental investigations. This paper deals with the behavior of the lateral load responses of MI-RC frames with and without polyester geo-fabric reinforcements at interfaces subjected to in-plane reversed cyclic lateral loads. This experimental investigation aims to describe the influence of brick MI with and without geo-fabric on the lateral-load-carrying capacity and in-elastic behavior of the MI-RC frame.

2. Experimental Program

2.1. Test Specimens

In the present experimental investigations, three geometrically half-scaled reinforced concrete frames have been considered. The specimen indicates the following three cases:

- Single-bay two-story conventional MI-RC frame.
- Single-bay two-story MI-RC frame with geo-fabric reinforcements at interfaces.
- Single-bay two-story MI-RC frame with geo-fabric reinforcements at interfaces with an open ground (soft) story.

The columns and beams in RC frames have 100 mm × 100 mm cross-sectional dimensions along with 8 mm high yield strength deformed bars with 6 mm mild steel stirrups at 50 mm center to center, as shown in Figure 1. The steel reinforcement in the framed section conformed to IS 1390-1993 [18], and using locally available table-molded bricks, brick masonry in cement mortar 1: 6 was used as the infill wall.

2.2. Characteristics of Materials Used in the Experimental Program

Due to the scaled structures, the RC-framed structures had closely spaced reinforcements and were also small in dimension. Hence, to facilitate the placement of concrete, self-compacting concrete (SCC) was used to avoid the vibrations of concrete. Ordinary Portland cement (OPC 53 grade) conforming to IS: 12269-1987 [19] requirements was used. Manufactured sand (M-sand) was used as fine aggregate, and ground granulated blast slag (GGBS) was used as filler material. The coarse aggregates used were 12.5 mm downsize, and a dosage of 1.15% super-plasticizer Master glenium-Sky 8233 was used to achieve the target strength of 40 MPa after 28 days of curing, as per IS 10262:2019 [20]. The GGBS was used as a partial replacement of cement at about 35%, and the workability of SCC flow was maintained at between 700 mm and 800 mm. A cement content of 300 kg/m^3 and water–cement ratio of 0.6 was adopted, as per IS 456:2000. This mix was used to prepare all three MI-RC models cast in this study. However, the concrete cubes which were tested at the time of the cyclic loading test of MI-RC frames indicated a cube strength crushing strength of 46.22 MPa, which was slightly more than the target strength.

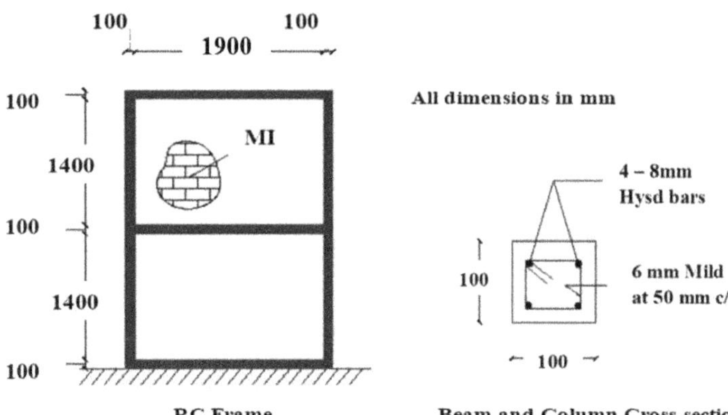

Figure 1. Details of test specimens.

Standard molded bricks available in the South Indian Bangalore region were used in this experimental study, having a unit strength of 7.6 MPa. Brick masonry was constructed using cement mortar 1:6 along with a water–cement ratio of 1.2. The brick masonry assemblages were cast, cured for 28 days and tested under compression to evaluate the properties of masonry assemblages. The SCC and cement mortar were tested, and the material characteristics used in the models are tabulated in Table 1.

Table 1. Material characteristics used in the models.

Sl. No.	Material	Compressive Strength/Tensile Strength (MPa)	Modulus of Elasticity (MPa) (Initial Tangent Modulus)
1	Brick masonry	$f''_m = 2.55$ (Corrected prism compressive strength)	1560 (Panel test)
2	Concrete	$f_{ck} = 46.22$ (Cube crushing strength)	27,109 (Compressive tests on cylinders)
3	Cement mortar	5.02 (Cube crushing strength)	6477 (Compressive tests on cylinders)
4	Reinforcing steel in columns	HYSD bars—425 (Tensile Strength) Mild steel—262 (Tensile strength)	HYSD bars—2,00,102 Mild steel—2,10,101
5	Polyester geo-fabric	129 (Tensile strength)	14,989 (Tensile test)

2.3. Estimation of Failure Lateral Load

The capacity moment of RC members can be computed for the test specimens with the achieved strength of concrete using the ultimate load method. Here, the bending compressive stress distribution of concrete is assumed to be parabolic with a maximum compressive strength of 0.67 fck for the achieved grade of concrete, and the strain was assumed to be linear across the depth of the section. The yield stress of corresponding steel used was taken as 415 MPa. Using these parameters, the ultimate moment capacity of the RC section under flexure was assessed to be 3249 N-m for the achieved strength of concrete grade which has been reported in Table 2.

Table 2. Estimated failure lateral load of the models.

S. No.	Type of Model	Ultimate Moment of Resistance of the Frame Section (N–m)	Max Bending Moment in the Frame as per Linear Static Analysis (N–m)	Base Shear Applied (N)
1	MI-RC frame	3249.0	3253.0	55,500.0
2	MI-RC frame with geo-fabric reinforcement at interface			
3	MI-RC Frame with geo-fabric reinforcement at interfaces with open ground story	3249.0	3250.0	6975.0

The prime focus of this present experimental investigation was to study the behavior of the lateral load response of MI-RC frames reinforced with geo-fabric at interfaces under in-plane lateral loads. The lateral load capacity of any model which was also equal to respective base shears was estimated using the equivalent lateral load method conforming to IS 1893:2016 [21]. For the present experimental models, including the MI-RC frames, MI-RC frames with geo-fabric at interfaces and MI-RC frames reinforced with geo-fabric interfaces with a soft story, the base shear and equivalent lateral loads were estimated using the fundamental period of vibration and response spectra of the models.

Eigen value analysis of the models was performed on three models; the fundamental period of vibrations was found to be 0.05 s, 0.045 s and 0.13 s for MI-RC, MI-RC with geo-fabric as interface reinforcement and MI-RC with geo-fabric as interface reinforcement with an open ground story (soft story), respectively. In this present study, only dead loads due to concrete and brick masonry infill have been considered, and the calculated seismic weight on the frame was found to be 13.1 kN. The corresponding spectral acceleration co-efficient (sa/g) for all three models as per IS 1893:2016 [22] was worked out as 2.5. The calculated design base shear of all the models was 55,000 N, and based on the design base shear, the calculated design load forces of the first and second stories were found to be in the ratio of 0.5:1.

Static lateral load analysis of the models subjected to the combination of loads (dead + lateral) was carried out. The lateral load was applied to the models in the ratio of 0.5:1 on the first and second stories of the frame. Further, through a trial-and-error method, the magnitude of the lateral load which could produce the maximum bending moment in beam–column joints equal to the ultimate moment of resistance of the sections was determined. Here, the frame was analyzed using standard FEM software where the column and beam were considered as line elements, MI was considered as a four-noded shell element and the interfacial element was considered as a rigid element. The theoretical evaluation of the base shear was conducted to have a rough estimation of the maximum lateral load that might cause failure of infilled frames, and no redistribution was considered.

The expected bending moments developed at the beam to column joints due to applications of lateral loads for the considered MI-RC frames are shown in Figures 2 and 3. The summations of lateral loads (base shear) applied to models are estimated as theoretical failure lateral loads for respective models, and the same has been represented as respective cycles of increasing magnitude, as shown in Figures 4–6.

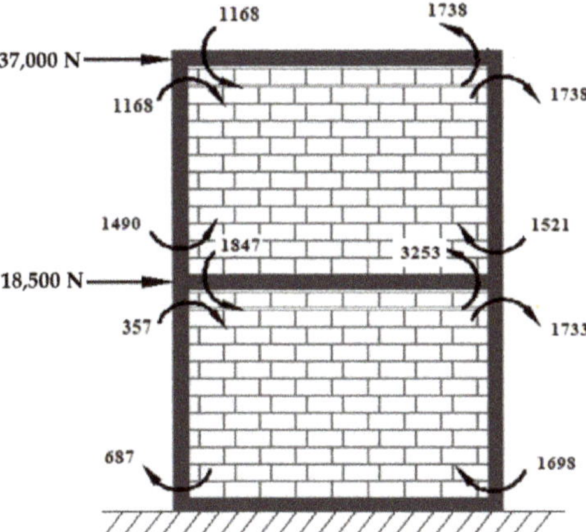

Figure 2. Bending moments (N-m) in MI-RC frame models with and without geo-fabric reinforcements at interfaces.

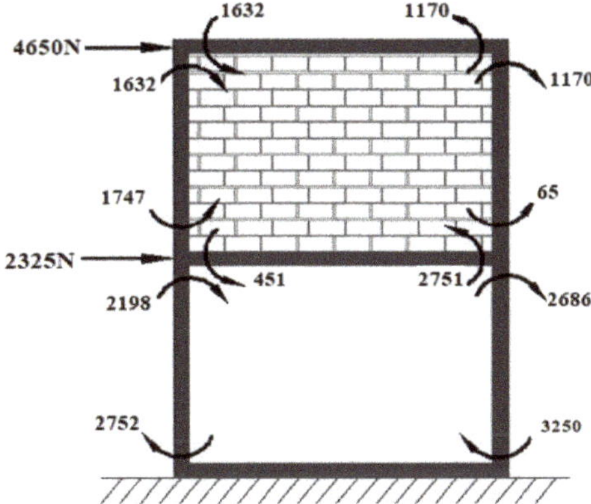

Figure 3. Bending moments (N-m) in MI-RC frame model with an open ground story along with geo-fabric reinforcement at interfaces.

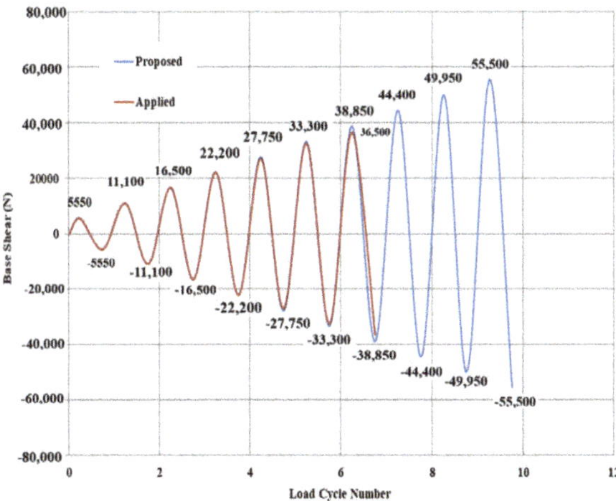

Figure 4. Loading history for the MI-RC frame model.

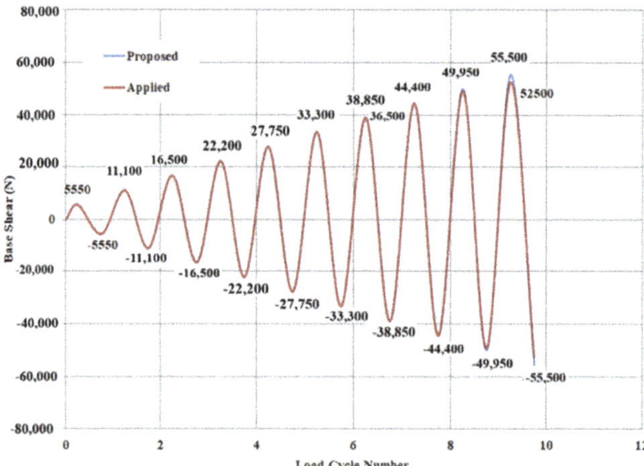

Figure 5. Loading history for the MI-RC frame model with geo-fabric reinforcements at interfaces.

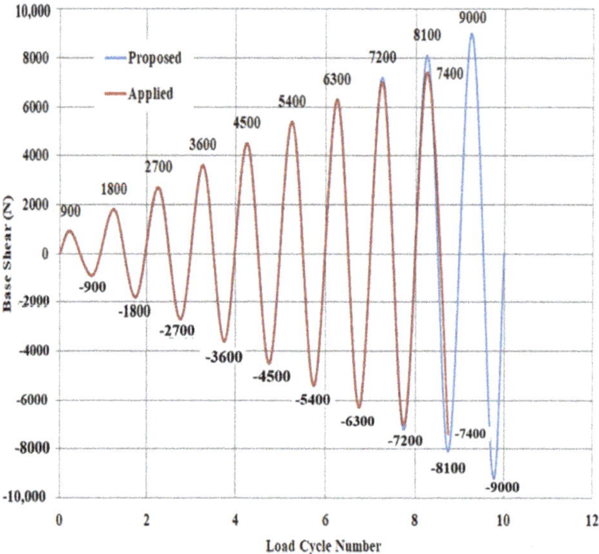

Figure 6. Loading history for the MI-RC frame model with an open ground story along with geo-fabric reinforcement at interfaces.

2.4. Experimental Setup and Proceedings

The RC frames were provided with RC flanges of 450 mm in length on either side of the column having a cross section of 100 mm × 100 mm to facilitate the fixity conditions at the base. The RC flanges were firmly fixed with the help of mild steel channels and plates which were welded at the base of the loading frame. The bottom flanges were clamped at the base using lipped channels, which were anchored using 12 mm diameter anchor bolts. Fixity was provided at the bottom of the RC frame. To ensure the proper application of in-plane lateral loads on the MI-RC frame, the frame was connected with suitable restraint arrangements with the help of an I-section and a roller hung at the top of the beam. The restraints helped to control the probable out-of-plane deformation of MI-RC models and also helped to ensure that the models were subjected to only in-plane lateral loads. All three models were checked in same way and tested under a 2000 kN loading frame for reversed cyclic lateral loads. The testing models along with the experimental setup are shown in Figures 7–9.

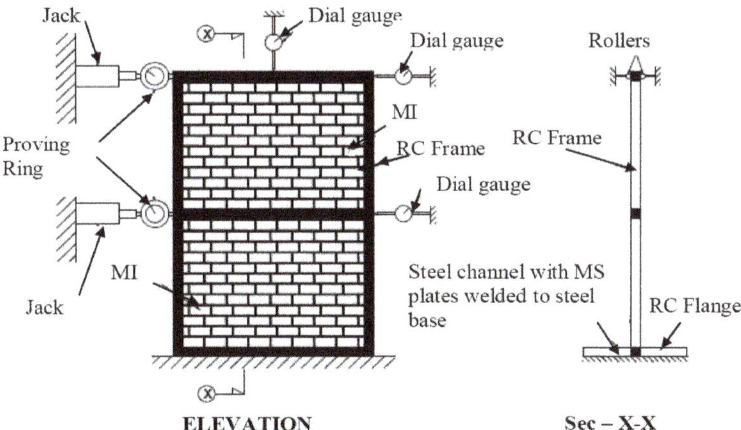

Figure 7. Test setup for the MI-RC model.

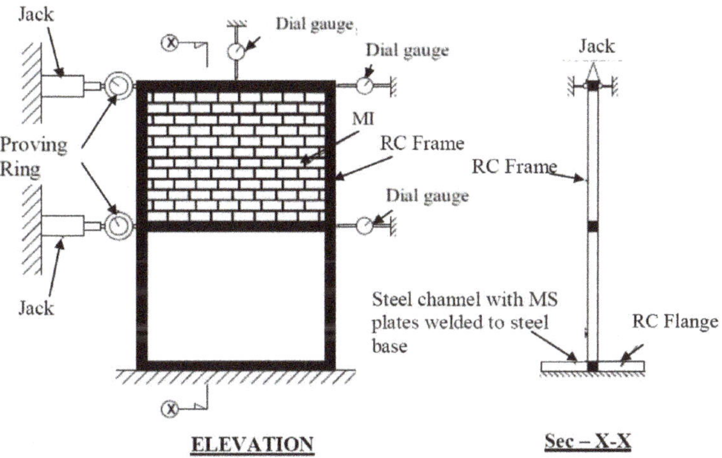

Figure 8. Test setup for the open ground (soft) story.

Figure 9. Providing column fixity with plates and a channel section.

The loads were applied to the RC frame with the help of manually operated jacks which were mounted at each story level. The jacks were mounted with proving rings at each story level, and the lateral loads were applied at the bottom and top stories at a rate of 0.5:1. The maximum base shear of each cycle shown as per loading history (Figures 4–6) was applied using the equivalent static lateral load method to each model. The loads were increased stepwise up to the maximum value, and later, the lateral load was unloaded in a similar fashion. During sequential loading, the corresponding story drift and deflection of beams were recorded using digital dial gauges. The directions of the in-plane lateral loads were reversed from the other end of the model, and the corresponding story drift and deflections were recorded for the applied loads at each story level. In each first half of the cycle of loading, the lateral loads were gradually applied up to the maximum value, and the same procedure was repeated in the second half of the cycle of loading by reversing the loading direction for the same magnitude of lateral loads.

The specimen fabrications were conducted under the 2000 kN loading frame. In the first stage, reinforcement skeletal frames were constructed, and SCC was used to cast the RC frames. In the case of conventional construction, brick masonry was constructed after water curing the RC frame for 28 days. All three specimens were allowed to cure for 28 days with the help of gunny bags after the completed construction of the MI-RC frames.

Before the construction of masonry panels, the locations on column faces connecting the geo-fabric were identified using a rebar locator. After identifying the locations, the geo-fabrics were connected with the help of washers and 3 mm diameter bolts using mechanical drill bits. At every course along the bed joints, 400 mm lengths of geo-fabrics were laid and connected at the interface of the column. The length of the geo-fabric was restricted because the provision of openings in infill may be possible practically, and the prime focus of this study is the interfacial study of RC and infill panels. The casting technique of geo-fabric reinforcement and the construction of the MI-RC model is shown in Figure 10a–f.

Flexible polyester geo-fabric is used as a reinforcing material at the interface of the MI and RC frame. Polyester-type geo-fabrics are very easy to handle, non-corrosive and extensively used in the field of soil reinforcements. Hence, an attempt is made in this present experimental investigation to study the behavior of an MI-RC frame using polyester geo-fabric as an interfacial reinforcing material.

(a)

(b)

Figure 10. *Cont.*

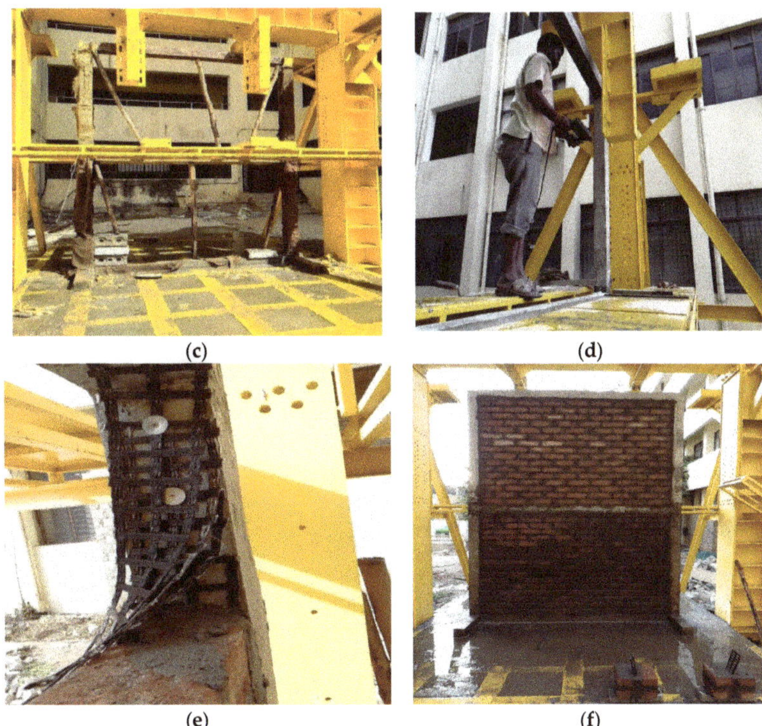

Figure 10. Frame Construction. (**a**) Reinforcement adopted for MI-RC frames; (**b**) column and beam with shuttering; (**c**) top and bottom of column and beam with shuttering; (**d**) using a drill bit to drill into columns in a safe region of the RC frames for connecting geo-fabric; (**e**) geo-fabric connected to RC frame interfaces using mechanical anchors; (**f**) MI-RC frame with geo-fabric reinforcement at interface model.

3. Results and Discussion

One-bay two-story MI-RC frames are considered in the present experimental investigation. An MI-RC frame reinforced at interfaces with geo-fabric on both stories was considered in the first model, and reinforcement at the interface of MI-RC with geo-fabric only on the top story and provided bottom soft story (open ground story) was considered in the second model. The third model is a conventional MI-RC frame with only brick MI on both the top and bottom stories. The three types of models are investigated for their story drift, stiffness degradation and ductility; energy dissipation and modes of failure when subjected to cyclic in-plane lateral loads have been discussed in the following section.

3.1. Microstructural Behavior of SCC Used in RC Frames

The SCC was prepared using GGBS as a filler material. The hydration of cement plays a major role in attaining target strength, and microstructural observations help us to know the actual behavior of the cement paste. To know the behavior of the cement paste in all three MI-RC frames, microstructural images were considered to observe the internal morphology of cement paste. The concrete cubes which were tested at the time of the cyclic loading test of the MI-RC frames were subjected to microstructural analysis after the test. The test specimens were extracted after the cube strength test. The size of the specimens considered for the analysis was 10 mm × 10 mm, and the microstructural images considered in each frame samples are shown in Figures 11–13.

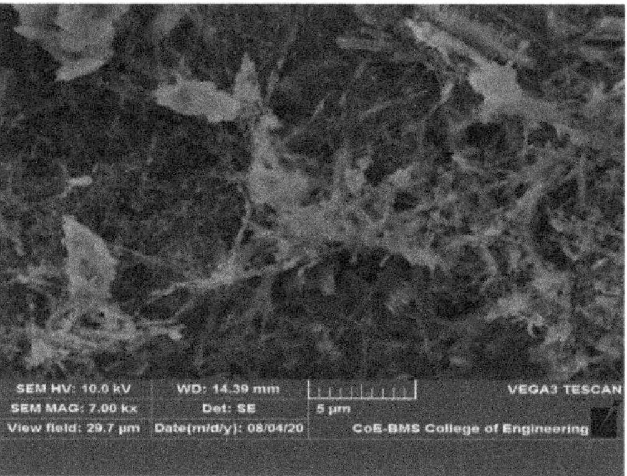

Figure 11. SEM image of MI-RC Frame-01 cube sample.

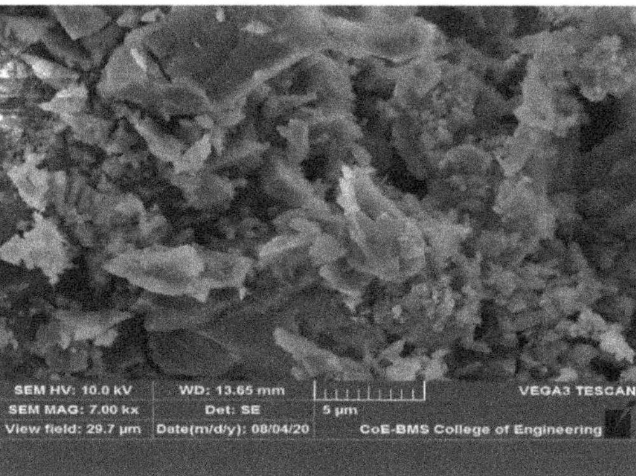

Figure 12. SEM image of MI-RC Frame-02 cube sample.

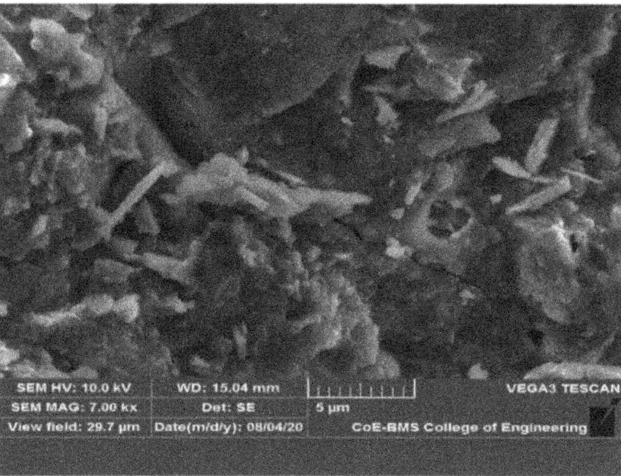

Figure 13. SEM image of MI-RC Frame-03 cube sample.

The hydrated cement paste microstructures of frames presented in the figures above are considered at a five-micron level for each frame. It can be observed that mix M40 has sufficient hydrated phases in its microstructure. There are a few un-hydrated components observed as bright spots in all of the microstructural images. In comparison with Figures 11 and 12 with concrete frames, the microstructure of concrete exhibits similar phases of hydrated and un-hydrated cement mass in its microstructure. All the three RC frame mixes resulted in the uniformity of microstructure with an observable presence of capillary voids.

The overall SEM studies showed consistency in the mix quality of the prepared grade of concrete (M40) in terms of its microstructure. There was not a considerable variation or presence of anomalies from the point of observation of all three RC frame concrete mixes.

3.2. Story Drift

The relative displacement of the frame with respect to one level to another level above or below the story is referred to as story drift. Generally, drift has been defined as the total lateral displacement at the top of the building, and the relative lateral displacement between the consecutive story levels is termed as inter-story drift. Top story drift plays a major role in the case of analysis and design of high-rise buildings, and they are subjected to this mainly under lateral loads.

The conventional unreinforced MI-RC frame was subjected to maximum base shear of 27,000 N in the sixth cycle. During the fourth cycle of loading, the MI lost contact with the RC frame at the interfaces of the bottom story. It was observed that in the fifth load cycle, the MI in the bottom story failed by indicating horizontal shearing of mortar bed joints. Further, larger values of lateral drifts were observed from the fourth to the sixth cycles of loading due to separation of the MI from the RC frame. After reaching maximum story shear in the sixth load cycle, the top story drift increased gradually, and the base shear reduced to around 50% of the maximum value in the subsequent cycles of loading.

The MI-RC frame with geo-fabric reinforcement at the interfaces model was subjected to a maximum of 11 load cycles with a maximum base shear of 44,000 N. The first crack appeared on the interfaces in the 5th cycle of loading, but the cracks widen after the 10th cycle. The geo-fabric's contribution to holding the MI and frame together as a single integral unit was observed between the 5th and 10th cycles of cyclic in-plane lateral loading of frames. Horizontal bed joint failure was observed in the MI in the 10th cycle of loading at the bottom story, and a horizontal crack on the face of the columns was also observed at the bottom story of the frame. Plastic hinge deformation was observed at the beam–column junction in the 10th cycle, and lateral drift was observed to be slightly increased between the 7th and 10th cycles of loading. The interface connection was lost in the 11th cycle of loading, and horizontal bed joint sliding was observed on the top story of MI in the same cycle of loading.

The MI-RC frame reinforced with geo-fabric at interfaces in the top story with open ground story model was subjected to a maximum base shear of 6050 N in the ninth cycle of loading. It has been observed that during the maximum cycle of loading, MI was not detached at interfaces of the top story, and only open ground story columns failed because of less stiffness. After the ninth cycle of loading, the model loses its lateral-load-resisting ability, and crushing of the column is observed at the bottom of the soft story due to the development of larger moments. A horizontal crack was also noticed on the face of the column at the bottom story, and X-type cracks were observed at the beam–column junction in the 10th cycle of loading.

Figure 14 shows the top story drifts of the MI-RC frames for the ultimate story shear with respective models. It indicates that the model with an open ground story exhibited the highest amount of top story drift, while the model reinforced with geo-fabric at interfaces of MI-RC exhibited the least. It can be noted that expectedly, the deformation of the open ground story model was two times that of the geo-fabric reinforced at interfaces model. Hence, from the experimental studies, it is noticed that geo-fabric reinforcement at interfaces of MI-RC helps to improve lateral-load-carrying capacity and controls lateral drift effectively under in-plane lateral loads.

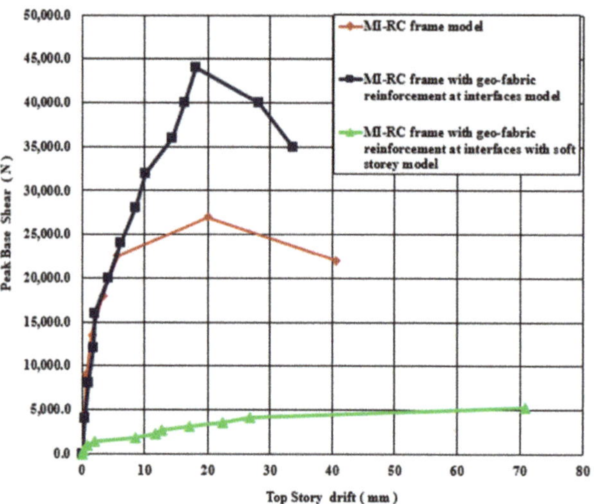

Figure 14. Comparison of peak story shear v/s top story drift.

3.3. Stiffness Degradation

The slope of the load deflection was used to measure the stiffness of the models. An initial stiffness of 10,000 N/mm was observed for the conventional MI-RC and reinforced MI-RC models, while the open ground story model exhibited an initial stiffness of only 2065 N/mm. The stiffness of the MI-RC model decreased gradually after the 3rd cycle, and the residual stiffness recorded at the end of the cycle was 1090 N/mm. The bottom story MI suffered with failure cracking of the horizontal bed joint and stepped crack failure in the fourth cycle of loading, and hence, the stiffness of the MI-RC model decreased because of the failure of the masonry infill.

Expectedly, the geo-fabric reinforced at interfaces MI-RC model performed better than the conventional MI-RC model. In the 5th cycle of loading, the interface failure mechanism was observed in both the models, but the geo-fabric-reinforced model showed better consistent performance in stiffness up to the 8th cycle, and even the residual stiffness was 1340 N/mm, which is high when compared with the MI-RC model. The presence of geo-fabric enhanced the stiffness behavior of the MI-RC frame. The bed joint failure mechanism was observed in the ninth cycle of loading at the bottom story in the unreinforced bed joint regions.

The MI-RC model reinforced with geo-fabric with open ground story model is relatively very flexible at the bottom story such that initial stiffness was 2065 N/mm, which is only 10.5% of the initial stiffness of the other 2 models. The stiffness steadily decreased up to the fifth cycle and thereafter at a faster rate because the RC frame had developed X-type cracks at the beam–column junction, and shear cracks were also noticed in the columns. The residual stiffness of the frame at the end of 10th cycle of loading was 90 N/mm, which is much lower than that of the other types of models.

Figure 15 shows the stiffness degradation of all the three types of MI-RC models tested corresponding to in-plane reverse cyclic loading. It can be noted that conventional MI-RC and geo-fabric-reinforced MI-RC models' initial stiffness is 80% more than that of the open ground story model. This clearly indicates that the infill plays a major role in contributing stiffness under lateral loads. The open ground story model exhibited lesser stiffness; even the top story MI was reinforced with geo-fabric at interfaces.

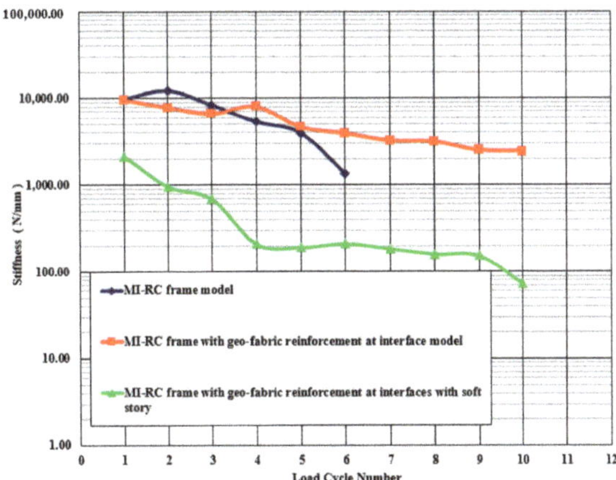

Figure 15. Comparison of stiffness degradation chart.

3.4. Ductility

Ductility is a measure of the material's ability to undergo permanent deformation without fracture. In all three MI-RC frame models, the ductility factor is considered for evaluation, and it is the ratio of failure displacement to yield displacement. In the present section, top story drift was considered for the evaluation of ductility factor, and calculated cumulative ductility factor is defined as the total sum of ductility factor at maximum base shear levels in each cycle up to the load cycles considered.

The conventional MI-RC frame model exhibited an appreciable top story drift during the 4th cycle of loading, and a drift of 3.37 mm is considered as the yield drift of the MI-RC frame. The load–displacement behavior of the MI-RC frame model is shown in Figure 16, and it is evident from the chart that the frame has displaced more after the fourth cycle of loading. In the same loading phase, the MI separated from the RC frame at interfaces, and crack sizes increased in the MI of the bottom story.

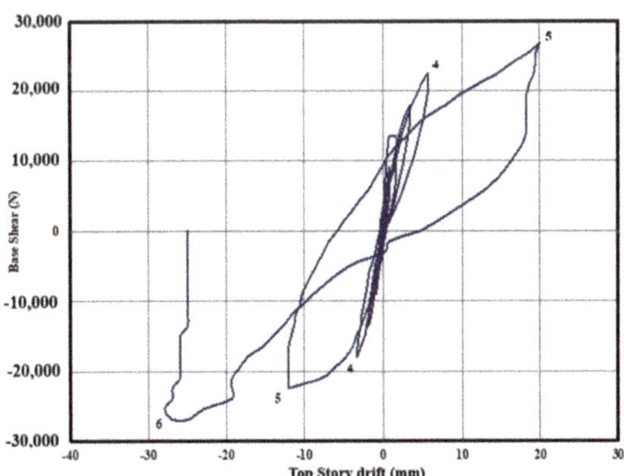

Figure 16. Load–displacement behavior of the conventional MI-RC frame model.

The MI-RC reinforced with geo-fabric at interfaces model had a yield drift of 1.9 mm in the 5th cycle of loading, which was much lower than that of the conventional MI-RC frame. Figure 17 shows the load–displacement behavior of the model subjected to in-plane cyclic loads. It is evident from the load–displacement chart that the geo-fabric helped to control the lateral drift of the model by making the MI and RC frame an integral unit by strongly connecting them at the interfaces.

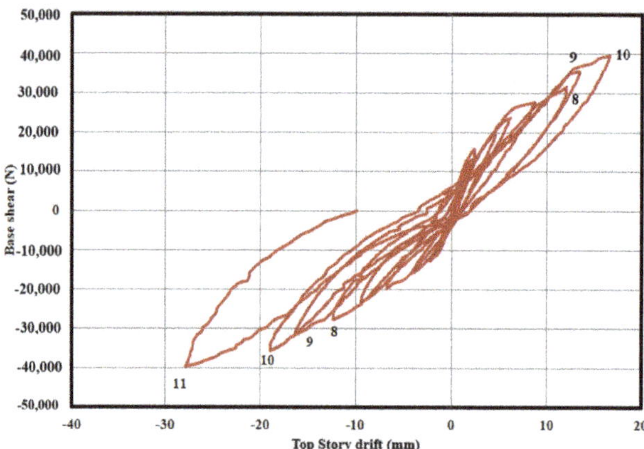

Figure 17. Load–displacement behavior of the geo-fabric-reinforced MI-RC frame model.

For the open ground story model, the yield drift was noticed in the 5th cycle of loading at a magnitude of 11.6 mm. Subsequently, cumulative ductility factors were evaluated, and the model lost its ability to resist deformation in the 10th cycle because of plastic deformation observed at beam–column junctions. It was observed that the presence of geo-fabric reinforcement at the interfaces of the top story and open ground story model did not have much impact on load-carrying capacity or on the prevention of lateral drift deformation. The load–displacement behavior of this case is presented in Figure 18.

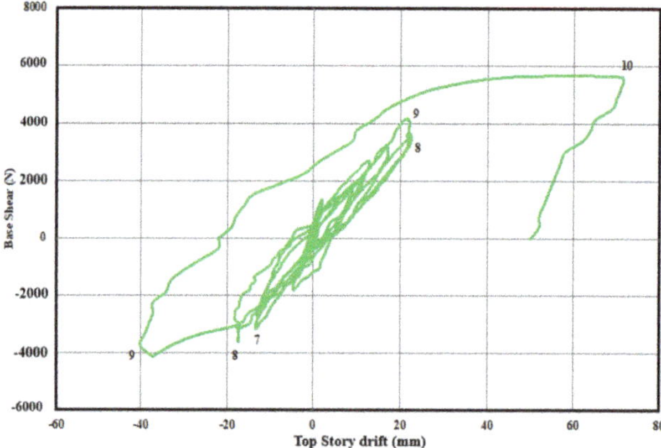

Figure 18. Load–displacement behavior of the geo-fabric-reinforced MI-RC frame with open ground story model.

A generic comparison using the hysteretic curves for the three specimens is presented in Figure 19. The cumulative ductility factor is presented in Figure 20, and it is observed that after the eighth cycle of loading, the model lost its ability gradually, and this resulted in maximum deformation/drift.

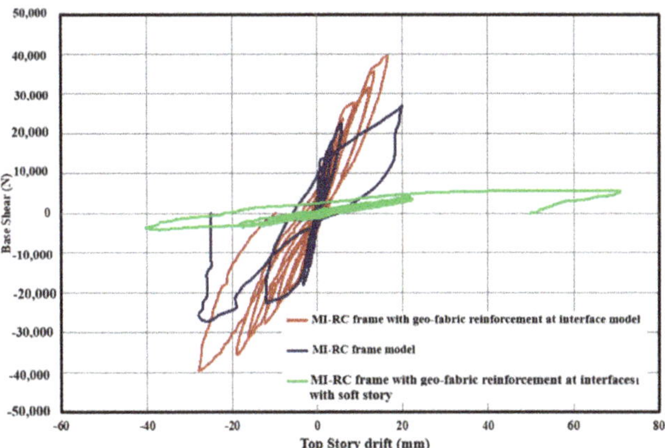

Figure 19. Comparison of hysteretic curves for the three specimens.

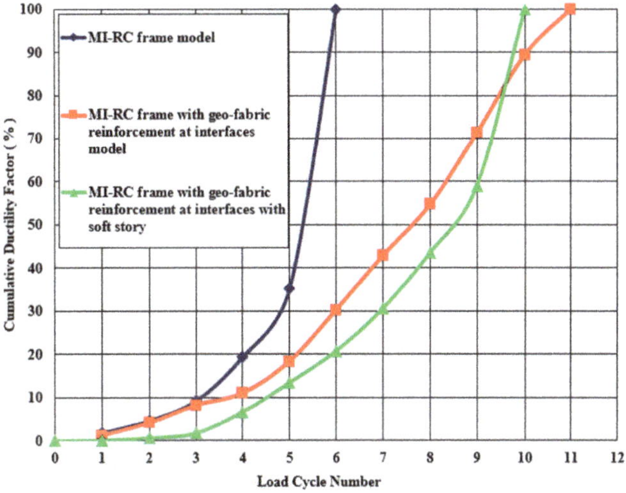

Figure 20. Cumulative ductility chart.

3.5. Energy Dissipation

The energy dissipation in the case of the models tested is defined as the area enclosed by the lateral load displacement of the hysteresis loop for the in-plane cyclic lateral loads. It is considered as an important aspect to study the behavior of MI-RC frames under in-plane lateral loads. The main parameter used to study the energy dissipation is the load–displacement relationship. The energy dissipated by the conventional MI-RC frame model in the initial 1st cycle was 0.96 kN-mm, while the cumulative energy dissipated in the final 6th cycle of loading was 1306.75 kN-mm. The energy absorbed by the geo-grid reinforced MI-RC model in first cycle was 0.38 kN-mm, and the final cumulative energy dissipated at the end of the 10th cycle was 3990.68 kN-mm. It is also noted that in the case of the open ground story model, the energy dissipation was much lower. It had an initial cumulative energy dissipation of 0.46 kN-mm in the 1st cycle and 934.39 kN-mm at the end of the 10th cycle of loading. The cumulative energy dissipation chart is presented in Figure 21.

The total energy dissipated by geo-fabric-reinforced MI-RC frames was about 2.5 times higher than that of the conventional MI-RC frames and 4.27 times more than that of the open ground story model. Geo-fabric reinforcement had no significance in the open ground story models.

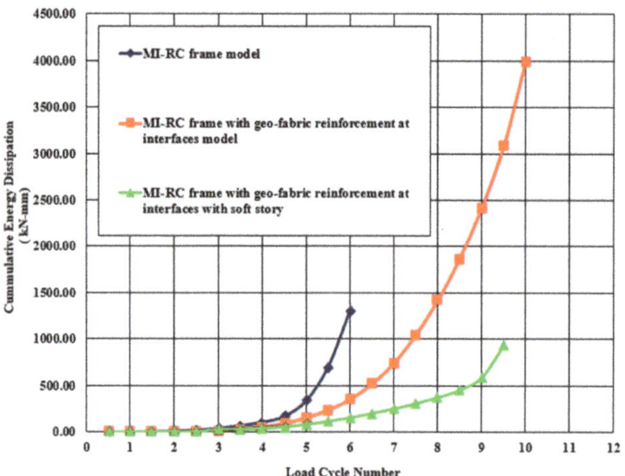

Figure 21. Cumulative energy dissipation chart.

3.6. Mode of Failure

The first crack was observed in the case of the MI-RC model between the interfaces of MI and RC at the bottom story of the frame under the base shear of 22,500 N during the 5th cycle of loading, as shown in Figure 22. After the interface detachment at the bottom story, the infill failed in a stepped way along with horizontal bed joint sliding. The various types of failures in MI such as sliding and bed joint failure were observed in the 6th cycle of loading at 25,600 N of base shear. In the same cycle of loading, the shear cracks were observed in columns and the X-type of failure was observed in the beam–column junction. On further loading, the frame experienced permanent deformation, and this resulted in a maximum drift of 20.01 mm.

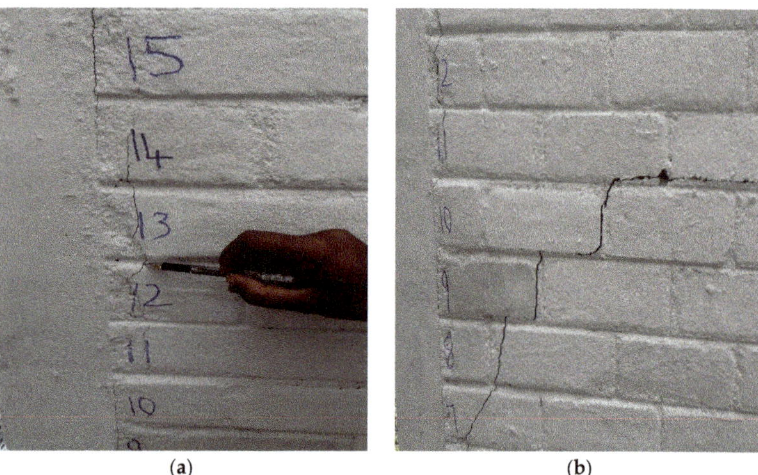

(**a**) (**b**)

Figure 22. *Cont.*

(c)   (d)

Figure 22. MI-RC model: (**a**) interfacial cracks appeared in the bottom story at a base shear of 22,500 N; (**b**) stepped cracks observed in the bottom story of the MI-RC frame model; (**c**) horizontal and stepped cracks in the MI-RC frame model; (**d**) final deformed shape of the MI-RC model.

The different failures that occurred in different cycles of loading are presented in Figure 22. During subsequent cycles of loading, the frame lost its lateral-load-resisting ability, and the cracks widened at the MI. This resulted in maximum drift and eventually failure of the frame. There was a sufficient increase in the lateral-load-resisting ability of the geo-fabric-reinforced MI-RC frame model. The model exhibited a better performance between the fifth and ninth cycles of loading. Because of interfacial strengthening, the MI resisted the lateral loads with better energy dissipation capacity. The first interfacial cracks were observed in the fifth cycle of loading, and after the ninth cycle of loading, horizontal bed sliding and stepped cracks developed at the bottom story of MI. The final deformed shapes along with the different failure patterns observed under different cycles of loading are shown in Figure 23.

(a)   (b)

Figure 23. *Cont.*

(c) (d)

Figure 23. Geo-fabric-reinforced MI-RC frame model: (**a**) interfacial crack appeared in the geo-fabric-reinforced MI-RC frame model; (**b**) horizontal and stepped cracks in the geo-fabric-reinforced MI-RC frame model; (**c**) shear cracks in columns of the geo-fabric-reinforced MI-RC frame model; (**d**) final deformed shape of the geo-fabric-reinforced MI-RC model.

The geo-fabric-reinforced MI-RC model with open ground story showed decreased lateral-load-resisting capacity compared with the other two types of models. The bottom story experienced the lowest stiffness due to the open ground story, and the upper story, which was reinforced with geo-fabric, exhibited rigid body displacement in the initial cycles of loading. Initially, a horizontal crack was observed in the bottom of the column when base shear was 3600 N in the 8th cycle of loading. Upon further loading, cracks developed at the beam–column junction, and shear cracks were observed at the bottom of the columns. It was also observed that there was no major crack observed in the geo-fabric-reinforced MI interfaces at the top story. The final deformed shape along with the different types of failures is shown in Figure 24.

(a) (b)

Figure 24. *Cont.*

(c)   (d)

Figure 24. Open ground story geo-fabric-reinforced MI-RC model: (**a**) initial cracks in columns; (**b**) shear cracks in columns; (**c**) final crack at beam–column junction; (**d**) final deformed shape.

In the MI-RC frame with an open ground story, geo-fabric reinforcement does not help to increase the lateral load capacity of the frame. The beam–column joints fail mainly because of a large moment developed because of MI at the junction. Further, these kinds of failures can be controlled by increasing stiffness in open ground story columns.

4. Summary and Conclusions

The three half-scaled models considered in this present study are the MI-RC frame model, the MI-RC frame reinforced with geo-fabrics at interfaces model and the MI-RC open ground story reinforced with a geo-fabric element. Based on experimental investigations, the following set of conclusions has been drawn.

The peak base-shear capacities of the conventional MI-RC frame model and the geo-fabric-reinforced MI-RC frame model were 27,000 N and 44,000 N, respectively. This indicates that the peak base shear capacity of the MI-RC frame with geo-fabric reinforcement is 163% of that of the conventional MI-RC frame model. The corresponding theoretically computed value is 55,000 N for both the models. Further, the peak base shear of the open ground story model reinforced with geo-fabric at the interfaces of the top story was 6050 N, compared to the theoretically computed value of 6975 N. It is clear that the presence of infill strengthened the interfaces of MI-RC, helping to enhance the lateral load capacity of MI-RC frames.

The maximum top story drift observed in the case of the conventional MI-RC frame model and geo-fabric-reinforced MI-RC frame model was 42 mm and 32 mm, respectively. It was observed that the maximum top story drift of the geo-fabric-reinforced MI-RC frame model was 76% of that of the conventional MI-RC frame model. Further, the maximum top story drift of the frame with an open ground story was 74 mm, which indicates the influence of masonry infill on the lateral drift of the model.

The initial stiffness of the MI-RC frame model and geo-fabric-reinforced MI-RC frame model showed a similar value of 10,000 N/mm, but after the 5th cycle of loading, the MI-RC frame completely lost its resistance against lateral loads. The geo-fabric-reinforced model showed consistent stiffness between the fifth and the eighth cycles of loading. Hence, this indicated that the presence of geo-fabric as a reinforcement at interfaces helps to enhance the stiffness and lateral-load-carrying ability of the MI-RC frame. However, in the case of the open ground story model, the initial stiffness was 2065 N/mm, which is 5 times less than that of the MI-RC frame model. This illustrates the influence of infill in enhancing the stiffness of the frame.

The geo-fabric-reinforced MI-RC frame model exhibited higher post-yield ductility, while the open ground story model exhibited that lowest post-yield ductility. Further, the energy dissipated in the geo-fabric-reinforced MI-RC frame model was 300% of that of the conventional MI-RC frame model. This is because of the performance of geo-fabric at interfaces, which helped to enhance the bonding behavior between the RC frame and the masonry infill. The energy dissipation capacity of the MI-RC frame model drastically decreased after the failure of the masonry infill.

In the case of the conventional MI-RC frame model, interfacial failure was observed in between the MI and RC frame. A brittle fracture of the masonry resulted in an increase in lateral loads on columns, which lead to shear failure at the lower story of the columns. In the case of the geo-fabric-reinforced MI-RC frame models, masonry failure was observed in a stepped pattern in unreinforced regions. Minor cracks were initiated at interfaces in the initial loading stages, but the geo-fabric reinforcement at interfaces helped to prevent the MI-RC interfacial failure. A horizontal shear crack was observed in the final cycles of loading due to large stiffness degradation of MI at the bottom story of columns.

The open ground story model failed at the bottom ground story because of lower stiffness. In the upper story, the MI did not develop any cracks, and it is to be noted that the upper story exhibited rigid body translation behavior. This mechanism led to shear–flexure cracks in the columns at the bottom story.

The presence of geo-fabric reinforcements clearly maintains the contribution of the infill wall to the RC structure, improving the global load-carrying capacity of the frame. One the main issues related to infilled structures is that the infilled walls can protect the structures for lower seismic demands; the use of geo-fabric reinforcements may increase this protection capacity.

Author Contributions: Conceptualization, K.S.S. and A.S.A.; methodology, K.S.S. and A.S.A.; validation, K.S.S., A.S.A. and H.R.; formal analysis, K.S.S., A.S.A. and H.R.; writing—original draft preparation, K.S.S. writing—review and editing, A.S.A. and H.R. All authors have read and agreed to the published version of the manuscript.

Funding: This research received no external funding.

Data Availability Statement: Data available on request.

Conflicts of Interest: The authors declare no conflict of interest.

References

1. Asteris, P.G. Lateral Stiffness of Brick Masonry Infilled Plane Frames. *J. Struct. Eng.* **2005**, *131*, 523–524. [CrossRef]
2. Furtado, A.; Rodrigues, H.; Arêde, A.; Varum, H. Experimental evaluation of out-of-plane capacity of masonry infill walls. *Eng. Struct.* **2016**, *111*, 48–63. [CrossRef]
3. Rodrigues, H. *Simplified Macro-Model for Infill Masonry Panels*; Taylor & Francis: Abingdon, UK, 2022; Available online: https://www.tandfonline.com/doi/abs/10.1080/13632460903086044 (accessed on 1 March 2023).
4. Furtado, A.; Rodrigues, H.; Arêde, A.; Varum, H. Simplified macro-model for infill masonry walls considering the out-of-plane behaviour. *Earthq. Eng. Struct. Dyn.* **2015**, *45*, 507–524. [CrossRef]
5. Sreekeshava, K.S.; Arunkumar, A.S. *Experimental Studies on Performance of Geo-Synthetic Strengthened Brick Masonry Infill*; Lecture Notes in Civil Engineering; Springer: Cham, Switzerland, 2021; Volume 97. [CrossRef]
6. Wang, X.; Li, S.; Wu, Z.; Bu, F.; Wang, F. Experimental Study on Seismic Strengthening of Confined Masonry Walls Using RPC. *Adv. Mater. Sci. Eng.* **2019**, *2019*, 5095120. [CrossRef]
7. Gopinath, S.; Madheswaran, C.K.; Prabhakar, J.; Devi, K.G.T.; Anuhya, C.L. Strengthening of Unreinforced Brick Masonry Panel Using Cast-in-Place and Precast Textile-Reinforced Concrete. *J. Earthq. Eng.* **2020**, *26*, 1209–1227. [CrossRef]
8. Sreekeshava, K.S.; Arunkumar, A.S.; Ravishankar, B.V. *Experimental Studies on Polyester Geo-Fabric Strengthened Masonry Elements*; Lecture Notes in Civil Engineering; Springer: Singapore, 2020; Volume 55. [CrossRef]
9. Sreekeshava, K.S.; Arunkumar, A.S. Effect of polypropylene (PP) geo-fabric reinforcement in brick masonry under axial loads. *Int. J. Recent Technol. Eng. Regular* **2019**, *8*, 369–373.
10. Sreekeshava, K.S.; Arunkumar, A.S.; Ravishankar, B.V. *Experimental Studies on Brick Masonry Elements with Geo-Fabric Bed Joint Reinforcement*; Lecture Notes in Civil Engineering; Springer: Singapore, 2020; Volume 68. [CrossRef]
11. Sadek, H.; Lissel, S. Seismic performance of masonry walls with GFRP and Geo-grid bed joint reinforcement. *Constr. Build. Mater.* **2013**, *41*, 977–989. [CrossRef]
12. Faella, C.; Camorani, G.; Martinelli, E.; Paciello, S.; Perri, F. Bond behavior of FRP strips glued on masonry: Experimental investigation and empirical formation. *Constr. Build. Mater.* **2012**, *31*, 353–363. [CrossRef]

13. Parisi, F.; Iovinella, I.; Balsamo, A.; Augenti, N.; Prota, A. In-plane behavior of tuff masonry strengthened with inorganic matrix-grid composites. *Compos. Part B* **2013**, *45*, 1657–1666. [CrossRef]
14. Saneinejad, A.; Hobbs, B. Inelastic design of in-filled frames. *J. Struct. Eng.* **1995**, *121*, 634–650. [CrossRef]
15. Madan, A.; Reinhorn, A.M.; Mander, J.B.; Valles, R.E. Modeling of Masonry Infill Panels for Structural Analysis. *J. Struct. Eng.* **1997**, *123*, 1295–1302. [CrossRef]
16. Del Vecchio, C.; Di Ludovico, M.; Verderame, G.; Prota, A. Pseudo-dynamic tests on full-scale two storeys RC frames with different infill-to-structure connections. *Eng. Struct.* **2022**, *266*, 114608. [CrossRef]
17. Molitierno, C.; Del Vecchio, C.; Di Ludovico, M.; Prota, A. Pseudodynamic Tests and Numerical Modelling for Damage Analysis of Infilled RC Frames. *J. Earthq. Eng.* **2023**, 1–26. [CrossRef]
18. Dymiotis, C.; Kappos, A.J.; Chryssanthopoulos, M.K. Seismic reliability of masonry in-filled RC frames. *J. Struct. Eng.* **2001**, *127*, 296–305. [CrossRef]
19. *IS:13920-1993*; Indian Standard Code of Practice for. Ductile Detailing of Reinforced Concrete Structures Subjected to Seismic Forces. Bureau of Indian Standards: Delhi, India, 1993.
20. *IS 12269*; 53 Grade Ordinary Portland Cement. Bureau of Indian Standards: Delhi, India, 2013.
21. *IS 10262*; 2019 Concrete Mix Proportioning—Guidelines. Bureau of Indian Standards: Delhi, India, 2019.
22. *IS 1893*; Part 1: 2016 Criteria for Earthquake Resistant Design of Structures—Part 1: General Provisions and Buildings. Bureau of Indian Standards: Delhi, India, 2016.

Article

The Comparison of Code-Based and Empirical Seismic Fragility Curves of Steel and RC Buildings

Mahnoosh Biglari [1,*], Behrokh Hosseini Hashemi [2] and Antonio Formisano [3]

[1] Civil Engineering Department, School of Engineering, Razi University, Kermanshah P.O. Box 67149-67346, Iran
[2] Department of Structures, International Institute of Earthquake Engineering and Seismology (IIEES), Tehran P.O. Box 19537-14453, Iran; behrokh@iiees.ac.ir
[3] Department of Structures for Engineering and Architecture, School of Polytechnic and Basic Sciences, University of Naples Federico II, 80138 Naples, Italy; antoform@unina.it
* Correspondence: m.biglari@razi.ac.ir

Abstract: Seismic codes were developed to reduce the structural vulnerability and risk associated with earthquakes in earthquake-prone regions of the world. The effectiveness of the code in preventing damage is dependent on the performance level defined and the construction technology employed. The seismic fragility curves for two recent versions of the seismic code of Iran are determined by using the hybrid method. The probability of damage levels is visualized by these curves. To develop these curves, only the assumptions of the code are taken into account. These curves are compared with the empirical fragility of the recent devastating earthquake in Iran. The results indicate that, despite a similar probability of damage to the different seismic-resistant systems, steel-braced frames pose a greater risk of collapse. Concerning earthquake damage, the steel and RC moment-resisting frames have shown higher damage probability than expected from the code.

Keywords: capacity curves; fragility curves; code-based approach; steel and RC buildings; Iranian earthquake code

Citation: Biglari, M.; Hosseini Hashemi, B.; Formisano, A. The Comparison of Code-Based and Empirical Seismic Fragility Curves of Steel and RC Buildings. *Buildings* **2023**, *13*, 1361. https://doi.org/10.3390/buildings13061361

Academic Editors: Rajesh Rupakhety and Dipendra Gautam

Received: 23 April 2023
Revised: 15 May 2023
Accepted: 20 May 2023
Published: 23 May 2023

1. Introduction

Seismic regulations are developed to ensure that all buildings that are at risk of being damaged by an earthquake are safe. Meanwhile, economic prosperity and technological advancement in construction are expected to have an impact on seismic code performance. Seismic regulations generally aim to minimize casualties from each earthquake event.

Other factors, such as complexity and diversity of site stratification, the quality of materials used, design and execution errors, past seismic experience, and maintenance throughout the structures, are also factors that determine earthquake damage to structures. Despite all of the above, seismic codes for each building accept damage even when all the design rules have been observed. Recommended assessment methods offered by different seismic codes were investigated in [1]. They compared the fragility curves of existing low- to mid-rise RC buildings designed with four seismic codes of TEC-2007 [2], TBEC-2018 [3], EC8/3 [4], and ASCE 41-17 [5] and showed that different code methods give remarkably different damage estimations under similar seismic demand levels. However, implementing the code design rules reduces the damage rate.

The seismic fragility curves provide an estimate of the likelihood of reaching or exceeding a specific damage level at each acceleration level for different limit states. The research focus is on determining seismic fragility curves for constructed buildings, rather than relying on assumptions in seismic codes. There are several ways to calculate seismic fragility curves: empirical, analytical, expert judgment, or a hybrid method. To propose the empirical method, an extensive database of peak ground motion parameters is required. The database is primarily accessible through the collection of information from strong earthquakes [6–9]. The empirical fragility curves are widely used in urban-scale seismic vulnerability assessment [10]. The analytical seismic fragility curves are developed by modeling and analyzing structures statically or dynamically. The ground motion records should exhibit a wide range of low to high PGA values in the dynamic analysis. During the pioneering research, incremental dynamic analysis methods were used by scaling a ground motion record to various peak accelerations [11]. Therefore, only the amplitude of the record changes, and other parameters such as frequency content and duration are not scaled. Other researchers used a set of unscaled natural ground motion records with a wide range of amplitudes for the dynamic analysis, e.g., [12]. In the absence of natural ground motion records corresponding to the seismo-tectonic conditions of a region, stochastic ground motion records have been used to propose probabilistic seismic fragility curves [13] and three-dimensional consistent hazard–fragility curves considering multiple capacity–demand uncertainties [14]. Modeling requires a good understanding of the material properties and geometry. Many researchers followed the analytical method for fragility curves, e.g., [15–18]. Expert judgment-based methods are based on human opinions and are used to replace the process of numerical modeling or observed data. However, this method should be verified by the observed data or analytical methods. The pioneer of this method was the Applied Technology Council (funded by the Federal Emergency Management Agency (FEMA)), as summarized in ATC-13 [19]. The hybrid method combines the analytical results with the empirical parameters to establish damage limit state definitions. The RISK-UE Project [20] employed this method to determine seismic fragility curves for the Unified Building Code [21].

This research utilizes hybrid fragility curves based on seismic code considerations, independent of earthquake-resistant configuration systems, retrofitting, and maintenance interventions. Therefore, seismic code users are informed of the code-based fragility curves used to evaluate the seismic vulnerability of the structure. This paper presents capacity and hybrid seismic fragility curves for residential buildings designed following Iranian seismic code IRSt2800 [22,23] based on the methodology presented in Figure 1. These curves are not included in seismic standards.

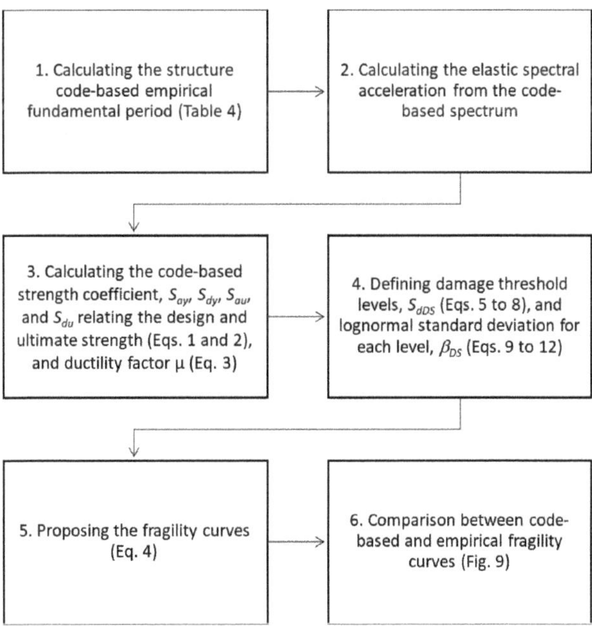

Figure 1. Methodology flowchart.

The Iranian seismic code has four versions. The third [22] and fourth [23] versions were considered in this research. These versions differ in the parameters used to calculate the base shear coefficient. The effects of these differences will be observed in the fragility curves later. The third version was considered because it was used for nine years to renovate residential, commercial, and administrative buildings in most major Iranian cities. Therefore, many existing steel (ST) buildings and reinforced concrete (RC) buildings were built based on this version of the seismic code. Furthermore, regarding the recent catastrophic earthquake in Iran, which occurred on 2017 November 12th in Sarpol-e-zahab, most of the exposed buildings were constructed according to the third version of the Iranian seismic code [22]. Therefore, the fragility curves of the third version can be compared with the existing empirical seismic fragility curves. Furthermore, the fragility curves provided in the fourth version of the seismic code [23], the most recent version, were utilized to assess the seismic performance of new buildings.

2. Building Typology and Damage Grades

The selected building typology matrix includes buildings built in the last 20 years based on the Iranian seismic code. The buildings' typologies are characterized by four factors: the seismic code, the building height, the construction frame material, and the seismic-resistant system.

Buildings are divided according to their seismic code version into two general categories: moderate code and high code. Buildings of the moderate code are constructed based on the third version (V3) of the Iranian seismic code [22], while high-code buildings are constructed based on the fourth (latest) version (V4) of the Iranian seismic code [23]. Tables 1 and 2 show the building typology matrices for moderate-code and high-code buildings, respectively. There are two sub-groups of buildings: low-rise (L) buildings (9 m in height or with three stories) and mid-rise (M) buildings (18 m in height or with six stories). In terms of construction materials, they are divided into two categories: steel (ST) buildings and reinforced concrete (RC) buildings. Each of these buildings has a different type based on the seismic-resistant system. Steel buildings are divided into three sub-categories: braced frames (including eccentric/concentric), moment-resisting frames (including special/intermediate), and a combination of moment-resisting and braced frames. The reinforced concrete buildings are divided into two sub-categories: moment-resisting frames (including special or intermediate) and a combination of moment-resisting frames and shear walls (including special or intermediate). The seismic-resistant system is named based on the seismic code system of Iran to be able to match with the code more easily. The last columns of Tables 1 and 2 contain the name of each building type. Subsequently, the studied structures are divided into 50 types (24 for moderate code and 26 for high code). For each of these 50 types, the fragility curves are presented based on the parameters introduced in the seismic code.

Table 1. The building typology matrix for moderate-code (V3) buildings.

Main Type	Description		Height	No. of Stories	Type Name
Steel (ST)	Braced frames	Eccentrically (EBF)	Low-rise (L)	3	B5-ST-L-V3
			Mid-rise (M)	6	B5-ST-M-V3
		Concentrically (CBF)	Low-rise (L)	3	B6-ST-L-V3
			Mid-rise (M)	6	B6-ST-M-V3
	Moment-resisting frame	Special (SMF)	Low-rise (L)	3	P4-ST-L-V3
			Mid-rise (M)	6	P4-ST-M-V3
		Intermediate (IMF)	Low-rise (L)	3	P5-ST-L-V3
			Mid-rise (M)	6	P5-ST-M-V3

Table 1. *Cont.*

Main Type	Description		Height	No. of Stories	Type Name
	Combination of moment-resisting frame and braced frame	SMF + EBF	Low-rise (L)	3	T4-ST-L-V3
			Mid-rise (M)	6	T4-ST-M-V3
		SMF + CBF	Low-rise (L)	3	T5-ST-L-V3
			Mid-rise (M)	6	T5-ST-M-V3
		IMF + EBF	Low-rise (L)	3	T6-ST-L-V3
			Mid-rise (M)	6	T6-ST-M-V3
		IMF + CBF	Low-rise (L)	3	T7-ST-L-V3
			Mid-rise (M)	6	T7-ST-M-V3
Reinforced concrete (RC)	Moment-resisting frame	Special (SMF)	Low-rise (L)	3	P1-RC-L-V3
			Mid-rise (M)	6	P1-RC-M-V3
		Intermediate (IMF)	Low-rise (L)	3	P2-RC-L-V3
			Mid-rise (M)	6	P2-RC-M-V3
	Combination of moment-resisting frame and RC shear wall	SMF + Special shear walls	Low-rise (L)	3	T1-RC-L-V3
			Mid-rise (M)	6	T1-RC-M-V3
		IMF + Intermediate shear walls	Low-rise (L)	3	T2-RC-L-V3
			Mid-rise (M)	6	T2-RC-M-V3

Table 2. The building typology matrix for high-code (V4) buildings.

Main Type	Description		Height	No. of Stories	Type Name
Steel (ST)	Braced frames	Eccentrically (EBF)	Low-rise (L)	3	B5-ST-L-V4
			Mid-rise (M)	6	B5-ST-M-V4
		Concentrically (CBF)	Low-rise (L)	3	B8-ST-L-V4
			Mid-rise (M)	6	B8-ST-M-V4
	Moment-resisting frame	Special (SMF)	Low-rise (L)	3	P4-ST-L-V4
			Mid-rise (M)	6	P4-ST-M-V4
		Intermediate (IMF)	Low-rise (L)	3	P5-ST-L-V4
			Mid-rise (M)	6	P5-ST-M-V4
	Combination of moment-resisting frame and braced frame	SMF + EBF	Low-rise (L)	3	T5-ST-L-V4
			Mid-rise (M)	6	T5-ST-M-V4
		IMF + EBF	Low-rise (L)	3	T6-ST-L-V4
			Mid-rise (M)	6	T6-ST-M-V4
		SMF + CBF	Low-rise (L)	3	T7-ST-L-V4
			Mid-rise (M)	6	T7-ST-M-V4
		IMF + CBF	Low-rise (L)	3	T8-ST-L-V4
			Mid-rise (M)	6	T8-ST-M-V4
Reinforced concrete (RC)	Moment-resisting frame	Special (SMF)	Low-rise (L)	3	P1-RC-L-V4
			Mid-rise (M)	6	P1-RC-M-V4
		Intermediate (IMF)	Low-rise (L)	3	P2-RC-L-V4
			Mid-rise (M)	6	P2-RC-M-V4

Table 2. *Cont.*

Main Type	Description		Height	No. of Stories	Type Name
Combination of moment-resisting frame and RC shear wall	SMF + Special shear walls		Low-rise (L)	3	T1-RC-L-V4
			Mid-rise (M)	6	T1-RC-M-V4
	IMF + Special shear walls		Low-rise (L)	3	T2-RC-L-V4
			Mid-rise (M)	6	T2-RC-M-V4
	IMF + Intermediate shear walls		Low-rise (L)	3	T3-RC-L-V4
			Mid-rise (M)	6	T3-RC-M-V4

The ground type is assumed to be rocky (type I based on IRSt2800 [22,23]). Similar to the area affected by the Sarpol-e-zahab earthquake, the hazard zone is considered a high-risk area (peak ground acceleration of PGA = 0.3 g based on IRSt2800 [22,23]). The results are therefore comparable with the existing empirical fragility curves presented in [9].

Based on the LM2 methodology of the RISK-UE project [20], the damage state is assessed following the FEMA/NIBS [24] guidelines. It uses four labels of DS (S = 1 to 4) which distinguish the no-damage building state from D0 (Table 3).

Table 3. Damage grading description.

Damage Grade	Damage Grade Label	Structural Damage	Non-Structural Damage	Description
None	D0	No	No	No
Minor	D1	No	Slight	Fine cracks in plaster over frame members or in walls at the base. Fine cracks in partitions and infills.
Moderate	D2	Slight	Moderate	Cracks in columns and beams of frames and structural walls. Cracks in partition and infill walls; fall of brittle cladding and plaster. Falling mortar from the joints of wall panels.
Severe	D3	Moderate	Heavy	Cracks in columns and beam–column joints of frames at the base and joints of coupled walls. Spilling of concrete cover, buckling of reinforced rods. Large cracks in partition and infill walls, failure of individual infill panels.
Collapse	D4	Heavy and very heavy	Very heavy and total collapse	Large cracks in structural elements with compression failure of concrete and fracture of rebars; bond failure of beam-reinforced bars; tilting of columns. The collapse of either a few columns or a single upper floor. The collapse of the ground floor or parts (e.g., wings) of buildings.

3. Capacity Spectra

It is necessary to obtain capacity curves from the code parameters for each type of building to develop code-based seismic fragility curves. This method estimates the expected first-mode peak response of a building at a given demand. The first mode of vibration of buildings is assumed to be dominant. Iranian seismic code-based bilinear capacity curves are based on yield and ultimate structural strength levels. The values are obtained from the prescribed values of the code for each type of seismic-resistant system.

Equations (1) and (2) define the coordinates for the yield capacity and the ultimate capacity points of the capacity curves, respectively. From Fajfar [25], period and ductility are assumed, and spectral acceleration and displacement are all determined in these two equations for four unknown quantities in the force-based design method.

$$\text{Yield capacity point } (S_{ay}, S_{dy}): \quad \begin{cases} S_{ay} = \Omega_0 \dfrac{C_s}{\alpha_1} g \\ S_{dy} = \dfrac{S_{ay}}{4\pi^2} T^2 = \Omega_0 \dfrac{C_s}{\alpha_1} \dfrac{T^2}{4\pi^2} g \end{cases} \quad (1)$$

$$\text{Ultimate capacity point } (S_{au}, S_{du}): \quad \begin{cases} S_{au} = \lambda S_{ay} = \lambda \Omega_0 \dfrac{C_s}{\alpha_1} g \\ S_{du} = \lambda \mu S_{dy} = \lambda \mu \Omega_0 \dfrac{C_s}{\alpha_1} \dfrac{T^2}{4\pi^2} g \end{cases} \quad (2)$$

where S_{ay} and S_{au} are spectral acceleration for yield and ultimate points, respectively. S_{dy} and S_{du} are spectral displacements for yield and ultimate points, respectively. Cs is the base shear coefficient corresponding to the design strength of the first plastic hinge (Figure 2). According to the seismic code of Iran [22,23], Cs is the ratio of ground spectral acceleration from the standard design spectra of the code (Figure 3) and the code strength reduction factor, R (Table 4) (i.e., $Cs = S_{ag}/R$). The code recommends that this coefficient be multiplied by the coefficient of the importance of structures. This coefficient is equal to 1 for residential buildings considered in this research. Ω_0 is an over-strength factor equal to 2.8 for the V3 of [22] and 2 for the V4 of [23]. $\lambda = 1 + k_{pl}$ where k_{pl} is the slope of the plastic branch considered here based on expert judgment equal to 20%. α_1 is an effective mass coefficient that Freeman [26] suggested adopting as between 0.75 to 0.83 for most multi-story buildings. Milutinovic and Trendafiloski [27] estimated 0.71 to 0.73 for RC frame and RC dual-system buildings. Here, α_1 is considered equal to 0.75 and T is a typical elastic period of the building that may be estimated using empirically developed formulas from the code presented in Table 4 for various structural typologies. μ is the ductility demand calculated by bilinear Equation (3) (e.g., [28,29]) from the representation of the code strength reduction factor R.

$$\begin{cases} \mu = (R-1)\dfrac{T_C}{T} + 1 & T < T_C \\ \mu = R & T \geq T_C \end{cases} \quad (3)$$

where T_C is a characteristic period of the ground motion, typically defined as the corner period at the beginning of the constant velocity range. A typical value of T_C, for IRSt2800 [22,23] for a rocky site, is equal to 0.4 s.

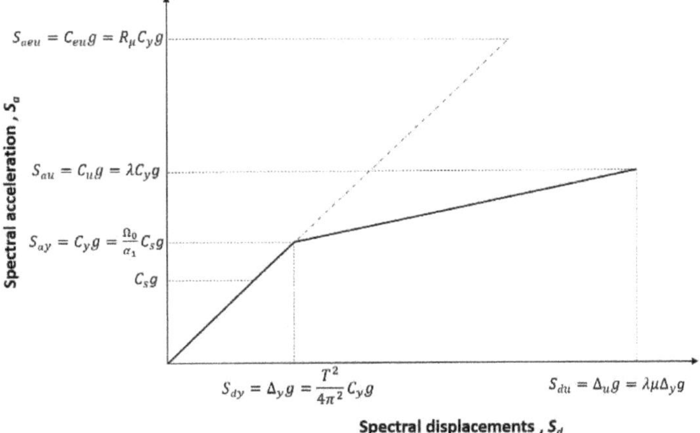

Figure 2. Bilinear idealization of capacity curve.

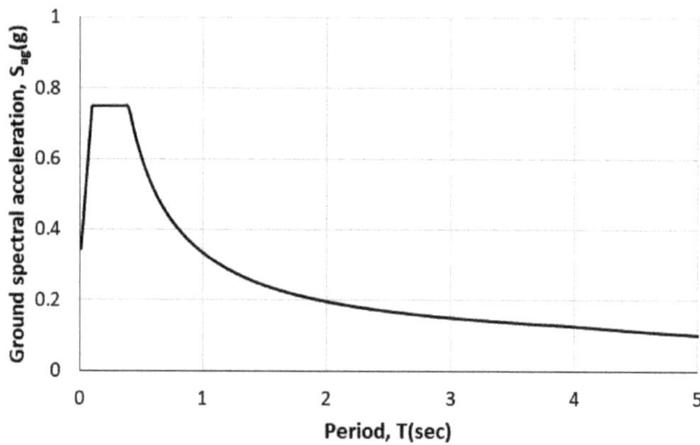

Figure 3. Standard design spectra of the codes [22,23] for a 475-year return period.

Table 4. Empirical formula for fundamental period of the building and the code strength reduction factor R of V3 [22] and V4 [23].

Typology		V3		V4	
		T (s)	R	T (s)	R
Steel	EBF	$0.05H^{0.75}$	7	$0.05H^{0.75}$	7
	CBF		6		5.5
	SMF	$0.08H^{0.75}$	10	$0.08H^{0.75}$	7.5
	IMF		7		5
	SMF + EBF	$0.05H^{0.75}$	10	$0.05H^{0.75}$	7.5
	SMF + CBF		9		7
	IMF + EBF		7		6
	IMF + CBF		7		6
Reinforced concrete	SMF	$0.07H^{0.75}$	10	$0.05H^{0.9}$	7.5
	IMF		7		5
	SMF + Special shear walls	$0.05H^{0.75}$	11	$0.05H^{0.75}$	7.5
	IMF + Special shear walls		-		6.5
	IMF + Intermediate shear walls		8		6

The capacity model parameters based on moderate code and high code for all 50 building typologies are presented in Tables 5 and 6, respectively. This presentation format of the demand spectrum is known as the acceleration–displacement response spectrum (ADRS) [30].

In both seismic codes, the capacity spectra for buildings with the same height show that steel moment-resisting frame buildings have the most significant spectral displacement (P4 and P5). Buildings with steel-braced frames (B5 and B8) have the lowest values of spectral displacement. However, values of spectral displacement in steel and RC combined frames are also close to those of steel-braced frames with a slight difference. The values presented in the capacity curves were determined by the parameters introduced in the seismic codes. If the drift of the floor surpasses the allowable value in the code, the designer will reduce the assumed ductility for the next attempt.

Table 5. Parameters of the capacity curves for moderate-code steel and RC buildings.

Building Typology	Yield Point		Ultimate Point	
	S_{dy} (cm)	S_{ay} (cm/s^2)	S_{du} (cm)	S_{au} (cm/s^2)
B5-ST-L-V3	0.94	549.36	11.54	659.23
B6-ST-L-V3	1.10	640.92	11.44	769.10
P4-ST-L-V3	1.68	384.55	20.20	461.46
P5-ST-L-V3	2.40	549.36	20.20	659.23
T4-ST-L-V3	0.66	384.55	11.72	461.46
T5-ST-L-V3	0.73	427.28	11.67	512.74
T6&T7-ST-L-V3	0.94	549.36	11.54	659.23
P1-RC-L-V3	1.29	384.55	16.85	461.46
P2-RC-L-V3	1.84	549.36	16.79	659.23
T1-RC-L-V3	0.60	349.59	11.76	419.51
T2-RC-L-V3	0.82	480.69	11.61	576.83
B5-ST-M-V3	2.66	549.36	22.32	659.23
B6-ST-M-V3	3.10	640.92	22.32	769.10
P4-ST-M-V3	3.28	265.03	39.37	318.04
P5-ST-M-V3	4.69	378.62	39.37	454.34
T4-ST-M-V3	1.86	384.55	22.32	461.46
T5-ST-M-V3	2.07	427.28	22.32	512.74
T6&T7-ST-M-V3	2.66	549.36	22.32	659.23
P1-RC-M-V3	2.75	289.71	32.95	347.65
P2-RC-M-V3	3.92	413.87	32.95	496.64
T1-RC-M-V3	1.79	369.29	23.57	443.15
T2-RC-M-V3	2.46	507.77	23.57	609.32

Table 6. Parameters of the capacity curves for high-code steel and RC buildings.

Building Typology	Yield Point		Ultimate Point	
	S_{dy} (cm)	S_{ay} (g)	S_{du} (cm)	S_{au} (g)
B5-ST-L-V4	0.48	280.29	5.89	336.34
B8-ST-L-V4	0.61	356.73	5.80	428.07
P4-ST-L-V4	1.72	392.40	15.46	470.88
P5-ST-L-V4	2.58	588.60	15.46	706.32
T5-ST-L-V4	0.56	327.00	7.38	392.40
T7-ST-L-V4	0.60	350.36	7.36	420.43
T6&T8-ST-L-V4	0.70	408.75	7.29	490.50
P1-RC-L-V4	1.30	392.40	12.76	470.88
P2-RC-L-V4	1.94	588.60	12.67	706.32
T1-RC-L-V4	0.59	345.42	7.80	414.51
T2-RC-L-V4	0.68	398.56	7.74	478.28
T3-RC-L-V4	0.74	431.78	7.71	518.13
B5-ST-M-V4	1.35	280.29	11.39	336.34
B8-ST-M-V4	1.72	356.73	11.39	428.07

Table 6. *Cont.*

Building Typology	Yield Point		Ultimate Point	
	S_{dy} **(cm)**	S_{ay} **(g)**	S_{du} **(cm)**	S_{au} **(g)**
P4-ST-M-V4	2.78	224.51	25.02	269.42
P5-ST-M-V4	4.17	336.77	25.02	404.13
T5-ST-M-V4	1.58	327.00	14.23	392.40
T7-ST-M-V4	1.69	350.36	14.23	420.43
T6&T8-ST-M-V4	1.98	408.75	14.23	490.50
P1-RC-M-V4	2.68	232.85	24.12	279.42
P2-RC-M-V4	4.02	349.27	24.12	419.13
T1-RC-M-V4	1.67	345.42	15.03	414.51
T2-RC-M-V4	1.93	398.56	15.03	478.28
T3-RC-M-V4	1.97	408.21	14.21	489.85

4. Fragility Curves

Seismic fragility curves indicate the probability of $P[\text{DS} \mid S_d]$ reaching or exceeding a specific damage state DS under a given ground motion parameter (e.g., peak ground acceleration PGA, spectrum displacement S_d, intensity I). The LM2 hybrid method introduced in RISK-UE [20] was employed to extract the seismic fragility curves for a given spectrum displacement S_d, as defined in Equation (4) [31]:

$$P[DS|S_d] = \Phi\left[\frac{1}{\beta_{DS}} \ln\left(\frac{S_d}{S_{dDS}}\right)\right] \tag{4}$$

where Φ is the standard normal cumulative distribution function, S_{dDS} is the median value of spectral displacement at which the building reaches the threshold of the damage state, DS, and β_{DS} is the lognormal standard deviation of spectral displacement for the damage state DS.

The median values introduced by LM2 of the RISK-UE method [20] for damage limit states, depending on S_{dy} and S_{du}, are used in Equations (5)–(8):

$$S_{dD1} = 0.7 S_{dy} \tag{5}$$

$$S_{dD2} = S_{dy} \tag{6}$$

$$S_{dD3} = S_{dy} + 0.25(S_{du} - S_{dy}) \tag{7}$$

$$S_{dD4} = S_{du} \tag{8}$$

The lognormal standard deviation of spectral displacement for damage state DS presented in LM2 of the RISK-UE method [20] as a function of ductility μ is as in Equations (9)–(12):

$$\beta_{D1} = 0.25 + 0.07 \ln(\mu) \tag{9}$$

$$\beta_{D2} = 0.20 + 0.18 \ln(\mu) \tag{10}$$

$$\beta_{D3} = 0.10 + 0.4 \ln(\mu) \tag{11}$$

$$\beta_{D4} = 0.15 + 0.5\ln(\mu) \tag{12}$$

Tables 7 and 8 show the parameters of hybrid fragility curves for the moderate-code buildings and the high-code buildings, respectively. Likewise, spectral displacement curves corresponding to moderate-code low-rise, high-code low-rise, and high-code mid-rise buildings are shown in Figures 4–7.

Table 7. Parameters of fragility curves for moderate-code steel and RC buildings.

| Building Properties | | Spectral Displacements, S_d (cm) | | | | | | | | |
| | | D1 | | D2 | | D3 | | D4 | |
Typology	Height (m)	Median	Beta	Median	Beta	Median	Beta	Median	Beta
B5-ST-L-V3		0.66	0.41	0.94	0.62	3.59	1.03	11.54	1.31
B6-ST-L-V3		0.77	0.40	1.10	0.59	3.68	0.96	11.44	1.23
P4-ST-L-V3		1.18	0.41	1.68	0.61	6.31	1.02	20.20	1.30
P5-ST-L-V3		1.68	0.39	2.40	0.55	6.85	0.88	20.20	1.12
T4-ST-L-V3		0.46	0.44	0.66	0.69	3.42	1.18	11.72	1.50
T5-ST-L-V3	9	0.51	0.43	0.73	0.67	3.47	1.14	11.67	1.44
T6&T7-ST-L-V3		0.66	0.41	0.94	0.62	3.59	1.03	11.54	1.31
P1-RC-L-V3		0.90	0.42	1.29	0.63	5.18	1.05	16.85	1.34
P2-RC-L-V3		1.29	0.39	1.84	0.56	5.58	0.91	16.79	1.16
T1-RC-L-V3		0.42	0.45	0.60	0.70	3.39	1.22	11.76	1.55
T2-RC-L-V3		0.57	0.42	0.82	0.64	3.52	1.09	11.61	1.38
B5-ST-M-V3		1.86	0.39	2.66	0.55	7.57	0.88	22.31	1.12
B6-ST-M-V3		2.17	0.37	3.10	0.52	7.90	0.82	22.32	1.05
P4-ST-M-V3		2.30	0.41	3.28	0.61	12.30	1.02	39.37	1.30
P5-ST-M-V3		3.28	0.39	4.69	0.55	13.36	0.88	39.37	1.12
T4-ST-M-V3		1.30	0.41	1.86	0.61	6.97	1.02	22.32	1.30
T5-ST-M-V3	18	1.45	0.40	2.07	0.59	7.13	0.98	22.32	1.25
T6&T7-ST-M-V3		1.86	0.39	2.66	0.55	7.57	0.88	22.32	1.12
P1-RC-M-V3		1.92	0.41	2.75	0.61	10.30	1.02	32.95	1.30
P2-RC-M-V3		2.75	0.39	3.92	0.55	11.18	0.88	32.95	1.12
T1-RC-M-V3		1.25	0.42	1.79	0.63	7.23	1.06	23.57	1.35
T2-RC-M-V3		1.72	0.40	2.45	0.57	7.73	0.93	23.57	1.19

Table 8. Parameters of fragility curves for high-code steel and RC buildings.

| Building Properties | | Spectral Displacements, S_d (cm) | | | | | | | | |
| | | D1 | | D2 | | D3 | | D4 | |
Typology	Height (m)	Median	Beta	Median	Beta	Median	Beta	Median	Beta
B5-ST-L-V4		0.33	0.41	0.48	0.62	1.83	1.03	5.89	1.31
B8-ST-L-V4		0.43	0.39	0.618	0.57	1.91	0.93	5.80	1.18
P4-ST-L-V4	9	1.20	0.39	1.72	0.56	5.15	0.91	15.46	1.16
P5-ST-L-V4		1.80	0.36	2.58	0.49	5.80	0.74	15.46	0.95
T5-ST-L-V4		0.39	0.42	0.56	0.63	2.27	1.06	7.38	1.35

Table 8. *Cont.*

| Building Properties | | Spectral Displacements, S_d (cm) | | | | | | | |
| Typology | Height (m) | D1 | | D2 | | D3 | | D4 | |
		Median	Beta	Median	Beta	Median	Beta	Median	Beta
T7-ST-L-V4		0.42	0.41	0.60	0.62	2.29	1.03	7.36	1.31
T6&T8-ST-L-V4		0.49	0.40	0.70	0.59	2.35	0.96	7.29	1.23
P1-RC-L-V4		0.91	0.40	1.30	0.58	4.16	0.94	12.76	1.20
P2-RC-L-V4		1.36	0.37	1.94	0.50	4.63	0.78	12.67	1.00
T1-RC-L-V4		0.41	0.42	0.59	0.63	2.39	1.06	7.80	1.35
T2-RC-L-V4		0.48	0.41	0.68	0.60	2.45	1.00	7.74	1.27
T3-RC-L-V4		0.52	0.40	0.74	0.59	2.48	0.96	7.71	1.23
B5-ST-M-V4		0.95	0.39	1.35	0.55	3.86	0.88	11.39	1.12
B8-ST-M-V4		1.21	0.37	1.72	0.51	4.14	0.78	11.39	1.00
P4-ST-M-V4		1.95	0.39	2.78	0.56	8.34	0.91	25.01	1.16
P5-ST-M-V4		2.92	0.36	4.17	0.49	9.38	0.74	25.02	0.95
T5-ST-M-V4		1.11	0.39	1.58	0.56	4.74	0.91	14.23	1.16
T7-ST-M-V4	18	1.186	0.39	1.69	0.55	4.83	0.88	14.23	1.12
T6&T8-ST-M-V4		1.38	0.37	1.98	0.52	5.04	0.82	14.23	1.05
P1-RC-M-V4		1.88	0.39	2.68	0.56	8.04	0.91	24.12	1.16
P2-RC-M-V4		2.81	0.36	4.02	0.49	9.04	0.74	24.12	0.95
T1-RC-M-V4		1.17	0.39	1.67	0.56	5.01	0.91	15.03	1.16
T2-RC-M-V4		1.35	0.38	1.93	0.54	5.20	0.85	15.03	1.09
T3-RC-M-V4		1.38	0.37	1.97	0.52	5.03	0.82	14.21	1.05

For an elastic system with a single degree of freedom, the spectral acceleration S_a associated with the first mode-dominated period T is converted into the corresponding spectral displacement S_d, as shown in Equation (13):

$$S_d(T) = \frac{S_a(T)}{4\pi^2} T^2 \tag{13}$$

The corresponding fragility curves can be proposed in terms of spectral acceleration and damage discrete distribution. Figure 8 shows the fragility curves and damage distribution for high-code mid-rise buildings at S_a = 300 cm/s^2 (high relative seismic hazard zone). This shows the accepted damage by the IRSt2800 seismic code [23] for mid-height engineered buildings.

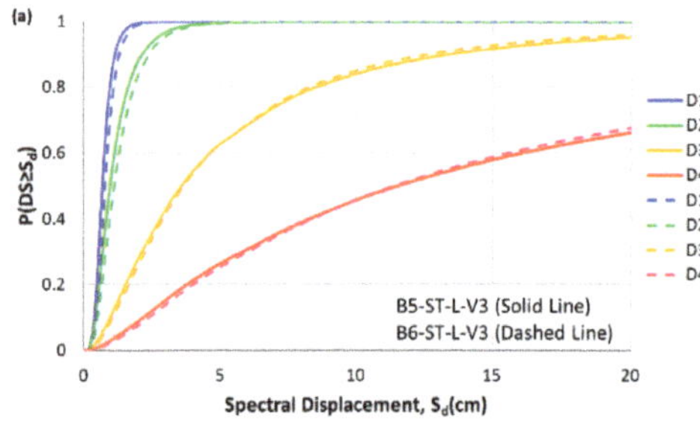

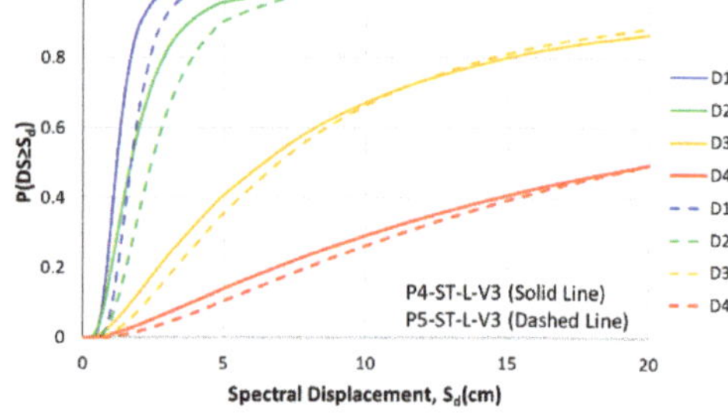

Figure 4. *Cont.*

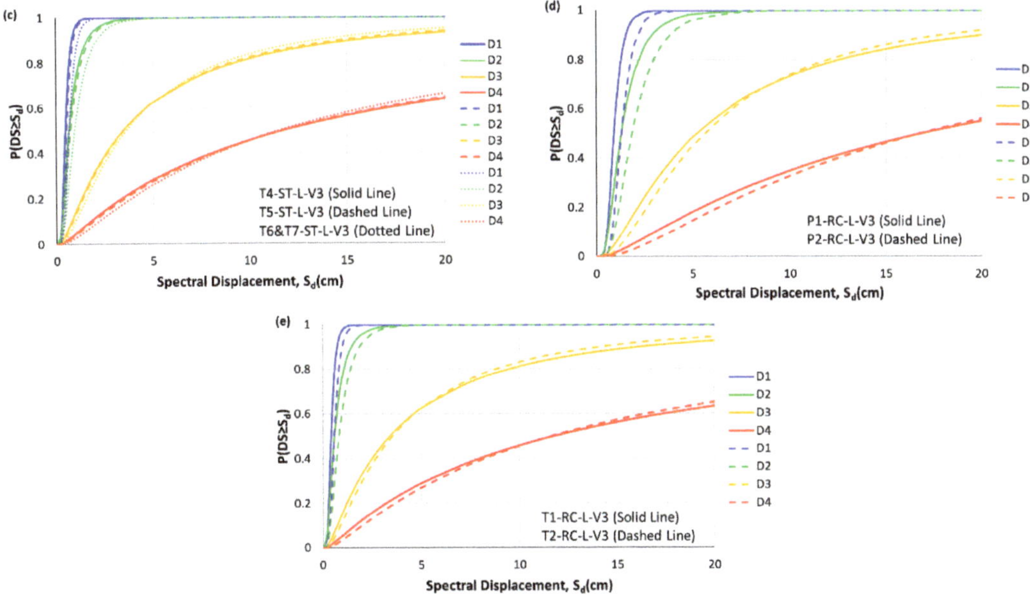

Figure 4. Fragility curves for moderate-code low-rise (9 m tall) steel and RC buildings: (**a**) steel-braced frame, (**b**) steel moment-resisting frame, (**c**) combination of moment-resisting frame and braced frame, (**d**) RC moment frame, and (**e**) combination of moment-resisting frame and RC shear wall.

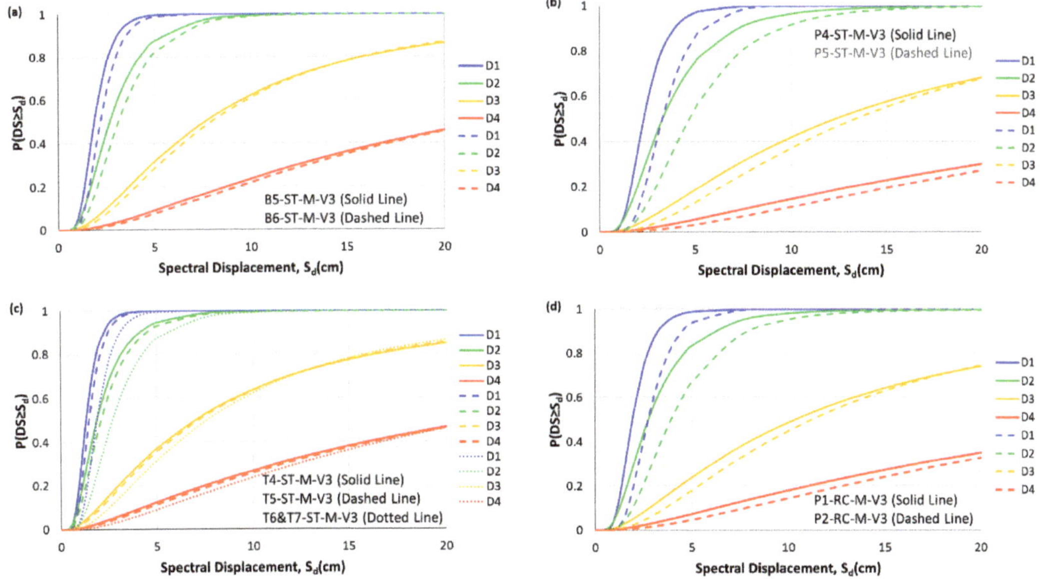

Figure 5. *Cont.*

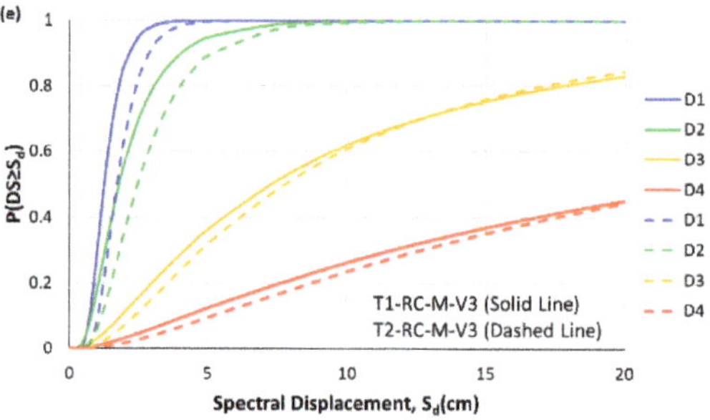

Figure 5. Fragility curves for moderate-code mid-rise (18 m tall) steel and RC buildings: (**a**) steel-braced frame, (**b**) steel moment-resisting frame, (**c**) combination of moment-resisting frame and braced frame, (**d**) RC moment frame, and (**e**) combination of moment-resisting frame and RC shear wall.

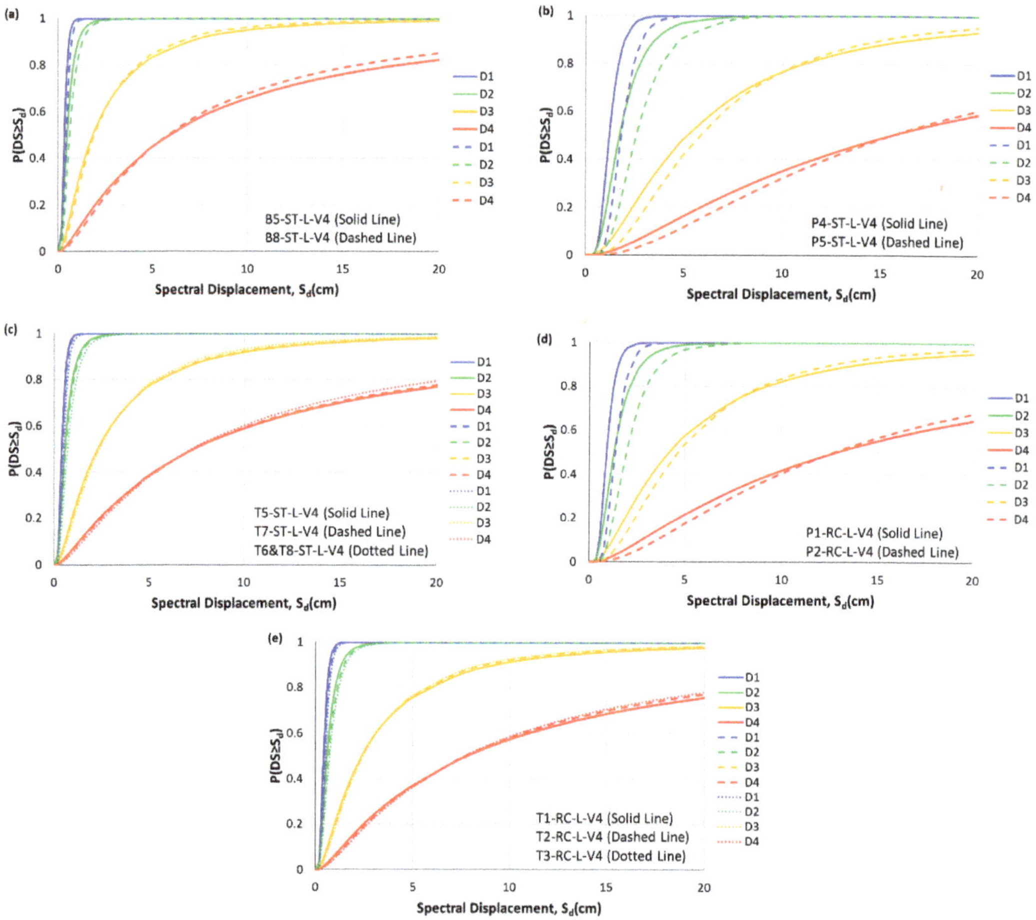

Figure 6. Fragility curves for high-code low-rise (9 m tall) steel and RC buildings: (**a**) steel-braced frame, (**b**) steel moment-resisting frame, (**c**) combination of moment-resisting frame and braced frame, (**d**) RC moment frame, and (**e**) combination of moment-resisting frame and RC shear wall.

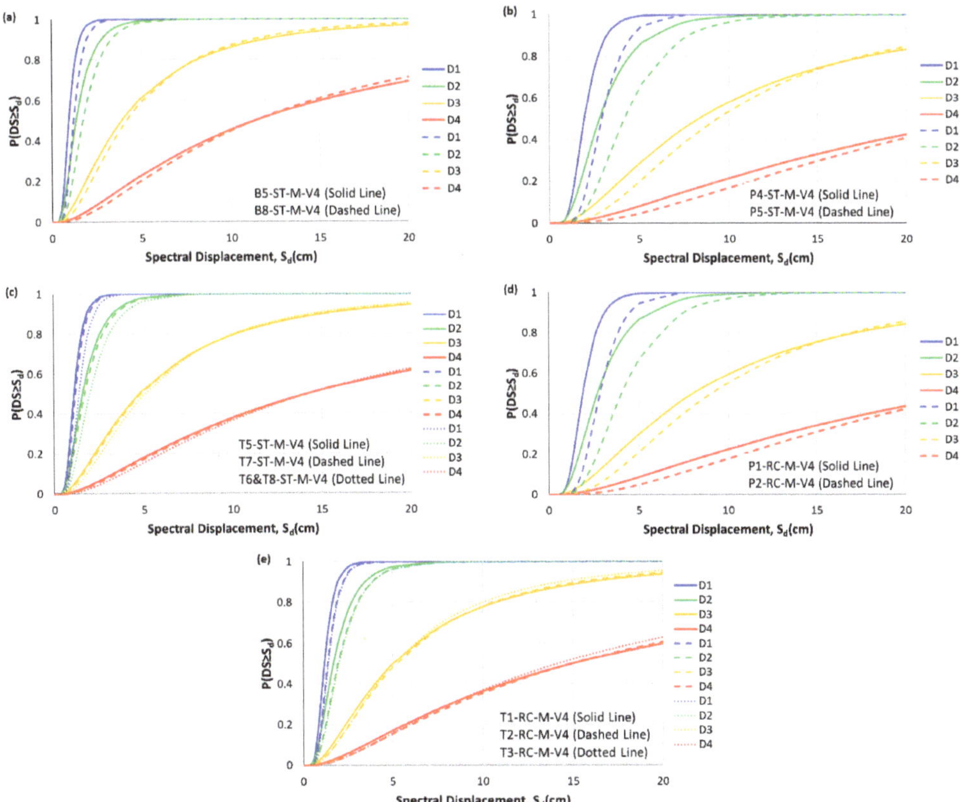

Figure 7. Fragility curves for high-code mid-rise (18 m tall) steel and RC buildings: (**a**) steel-braced frame, (**b**) steel moment-resisting frame, (**c**) combination of moment-resisting frame and braced frame, (**d**) RC moment frame, and (**e**) combination of moment-resisting frame and RC shear wall.

Different types of damage probability curves were evaluated by comparing their D4 damage level (collapse). For the low-rise and mid-rise buildings of the third version of the code, the highest probability of D4 damage in constant spectral displacement is related to the steel-braced frames and the combined steel and RC frames (Figure 4a,c,e and Figure 5a,c,e). However, according to the assumptions of the fourth version of the code, steel-braced frame buildings (Figures 6a and 7a) have the highest probability of D4-level damage in constant spectral displacement. Furthermore, the steel moment-resisting frame building type has the lowest probability of D4 damage in constant spectral displacement in both code versions and heights (Figures 4b, 5b, 6b and 7b).

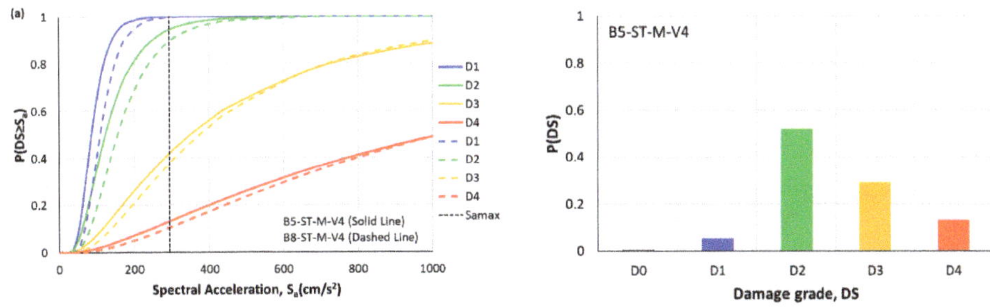

Figure 8. *Cont.*

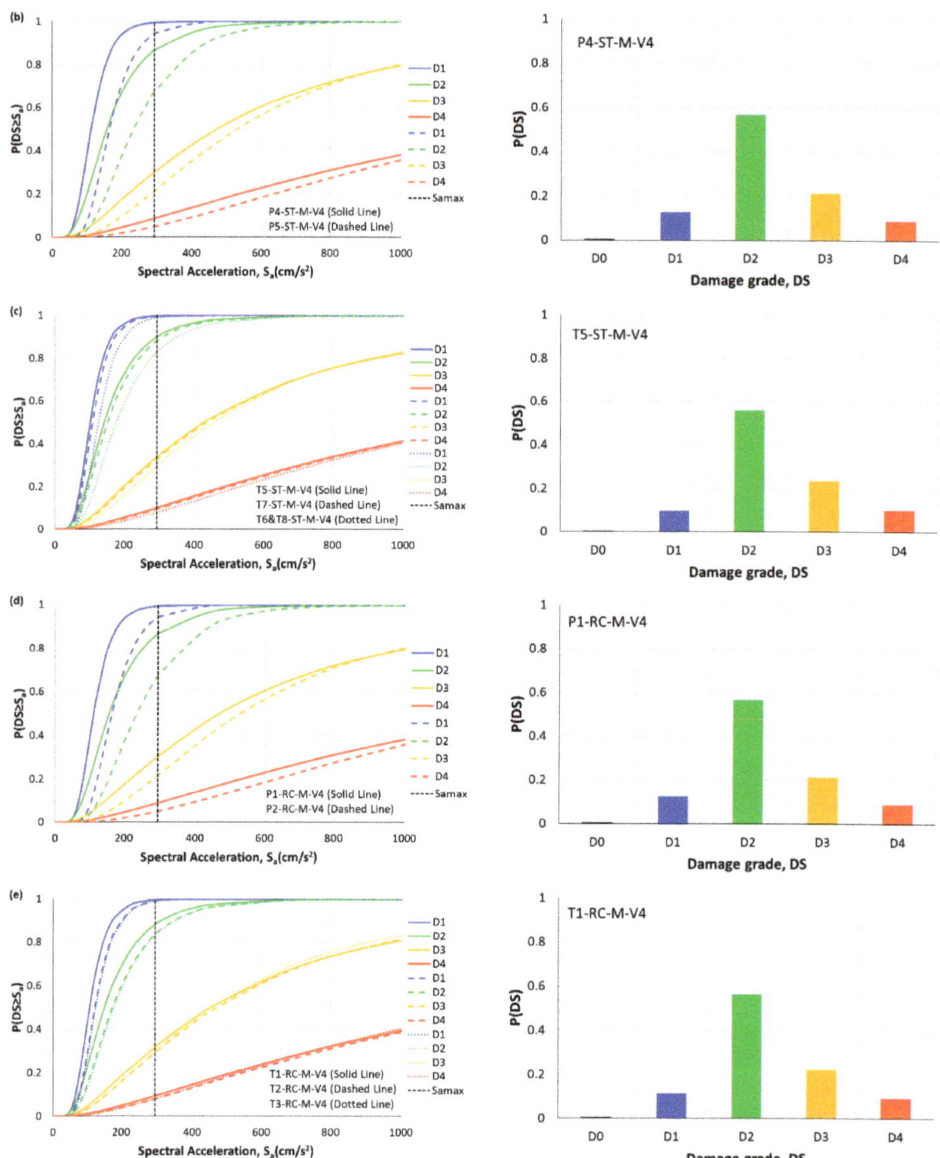

Figure 8. Fragility curves in terms of spectral acceleration and the discrete damage distribution referring to $S_a = 300$ cm/s^2 for high-code mid-rise (18 m tall) steel and RC buildings: (**a**) steel-braced frame, (**b**) steel moment-resisting frame, (**c**) combination of moment-resisting frame and braced frame, (**d**) RC moment frame, and (**e**) combination of moment-resisting frame and RC shear wall.

The results of the discrete damage distribution at constant spectral acceleration in mid-rise buildings of the fourth version of the code show that although the damage of different types is close at each level, the probability of D4 damage is slightly higher in the steel-braced frame. As shown in Figure 8, at a constant spectral acceleration of 0.3 g for different systems, the damage probability for D1 is about 5% to 13%, that of D2 is about 52% to 57%, that of D3 is about 21% to 29%, and that of D4 is estimated at 9% to 13%.

5. Comparison with Empirical Fragility Curves

The proposed code-based fragility curves (solid lines) are compared to the empirical fragility curves proposed in [9] (dashed lines) for buildings with the same seismic-resistant system, that were damaged during the 2017 Sarpol-e-zahab earthquake, in Figure 9. The proposed empirical fragility curves, based on the beta distribution of collected data, are regenerated to facilitate better comparisons since the empirical fragility curves are based on five levels of damage (D1 to D5) for LM1 of the RISK-UE method (Milutinovic and Trendafiloski, 2003). To achieve this, we assume that the D4 and D5 damage grades for LM1 of RISK-UE [20] will be the same as the collapse level (or D4 damage level) of LM2 of the method, based on the damage scale correlation of LM2 of the RISK-UE method [20]. It is also important to note that the curves at low spectral accelerations do not match up because the probability distribution functions are different (i.e., the beta distribution for the empirical fragility curves and the normal distribution for the code-based fragility curves). Furthermore, the height of the structures in the database of empirical fragility curves was not necessarily 18 m. According to a previous study [32] the demand estimation approaches can affect the performance of structures as well as the capacity estimation methods. Therefore, the observed discrepancies may also be attributed to assumptions underlying capacity and demand determinations.

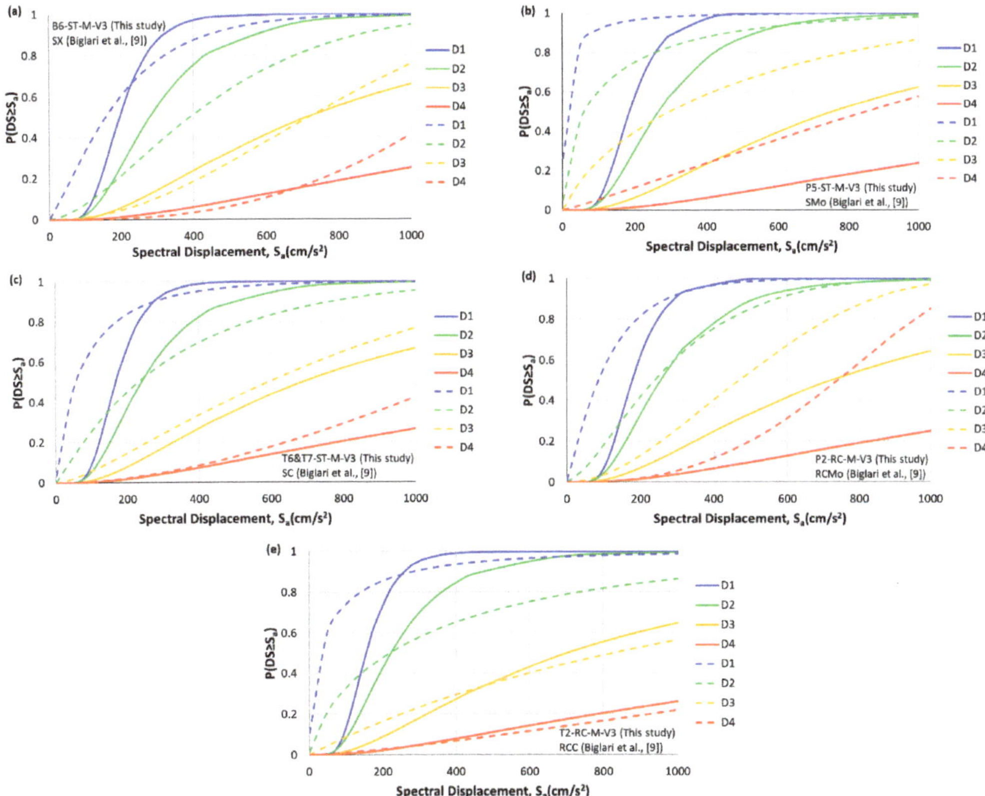

Figure 9. Comparison between moderate-code fragility curves proposed herein (solid line) and the empirical fragility curves developed by Biglari et al. [9] (dashed line) in terms of PGA for steel and RC buildings: (**a**) steel-braced frame, (**b**) steel moment-resisting frame, (**c**) combination of moment-resisting frame and braced frame, (**d**) RC moment frame, and (**e**) combination of moment-resisting frame and RC shear wall.

The results show that the damage probability at a high damage level (i.e., D4) in the real condition is higher than what the seismic code would expect. This poor performance is due to weaknesses in construction technologies and materials, as well as neglect of maintenance of structures. In RC buildings with a combination of moment-resisting frames and RC shear walls (Figure 9e) and steel-braced frames (Figure 9a), this difference is less than in the other seismic-resistant systems. This shows that these types of seismic-resistant systems work better with construction technologies in Iran. In the meantime, buildings with steel and RC moment frames (Figure 9b,d) have greater inconsistencies with the probability of code-expected damage. It is recommended to increase the technical controls of these buildings due to the increased demand for this seismic-resistant system in Iran.

6. Conclusions

The default values in seismic codes can be seen as a fragility curve, which gives seismic code developers a visual way to understand the seismic performance of different buildings and adjust them to meet real social and economic needs. These curves also help to plan seismic risk mitigation interventions for new code-based buildings, which are generally assumed to be invulnerable. This is specifically a comparison of codes and other parameters, such as the quality of materials, the construction methodology, the supervision of construction, and the maintenance during operation, which are other influential factors that will control the final seismic behavior of the structure.

The methodology employed in this study relies exclusively on the assumptions contained in the two latest Iranian seismic codes. These pure code-based seismic fragility curves have not been proposed to date. However, it is essential to consider the possibility of simulated damage in seismic codes. The limitation of this method is that it ignores design changes during deformation control. This issue occurs mostly with tall structures, and this effect cannot be considered in this method.

These two versions of the seismic design code recommended values for the minimum amount of nonlinear deformation and, therefore, the ductility factor for steel buildings with braced frames. However, steel buildings with special moment-resisting frames had the highest seismic capacity. Furthermore, the hybrid fragility curves for all typologies of the studied buildings were proposed. Seismic fragility curves are useful for figuring out how good codes are and how strong they are. These curves are not presented in seismic codes. Upon comparing the fragility curves of the third and fourth editions of the seismic code of Iran, it was found that the severe damage probability (D4) of the fourth version is higher than that of the third version. This comparison used a similar approach and only considered the assumptions included in the standard. The curves showed that steel buildings with braced frames had the highest probability of collapsing at constant spectral displacement, and steel buildings with moment-resistant frames had the lowest probability. Steel-braced frame buildings have the highest collapse probability in constant spectral acceleration among mid-rise buildings.

A comparison of results with empirical values showed that the steel-braced frames and RC buildings with a combination of moment-resisting frames and RC shear walls are more compatible with construction technologies in Iran. However, the steel and RC moment frames of buildings are incompatible. It is recommended to increase the technical controls on these buildings.

Determining these curves for other seismic codes and consciously changing the seismic codes could reduce the probability of D3 and D4 damage and, consequently, reduce the vulnerability of newly built structures. Further investigation is needed to find the most effective seismic code parameters to reduce the probability of high damage levels.

Author Contributions: Conceptualization, M.B.; methodology, M.B., B.H.H. and A.F.; software, M.B.; validation, M.B. and B.H.H.; formal analysis, M.B.; investigation, M.B.; resources, M.B.; writing—original draft preparation, M.B.; writing—review and editing, M.B., B.H.H. and A.F.; visualization, M.B.; supervision, M.B., B.H.H. and A.F.; project administration, M.B. All authors have read and agreed to the published version of the manuscript.

Funding: This research received no external funding.

Data Availability Statement: The text included all the presented research data.

Conflicts of Interest: The authors declare no conflict of interest.

References

1. Karakas, C.C.; Palanci, M.; Senel, S.M. Fragility based evaluation of different code based assessment approaches for the performance estimation of existing buildings. *Bull. Earthq. Eng.* **2022**, *20*, 1685–1716. [CrossRef]
2. Ministry of Public Works and Settlement of Turkey. *TEC 2007, Specification for Buildings to be Built in Seismic Zones*; Ministry of Public Works and Settlement of Turkey: Ankara, Turkey, 2007.
3. Disaster and Emergency Management Presidency of Turkey. *TBEC 2018, Turkey Earthquake Building Regulations*; Disaster and Emergency Management Presidency of Turkey: Ankara, Turkey, 2018.
4. CEN. *Eurocode 8: Design of Structures for Earthquake Resistance—Part 3: Assessment and Retrofitting of Buildings*; CEN: Brussels, Belgium, 2005.
5. American Society of Civil Engineers. *ASCE 41-17, Seismic Evaluation and Retrofit of Existing Buildings*; American Society of Civil Engineers: Reston, VA, USA, 2017.
6. Omidvar, B.; Gatmiri, B.; Derakhshan, S. Experimental vulnerability curves for the residential buildings of Iran. *Nat. Hazards* **2011**, *60*, 345–365. [CrossRef]
7. Del Gaudio, C.; De Martino, G.; Di Ludovico, M.; Manfredi, G.; Prota, A.; Ricci, P.; Verderame, G.M. Empirical fragility curves from damage data on RC buildings after the 2009 L'Aquila earthquake. *Bull. Earthq. Eng.* **2017**, *15*, 1425–1450. [CrossRef]
8. Gautam, D.; Fabbrocino, G.; de Magistris, F.S. Derive empirical fragility functions for Nepali residential buildings. *Eng. Struct.* **2018**, *171*, 617–628. [CrossRef]
9. Biglari, M.; Formisano, A.; Hashemi, B.H. Empirical fragility curves of engineered steel and RC residential buildings after Mw 7.3 2017 Sarpol-e-zahab earthquake. *Bull. Earthq. Eng.* **2021**, *19*, 2671–2689. [CrossRef]
10. Biglari, M.; Formisano, A. Urban seismic scenario-based risk analysis using empirical fragility curves for Kerendegharb after Mw 7.3, 2017 Iran earthquake. *Bull. Earthq. Eng.* **2022**, *20*, 6487–6503. [CrossRef]
11. Vamvatsikos, D.; Cornell, C.A. Incremental dynamic analysis. *Earthq. Eng. Struct. Dyn.* **2001**, *31*, 491–514. [CrossRef]
12. Réveillère, A.; Gehl, P.; Seyedi, D.; Modaressi, H. Development of seismic fragility curves for damaged reinforced concrete structures. In Proceedings of the 15th World Conference on Earthquake Engineering, Lisboa, Portugal, 24–28 September 2012; p. 999.
13. Cao, X.-Y.; Feng, D.-C.; Li, Y. Assessment of various seismic fragility analysis approaches for structures excited by non-stationary stochastic ground motions. *Mech. Syst. Signal Process.* **2023**, *186*, 109838. [CrossRef]
14. Cao, X.-Y.; Feng, D.-C.; Beer, M. Consistent seismic hazard and fragility analysis considering combined capacity-demand uncertainties via probability density evolution method. *Struct. Saf.* **2023**, *103*, 102330. [CrossRef]
15. Kappos, A.J.; Panagopoulos, G.; Panagiotopoulos, C.; Penelis, G. A hybrid method for the vulnerability assessment of R/C and URM buildings. *Bull. Earthq. Eng.* **2006**, *4*, 391–413. [CrossRef]
16. D'Ayala, D.; Meslem, A.; Vamvatsikos, D.; Porter, K.; Rossetto, T.; Crowley, H.; Silva, V. *Guidelines for Analytical Vulnerability Assessment of Low-Mid-Rise Buildings—Methodology*; Vulnerability Global Component Project; GEM Foundation: Pavia, Italy, 2014.
17. Simoncelli, M.; Aloisio, A.; Zucca, M.; Venturi, G.; Alaggio, R. Intensity and location of corrosion on the reliability of a steel bridge. *J. Constr. Steel Res.* **2023**, *206*, 107937. [CrossRef]
18. Suzuki, A.; Iervolino, I. Seismic Fragility of Code-conforming Italian Buildings Based on SDoF Approximation. *J. Earthq. Eng.* **2019**, *25*, 2873–2907. [CrossRef]
19. Applied Technology Council (ATC). *Earthquake Damage Evaluation Data for California, ATC-13*; Applied Technology Council (ATC): Redwood, CA, USA, 1985.
20. Milutinovic, Z.V.; Trendafiloski, G.S. *RISK-UE. An Advanced Approach to Earthquake Risk Scenarios with Applications to Different European Towns*; WP4: Vulnerability of Current Buildings; European Commission: Brussels, Belgium, 2004.
21. International Conference of Building Officials (ICBO). *Uniform Building Code*; International Conference of Building Officials (ICBO): Whittier, CA, USA, 1994.
22. Building and Housing Research Center. *IRSt2800, Iranian Code of Practice for Seismic Resistant Design of Buildings*, 3rd ed.; Building and Housing Research Center: Tehran, Iran, 2005.
23. Building and Housing Research Center. *IRSt2800, Iranian Code of Practice for Seismic Resistant Design of Buildings*, 4th ed.; Building and Housing Research Center: Tehran, Iran, 2014.
24. FEMA/NIBS. *HAZUS—Earthquake Loss Estimation Methodology*; Department of Homeland Security Federal Emergency Management Agency Mitigation Division: Washington, DC, USA, 1998; Volume 1.
25. Fajfar, P. Capacity spectrum method based on inelastic demand spectra. *Earthq. Eng. Struct. Dyn.* **1999**, *28*, 979–993. [CrossRef]
26. Freeman, S.A. Development and use of capacity spectrum method. In Proceedings of the 6th US National Conference on Earthquake Engineering, Seattle, WA, USA, 31 May–4 June 1998.
27. Milutinovic, Z.V.; Trendafiloski, G.S. *WP4: Level 2 Methodology—Code Based Approach, Case Study: Aseismic Design Codes in Macedonia*; RISK-UE WP4 L2 Report; IZIIS: Skopje, North Macedonia, 2002.

28. Vidic, T.; Fajfar, P.; Fischinger, M. Consistent inelastic design spectra: Strength and displacement. *Earthq. Eng. Struct. Dyn.* **1994**, *23*, 507–521. [CrossRef]
29. Fajfar, P. A nonlinear analysis method for performance-based seismic design. *Earthq. Spectra* **2000**, *16*, 573–592. [CrossRef]
30. Mahaney, J.A.; Paret, T.F.; Kehoe, B.E.; Freeman, S.A. The capacity spectrum method for evaluating structural response during the Loma Prieta earthquake. In Proceedings of the 1993 National Earthquake Conference, Memphis, TN, USA, 2–5 May 1993.
31. Department of Homeland Security, FEMA. *Multi-Hazard Loss Estimation Methodology: Earthquake Model*; Technical Report; Department of Homeland Security, FEMA: Washington, DC, USA, 2003.
32. Palanci, M.; Kalkan, A.; Sene, S.M. Investigation of shear effects on the capacity and demand estimation of RC buildings. *Struct. Eng. Mech.* **2016**, *60*, 1021–1038. [CrossRef]

Article

Kahramanmaraş—Gaziantep, Türkiye M_w 7.8 Earthquake on 6 February 2023: Strong Ground Motion and Building Response Estimations

George Papazafeiropoulos [1] and **Vagelis Plevris** [2,*]

1   School of Civil Engineering, National Technical University of Athens, Zografou Campus, 15780 Athens, Greece; gpapazafeiropoulos@yahoo.gr
2   Department of Civil and Architectural Engineering, Qatar University, Doha P.O. Box 2713, Qatar
*   Correspondence: vplevris@qu.edu.qa

Abstract: The effects on structures of the earthquake with the magnitude of 7.8 on the Richter scale (moment magnitude scale) that took place in Pazarcık, Kahramanmaraş, Türkiye at 04:17 a.m. local time (01:17 UTC) on 6 February 2023, are investigated by processing suitable seismic records using the open-source software OpenSeismoMatlab. The earthquake had a maximum Mercalli intensity of XI (Extreme) and it was followed by a M_w 7.5 earthquake nine hours later, centered 95 km to the north–northeast from the first. Peak and cumulative seismic measures as well as elastic response spectra, constant ductility (or isoductile) response spectra, and incremental dynamic analysis curves were calculated for two representative earthquake records of the main event. Furthermore, the acceleration response spectra of a large set of records were compared to the acceleration design spectrum of the Turkish seismic code. Based on the study, it is concluded that the structures were overloaded far beyond their normal design levels. This, in combination with considerable vertical seismic components, was a contributing factor towards the collapse of many buildings in the region. Modifications of the Turkish seismic code are required so that higher spectral acceleration values can be prescribed, especially in earthquake-prone regions.

Keywords: earthquake; Türkiye; design; collapse; ductility; reinforcement; concrete

Citation: Papazafeiropoulos, G.; Plevris, V. Kahramanmaraş—Gaziantep, Türkiye M_w 7.8 Earthquake on 6 February 2023: Strong Ground Motion and Building Response Estimations. *Buildings* **2023**, *13*, 1194. https://doi.org/10.3390/buildings13051194

Academic Editors: Rajesh Rupakhety and Dipendra Gautam

Received: 30 March 2023
Revised: 23 April 2023
Accepted: 28 April 2023
Published: 30 April 2023

Correction Statement: This article has been republished with a minor change. The change does not affect the scientific content of the article and further details are available within the backmatter of the website version of this article.

1. Introduction

An earthquake with the magnitude (M_w) of 7.8 took place in Pazarcık, Kahramanmaraş, Türkiye at 04:17 a.m. local time (01:17 UTC) on 6 February 2023. The earthquake had a maximum Mercalli intensity of XI (Extreme) and it was followed by a M_w 7.5 earthquake nine hours later, centered 95 km to the north–northeast from the first. According to information available as of 22 February 2023, and a press release of the Turkish government [1], 42,310 people lost their lives in Kahramanmaraş, Gaziantep, Şanlıurfa, Diyarbakır, Adana, Adıyaman, Osmaniye, Hatay, Kilis, Malatya, and Elazığ, and 448,010 people have been evacuated from the earthquake zone. A total of 7184 aftershocks occurred, and a total of 5606 buildings have reportedly been destroyed in Türkiye [2]. In Türkiye and Syria combined, more than 6500 buildings have collapsed due to the two main shocks. As of 6 February 2023, a three-month state of emergency is in place in provinces directly affected by the earthquake in Türkiye [3]. The details of the earthquake event (from now on referred to as the M_w 7.8 event) are shown in Table 1 [4].

Table 1. Details of the main M_w 7.8 earthquake event [4].

Magnitude	7.8 (M_w)
Location	Pazarcık (Kahramanmaraş), 26 km ENE of Nurdağı, Türkiye
Date and time	6 February 2023, 01:17:34 UTC
Latitude	37.225° N
Longitude	37.021° E
Depth	10.0 km

2. Literature Review

Many studies in the literature have attempted to quantify the impact of earthquakes on structures. One of the most important quantities that can describe the destructive effects of an earthquake is the seismic energy that is absorbed from the structures which respond to seismic excitation, as well as its distribution inside the structure, e.g., among its stories in the case of buildings [5]. In addition, there is strong evidence that taking the seismic energy absorbed from structures into account leads to the development of more realistic acceleration time histories that can be used for more rigorous structural design [6]. More recently, attempts have been made to use computational intelligence methods for predicting seismic damage on structures through energy-related measures for quantification of the seismic damage, such as the Park–Ang index [7]. Apart from the seismic energy considerations above that apply to all earthquakes in general, a more specific presentation regarding particularly the 6 February 2023 earthquakes in Türkiye are pursued in the following paragraphs, and the results of the various studies are emphasized.

Lu [8] attempted a preliminary assessment of the damage of the 6 February 2023 Türkiye earthquake on buildings. The author tried to explain why building damages due to this earthquake were so severe and whether Chinese structures would be strong enough to resist an earthquake of this magnitude. The RED-ACT system [9] was used to analyze the recorded ground motion of the 1st (M_w 7.8) earthquake event which gave "collapse" results for the ground motion input of Station 3138. Following this, the earthquake ground motions were input to typical individual and urban buildings in Türkiye to assess the damage. Low- and high-rise reinforced concrete frame models were analyzed and subjected to the ground motion recorded at station 3123, and the results showed that there were large inter-story drifts at the lower stories in all cases, resulting in the collapse of all three frames analyzed.

A recent study has pointed out some noticeable facts about the buildings' design and construction and the impact that they had on the devastation in the aftermath of the earthquake [10], for example, (i) there was low concrete quality, inadequate reinforcement, and/or poor detailing, which led to either heavily damaged or collapsed structures [11]; (ii) the extensive building collapse can be attributed to a variety of reasons, including the old age of buildings, some of which were built according to previous versions of the Turkish seismic code, and the improper application of the current Turkish seismic code [10]. In addition, some peculiar phenomena have been observed, such as the fact that a newly constructed housing estate in Antakya municipality extensively collapsed killing many people, yet in nearby Erzin there were no collapses at all [12]. It has also been argued that the ongoing war in Syria may have played its role, where conflicts have made building standards impossible to enforce and some war-damaged buildings have been rebuilt using low-quality materials or "whatever materials are available" [13].

In another study, researchers have tried to identify the main features of the earthquake sequence and also to present various spectra of near-source ground motion data and ground motion data with pulse-like features [14]. In this study, it was concluded that the seismic actions were generally challenging for the structures, based on the response spectra of stations close to the source. Relatively large near-source pulses were also identified, whereas their attribution to rupture phenomena (e.g., forward directivity) needs further nvestigation.

The generalized continuous wavelet transform (GCWT) method proposed by Chen et al. [15] has been applied to identify pulse-like ground motions in the February 2023 earthquakes in Türkiye [16]. This method applies three criteria for the identification of pulse components which establish lower thresholds for the following quantities: (1) PGV, (2) the ratio of the energy of the pulse part of the ground motion to the energy of the original ground motion, and (3) the Pearson correlation coefficient between the pulse part of the ground motion and the time history of the pulse model. A total of 99 pulse-like ground motions were discovered that occurred in the February 2023 Türkiye earthquakes, with two of them containing more than one pulse. In addition, by examining four intensity measures, i.e., Arias intensity, cumulative Fourier spectrum, 5% damped pseudo-spectral velocity, and 5% damped pseudo-spectral acceleration, it was concluded in the same study that the response spectra, as well as the PGA, PGV and energy-frequency parameter (defined in [17]) values of pulse-like motions, are larger than those of non-pulse like motions.

A preliminary analysis of the acceleration time histories of the first earthquake of 6 February 2023 has appeared in [18]. It was shown there that the different rupture episodes appeared distinctly at the recordings of the stations closest to the epicenter (e.g., TK.2712 and TK.4615), whereas at stations farther away, such as TK.4406, the different rupture episodes are no longer distinguishable, and the PGA is lower. Far-field stations, such as TK.3145 sensed the rupture episodes at a considerable time delay from the time point of their occurrence.

Another preliminary report was presented by Garini and Gazetas [19]. Seismological and acceleration time history information was presented in this report for both earthquakes of the doublet, followed by an extensive analysis of the main reasons for the collapses of the buildings that happened. The acceleration time histories of the TK.4614 and TK.3123 recorded during the first mainshock were analyzed. In addition, their acceleration spectra were extracted, where it was observed that the TK.4614 record has increased spectral acceleration values for periods lower than roughly 0.6 s, whereas the TK.3123 record has increased spectral acceleration for periods larger than 0.6 s.

A preliminary analysis of strong ground motion characteristics has been made in [20]. In this report, a variety of aspects of ground motions recorded during the Türkiye earthquakes of February 2023 have been addressed as follows: (1) procedures for strong motion data processing have been presented, along with the types of nonstandard errors usually met in real raw acceleration data; (2) ground motion intensity measures, such as peak ground motion amplitudes, significant duration, and Arias as well as Housner intensity values of the recordings within distance from rupture lower than 100 km, are provided. Furthermore, acceleration time histories, Fourier amplitude spectra, and the 5%-damped acceleration response spectra of the recorded accelerations at the stations TK.2708, TK.3126, TK.3138, TK.4615, and TK.4624 are presented and discussed. Besides the aforementioned ground motion data, the spatial distribution of peak and spectral accelerations and soil amplifications and Horizontal-to-Vertical Spectral Ratio (HVSR) analyses are discussed in detail.

The characteristics of the soil and geological conditions where a building is located can greatly influence the amount of shaking that the building experiences during an earthquake [21,22]. Parameters such as soil type (rock foundations, earth foundations, etc.), soil depth, geological features, and site amplification can significantly affect the response of a building in an earthquake event. In the case of a soil foundation, different grain gradation [23], pore ratio, and water content [24] will also lead to a different structural response. Sun and Huang [23] presented a particle discrete element method that can be used to calculate the soil particle gradation, while Sun [24] investigated soil porosity and attempted to calculate the permeability of particle soils under soil pressure.

Apart from the aforementioned factors, various additional parameters mainly of seismological nature can be considered for the evaluation of the Kahramanmaraş–Gaziantep, Türkiye M_w 7.8 earthquake on 6 February 2023. For example, in [25], kinematic rupture models from a joint inversion of High Rate Global Navigation Satellite System (HR-GNSS) and strong motion data sets of the two events in the 6 February 2023 Türkiye earthquake doublet have been developed, and it is shown that the M_w 7.8 earthquake nucleated on a previously unmapped fault before transitioning to the East Anatolian Fault (EAF), rupturing for ~350 km; the maximum rupture speeds were estimated to be 3.2 km/s for the same event. In another study [26], a long-period coda moment magnitude method was used to measure the moment magnitudes of the two large mainshocks. It was found that the magnitude of the first event (with one standard error of the magnitude estimation) is 7.95 ± 0.013. Results about the tectonic setting, the seismicity and its temporal evolution as well as the seismic moment release rate in the region have been presented in [27], whereas various issues about seismic forecasting are discussed in [28].

In the present study, various seismic parameters of the M_w 7.8 earthquake are calculated and assessed in an effort to provide some explanations about the large destructiveness of the earthquake and the devastating effects it had on buildings. A building-oriented evaluation of the earthquake impact is performed not only by extracting its various peak and cumulative parameters but also by calculating various types of linear and nonlinear (isoductile) seismic spectra. Furthermore, incremental dynamic analysis (IDA) [29] is performed for various simplified cases of buildings in an effort to estimate the response that they would exhibit during the earthquake. To the best of the authors' knowledge, such a detailed investigation has not been conducted in the literature for the M_w 7.8 earthquake event. It is examined if the isoductile seismic spectra, the IDA curves, and the other calculated parameters can provide some hints about the destructiveness of the earthquake and how the buildings could be designed to be able to resist such earthquakes in the future. The objective of the study is to highlight various particular characteristics of the M_w 7.8 event through the investigation of all the aforementioned strong ground motion data processing results.

3. Record Data

There are two main earthquake monitoring networks currently operating in Türkiye: (i) the Turkish Civil Defense Network (AFAD, Republic of Türkiye, Ministry of Interior Disaster And Emergency Management Presidency, code TK) and (ii) the Kandilli Observatory and Earthquake Research Institute (Boğaziçi University, code KO). Two representative seismic recording stations were selected for obtaining acceleration time history data of the M_w 7.8 earthquake event, one from each network: (i) Station No 3137, TK Network (Lat.: 36.69293°, Long.: 36.48852°) and (ii) KHMN Station, KO Network KO (Lat.: 37.3916°, Long.: 37.1574°); both are shown in Figure 1, together with the epicenter of the earthquake event. For the processing of the acceleration time histories, the open-source Matlab code OpenSeismoMatlab [30] was used, which has been developed by the authors and is quite reliable since it has been successfully verified in several cases in the literature [31–33]. The software uses an advanced time integration algorithm first presented in [34].

Figure 2 shows the two horizontal acceleration components of the TK.3137 station record, while Figure 3 shows the corresponding vertical acceleration component. Similarly, Figure 4 shows the two horizontal acceleration components of the KO-KHMN station record, and Figure 5 shows its vertical acceleration component. The severity of the ground motions is obvious since the TK.3137 record reached a peak ground acceleration (PGA) of 0.75 g during ground shaking, whereas the KO-KHMN record reached PGA values as high as 0.60 g. Another observation is that the duration of the shaking was quite large, reaching about 1.5 min in the case of the TK.3137 station.

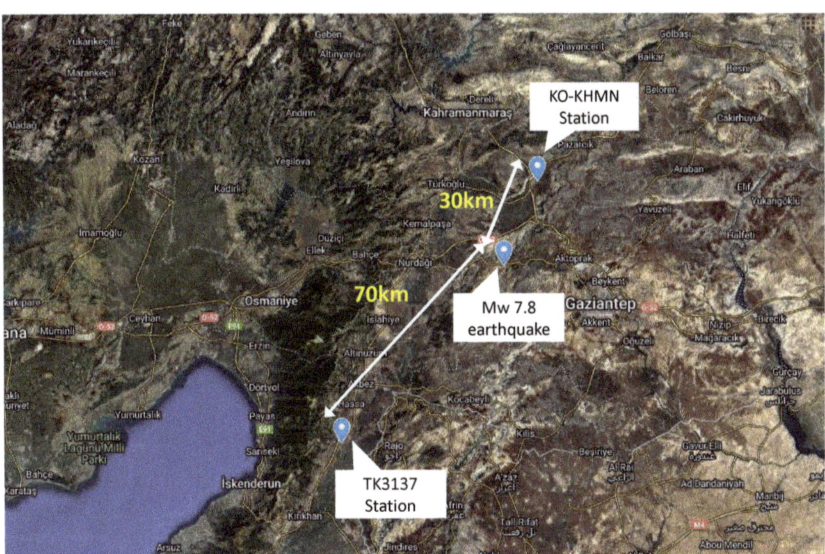

Figure 1. Locations of the two stations (No 3137 of TK Network and KHMN of KO Network) and the epicenter of the M_w 7.8 earthquake.

Table 2 shows the peak values recorded for each station, namely the PGA as well as the peak ground velocity (PGV) and the peak ground displacement (PGD) for both records, for the horizontal (EW, NS) and the vertical (UD) components.

Table 2. Peak seismic parameters of the M_w 7.8 earthquake based on the two recordings.

Station	TK.3137			KO-KHMN		
Component	PGA (m/s^2)	PGV (m/s)	PGD (m)	PGA (m/s^2)	PGV (m/s)	PGD (m)
EW Horizontal	7.47	0.75	0.50	5.09	1.08	0.61
NS Horizontal	4.26	0.76	1.15	6.06	0.89	0.50
UD Vertical	4.46	0.40	0.16	4.79	0.45	0.34

3.1. Cumulative Energy, Arias Intensity, and Significant Duration Data

A significant measure of the destructive effect of an earthquake on structures is the amount of energy that it releases. This energy, after having been released by the earthquake, is partly absorbed by the structures in the area and this results in their gradual damage or even collapse. Energy-based design theory (EBDT), which introduces energy demand as the critical parameter to evaluate structural damage, has gained attention around the world in recent decades. According to this approach, each structure must be designed to be capable of absorbing a certain amount of energy. It is noted that the EBDT concept accounts for the cumulative damage and the effective duration of earthquakes in a theoretically sound manner. The effects of the latter factors are not taken sufficiently into account in traditional force-based design approaches. In this section, the data for the total cumulative energy, Arias intensity, and significant duration are presented for the two records.

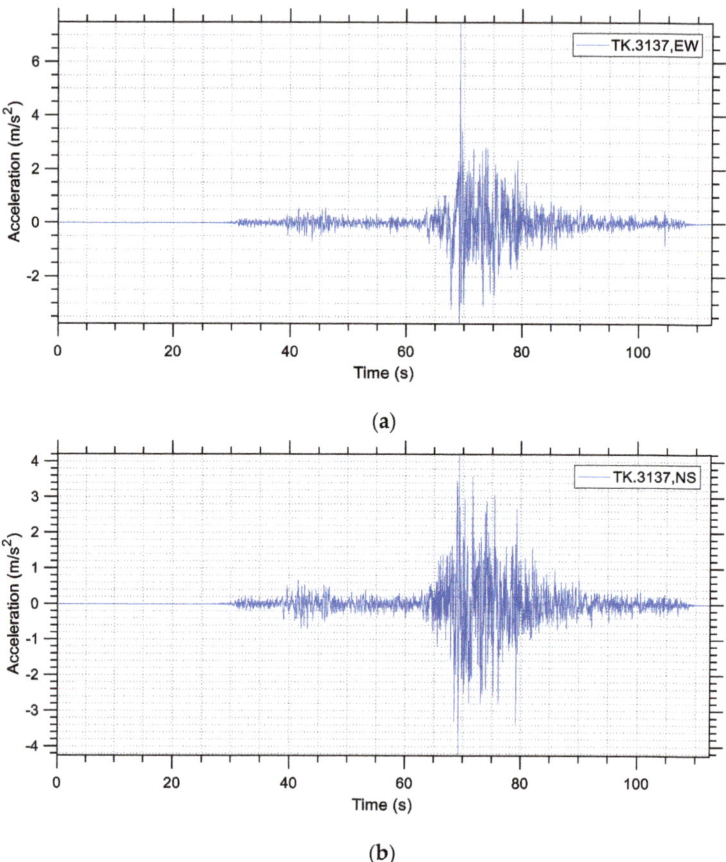

(**a**)

(**b**)

Figure 2. Earthquake acceleration time histories of the M_w 7.8 earthquake recorded at the TK.3137 station: (**a**) east–west component, (**b**) north–south component.

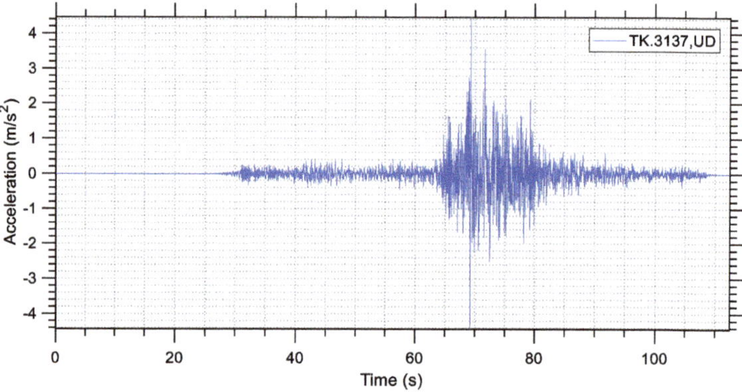

Figure 3. Earthquake acceleration time history of the M_w 7.8 earthquake recorded at the TK.3137 station: vertical component.

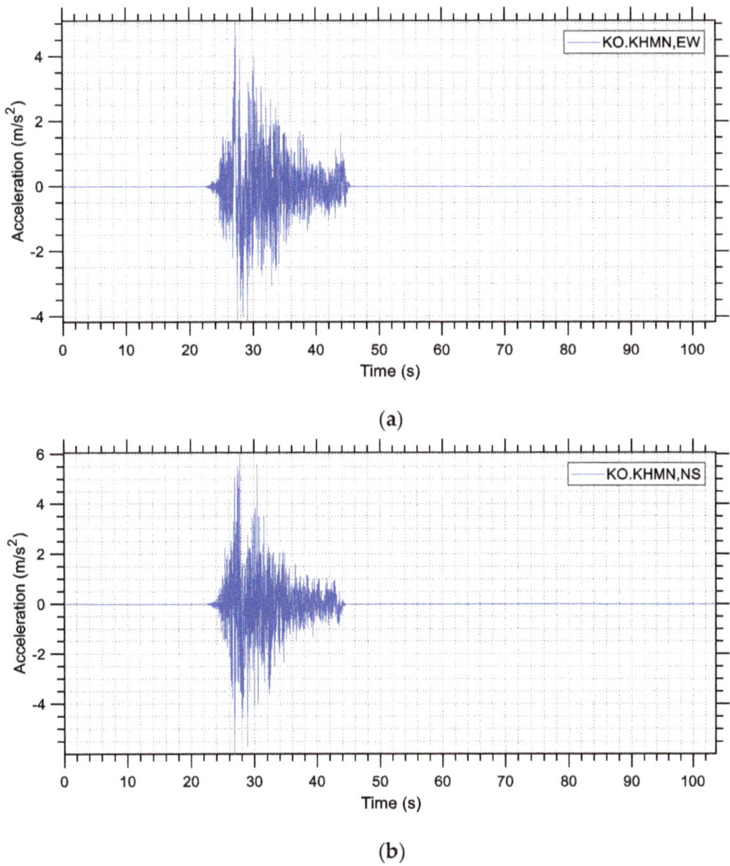

Figure 4. Earthquake acceleration time histories of the M_w 7.8 earthquake recorded at the KHMN Station: (**a**) east–west component, (**b**) north–south component.

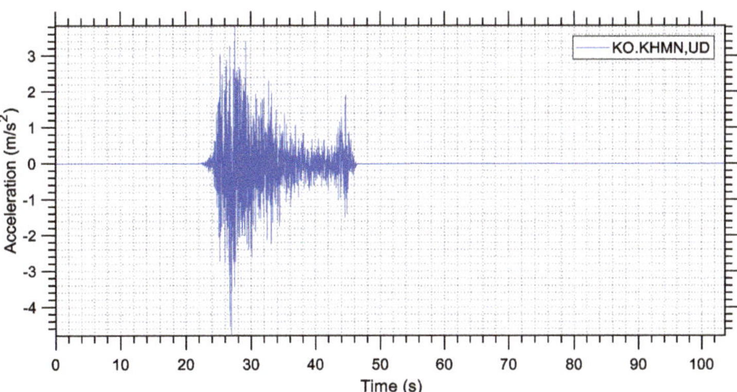

Figure 5. Earthquake acceleration time histories of the M_w 7.8 earthquake recorded at the KHMN Station: vertical component.

A summary report is given in Table 3, where *Ecum* denotes the cumulative energy, *Arias* denotes the Arias intensity, $td_{5\text{-}95}$ denotes the time needed for the 90% of the seismic energy to be released (5–95% interval), and td_{5-75} denotes the time needed for the 70% of the seismic energy to be released (5–75% interval). These quantities are calculated using the following equations:

$$E_{cum} = \int_0^{td} a(t)^2 dt \tag{1}$$

$$Arias = \frac{\pi}{2g} \int_0^{td} a(t)^2 dt \tag{2}$$

where td stands for the time duration of the strong motion record of the earthquake. The time durations are given as follows:

$$td_{5-95} = td_{95} - td_5 \tag{3}$$

$$td_{5-75} = td_{75} - td_5 \tag{4}$$

where for td_5, td_{75}, and td_{95} the following conditions hold:

$$\int_0^{td_5} a(t)^2 dt = 0.05 E_{cum} \tag{5}$$

$$\int_0^{td_{75}} a(t)^2 dt = 0.75 E_{cum} \tag{6}$$

$$\int_0^{td_{95}} a(t)^2 dt = 0.95 E_{cum} \tag{7}$$

Table 3. Cumulative seismic parameters of the M_w 7.8 earthquake event.

Station	TK.3137				KO-KHMN			
Component	Ecum (m^2/s^3)	td_{5-95} (s)	td_{5-75} (s)	Arias (m/s)	Ecum (m^2/s^3)	td_{5-95} (s)	td_{5-75} (s)	Arias (m/s)
EW Horizontal	22.755	16.27	8.26	3.6	21.220	12.145	5.88	3.4
NS Horizontal	22.232	16.79	9.58	3.6	28.743	10.28	4.64	4.6
UD Vertical	13.932	16.68	9.71	2.2	13.731	16.59	5.44	2.2

The main trend, apart from the relatively high Arias intensity and cumulative energy values, is that the seismic energy was released in a small fraction of the total duration of the earthquake. For example, for station TK.3137, it took only 16 s for 90% of the seismic energy to be released, whereas only roughly 8 s were needed for 70% of it to be released. If these time durations are compared to the total duration of the earthquake, which is larger than 80 s, it becomes apparent that the M_w 7.8 event was an event of large seismic power. This fact played a critical role in the intensity of the shaking that was experienced by structures and could provide some indirect hints explaining the large number of structural collapses. The aforementioned points become obvious by observing the normalized cumulative energy time histories shown in Figure 6 [35].

3.2. Elastic Response Spectra

A response spectrum is a plot that shows the maximum response of a structure to a ground motion as a function of frequency (or equivalently the period). The elastic response spectra simulate a single degree of freedom (SDOF) system and the way that it would respond to the time history of a given earthquake. We are interested both in the peak and the cumulative spectral response of an SDOF system for an earthquake. Typical spectral quantities of the first category are spectral acceleration and pseudo-acceleration, spectral velocity and pseudo-velocity, and spectral displacement. Typical spectral quantities of the latter category are the seismic input energy equivalent velocity [36], absolute and relative. It is noted that the seismic input energy equivalent velocity is a measure of the seismic energy that is input to a structure by an earthquake, not the energy that is released by the earthquake itself. In structural analysis, response spectra are utilized in response spectrum modal analysis (RSMA), a method commonly used in the design of buildings, bridges, and other structures that are critical to public safety. By analyzing the structure's response to different ground motion records, given the uncertainties involved, the engineers can hope that the structure will not experience excessive deformation or failure during an earthquake.

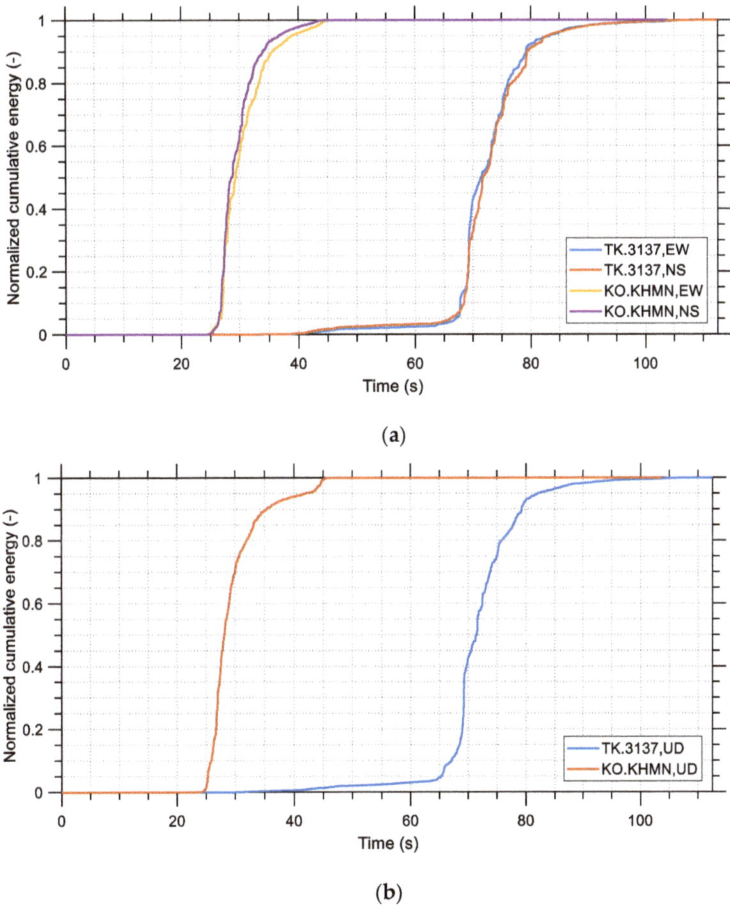

(a)

(b)

Figure 6. Normalized cumulative energy that was released by the recorded M_w 7.8 earthquake, as recorded at the TK.3137 and the KO–KHMN stations: (**a**) Horizontal components, (**b**) Vertical components.

The spectral displacement, velocity, and acceleration are shown in Figures 7–9, respectively, for the two records and their different components (the two horizontal and the vertical). Similarly, Figures 10 and 11 show the spectral pseudo-acceleration and the spectral pseudo-velocity, respectively. All diagrams have been generated for damping ratio ζ equal to 5% and their horizontal axis is in logarithmic scale. In Figure 9, it can be observed that the NS component of the KO-KHMN station record reached very high spectral accelerations, around 2.35 g, whereas the vertical component of the same record reached a maximum spectral acceleration close to 1.7 g. These spectral acceleration values are extraordinarily high. On the other hand, it is seen from Figure 8 that at the high period range the spectral velocity of the record of the TK.3137 station is generally larger, with some exceptions, for both its horizontal and vertical components. This is an important observation that may explain the large casualties that occurred in the Hatay region, although it is far away from the epicenter of the main earthquake, especially in comparison to Kahramanmaraş. The TK.3137 station, which gave higher spectral velocities than the KO-KHMN station, is much closer to the Hatay region, as shown in Figure 1. This may imply a stronger relationship between the destructive effects of an earthquake and its spectral velocity rather than its spectral acceleration. While the spectral acceleration has been traditionally taken into account for the design of buildings according to various seismic norms worldwide (including the Turkish seismic code), the spectral velocity is generally ignored, although it may be a better index in determining seismic hazard for taller buildings and it can serve as a parameter from which to estimate the macroseismic intensity and structural damage [37]. This is a well-known issue, and the M_w 7.8 earthquake may provide an incentive for further improvements to the seismic codes in this direction [35].

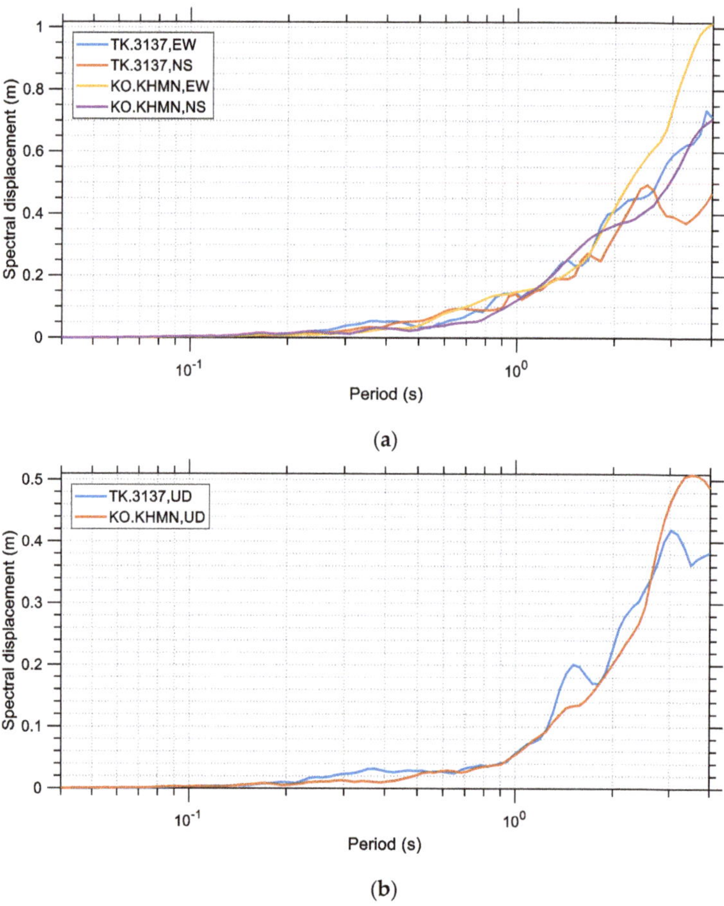

Figure 7. Spectral displacement (ζ = 5%) for the M_w 7.8 earthquake, as recorded at the TK.3137 and the KO–KHMN stations: (**a**) horizontal components, (**b**) vertical components.

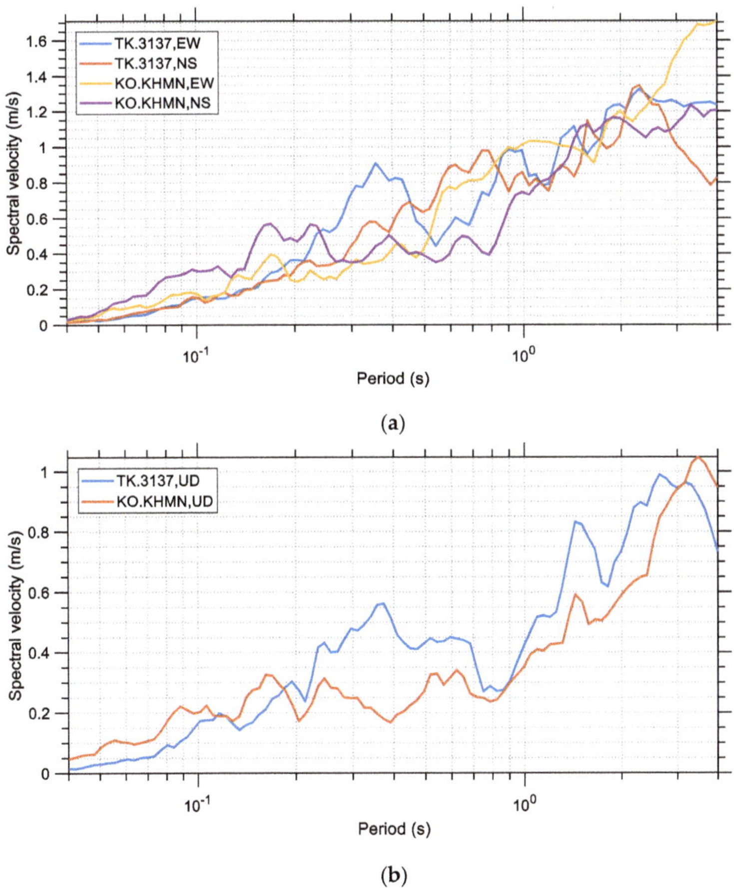

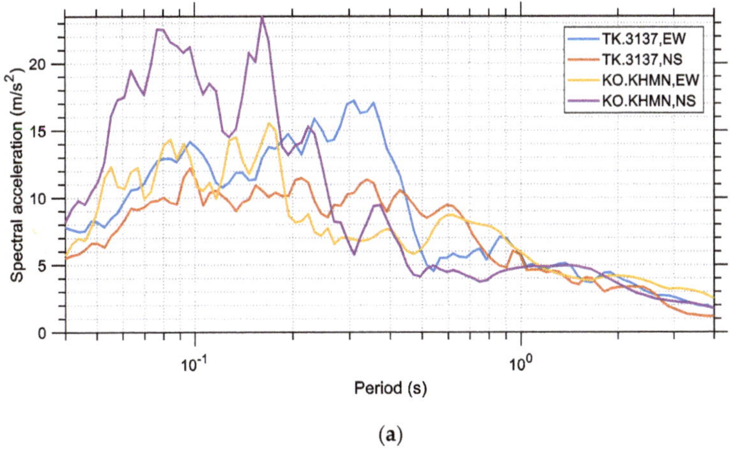

Figure 8. Spectral velocity ($\zeta = 5\%$) for the M_w 7.8 earthquake, as recorded at the TK.3137 and the KO-KHMN stations: (**a**) horizontal components, (**b**) vertical components.

Figure 9. *Cont.*

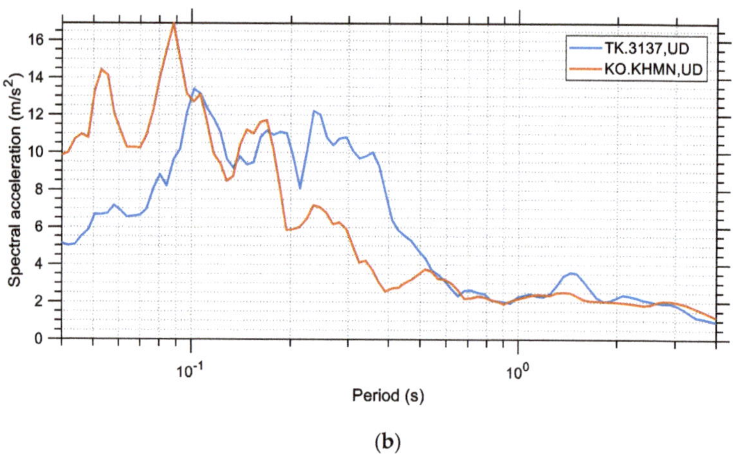

(**b**)

Figure 9. Spectral acceleration (ζ = 5%) for the M_w 7.8 earthquake, as recorded at the TK.3137 and the KO–KHMN stations: (**a**) horizontal components, (**b**) vertical components.

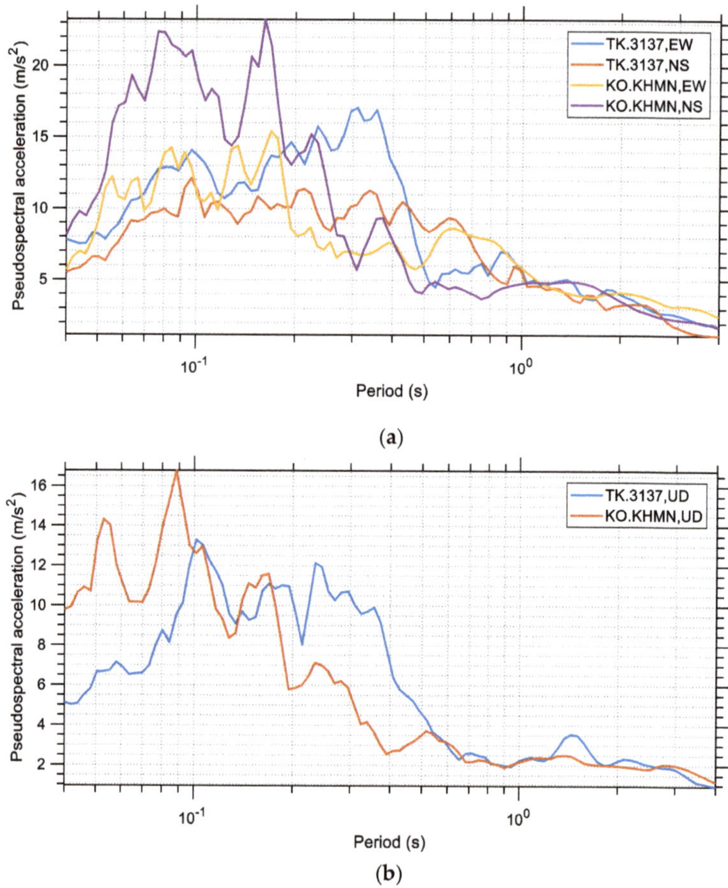

(**a**)

(**b**)

Figure 10. Spectral pseudo-acceleration (ζ = 5%) for the M_w 7.8 earthquake, as recorded at the TK.3137 and the KO–KHMN stations: (**a**) horizontal components, (**b**) vertical components.

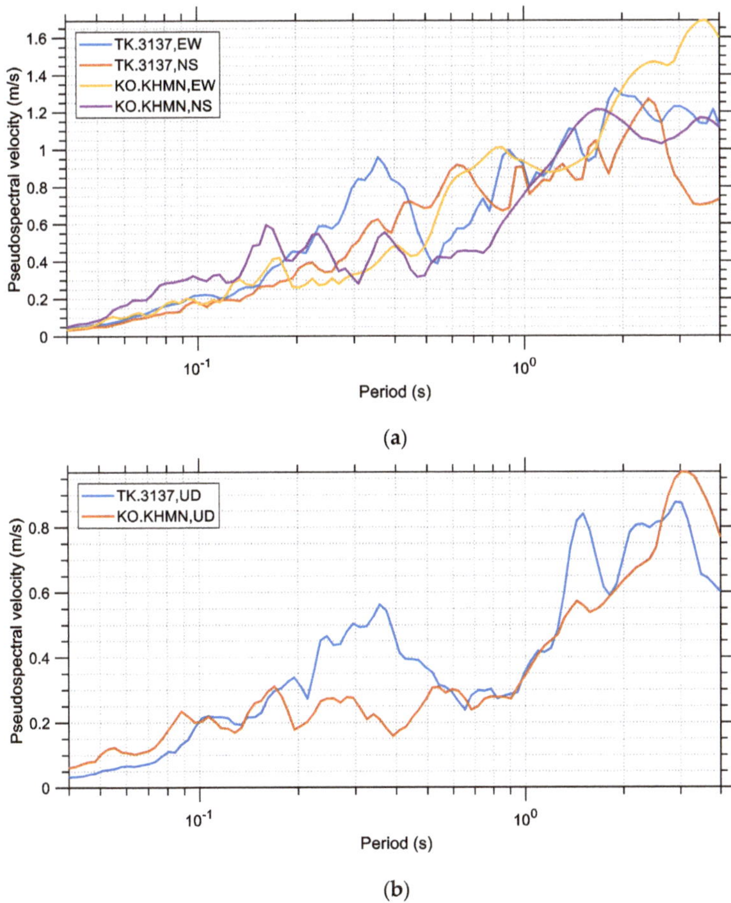

Figure 11. Spectral pseudo-velocity ($\zeta = 5\%$) for the M_w 7.8 earthquake, as recorded at the TK.3137 and the KO–KHMN stations: (**a**) horizontal components, (**b**) vertical components.

3.3. Isoductile response spectra

Seismic energy can be absorbed in structures in two main forms: (i) as elastic strain energy and (i) as hysterically dissipated energy. The former approach requires that the structure remains in the elastic region and resists the entire earthquake load, as high as its peak value, elastically. The latter approach permits the design of structures based on reduced earthquake loads (lower than those of the former case, which correspond to the maximum acceleration that the structure experiences during the seismic event) by relying on ductility and over-strength of the materials used. A structure that responds in an elastoplastic way through its ductile character experiences lower acceleration during an earthquake and its design becomes more economical, although damages may be expected in case of high accelerations. The basic ductile design philosophy is that the structure should survive the main shock through controlled damage but without collapse. The constant ductility (or isoductile) spectra assume that an SDOF system responds in an elastoplastic (i.e., elastic—perfectly plastic) way with constant ductility, which is defined as the ratio of the maximum displacement to the yielding displacement. The yielding displacement is the displacement that corresponds to the yield limit of the SDOF system. In other words, we are interested in the response of SDOF systems for varying eigenperiods, similar to the elastic spectra considered in the previous section, but for a specific ductility ratio. The isoductile spectral displacement, velocity, and acceleration are shown in Figures 12–14, respectively, for the two records and their two components.

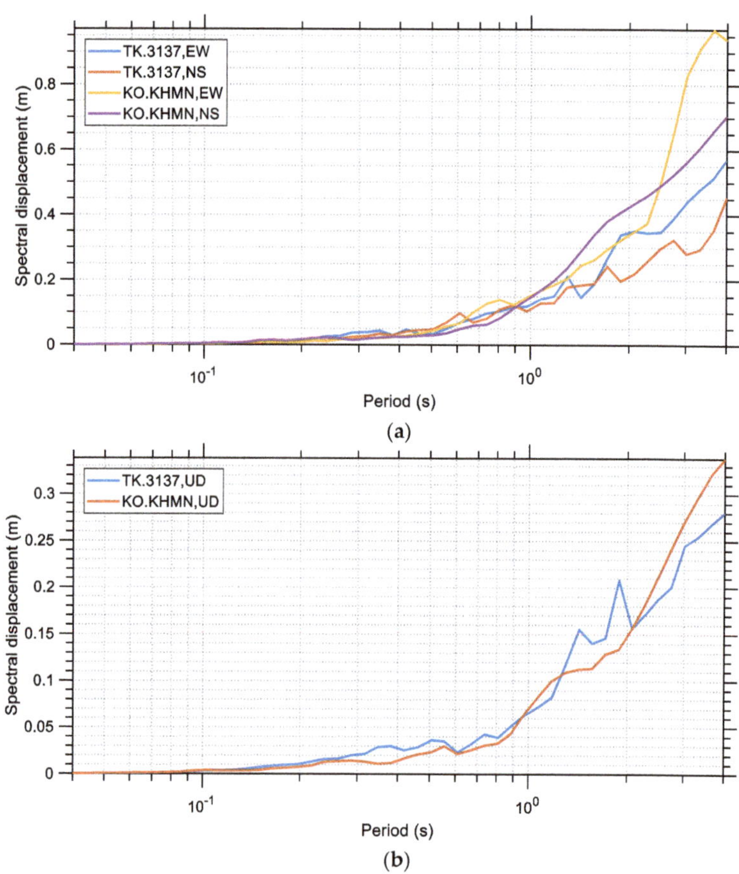

Figure 12. Isoductile spectral displacement ($\zeta = 5\%$, $\mu = 2$) for the M_w 7.8 earthquake, as recorded at the TK.3137 and the KO-KHMN stations: (**a**) horizontal components, (**b**) vertical components.

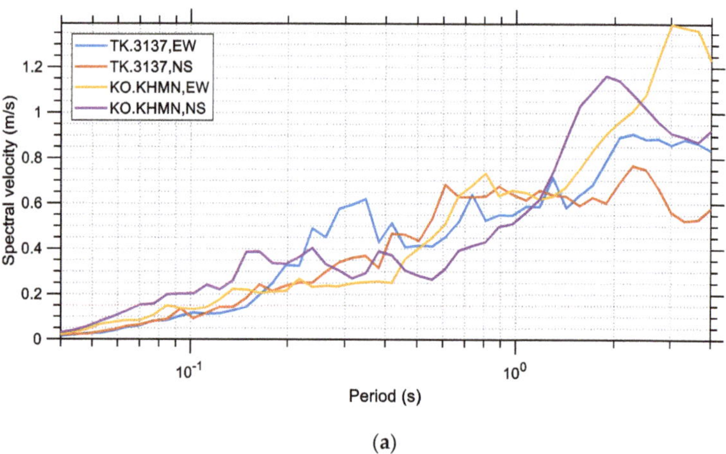

Figure 13. *Cont.*

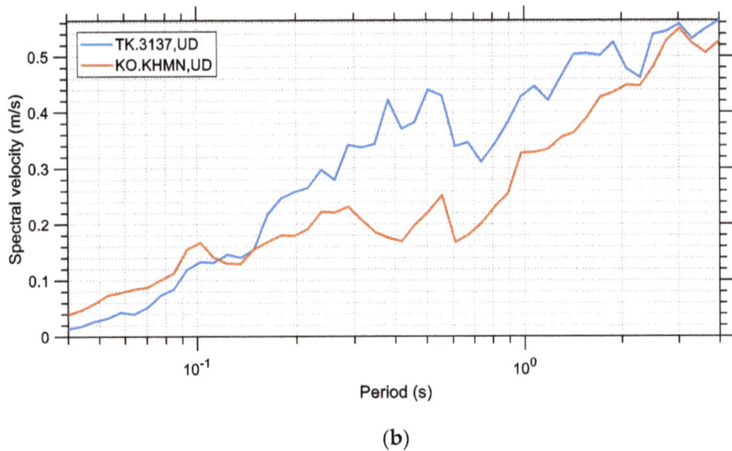

(b)

Figure 13. Isoductile spectral velocity ($\zeta = 5\%$, $\mu = 2$) for the M$_w$ 7.8 earthquake, as recorded at the TK.3137 and the KO–KHMN stations: (**a**) horizontal components, (**b**) vertical components.

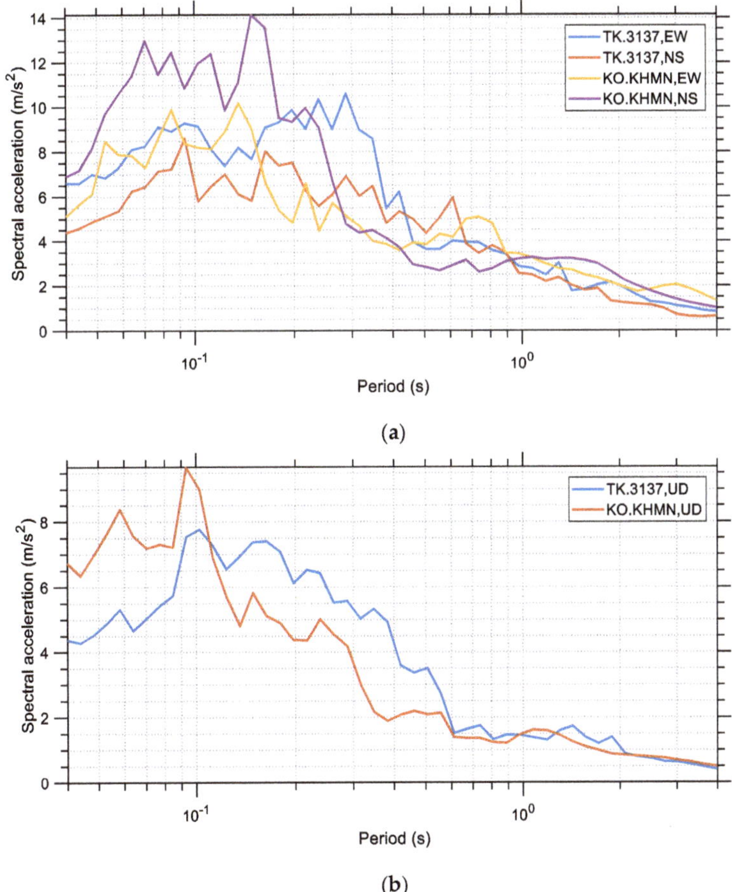

(a)

(b)

Figure 14. Isoductile spectral acceleration ($\zeta = 5\%$, $\mu = 2$) for the M$_w$ 7.8 earthquake, as recorded at the TK.3137 and the KO–KHMN stations: (**a**) horizontal components, (**b**) vertical components.

Similarly, Figures 15 and 16 show the isoductile spectral pseudo-acceleration and spectral pseudo-velocity, respectively. All diagrams correspond to damping ratio ζ equal to 5% and ductility μ equal to 2, and their horizontal axis is in logarithmic scale. As expected, the comparison between the linear elastic and the isoductile spectra reveals that the maximum responses in the isoductile spectra are generally lower than those in the elastic spectra. For example, the maximum spectral acceleration of the horizontal components of the M_w 7.8 event is equal to 2.35 g as shown in Figure 9, whereas the corresponding value for the isoductile spectra is equal to 1.4 g as shown in Figure 14. The difference in the maximum spectral acceleration implies a substantial difference in the applied seismic forces, and this shows the importance of structural ductility. The collapses due to the M_w 7.8 earthquake showed in many cases a nonductile, brittle behavior, which in the case of reinforced concrete (RC) structures is closely related to under-reinforced structural elements. These structures, having limited ductility, responded in a more linear elastic-wise manner, and thus experienced much larger accelerations, which explains many of the building collapses [35].

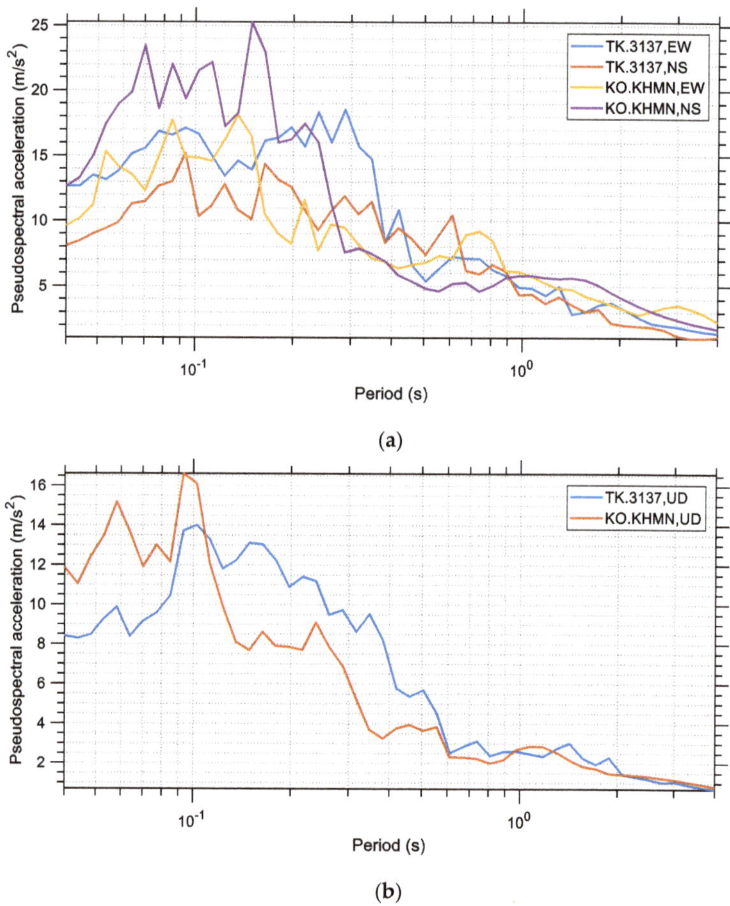

Figure 15. Isoductile spectral pseudo-acceleration ($\zeta = 5\%$, $\mu = 2$) for the M_w 7.8 earthquake, as recorded at the TK.3137 and the KO–KHMN stations: (**a**) horizontal components, (**b**) vertical components.

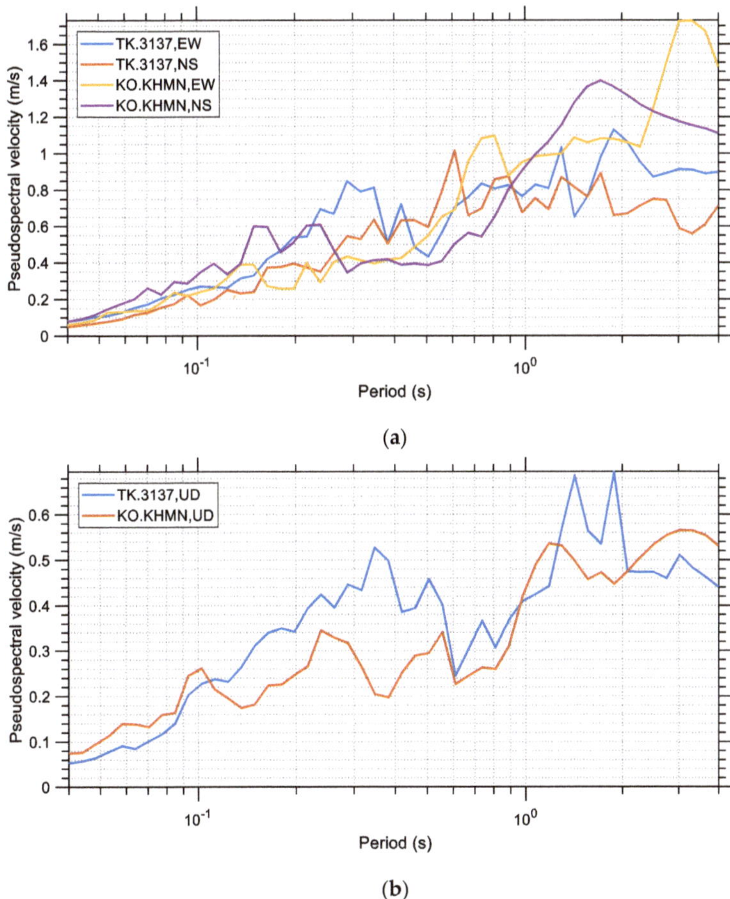

Figure 16. Isoductile spectral pseudo-velocity (ζ = 5%, μ = 2) for the M_w 7.8 earthquake, as recorded at the TK.3137 and the KO-KHMN stations: (**a**) horizontal components, (**b**) vertical components.

An important observation can be made based on the elastic and isoductile seismic response spectra: the maximum acceleration which is experienced by the elastoplastic ductile structures is substantially lower than that experienced by elastic structures. For example, in Figure 14, the maximum isoductile spectral acceleration for ductility μ = 2 is roughly 1.4 g and 0.97 g for the horizontal and vertical components, respectively. The corresponding values for linear elastic nonyielding structures can be seen in Figure 9, which are 2.35 g and 1.7 g for the horizontal and vertical components, respectively. In the horizontal direction, an increase from 1.4 g to 2.35 g accounts for 68% higher horizontal acceleration values for the nonductile structures. Low accelerations are directly related to low seismic forces, through Newton's second law, and thus lower requirements on behalf of the structure to resist these forces. Therefore, it becomes evident that ductility is an important aspect of seismic design since reduced seismic loads result in more economical designs. Apart from this, a structure of increased ductility is generally safer since it can accommodate large deformations which cannot go unnoticed by the occupants and act as a warning for the imminent failure of the structure. This can save some critical time in difficult situations for the occupants when evacuation is required and could potentially save their lives.

3.4. What Does the Turkish Seismic Code Provide?

It is interesting to compare the effect of the earthquake event on the structures with the requirements of the Turkish seismic code in the region. The comparison is made with reference to the records of the TK.3137 and the KO-KHMN stations and it is shown in Figure 17 for the cases of linear elastic response spectra. Based on the comparison, the major conclusion is that the earthquake struck mainly at the low period range, where the design acceleration was 1.4 g and the maximum acceleration observed was roughly equal to 2.4 g. This is a significant difference, not only in the acceleration magnitude but also in its period content. Even for site class I, which contains the lowest period content (i.e., corresponds to stiff rock), there was a significant acceleration below the lowest reference period. Therefore, suitable adjustments need to be made in the design spectrum of the Turkish code so that rare events, such as the one (M_w 7.8) considered in this study, can be taken into account.

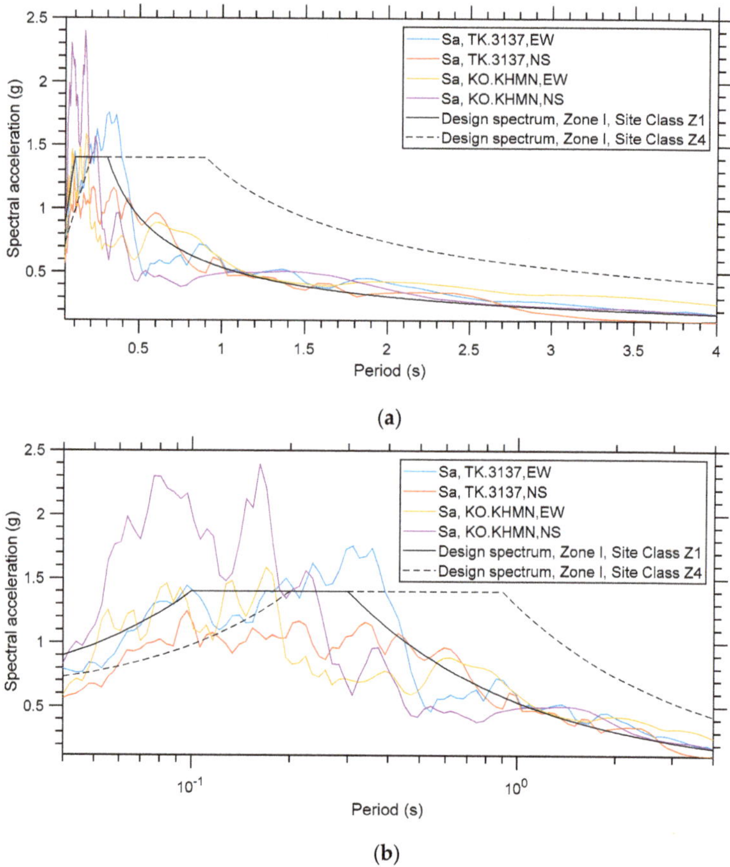

Figure 17. Design spectrum of the Turkish seismic code vs. actual acceleration response spectra for the M_w 7.8 earthquake (ζ = 5%), as recorded at TK.3137 and KO–KHMN stations: (**a**) linear scale, (**b**) logarithmic scale.

3.5. Does the High Spectral Acceleration in the Low Period Range Occur Only for the Two Examined Records or Is It a General Trend?

At this point, we need to check whether the trend that appears in Figure 17 is a general trend or if it is specific only to these two recordings. For this purpose, more earthquake acceleration records need to be taken into account. In Figure 18 the acceleration spectra of several earthquake records are shown and compared to the provisions of the Turkish seismic code (linear elastic response spectra). It is shown that there are higher spectral acceleration values for a broader range of eigenperiods for many of the recordings. Based on the envelope spectrum, a maximum spectral acceleration of 5.35 g is observed, which is extremely high and is responsible for the many collapses due to the earthquake event. A need for a revision of the seismic code standards seems to exist, i.e., higher acceleration values for the design spectra must be proposed [35].

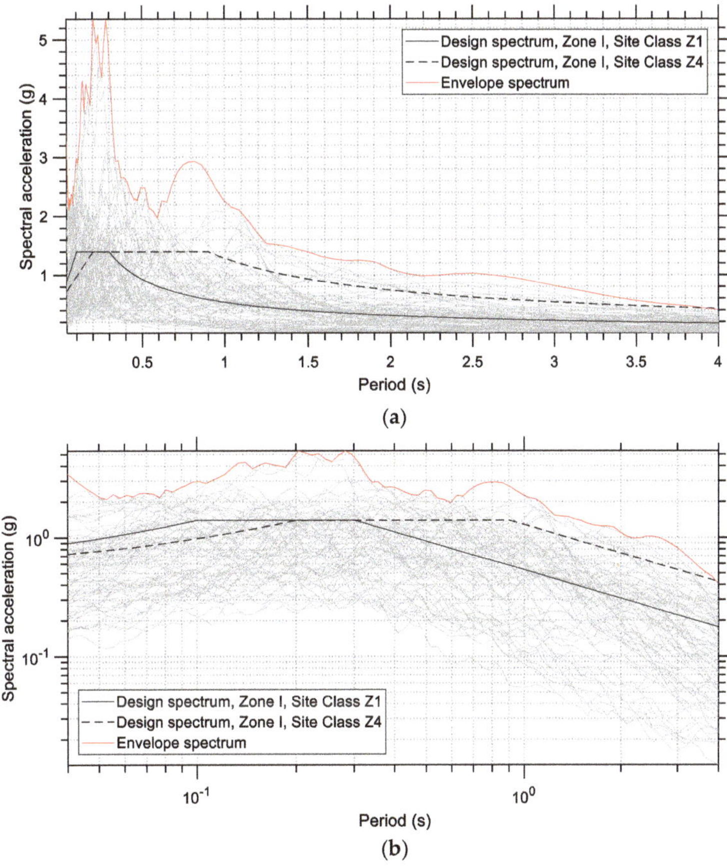

Figure 18. Design spectrum according to the Turkish seismic code vs. actual acceleration response spectra ($\zeta = 5\%$) of various records for the M_w 7.8 earthquake: (**a**) linear scale, (**b**) logarithmic scale.

In Figure 18, 76 recordings have been taken into account in total, namely the two horizontal components (EW and NS) from the following 38 stations: TK_0201, TK_0213, TK_1213, TK_2308, TK_2708, TK_2715, TK_2718, TK_3115, TK_3117, TK_3123, TK_3124, TK_3125, TK_3126, TK_3129, TK_3131, TK_3132, TK_3133, TK_3134, TK_3135, TK_3136, TK_3137, TK_3138, TK_3139, TK_3141, TK_3142, TK_3143, TK_3145, TK_3146, TK_4617, TK_2703, TK_2712, TK_4615, TK_4616, TK_4624, TK_4629, TK_4630, TK_4632, and TU_NAR.

4. Structural Incremental Dynamic Analysis

The effect of an earthquake on structures can be quantified in various forms such as by using the various peak and cumulative seismic parameters as well as the response spectra that were described in the previous sections. However, the engineer is often interested in monitoring the peak or cumulative structural response due to a seismic record, while varying a suitable intensity measure which is taken by appropriately scaling an earthquake record. This procedure is called incremental dynamic analysis (IDA) and it involves performing multiple nonlinear dynamic analyses of a structural model under a ground motion record scaled to several levels of seismic intensity. The scaling levels are appropriately selected to force the structure through the entire range of behavior, from elastic to inelastic [29]. OpenSeismoMatlab [30] is capable of performing IDA analysis for a single record and an SDOF structure. Such IDA curves contain useful information about a seismic record, from a structural point of view.

Normally, IDA involves performing multiple nonlinear dynamic analyses of a structural model under a suite of ground motion records, each scaled to several levels of seismic intensity. The scaling factors are selected in a way that the structural model being analyzed experiences the entire range of behavior, from linear elastic to inelastic global dynamic instability, where the structure collapses. The procedure is intended to provide an estimation of the seismic risk for a given structure. However, various approximate methods define SDOF systems with which the static pushover curves of MDOF systems are calculated to reduce the computational effort required for the IDA to calculate various seismic demand measures for the structures [38–40]. The last approximation is attractive in cases of a large variety of structures existing in urban densely populated areas, for the evaluation of the seismic risk. In addition to this, the various effects of a given (actual) earthquake on buildings are more easily conceivable when applied in the context of an SDOF system rather than MDOF systems, since the former is defined in terms of a much smaller number of independent parameters. Finally, performing IDA analysis in SDOF systems is a usual practice, as reflected in the relevant literature, e.g., [41,42].

4.1. Spectral Acceleration–Ductility Curves

In Figures 19 and 20 the IDA curves of an SDOF system for the horizontal components of the records TK.3137 and KO-KHMN, respectively, are shown. These curves plot the spectral acceleration as an intensity measure (IM) and the ductility demand of the structure as a damage measure (DM). It is shown in Figure 19 that for structures with low-yielding deformation, there is a higher ductility demand to withstand the same spectral acceleration (if possible). This trend is also observed in Figure 20. In addition, stiffer structures (with lower eigenperiods) seem to be more capable of withstanding higher spectral acceleration, as engineering intuition dictates. An important observation is the possibility of the fact that a structure can withstand more than one spectral acceleration for a single value of ductility (for example for $T = 1.5$ s and $u_y = 0.1$ m in Figure 19a, or $T = 1$ s and $u_y = 0.1$ m in Figure 19b). This is a common observation for structures responding in the elastoplastic regime. A general trend of the IDA curves is that with increasing intensity measure (spectral acceleration), the damage measure generally increases as well [35].

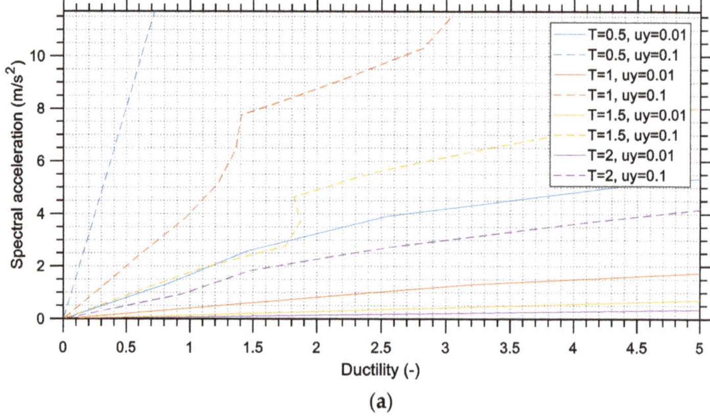

(a)

Figure 19. *Cont.*

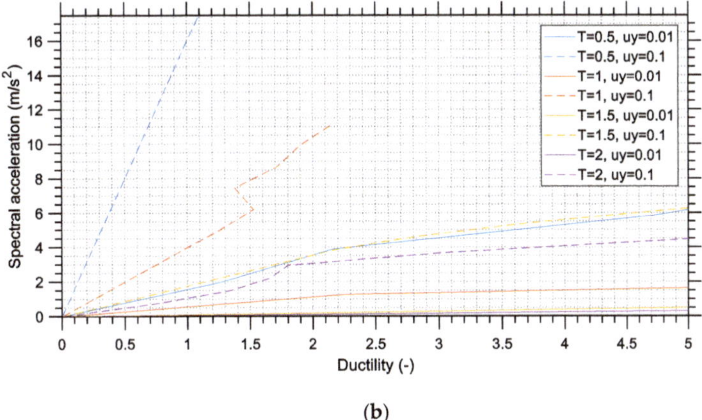

(**b**)

Figure 19. IDA—spectral acceleration vs. ductility curves for the TK.3137 record and for various combinations of yield displacement (u_y) and small strain eigenperiods (T) for the M_w 7.8 earthquake: (**a**) EW component, (**b**) NS component.

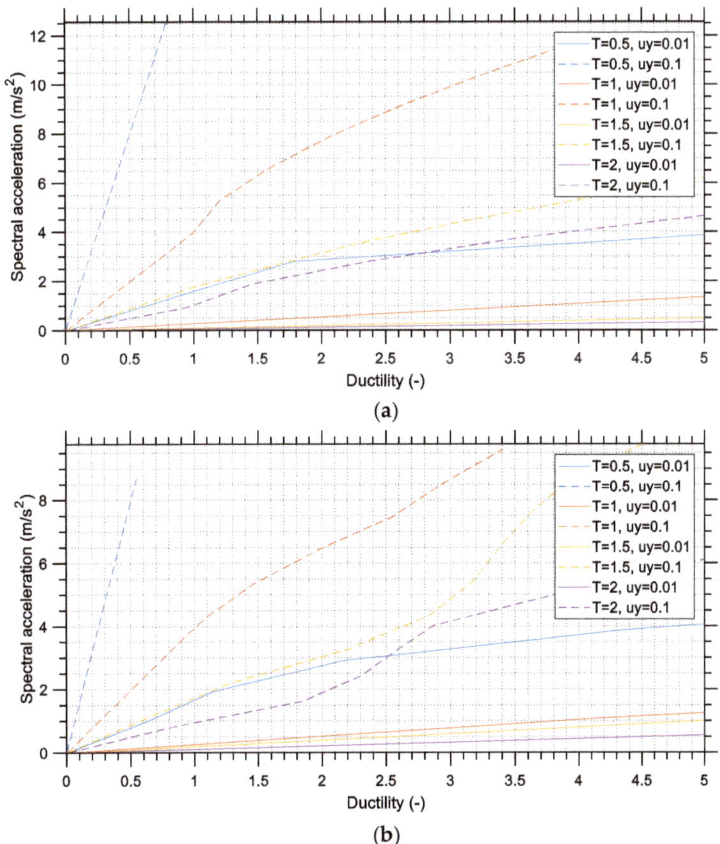

(**a**)

(**b**)

Figure 20. IDA—spectral acceleration vs. ductility curves for the KO–KHMN record and for various combinations of yield displacement (u_y) and small strain eigenperiods (T) for the M_w 7.8 earthquake: (**a**) EW component, (**b**) NS component.

4.2. Peak Ground Acceleration–Ductility Curves

In Figures 21 and 22, IDA curves are presented regarding PGA versus ductility of an SDOF system for the horizontal components of the records TK.3137 and KO-KHMN, respectively. The same trends as in the previous section can be observed. In addition, it is noted that in Figure 21a, the curves for $T = 0.5$ s and $u_y = 0.01$ m and $T = 2$ and $u_y = 0.1$ m are nearly identical. This means that a stiff structure with low yield deformation can be equivalent to a flexible structure with moderate yield deformation. This has important implications for structural design. The latter type of structure is preferable since it is more economical. Therefore, reduced stiffness should be accompanied by moderate levels of yield deformation to ensure that a structure will be able to withstand high earthquake acceleration levels.

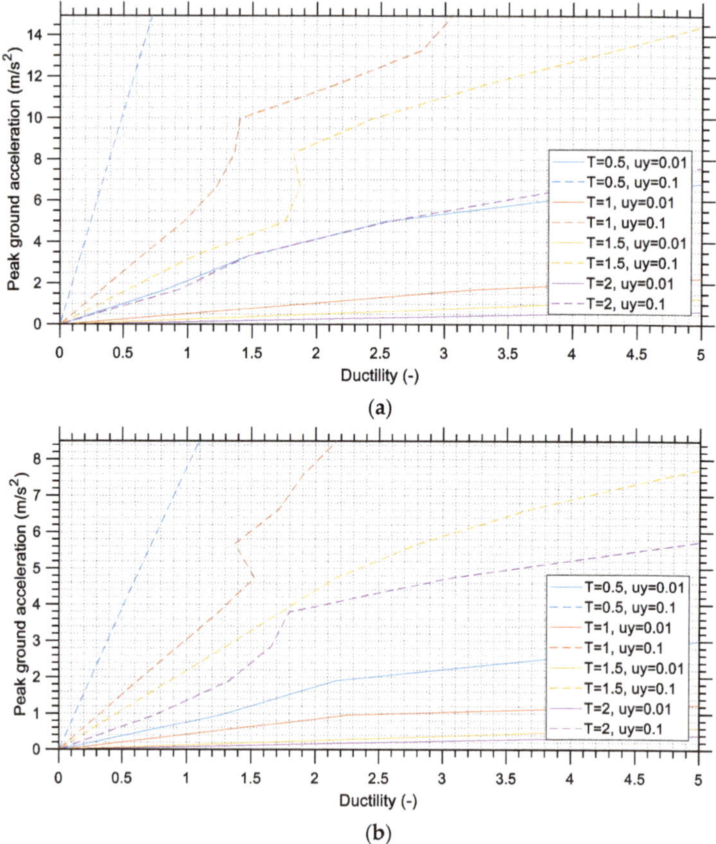

Figure 21. IDA—PGA vs. ductility curves for the TK.3137 record and for various combinations of yield displacement (u_y) and small strain eigenperiods (T) for the M_w 7.8 earthquake: (**a**) EW component, (**b**) NS component.

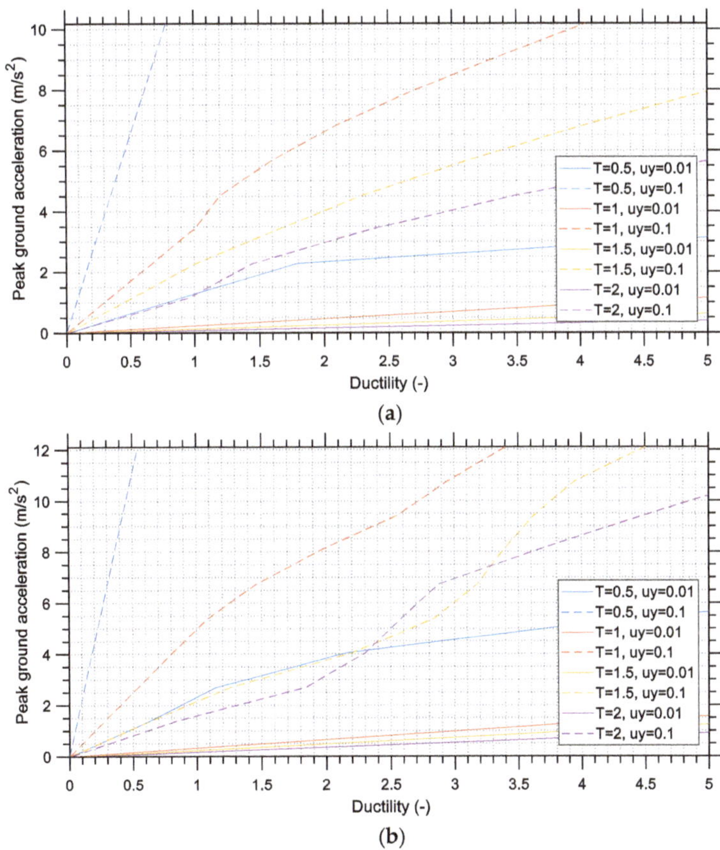

Figure 22. IDA—PGA vs. ductility curves for the KO–KHMN record and for various combinations of yield displacement (u_y) and small strain eigenperiods (T) for the M_w 7.8 earthquake: (**a**) EW component, (**b**) NS component.

5. Distributions of Several Earthquake Characteristics

The distributions of several characteristic seismic parameters are investigated in this section. For this purpose, the previously mentioned 76 acceleration time histories (i.e., the two horizontal components from 38 stations) are considered and their seismic parameters are calculated and then plotted in the form of statistical distributions. These distributions demonstrate the severity of the earthquakes that occurred on 6 February 2023 in Türkiye and indirectly may provide some explanations about the increased number of buildings that collapsed. The seismic parameters that are plotted include peak measures as well as cumulative measures. These are the PGA (plotted in Figure 23), effective PGA (EPGA, according to [43], plotted in Figure 24), PGV (plotted in Figure 25), spectral intensity defined according to [44] (plotted in Figure 26), spectral intensity according to [45] (plotted in Figure 27), Arias intensity (plotted in Figure 28), and significant duration (plotted in Figure 29). All the aforementioned parameters have been calculated using the OpenSeismoMatlab software [30], only for horizontal strong ground motion components. It is noted that the μ_{LN} and σ_{LN} parameters of the lognormal distribution that appear in the legends of the histogram plots are different from the mean value and standard deviation of the data being plotted. The increased mean values of the plotted seismic parameters denote the increased impact of the 6 February 2023 Türkiye M_w 7.8 earthquake event.

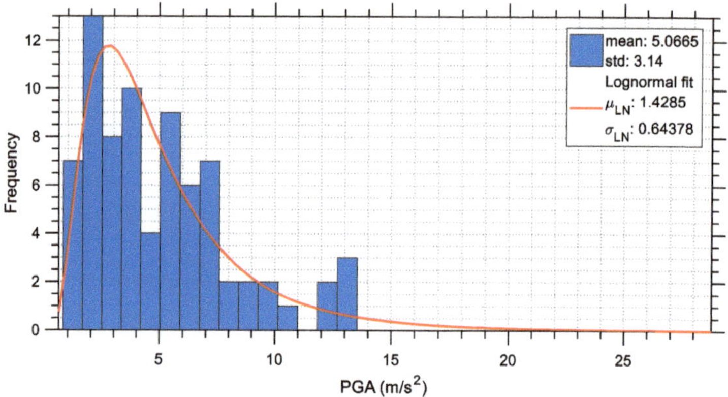

Figure 23. Histogram plot of the PGA values of the various acceleration time histories of the M_w 7.8 earthquake and lognormal distribution fit.

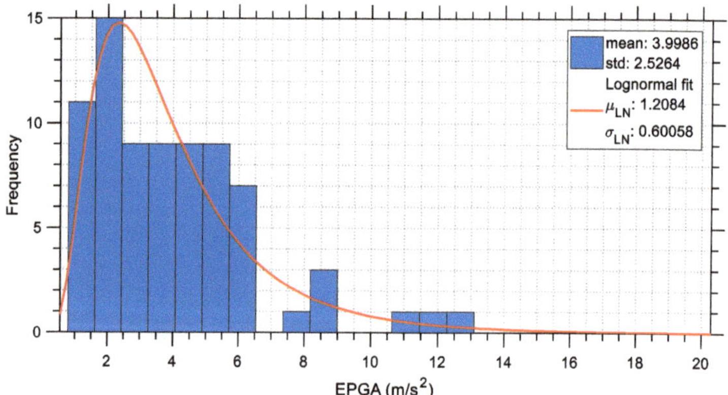

Figure 24. Histogram plot of the EPGA values [45] of the various acceleration time histories of the M_w 7.8 earthquake and lognormal distribution fit.

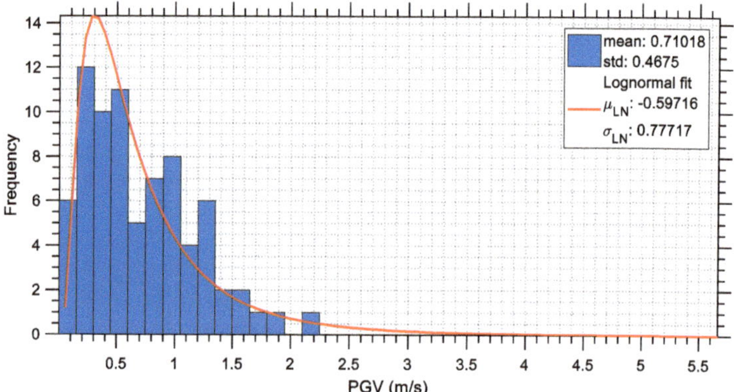

Figure 25. Histogram plot of the PGV values of the various acceleration time histories of the M_w 7.8 earthquake and lognormal distribution fit.

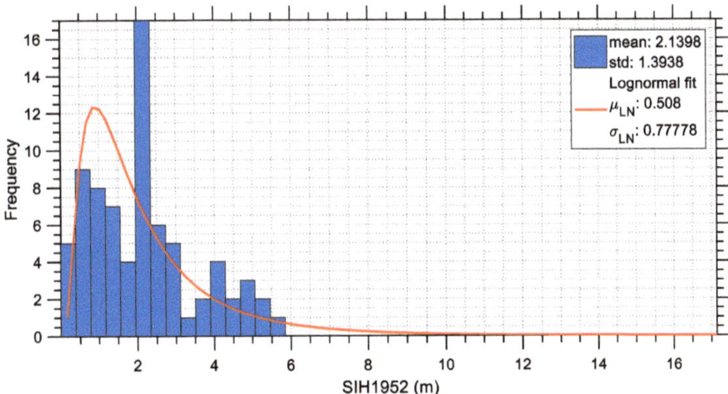

Figure 26. Histogram plot of the spectral intensity according to Housner (1952) [43] of the various acceleration time histories of the M_w 7.8 earthquake and lognormal distribution fit.

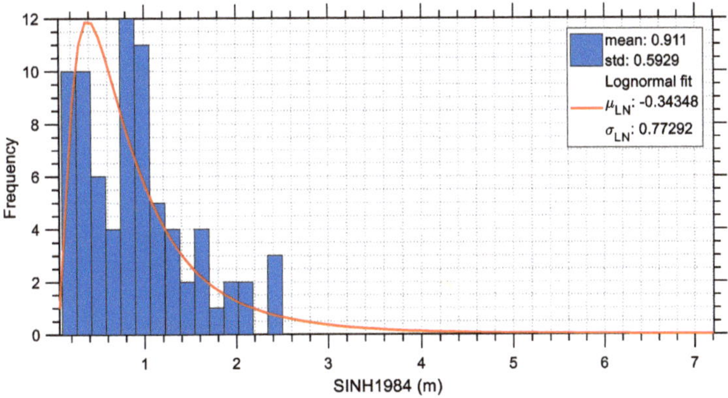

Figure 27. Histogram plot of the spectral intensity according to Nau and Hall (1984) [44] of the various acceleration time histories of the M_w 7.8 earthquake and lognormal distribution fit.

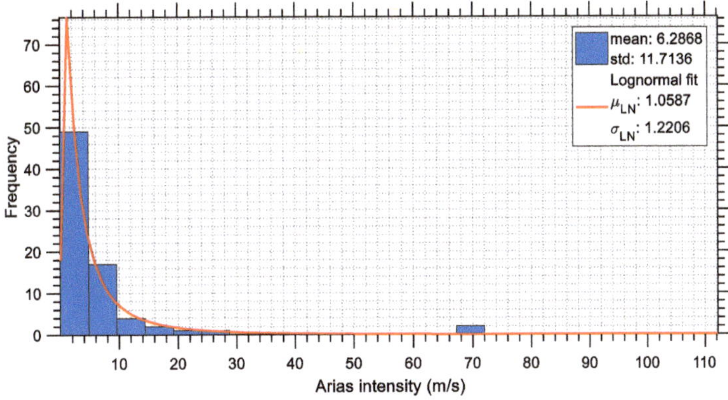

Figure 28. Histogram plot of the Arias intensity of the various acceleration time histories of the M_w 7.8 earthquake and lognormal distribution fit.

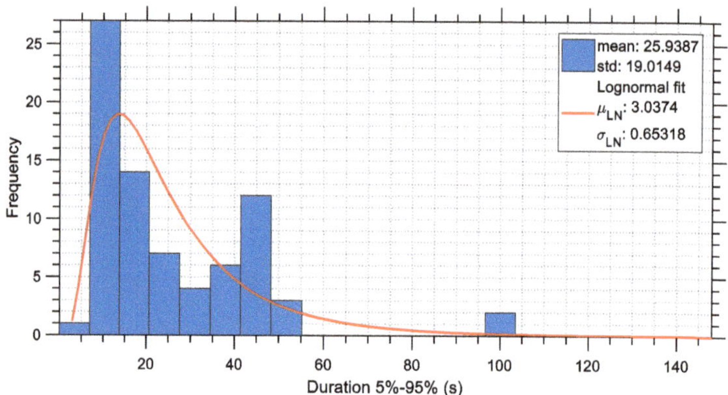

Figure 29. Histogram plot of the significant duration (5–95%) of the various acceleration time histories of the M_w 7.8 earthquake and lognormal distribution fit.

More specifically, it is apparent from Figure 26 that the spectral intensity calculated according to [43] has average and maximum values equal to 2.14 m and 6 m, respectively. Comparison of these values with the Housner intensities that are calculated for major earthquake events in the past reveals the severity of the M_w 7.8 earthquake event that occurred in Türkiye on 6 February 2023. For example, from Figure 5 of the work of Garini and Gazetas [46], where 99 recorded ground motions are selected to cover many of the well-known accelerograms from earthquakes of the last 30 years, and to include motions bearing near-fault characteristics (directivity and fling effects), it can be deduced that the Housner intensity values ranges roughly from 1 m to 6 m. In the study of Massumi and Gholami [47], a set of 85 far-field ground motion records from 17 earthquake events with moment magnitudes ranging from 5.9 to 7.6 and recorded for type II soil (Vs = 360–750 m/s) were processed. Figure 2 of this study shows the Housner spectral intensity ranging from 0 to 2.5 m. Since the range of the Housner intensities of the earthquake considered in this study is closer to that of [46], it is concluded that it can be highly possible that the earthquake considered in this study contains directivity and fling step phenomena, whereas it is confirmed that it was a severely strong event.

In addition, when the Arias intensities of the earthquake under consideration in this study and of various other major earthquakes are compared, similar observations can be made, as described above. For example, in Figure 2 of [46], it is observed that the Arias intensity values range roughly from 0.5 m/s to 12 m/s, whereas the mean and average values of the Arias intensity of the 6 February 2023 M_w 7.8 Türkiye earthquake are noted from Figure 28 to be equal to 6.29 m/s and 70 m/s, respectively. The fact that the average value falls into the middle of the aforementioned range shows the severity of the M_w 7.8 earthquake once again.

Similar observations can be made about the other seismic measures that are presented in this section.

6. Conclusions

The earthquake with the magnitude of M_w 7.8 that hit the Kahramanmaraş–Gaziantep regions in southern Türkiye on 6 February 2023, was a rare event of extremely large seismic power; as shown in Section 3.1, where the total cumulative energy as well as its time history are calculated. This fact played a critical role in the intensity of the shaking that was experienced by structures and could provide some indirect hints explaining the large number of structural collapses. The acceleration spectral values of the seismic records that are calculated in Section 3.2 were found to be substantially larger than the design acceleration spectrum values according to the Turkish seismic code. Moreover, this difference between the design and the actual response spectra covered a large interval of periods, which includes the eigenperiods of most common buildings.

It has been shown in this study that the maximum spectral acceleration of isoductile spectra is much lower than that of the corresponding linear elastic spectra, which implies a substantial difference in the seismic forces experienced by the structures and shows the importance of structural ductility for proper seismic design. Many post-earthquake surveys have shown that many concrete buildings were under-reinforced, which denotes that they had low ductility and therefore responded in a more linear elastic-wise manner, thus experiencing much higher accelerations and forces, which may have led to their collapse.

A close examination of the IDA curves that exhibit the seismic demand of the M_w 7.8 earthquake on SDOF structures can provide strong evidence that a relatively stiff structure with low yield deformation could be equivalent to a flexible structure with moderate yield deformation. This implies that reduced structural stiffness, which is the usual outcome of pursuing a more economical design, should be accompanied by moderate and not low levels of yield deformation to ensure that the structure will be able to withstand high earthquake acceleration levels. Under-reinforced concrete structures possess a low lateral stiffness combined with a low yield deformation, and this could explain the large number of collapses during the earthquake event when considering the much larger seismic damage imposed on SDOF systems with similar characteristics that are obvious in the various IDA curves presented in Section 4.

Spectral velocity appears to be an important parameter describing the destructive effects of an earthquake, as shown in Section 3, apart from the spectral acceleration which is adopted in most seismic codes worldwide (including the Turkish seismic code) for structural design against earthquake loading.

In Section 5 of this study, it has been shown that the M_w 7.8 earthquake was indeed severely strong by comparing its various peak and cumulative seismic parameters to the corresponding seismic parameters of strong motion datasets from other major earthquakes in the past.

Author Contributions: Conceptualization, G.P. and V.P.; methodology, G.P. and V.P.; software, G.P.; validation, G.P. and V.P.; formal analysis, V.P.; investigation, G.P. and V.P.; resources, G.P. and V.P.; data curation, G.P. and V.P.; writing—original draft preparation, G.P.; writing—review and editing, G.P. and V.P.; visualization, G.P. and V.P.; supervision, V.P.; project administration, G.P. and V.P. All authors have read and agreed to the published version of the manuscript.

Funding: This research received no external funding.

Institutional Review Board Statement: Not applicable.

Informed Consent Statement: Not applicable.

Data Availability Statement: The data presented in this study are available on request from the corresponding author.

Conflicts of Interest: The authors declare no conflict of interest.

References

1. Government of Türkiye. AFAD Press Bulletin about the Earthquake in Kahramanmaraş-34. 2023. Available online: https://reliefweb.int/report/turkiye/afad-press-bulletin-about-earthquake-kahramanmaras-34-entr (accessed on 23 February 2023).
2. Government of Türkiye. AFAD Press Bulletin about the Earthquake in Kahramanmaraş-8. 2023. Available online: https://reliefweb.int/report/turkiye/afad-press-bulletin-about-earthquake-kahramanmaras-8-entr (accessed on 23 February 2023).
3. UN High Commissioner for Refugees. UNHCR Türkiye Emergency Response to Earthquake (23 February 2023). 2023. Available online: https://reliefweb.int/report/turkiye/unhcr-turkiye-emergency-response-earthquake-23-february-2023 (accessed on 23 February 2023).
4. USGS Earthquake Hazards Program. M 7.8-26 km ENE of Nurdağı, Turkey. 2023. Available online: https://earthquake.usgs.gov/earthquakes/eventpage/us6000jllz/executive (accessed on 23 February 2023).
5. Papazafeiropoulos, G.; Plevris, V.; Papadrakakis, M. A new energy-based structural design optimization concept under seismic actions. *Front. Built Environ.* **2017**, *3*, 44. [CrossRef]

6. Papazafeiropoulos, G.; Georgioudakis, M.; Papadrakakis, M. Selecting and Scaling of Energy-Compatible Ground Motion Records. *Front. Built Environ.* **2019**, *5*, 140. [CrossRef]

7. Gharehbaghi, S.; Gandomi, M.; Plevris, V.; Gandomi, A.H. Prediction of seismic damage spectra using computational intelligence methods. *Comput. Struct.* **2021**, *253*, 106584. [CrossRef]

8. Lu, X. Preliminary assessment of the damage of the Feb. 6 Turkey earthquake on Chinese and Turkish buildings. 2023. Available online: https://www.researchgate.net/publication/368335285_Preliminary_assessment_of_the_damage_of_the_Feb_6_Turkey_earthquake_on_Chinese_and_Turkish_buildings (accessed on 23 February 2023). [CrossRef]

9. Lu, X.; Cheng, Q.; Xu, Z.; Xu, Y.; Sun, C. Real-Time City-Scale Time-History Analysis and Its Application in Resilience-Oriented Earthquake Emergency Responses. *Appl. Sci.* **2019**, *9*, 3497. [CrossRef]

10. Ahmed, I. Key Building Design and Construction Lessons from the 2023 Türkiye–Syria Earthquakes. *Architecture* **2023**, *3*, 104–106. [CrossRef]

11. Erdik, M.; Tümsa, M.B.D.; Pınar, A.; Altunel, E.; Zülfikar, A.C. A Preliminary Report on the February 6, 2023 Earthquakes in Türkiye. *Temblor.* 2023. Available online: https://temblor.net/temblor/preliminary-report-2023-turkey-earthquakes-15027/ (accessed on 23 February 2023). [CrossRef]

12. The Economist. Turkey's Earthquakes Show the Deadly Extent of Construction Scams. Available online: https://www.economist.com/europe/2023/02/12/turkeys-earthquakes-show-the-deadly-extent-of-construction-scams (accessed on 24 March 2023).

13. Naddaf, M. Turkey–Syria earthquake: What scientists know. *Nature* **2023**, *614*, 398–399. [CrossRef]

14. Baltzopoulos, G.; Baraschino, R.; Chioccarelli, E.; Cito, P.; Iervolino, I. Preliminary Engineering Report. on Ground Motion Data of the Feb. 2023 Turkey Seismic Sequence V2.5–25/02/2023. 2023. Available online: https://www.researchgate.net/publication/368330473_PRELIMINARY_ENGINEERING_REPORT_ON_GROUND_MOTION_DATA_OF_THE_FEB_2023_TURKEY_SEISMIC_SEQUENCE (accessed on 23 February 2023).

15. Chen, G.; Liu, Y.; Beer, M. Identification of near-fault multi-pulse ground motion. *Appl. Math. Model.* **2023**, *117*, 609–624. [CrossRef]

16. Chen, G. Report. on Pulse-Like Ground Motions in the Feb 2023 Turkey Earthquakes. 2023. Available online: https://www.researchgate.net/publication/368389270_Report_on_pulse-like_ground_motions_in_the_Feb_2023_Turkey_earthquakes?channel=doi&linkId=63e52dfbe2e1515b6b82d7c4&showFulltext=true (accessed on 23 February 2023). [CrossRef]

17. Chen, G.; Yang, J.; Liu, Y.; Kitahara, T.; Beer, M. An energy-frequency parameter for earthquake ground motion intensity measure. *Earthq. Eng. Struct. Dyn.* **2023**, *52*, 271–284. [CrossRef]

18. Luzi, L.; Felicetta, C.; D'Amico, M.; Ramadan, F.; Mascandola, C.; Pacor, F.; Sgobba, S.; Lanzano, G.; Brunelli, G.; Colavitti, L.; et al. *Preliminary Analysis of the accelerometric Registrations of the February 6th 2023 Turkish Earthquake (Mw 7.9)*; Istituto Nazionale di Geofisica e Vulcanologia: Via di Vigna Murata, Rome, 2023.

19. Garini, E.; Gazetas, G. The 2 Earthquakes of February 6th 2023 in Turkey, Preliminary Report. Available online: https://learningfromearthquakes.org/images/earthquakes/2023_02_05_Turkey/M7.8_Turkey-Syria_EQ_6_February_2023.pdf (accessed on 23 February 2023).

20. Gülerce, Z.; Askan, A.; Kale, Ö.; Sandıkkaya, A.; Işık, N.S.; İlhan, O.; Can, G.; Ilgaç, M.; Ozacar, A.A.; Sopacı, E.; et al. February 6, 2023 Kahramanmaraş-Pazarcik (Mw = 7.7) and Elbistan (Mw = 7.5) Earthquakes: Preliminary Analysis of Strong Ground Motion Characteristics. 2023. Available online: https://eerc.metu.edu.tr/en/system/files/documents/CH4_Strong_Ground_Motion_Report_2023-02-20.pdf (accessed on 23 February 2023).

21. Shou, K.J.; Wu, C.C.; Fei, L.Y.; Lee, J.F.; Wei, C.Y. Dynamic environment in the Ta-Chia River watershed after the 1999 Taiwan Chi-Chi earthquake. *Geomorphology* **2011**, *133*, 190–198. [CrossRef]

22. Shou, K.-J.; Wang, C.-F.; Chen, Y.-L. Risk estimation of earthquake induced rock sliding in Chiufengershan Taiwan. In Proceedings of the 2004 ISRM International Symposium-3rd Asian Rock Mechanics Symposium (ARMS), Kyoto, Japan, 30 November–2 December 2004.

23. Sun, J.; Huang, Y. Modeling the Simultaneous Effects of Particle Size and Porosity in Simulating Geo-Materials. *Materials* **2022**, *15*, 1576. [CrossRef] [PubMed]

24. Sun, J. Permeability of Particle Soils Under Soil Pressure. *Transp. Porous Media* **2018**, *123*, 257–270. [CrossRef]

25. Melgar, D.; Taymaz, T.; Ganas, A.; Crowell, B.; Öcalan, T.; Kahraman, M.; Tsironi, V.; Yolsal-Çevikbilen, S.; Valkaniotis, S.; Irmak, T.S.; et al. Sub- and super-shear ruptures during the 2023 Mw 7.8 and Mw 7.6 earthquake doublet in SE Türkiye. *Seismica* **2023**, *2*, 1–10. [CrossRef]

26. Jiang, X.; Song, X.; Li, T.; Wu, K. Special focus/Rapid Communication Moment magnitudes of two large Turkish earthquakes on February 6, 2023 from long-period coda. *Earthq. Sci.* **2023**, *36*, 169–174. [CrossRef]

27. Dal Zilio, L.; Ampuero, J.-P. Earthquake doublet in Turkey and Syria. *Commun. Earth Environ.* **2023**, *4*, 71. [CrossRef]

28. Hall, S. What Turkey's earthquake tells us about the science of seismic forecasting. *Nature* **2023**, *615*, 388–389. [CrossRef]

29. Vamvatsikos, D.; Cornell, C.A. Incremental dynamic analysis. *Earthq. Eng. Struct. Dyn.* **2002**, *31*, 491–514. [CrossRef]

30. Papazafeiropoulos, G.; Plevris, V. OpenSeismoMatlab: A new open-source software for strong ground motion data processing. *Heliyon* **2018**, *4*, e00784. [CrossRef] [PubMed]

31. Yu, C.-C.; Mir, F.U.H.; Whittaker, A.S. Validation of numerical models for seismic fluid-structure-interaction analysis of nuclear, safety-related equipment. *Nucl. Eng. Des.* **2021**, *379*, 111179. [CrossRef]

32. Marchisella, A.; Muciaccia, G. Comparative Assessment of Shear Demand for RC Beam-Column Joints under Earthquake Loading. *Appl. Sci.* **2022**, *12*, 7153. [CrossRef]
33. Acharjya, A.; Roy, R. Estimating seismic response to bidirectional excitation per unidirectional analysis: A reevaluation for motions with fling-step using SDOF systems. *Soil. Dyn. Earthq. Eng.* **2023**, *164*, 107563. [CrossRef]
34. Papazafeiropoulos, G.; Plevris, V.; Papadrakakis, M. A generalized algorithm framework for non-linear structural dynamics. *Bull. Earthq. Eng.* **2017**, *15*, 411–441. [CrossRef]
35. Papazafeiropoulos, G.; Plevris, V. Kahramanmaraş-Gaziantep, Türkiye Mw 7.8 Earthquake on February 6, 2023: Preliminary Report on Strong Ground Motion and Building Response Estimations. *ArXiv E-Prints* **2023**. [CrossRef]
36. Uang, C.-M.; Bertero, V.V. Evaluation of seismic energy in structures. *Earthq. Eng. Struct. Dyn.* **1990**, *19*, 77–90. [CrossRef]
37. Bommer, J.J.; Alarcon, J.E. The Prediction and Use of Peak Ground Velocity. *J. Earthq. Eng.* **2006**, *10*, 1–31. [CrossRef]
38. Han, S.W.; Chopra, A.K. Approximate incremental dynamic analysis using the modal pushover analysis procedure. *Earthq. Eng. Struct. Dyn.* **2006**, *35*, 1853–1873. [CrossRef]
39. Dolšek, M.; Fajfar, P. Simplified non-linear seismic analysis of infilled reinforced concrete frames. *Earthq. Eng. Struct. Dyn.* **2005**, *34*, 49–66. [CrossRef]
40. Vamvatsikos, D.; Cornell, C.A. Direct Estimation of Seismic Demand and Capacity of Multidegree-of-Freedom Systems through Incremental Dynamic Analysis of Single Degree of Freedom Approximation1. *J. Struct. Eng.* **2005**, *131*, 589–599. [CrossRef]
41. Azarbakht, A.; Dolšek, M. Progressive Incremental Dynamic Analysis for First-Mode Dominated Structures. *J. Struct. Eng.* **2011**, *137*, 445–455. [CrossRef]
42. Léger, P.; Kervégant, G.; Tremblay, R. Incremental dynamic analysis of nonlinear structures: Selection of input ground motions. In Proceedings of the 9th U.S. National and 10th Canadian Conference on Earthquake Engineering, Toronto, ON, Canada, 25–29 July 2010.
43. Musson, R.M.W. Effective Peak Acceleration as a Parameter for Seismic Hazard Studies. In Proceedings of the 12th the European Conference on Earthquake Engineering, London, UK, 9–13 September 2002.
44. Housner, G.W. *Intensity of Ground Motion during Strong Earthquakes*; California Institute of Technology, Earthquake Research Laboratory: Pasadena, CA, USA, 1952.
45. Nau, J.M.; Hall, W.J. Scaling Methods for Earthquake Response Spectra. *J. Struct. Eng.* **1984**, *110*, 1533–1548. [CrossRef]
46. Garini, E.; Gazetas, G. Destructiveness of earthquake ground motions: "Intensity Measures" versus sliding displacement. In Proceedings of the 2nd International Conference on Performance-Based Design in Earthquake Geotechnical Engineering, Taormina, Italy, 27–31 March 2012.
47. Massumi, A.; Gholami, F. The influence of seismic intensity parameters on structural damage of RC buildings using principal components analysis. *Appl. Math. Model.* **2016**, *40*, 2161–2176. [CrossRef]

 *buildings*

Article

Finite Element Analysis of Shear Reinforcing of Reinforced Concrete Beams with Carbon Fiber Reinforced Polymer Grid-Strengthened Engineering Cementitious Composite

Mohammadsina Sharifi Ghalehnoei [1,*], Ahad Javanmardi [2,*], Mohammadreza Izadifar [3], Neven Ukrainczyk [3] and Eduardus Koenders [3]

1   Department of Civil Engineering, University of Esfahan, Esfahan 8174673441, Iran
2   College of Civil Engineering, Fuzhou University, University Town, 2 Xueyuan Road, Fuzhou 350108, China
3   Institute of Construction and Building Materials, Technical University of Darmstadt, Franziska-Braun-Str. 3, 64287 Darmstadt, Germany
*   Correspondence: sharifisina10@yahoo.com or sharifisina10@sci.iaun.ac.ir (M.S.G.); ahadjavanmardi@gmail.com or ahad@fzu.edu.cn (A.J.)

Abstract: This study investigates the shear behavior of reinforced concrete (RC) beams that have been strengthened using carbon fiber reinforced polymer (CFRP) grids with engineered cementitious composite (ECC) through finite element (FE) analysis. The analysis includes twelve simply supported and continuous beams strengthened with different parameters such as CFRP sheets, CFRP grid cross-sectional area, and CFRP grid size. To conduct the analysis, FE models of the RC beams were created and analyzed using ABAQUS software. Research results show that the strengthened RC beams with CFRP grids and ECC had approx. 30–50% higher shear capacity than reference RC beams. The composite action of CFRP grids with the ECCs also showed a significant ability to limit diagonal cracks and prevent the degradation of the bending stiffness of the RC beams. Furthermore, this study calculated the shear capacity of the strengthened beams using an analytical model and compared it with the numerical analysis results. The analytical equations showed only a 4% difference from the numerical results, indicating that the analytical model can be used in practice.

Keywords: shear strengthening; CFRP grid; ECC; RC beams; finite element analysis

Citation: Sharifi Ghalehnoei, M.; Javanmardi, A.; Izadifar, M.; Ukrainczyk, N.; Koenders, E. Finite Element Analysis of Shear Reinforcing of Reinforced Concrete Beams with Carbon Fiber Reinforced Polymer Grid-Strengthened Engineering Cementitious Composite. *Buildings* **2023**, *13*, 1034. https://doi.org/10.3390/buildings13041034

Academic Editors: Rajesh Rupakhety and Dipendra Gautam

Received: 14 December 2022
Revised: 2 April 2023
Accepted: 11 April 2023
Published: 14 April 2023

1. Introduction

The deterioration of reinforced concrete (RC) structures has become increasingly severe in recent years [1,2] The deterioration not only compromises the serviceability of the structure but also poses serious safety risks for humans [3–5]. Corrosion, overload, fatigue, and other negative conditions have a direct impact on the performance of RC structures. Under extreme circumstances, traditional RC beams may be prone to damage and collapse, resulting in significant financial losses. Retrofitting damaged or strengthening weak RC structures with shear defects has become a major challenge in the construction sector [6–9], as it is crucial to enhance the durability and safety of these structures. As a result, developing techniques for strengthening damaged structures to increase the sustainability and durability of existing RC structures is of great practical and scientific importance [10]. In recent years, the application of Fiber-Reinforced Polymer (FRP) for repairing, strengthening, and retrofitting RC structural elements such as beams and columns has been increased [7,11–17] due to its lightweight, high-strength, strong corrosion resistance, and high-durability features. Bonding FRP laminates with adhesive to concrete surfaces is a common solution for external strengthening. An adhesive such as epoxy resin serves as a connector at the interface between the concrete surfaces and FRP laminates that transfers stress to the FRP laminates.

Shear strengthening of RC beams can be achieved by using different materials and methods. In recent decades, a great deal of attention has been paid to the shear strengthening of RC beams using FRP materials by either wrapping FRP or attaching the CFRP to the side of beams using adhesives such as epoxy resin [18–22]. However, epoxy resin has a few drawbacks including poor fire resistance, low glass transition temperature (T_g), sensitivity to UV radiation, and quick degradation in wet environments [6,23]. To overcome these shortcomings, the carbon-FRP (CFRP) grid with different cementitious materials such as UHPC for RC structures is a new strengthening method [6,24–26]. For example, the FRP grid improves the ultimate loading capacity and ultimate tensile capacity of the UHPC beams [24]. The FRP grid also enhances the shear and ductility of the UHPC beams [26].

Generally, several factors affect the shear strengthening of RC beams using FRP materials, as listed in Table 1. The failure mode of RC beams strengthened by FRP sheets is generally debonding, while BFRP grids prevent premature debonding failure at the interface of the concrete [6]. Meanwhile, the epoxy resin provides better bond performance at the interface of the BFRP gird with the concrete as compared with PCM. Azam et al.'s [27] comparative study concluded that the CFRP grid with cement-based mortar was slightly more effective than the CFRP sheets with epoxy resin in the shear strengthening of RC beams. Moreover, the grid size of the CFRP significantly influences the shear strength of the RC beams [28]. Guo et al. [29] investigated the application of CFRP grids with polymer-cement-mortar (PCM) shotcrete for the shear strengthening of RC beams. The shear capacity of the beam strengthened by the CFRP grids was 30–40% larger than that of the PCM-reinforced and unreinforced beams. Therefore, the effect of PCM shotcrete on the shear strength was insignificant and the CFRP grid sufficiently improved the shear capacity of the beam. Generally, the CFRP grids primarily contributed to improving the shear strength [30]. Cai et al. [31] used CFRP grids with epoxy mortar to improve the shear behavior of RC beams. The study results showed that the shear capacity of the RC beams increased up to 56%. The vertical grids of CFRP primarily contributed to the shear capacity improvement while the horizontal grids controlled the crack propagation and improved the bond behavior. Nevertheless, the RC beams strengthened with smaller grid dimensions have a better shear performance than the beams strengthened with larger grid dimensions [29]. In addition, the shear strength depends on the concrete strength and the shear span ratio of the beam [6,29]. Chen et al. [32] used CFRP meshes with UHPC for shear strengthening of corroded RC beams. The composite action between the CFRP and UHPC layer was very strong as there was no debonding during the loading test. The experimental results showed that this method substantially improves the shear capacity and crack resistance of the corroded RC beams. Nevertheless, increasing the number of layers does not necessarily increase the shear strength of the beam but leads to better ductile behavior [32].

In recent years, a new composite material named Engineered Cementitious Composite (ECC) has been developed by the mixing of cement, fine sand, other admixtures, and shortcut fibers randomly distributed in the matrix. The ECC has excellent tensile strength, high ductility, toughness performance, and durability. Studies on the mechanical properties of ECC showed that the ultimate tensile strain of ECC is much greater than 2%, which is about 200 times larger than normal concrete [33–36]. Further, the ultimate compressive strain of the ECC is twice larger than normal concrete. Meanwhile, the compressive strength of ECC is similar to normal concrete; however, the elastic modulus of ECC is about half of that of normal concrete because of the absence of coarse aggregates. Therefore, the ECC as a shear-strengthening material for RC beams is ineffective or has very low efficiency as compared with other methods.

Given the fact that FRP grids are able to improve the shear capacity of the RC beams, and to take the advantage of features of ECC, a new composite material was proposed in this study for the shear strengthening of RC beams. In this method, FRP grids were attached on both sides, and the soffit of the RC beam was within layers of ECC, as shown in Figure 1. The ECC layers have two purposes: (i) to make a much stronger bond between the FRP grids to the substrate concrete, and (ii) to strengthen the RC beam. For this purpose, a simply supported beam and a continuous RC beam were selected from the literature [37] for shear strengthening through numerical analysis. The selected RC beams were modeled with ABAQUS Finite Element (FE) software [38]. The models were then analyzed and validated with the results of the experiments [37]. Thereafter, the effectiveness of CFRP grids with ECC for shear strengthening of the RC beams was studied using the validated models. Further, the effect of different parameters including CFRP grid size, and CFRP grid cross-sectional area were studied and compared with the reference beams and beams strengthened with CFRP sheets. Lastly, calculation formulae for the shear capacity of the RC beam strengthened with the proposed composite material were given and compared with the numerical analysis.

Table 1. General parameters and their effects using FRP materials for the shear strengthening RC beams.

Ref.	Type of Shear Strengthening of Beams	Parameters	Effect
[6]	BFRP grid with epoxy resin	Epoxy resin	• Improve the bonding interface between the BFRP gird and the concrete interface • Control the diagonal cracks
[6]	BFRP sheets or BFRP gird with epoxy resin	BFRP sheets BFRP gird	• BFRP sheets fail due to premature failure debonding, while BFRP grids prevent the premature failure
[6]	BFRP grid	Grid orientation (0° and 45°)	• Increasing the grid angle to 45° significantly improves the shear performance in terms of crack resistance and maximum shear capacity
[29]	CFRP grid with PCM shotcrete	CFRP Grid size	• Smaller grid interval leads to higher shear strength
[27]	CFRP sheet with epoxy and CFRP gird with mortar	CFRP sheet CFRP grid Mortar and epoxy	• Mortar able to better control the diagonal cracks • CFRP gird with mortar offers better ultimate shear strength
[28]	CFRP grid with MBC	Grid size	• Smaller grid size leads to higher ultimate shear strength

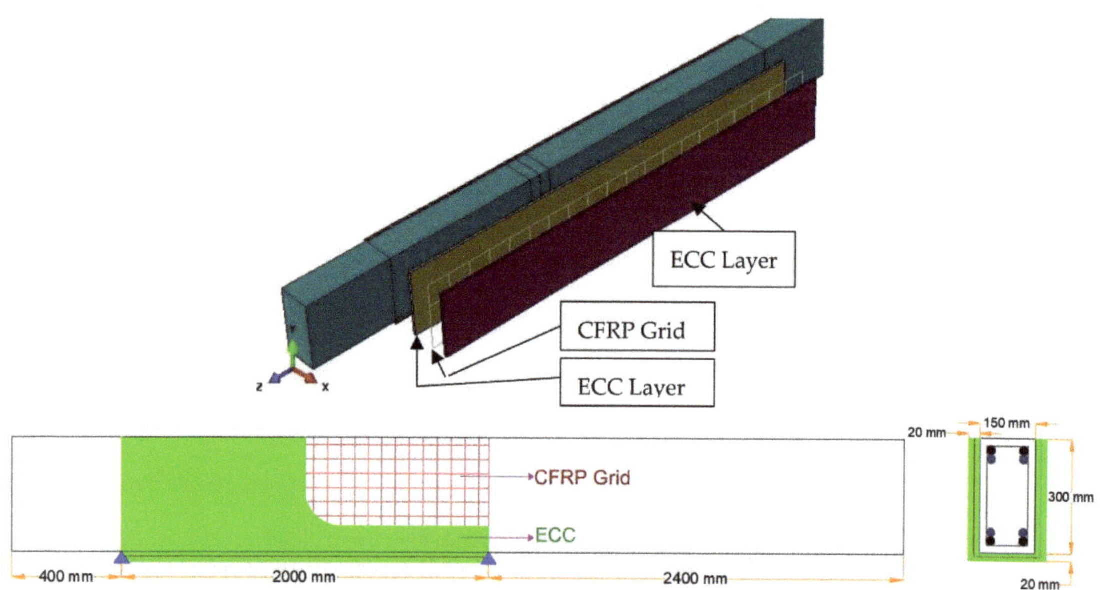

Figure 1. Details of the strengthening beams by CFRP grids with the ECC method.

2. Methodology

2.1. Experimental Study [37]

Pellegrino and Modena [37] investigated the shear behavior of RC beams strengthened with CFRP wraps. In total, 12 RC beams were tested in the laboratory including 3 reference beams and 8 beams with externally bonded U-wrapped CFRP. The CFRP was wrapped on both sides and the soffit of the beams. Figure 2 shows the detailed dimensions of the beams, steel reinforcements, and stirrups as well as the location of loads and supports. Four steel bars of 30 mm diameter were used for the tensile and compressive reinforcement of the beam. Steel bars of 8 mm diameter were used for stirrups with a spacing of 170 mm for both types of beams, as illustrated in Figure 2c. As shown in Figure 2, two configurations, i.e., a simply supported beam (Figure 2b) and a continuous beam (Figure 2a) were considered for shear testing of the beams in order to maximize the shear and bending moment acting at support. In other words, the beams were designed to fail due to shear before flexural failure. A 500 kN hydraulic jack was used for the three- or four-point flexural tests. Two strain gauges were installed on the beams in vertical and horizontal directions and one strain gauge was installed at a 45-degree inclination with respect to the horizontal direction, as shown in Figure 2a,b. More details of the experimental work can be found in the reference [37]. In the current study, eight of the twelve beams with the details shown in Table 2 were selected for FE modeling, validation, and further comparisons. These beams were further used for strengthening by the CFRP grid with ECC.

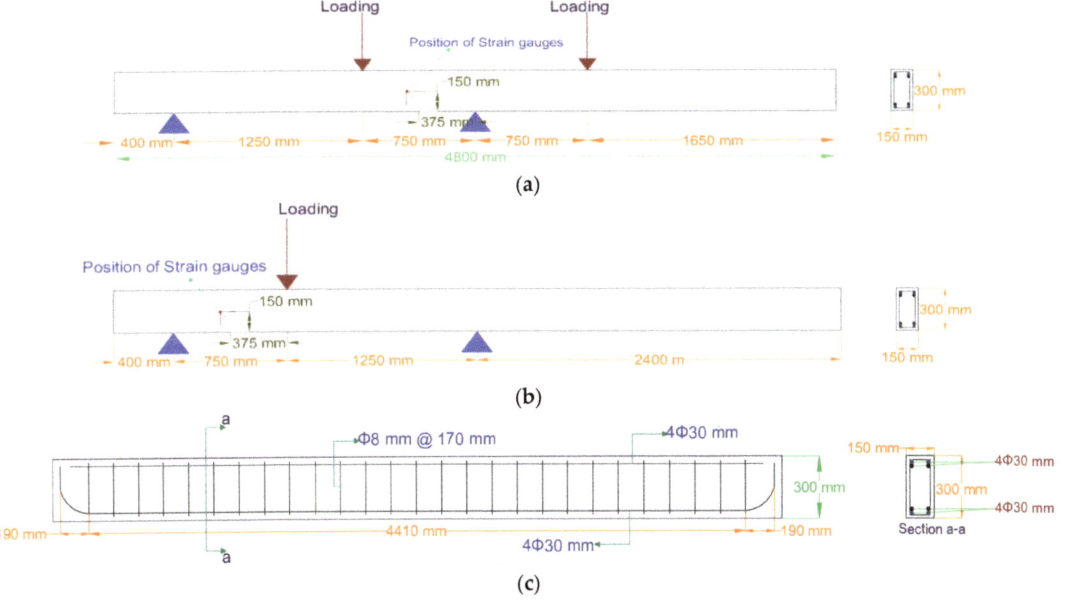

Figure 2. Schematic diagram of the loading position, support conditions, and the reinforcement detailing of the beams [37].

Table 2. Details of beams in the experimental work (adapted with permission from [37]; copyright 2006 American Concrete Institute). (Unit: mm.)

No.	Beam	Tension Reinforcement	ρ	Stirrup Diameter	Stirrup Spacing	ρ_w	Strengthening Layer	a/d	Beam Configuration
1	A-C-17	4φ30 mm	0.075	8	170	0.00392	-	3	Continuous
2	A-S-17	4φ30 mm	0.075	8	170	0.00392	-	3	Simply supported
3	CR-C-17	4φ30 mm	0.075	8	170	0.00392	CFRP	3	Continuous
4	CR-S-17	4φ30 mm	0.075	8	170	0.00392	CFRP	3	Simply supported

2.2. Beams' Details

In this research, a total of 12 RC beams were used for shear behavior investigations, as listed in Table 3. Two of these beams, namely, A-C-17 and A-S-17 were the reference beams, which had the same detailing as the experimental work (see Table 2). Two beams, namely, CR-C-17 and CR-S-17 were strengthened with CFRP sheets, which also had the same detailing as the experimental work (see Table 2). For shear strengthening of RC beams using CFRP grids with ECC, two parameters, i.e., CFRP grid size and cross-sectional area of CFRP grids were considered, as shown in Table 3. The procedure for strengthening the RC beams using the CFRP grid with ECC is as follows: (i) after cleaning the surface of the RC beams from the dirt and dust, a 10 mm thick layer of ECC was applied to the surface of the concrete; (ii) then, the CFRP grid was cut based on the dimension of the beam and placed on the beam; and (iii) another 10 mm thick layer of ECC was applied on the beam, as shown in Figure 1 [23,29]. Therefore, in total, 8 beams were strengthened by a CFRP grid with ECC and categorized as Series II to Series V in Table 3. It should be noted that the same strengthening scheme was used for both the continuous beams and the simply supported beam. The CFRP grid having square meshes with untwisted yarns, continuous carbon fiber, and impregnated with thermoset epoxy resin was used for strengthening. The nominal dimensions of the CFRP grid were 50 mm × 50 mm and 100 mm × 100 mm.

2.3. Numerical Modeling

The general purpose of ABAQUS FE software [38] was used for FE modeling and analysis. This section describes the numerical modeling of the beams in detail.

2.3.1. Geometric Modeling and Boundary Conditions

The concrete beams with dimensions shown in Figure 1 were modeled with three-dimensional solid elements. The steel reinforcements, stirrups, and CFRP grids were modeled with two-dimensional wire elements. The CFRP sheets with a thickness of 0.165 mm were modeled with three-dimensional shell elements. In addition, the ECC layer with a thickness of 20 mm was modeled with three-dimensional solid elements. The same loading and boundary conditions of the experiment were used in the FE modeling, as shown in Figure 3.

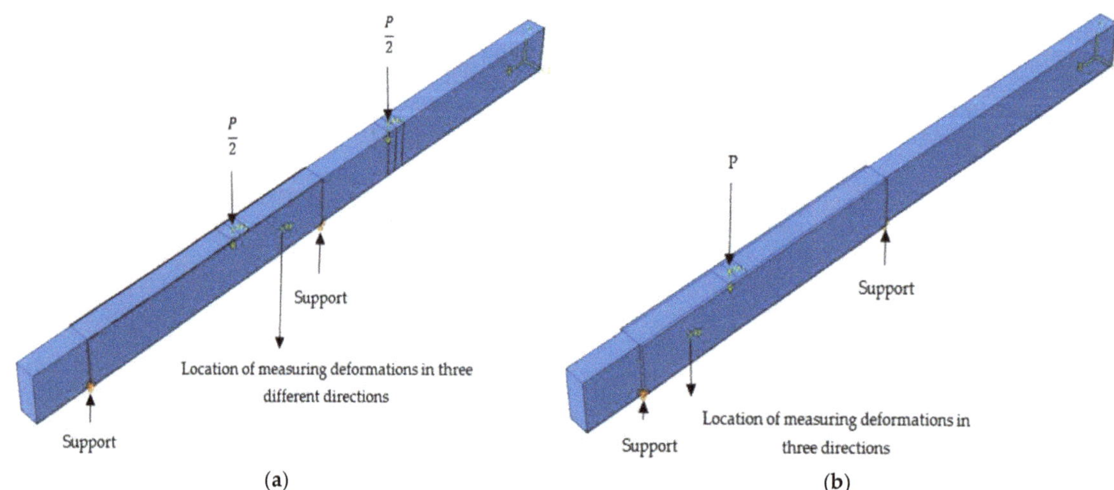

Figure 3. Loading and boundary conditions of (**a**) continuous beam, and (**b**) simply supported beam models in the FE software.

Table 3. Details of the beams (adapted with permission from [37]; copyright 2006 American Concrete Institute). (Unit: mm.)

Series	No.	Beam	Tension Reinforcement	ρ	Stirrup Dia.	Stirrup Spacing	ρ_w	Strengthening Layer	CFRP Grid Type	Grid Size, S (mm × mm)	a/d	Beam Configuration
Reference beams	1	A-C-17	4φ30 mm	0.075	8	170	0.00392	-	-	-	3	Continuous
	2	A-S-17	4φ30 mm	0.075	8	170	0.00392	-	-	-	3	Simply supported
Series I	3	CR-C-17	4φ30 mm	0.075	8	170	0.00392	CFRP	-	-	3	Continuous
	4	CR-S-17	4φ30 mm	0.075	8	170	0.00392	CFRP	-	-	3	Simply supported
Series II	5	CRG5-C-17	4φ30 mm	0.075	8	170	0.00392	CFRPG5 + ECC	CR5	50 × 50	3	Continuous
	6	CRG5-S-17	4φ30 mm	0.075	8	170	0.00392	CFRPG5 + ECC	CR5	50 × 50	3	Simply supported
Series III	7	CRG8-C-17	4φ30 mm	0.075	8	170	0.00392	CFRPG8 + ECC	CR8	50 × 50	3	Continuous
	8	CRG8-S-17	4φ30 mm	0.075	8	170	0.00392	CFRPG8 + ECC	CR8	50 × 50	3	Simply supported
Series IV	9	CRG′5-C-17	4φ30 mm	0.075	8	170	0.00392	CFRPG5 + ECC	CR5	100 × 100	3	Continuous
	10	CRG′5-S-17	4φ30 mm	0.075	8	170	0.00392	CFRPG5 + ECC	CR5	100 × 100	3	Simply supported
Series V	11	CRG′8-C-17	4φ30 mm	0.075	8	170	0.00392	CFRPG8 + ECC	CR8	100 × 100	3	Continuous
	12	CRG′8-S-17	4φ30 mm	0.075	8	170	0.00392	CFRPG8 + ECC	CR8	100 × 100	3	Simply supported

Note: CFRPG is the CFRP grids.

2.4. Material Modeling

The Concrete Damage Plasticity (CDP) was employed to model the concrete behavior under tension and compression [38,39]. The Poisson's ratio and Young's modulus were used for determining the linear elastic and isotropic behaviors of concrete in tension and compression. Nonlinear behavior was definable concerning inelastic strains as well as the associated yield stresses. Figure 4a,b shows the stress–strain relation of concrete used for the material modeling. The compressive and tensile strengths of the concrete were 41.4 MPa and 3.76 MPa, respectively [37]. As shown in Figure 4c, the bilinear isotropic hardening model was used to simulate the behavior of steel reinforcements, which is ideal for elastic-plastic materials. Steel material with the mechanical properties given in Table 4 was used for modeling the steel sections.

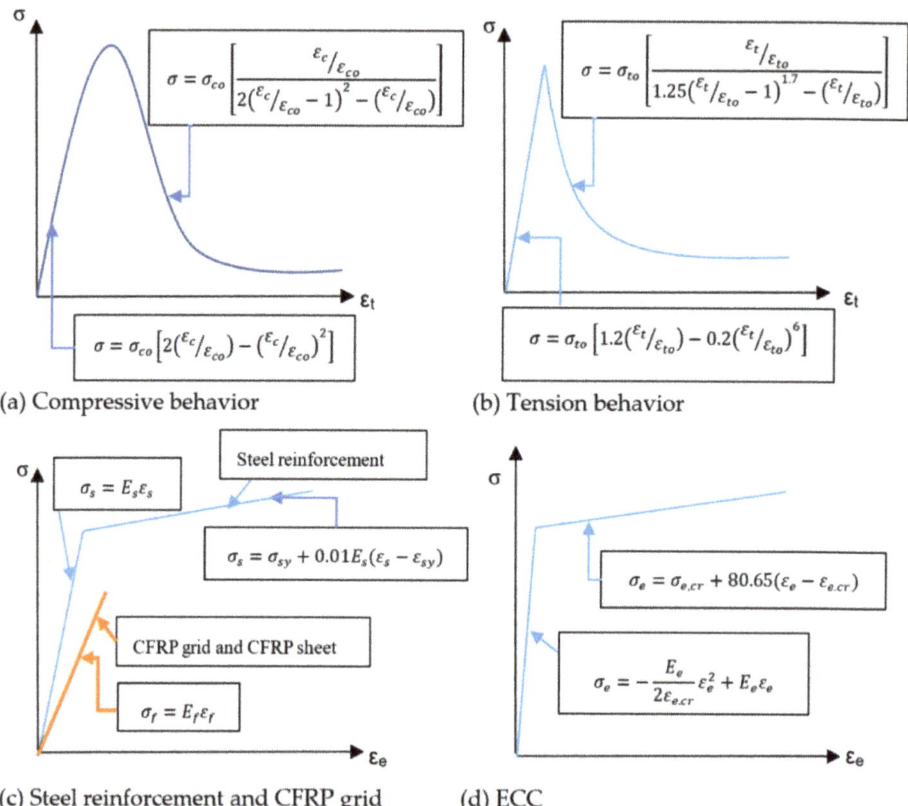

Figure 4. Constitutive material models used in FE modeling [40,41].

Table 4. Mechanical properties of steel reinforcement.

Bar Dia. (mm)	Area (mm^2)	Yield Stress (MPa)	Ultimate Strain (%)	Ultimate Stress (MPa)	Modulus of Elasticity (GPa)
30	706.5	534	0.15	717	193
8	50.24	534	0.15	717	193

The material of CFRP sheets had an elastic modulus of 230 GPa, an ultimate strength of 3.45 GPa, and a Poisson's ratio of 0.26 [37]. Two grades of CFRP grids CR5 and CR8 were selected based on the product datasheet provided by the manufacturer [42]. The difference between the two grades is the area of the cross-section of the CFRP grids. The linear elastic model, as shown in Figure 4c, was used for the CFRP sheets and girds. Both grades of the CFRP grids had an elastic modulus of 100 GPa and a tensile strength of 1400 MPa. The cross-sectional area of CR5 was half of CR8. The mechanical properties listed in Table 5 were used to model the CFRP grid materials in the FE software [23,43].

Table 5. Mechanical properties of CFRP girds used in the simulation [23,43].

Grade	Cross-Sectional Area of Bar (mm^2)	Interval of Grids, S (mm $\times$ mm)	E_{11} (GPa)	G_{12} (GPa)	v_{12}	X_T (MPa)	X_C (MPa)	Y_T (MPa)	Y_C (MPa)	S_L (MPa)	S_T (MPa)
CR5	13.2	50 $\times$ 50 100 $\times$ 100	100	4	0.29	2000	600	1200	150	50	50
CR8	26.4	50 $\times$ 50 100 $\times$ 100	100	4	0.29	2000	600	1200	150	50	50

Notes: E_{11}: longitudinal elasticity modulus, G_{12}: shear modulus, v_{12}: Poisson's ratio, X_T: tensile strength in fiber direction, X_C: compressive strength in fiber direction, Y_T: tensile strength in matrix direction, Y_C: compressive strength in matrix direction, S_L: longitudinal shear strength, and S_T: transverse shear strength.

As a cement-based material, ECC also has the same compressive behavior as a concrete material. However, no unified relation is currently available to define the stress–strain relationship of the ECC. Therefore, the compressive stress–strain relation of concrete was chosen for modeling the ECC material. As Figure 4d shows, a bilinear stress–strain curve was used for modeling the ECC based on the mechanical properties given in Table 6 [40,41].

Table 6. Mechanical properties of ECC (adapted with permission from [23]; copyright 2018 Elsevier).

Material	Tensile Strength (MPa)	Elastic Modulus (GPa)	Maximum Strain (%)	Compressive Strength (MPa)
ECC	3.8	16.9	3.04	30.2

2.4.1. Mesh Discretization

An eight-node linear brick element with reduced integration and hourglass control (C3D8R) was used for solid elements, i.e., concrete and ECC. A two-node linear truss element with three degrees of freedom in each node (T3D2) was used for the wire elements, i.e., rebars, stirrups, and CFRP grids. In addition, the quadrilateral shell element (S4R) with reduced integration was used for the shell element, i.e., CFRP sheets. In order to maximize the convergence rate and minimize the computational time, a convergence study was performed. For this purpose, one of the reference beam models (A-S-17) was selected and its ultimate load at vertical, horizontal, and 45-degree inclination was compared with the experimental results [37]. As a result, the mesh size of 30 mm was found to be optimal and used for all the beam models for consistency. Figure 5 shows the general view of the mesh detailing of beam models in the FE software.

Figure 5. Mesh detailing of the reference beam model in FE software.

2.5. Validation of the FE Models

In order to evaluate the accuracy numerical model, the force-deformation curves and tensile crack pattern of the FE analysis were compared and validated with the experimental results [37]. Figure 6 compares the load-axial deformation curves of one of the reference beams (A-S-17) obtained from the FE analysis with the experimental results. In the experiment, the axial deformations were directly collected from the strain gauges (See Figure 2). As Figure 6 demonstrates, numerical curves of the load and axial deformations closely followed the experimental curves. The obtained yield loads from the numerical analysis in three directions of deformations, i.e., 45° deformation (ε_{45}), horizontal deformation (ε_h), and vertical deformation (ε_v) were very close to that of the experimental tests. The yield forces at the inclined deformation angle of 45 of this beam from the experiment and FE analysis were 195 kN and 182 kN, respectively. The ultimate load of the numerical model at the 45-degree inclined deformation was 314 kN, which was close to 308 kN obtained from the experimental results. The obtained yield loads at the horizontal direction from the FE analysis and experiment were 226 kN and 227 kN, respectively. Meanwhile, the ultimate loads at the horizontal direction obtained from the FE analysis and experiment were 326 kN and 317 kN, respectively, which shows about 2.8% differences. Further, the yield forces in the vertical direction from the numerical and experimental analyses were 185 kN and 180 kN, respectively (about a 3% difference). The obtained ultimate load at the vertical direction from the numerical and experimental analyses were 325 kN and 316 kN, respectively, which indicates a 2.9% difference.

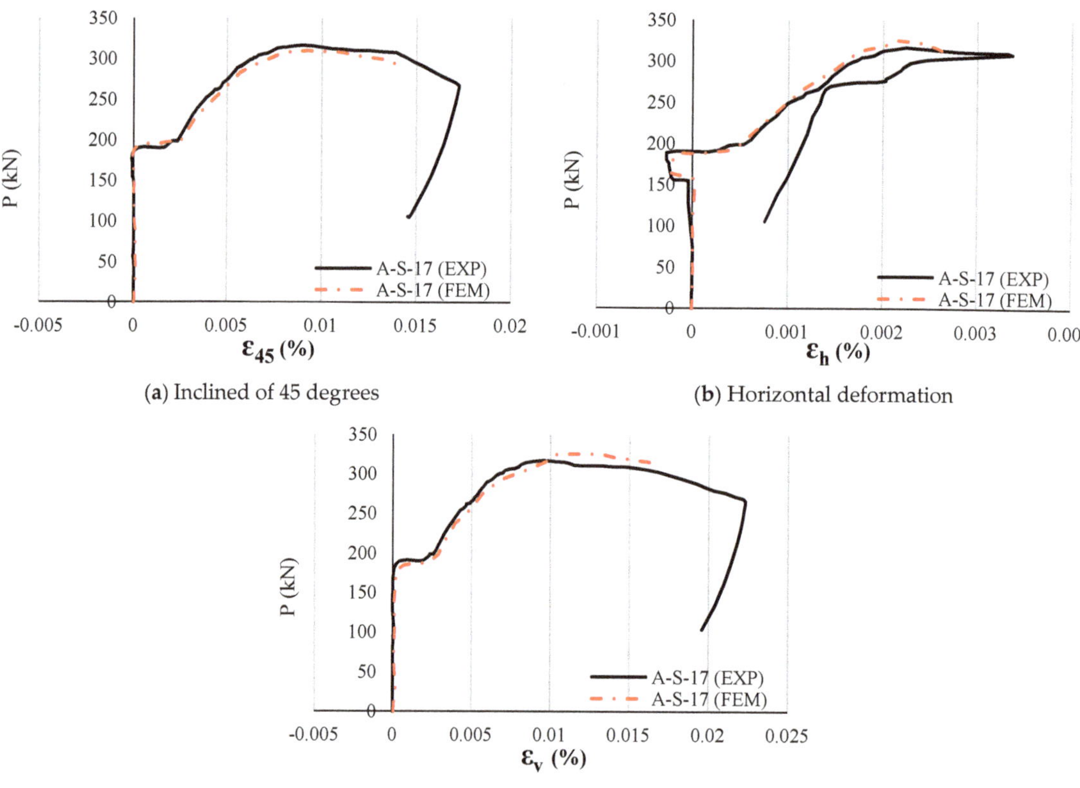

(a) Inclined of 45 degrees

(b) Horizontal deformation

(c) Vertical deformation

Figure 6. Comparison of the load-axial deformation curves of the reference beam (A-S-17) obtained from numerical analysis and the experiment [37].

Figure 7 compares the curves for the shear force against the shear deformation of the A-S-17 beam obtained from the FE and the experimental analyses. The shear deformation was calculated using the following equation [37]

$$\gamma = 2\varepsilon_{45} - \varepsilon_h - \varepsilon_v \tag{1}$$

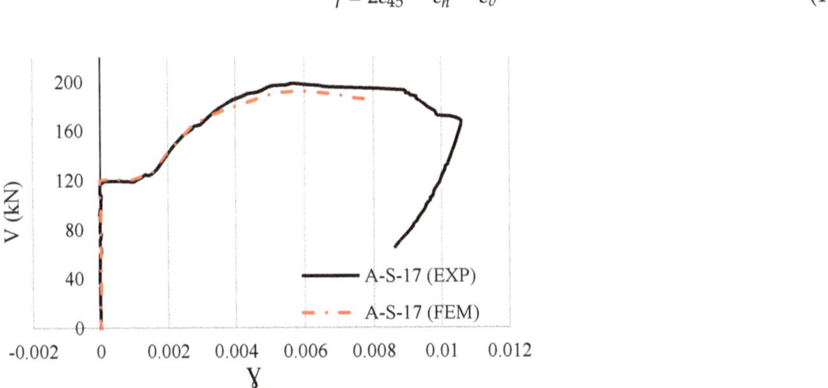

Figure 7. Comparison of the shear force-shear deformation curves of the reference beam (A-S-17) obtained from numerical analysis and the experiment [37].

It can be seen from Figure 7 that the curve trend of the shear force deformations calculated based on numerical and experimental analyses was almost the same. The yield shear force obtained from the experiment and FE analysis was 165 kN. Further, the maximum ultimate shear strengths calculated by the FE analysis and experimental results were 198 kN and 193 kN, respectively, which only indicate a 3% difference.

The results of the shear strength of the reference beams and beams strengthened with CFRP sheets obtained from the FE analysis were also compared with the experimental results, as shown in Table 7. As this table shows, the difference between the shear capacity of the beams obtained from the FE analysis and the experimental study was relatively small.

Table 7. Comparison of the shear strength of the beams from the numerical and experimental [37].

Beams	V_{FEM} (kN)	V_{EXP} (kN)	Difference (%)
A-C-17	183.57	185.2	0.9
A-S-17	192.74	198.1	2.71
CR-C-17	231.50	238.1	3
CR-S-17	253.02	247.3	2.26

Figure 8 compares the crack patterns of the FE analysis and the experimental results for the A-C-17 and A-S-17 beams (reference beams). In the experimental study, the A-C-17 beam with continuous beam configuration had a major diagonal crack and several other diagonal (with 45° angle) cracks originated from the loading positions to the nearest support [37]. This failure mechanism is a typical shear-tensile failure in RC beams. Similarly, in the FE analysis diagonal cracks were also observed between the support and loading locations, as shown in Figure 8a. For the A-S-17 beam having the simply supported configuration, the diagonal cracks were almost the same and were mainly started from loading to the nearest support in the experimental test. The crack pattern of this beam in the FE analysis was also the same as the experiment, where several closed space diagonal cracks were started from the support and extended to the loading location, as shown in Figure 8b. Therefore, it can be concluded from the load-deformation curves, shear strength, and failure mechanism of the beams that there is a good agreement between the FE analysis and the experimental tests.

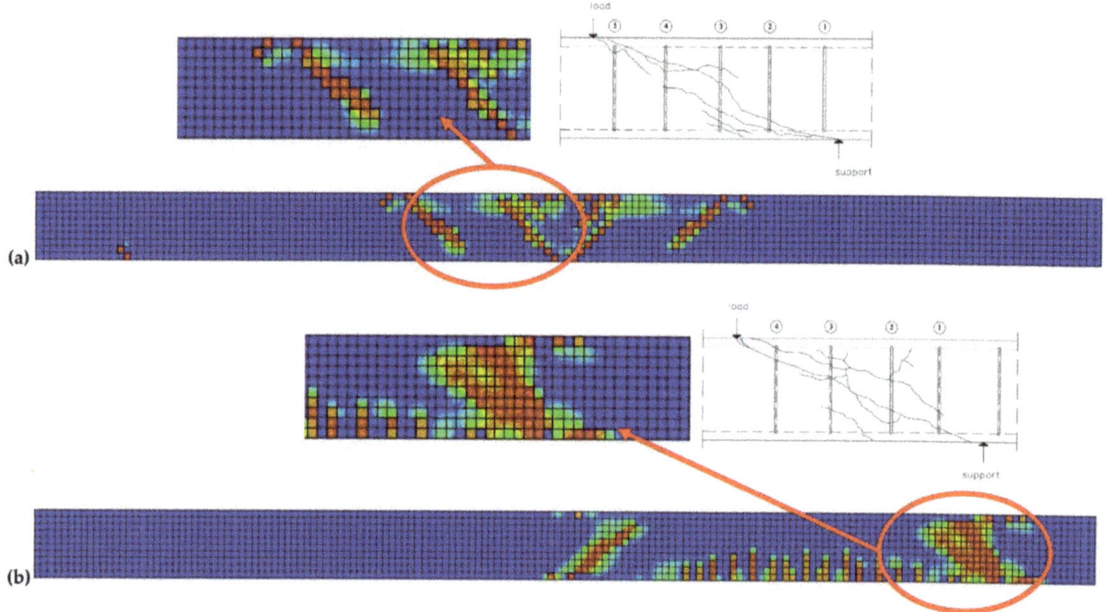

Figure 8. Concrete crack patterns of (**a**) A-C-17, and (**b**) A-S-17 beams.

3. Results and Discussion

This section explains the results for the reference beams and five types of shear-strengthened beams (see Table 3) to evaluate the effectiveness of the shear strengthening by the CFRP grids and ECC. The results include the failure mechanism, load-axial deformations curves, and shear force-shear deformation curves. In addition, the shear capacity calculation formulas for RC beams strengthened with CFRP grids and ECC are given.

3.1. Failure Mechanism

In order to comprehensively study the failure mechanism of the beams, the crack patterns (tensile crack) and stress results of the numerical analysis are briefly discussed in this section. Since the shear behavior of simply supported beams was more profound, only the results of the simply supported beams are presented here. Figure 9 shows the crack patterns of a reference beam and beams strengthened with different methods at the ultimate load. The CFRP sheets slightly reduced the shear diagonal cracks and some of the flexural cracks, as shown in Figure 9b. In general, CFRP grids and ECC not only reduced the shear cracks but also reduced the flexural cracks of the beams, as shown in Figure 9c–f. By comparing the CRG5-S-17 beam with the CRG′5-S-17 beam (Figure 9c,e) and the CRG8-S-17 beam with the CRG′8-S-17 beam (Figure 9d,f), it can be seen that the beams with the larger CFRP grid size had a lesser number of shear cracks and smaller crack widths. Further, the beams strengthened with the larger cross-sectional area of CFRP grids (CRG8-S-17 and CRG′8-S-17) had smaller shear cracks as compared with those beams strengthened with the smaller cross-sectional area of CFRP grids (CRG5-S-17 and CRG′5-S-17). It is worth mentioning that the flexural cracks significantly reduced for the beam with the largest cross-sectional area of CFRP grids and smallest grid size (CRG8-S-17) as compared to other strengthened beams.

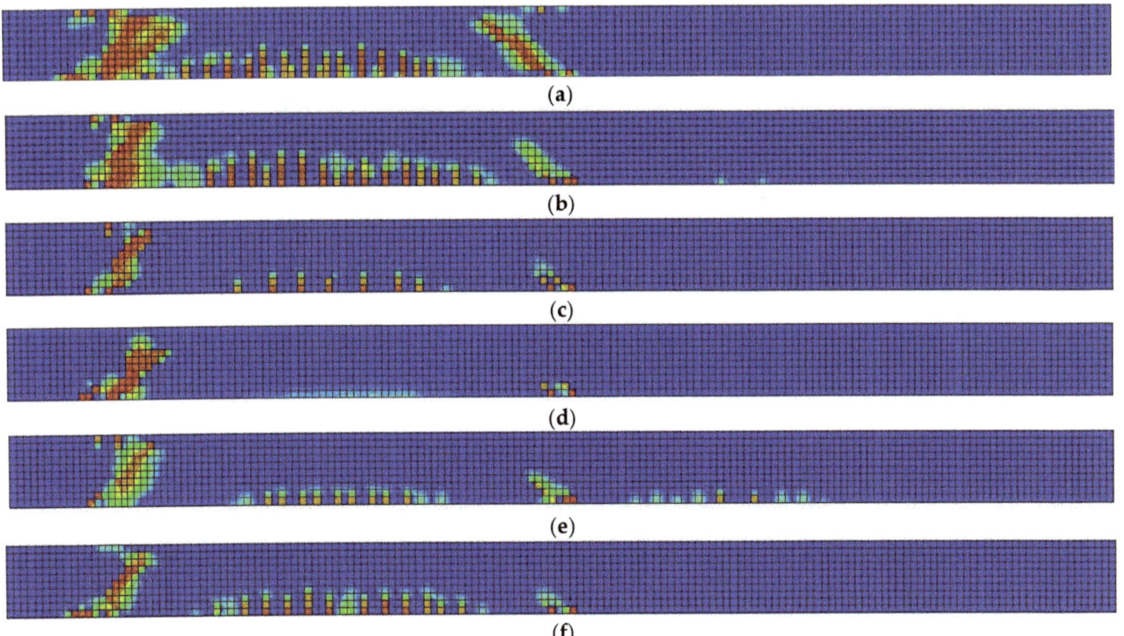

Figure 9. Concrete crack patterns of (**a**) A-S-17, (**b**) CR-S-17, (**c**) CRG5-S-17, (**d**) CRG8-S-17, (**e**) CRG′5-S-17, and (**f**) CRG′8-S-17 beams.

Figure 10 shows the stress distributions of the reference and strengthened simply supported beams in detail. For the reference beam, the stresses in the concrete diagonally span from the loading position to the supports (in both shear and moment areas), as shown in Figure 10a. Similarly, the stress concentration for reinforcements of the reference beam was at the stirrups and tensile bars below the loading position, as shown in Figure 10e. However, for the beam strengthened with CFRP sheets (CR-S-17), the stress distributions in the concrete (Figure 10b) and reinforcements (Figure 10f) were significantly reduced in shear and moment areas. The stress distributions in concrete and steel bars of the beams with CFRP grids and ECC (CRG5-S-17 and CRG8-S-17) were further reduced as compared to the reference beam (A-S-17) and beam with the CFRP sheet (CR-S-17). This indicated that the combined action of the CFRP grids with the ECC layers enhanced the shear capacity of the reinforced beams, and effectively limited the spread of concrete cracks. For the CFRP sheet of the CR-S-17 beam, the stress was diagonally distributed from the loading to the support, and the maximum stress was located in the vicinity of the support, as shown in Figure 10i. Further, the stress distributions of the CFRP grids of CRG5-S-17 and CRG8-S-17 beams were relatively similar; however, the maximum stress of CFRP grids of the CRG8-S-17 beam was about 12% larger than the CRG5-S-17 beam, as shown in Figure 10j,k. This is because the cross-sectional area of the CFRP grids of the CRG8-S-17 beam was larger than the CRG5-S-17 beam, which led to higher loading bearing capacity and stresses.

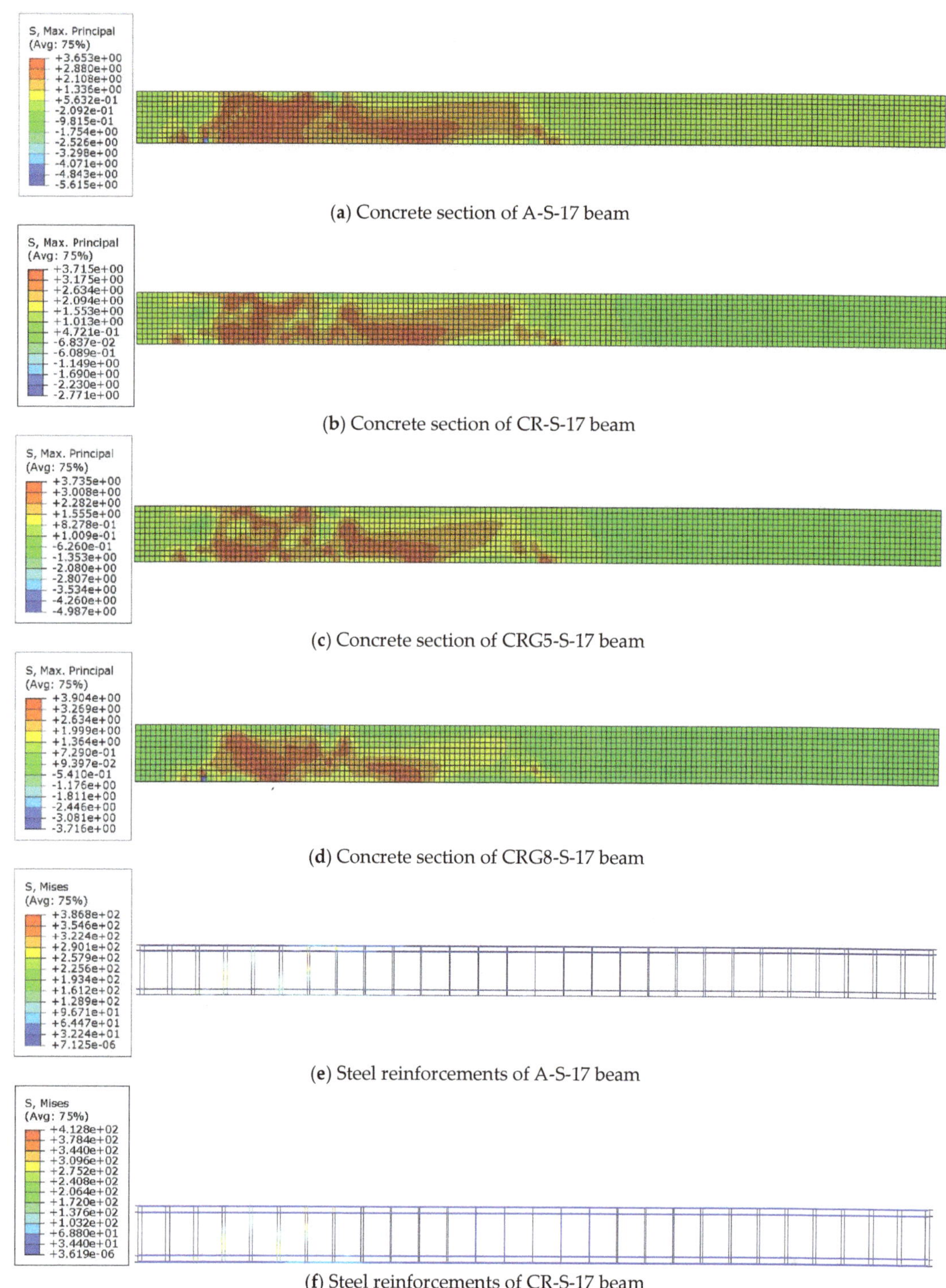

(**a**) Concrete section of A-S-17 beam

(**b**) Concrete section of CR-S-17 beam

(**c**) Concrete section of CRG5-S-17 beam

(**d**) Concrete section of CRG8-S-17 beam

(**e**) Steel reinforcements of A-S-17 beam

(**f**) Steel reinforcements of CR-S-17 beam

Figure 10. *Cont.*

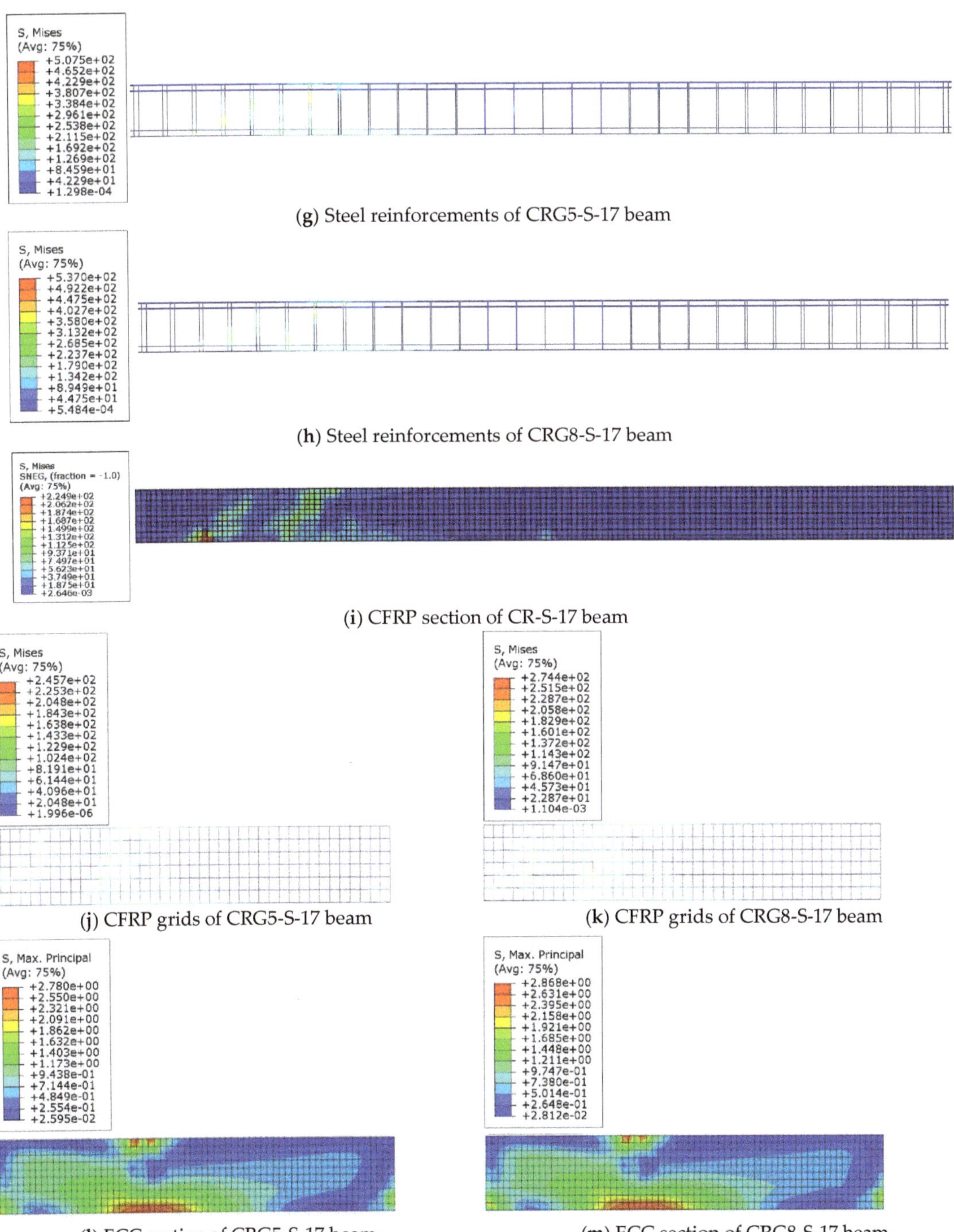

(**g**) Steel reinforcements of CRG5-S-17 beam

(**h**) Steel reinforcements of CRG8-S-17 beam

(**i**) CFRP section of CR-S-17 beam

(**j**) CFRP grids of CRG5-S-17 beam

(**k**) CFRP grids of CRG8-S-17 beam

(**l**) ECC section of CRG5-S-17 beam

(**m**) ECC section of CRG8-S-17 beam

Figure 10. Stress distribution of reference and strengthened beams.

3.2. Load-Deformation Response

Figure 11 shows the load-deformation curves of reference and strengthened beams. It can clearly be seen that all of the strengthened beams had higher strength than the reference beam. The CRG8-S-17 and CRG5-S-17 beams had better load-deformation responses as compared with other beams. This means that the CFRP grid size is a governing factor in this shear-strengthening method. The load-deformation response of the beam improved as the CFRP grid size was reduced. Nevertheless, the load-deformation response of the CRG8-S-17 beam outperformed the CRG5-S-17 beams due to the larger stiffness of the CFRP girds. The maximum ultimate load of the CRG5-S-17 beam was 46.2%, 37.68%, and 37.68% larger than the reference beam at 45°, horizontal, and vertical deformations, respectively. Comparatively, the ultimate load of the CRG8-S-17 beam at 45°, horizontal, and vertical deformations was 2.5%, 2.4%, and 2.4% larger than the CRG5-S-17 beam, respectively, while the cross-sectional area of the CFRP gird of CRG8-S-14 was double that of CRG5-S-17. Although the load-deformation responses of the beam strengthened by CFRP sheets (CR-S-17) were better than the reference beam, its responses were less significant as compared with beams strengthened with CFRP grids and ECC. By comparing Figure 11a with Figure 11b,c, it can be seen that the effect of the gird size of CFRP girds was more profound for the load-deformation response at a 45-degree inclination than the horizontal and vertical deformation. The ultimate load of the CRG5-S-17 beam at a 45-degree inclination, horizontal, and vertical deformation was 7%, 4%, and 3%, respectively.

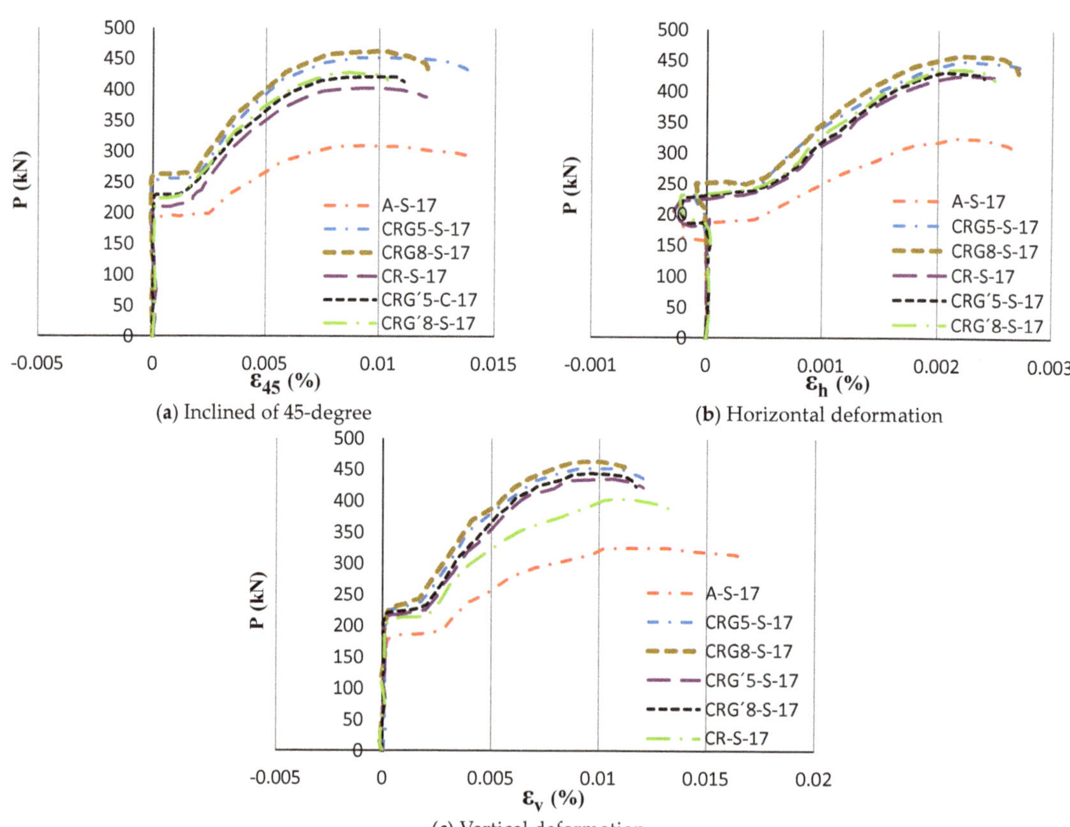

Figure 11. Load-axial deformation curves of the reference and strengthened beams.

Table 8 summarized the results of the flexural cracking load (P_{cr}), the shear cracking load (V_{cr}), the ultimate load at an incline of 45 degrees, vertical, and horizontal deformations, and their corresponding strain values. For the continuous beam configuration (A-C-17), the flexural and shear cracking loads were slightly smaller than the simply supported beam (A-S-17). In addition, the ultimate shear loads of the continuous beam were smaller but with larger strains (in three directions) than the simply supported beam. In other words, the failure mode of the continuous beam failed flexural-shear failure. After strengthening the continuous beam with different configurations, the flexural and shear cracking loads were found to be larger than that of the simply supported strengthened beams. On the other hand, after strengthening beams with different schemes, the ultimate shear forces (in three directions) of the simply supported beam were larger than that of the continuous beam. This means that the shear strengthening of the continuous beam improves its serviceability state better than the ultimate state. Among all the strengthen schemes, the flexural and shear cracking loads as well as the ultimate loads were the largest for the strengthened beams with the largest cross-sectional area of CFRP grids and the smallest CFRP grid size (CRG8-C-17 and CRG8-S-17 beams). When the CFRP grid size increased (from 50 mm to 100 mm), the flexural and shear cracking loads as well as the ultimate loads of the beam decreased. However, when the cross-sectional area of the CFRP grid increased, the loading response of the beam also increased.

Table 8. Summary of loads and strain of the beams.

Beam	P_{cr} (kN)	V_{cr} (kN)	P_{u45} (kN)	ε_{u45}	P_{uv} (kN)	ε_{uv}	P_{uh} (kN)	ε_{uh}
A-C-17	113.00	157.00	293.03	0.00987	311.09	0.01128	314.47	0.00236
A-S-17	119.00	165.00	308.46	0.00940	326.36	0.01074	326.33	0.00225
CR-C-17	130.78	218.30	327.06	0.00921	342.04	0.01101	336.34	0.00221
CR-S-17	128.67	210.47	403.00	0.00923	418.86	0.01010	426.69	0.00225
CRG5-C-17	197.50	300.40	380.79	0.00789	404.05	0.00900	367.48	0.00190
CRG5-S-17	168.00	257.00	450.96	0.00725	449.33	0.00829	449.29	0.00175
CRG8-C-17	215.00	318.50	383.46	0.00781	405.85	0.00889	406.91	0.00186
CRG8-S-17	183.00	270.00	462.19	0.00710	460.53	0.00809	460.48	0.00169
CRG'5-C-17	180.00	274.00	346.89	0.00823	369.51	0.00940	361.27	0.00199
CRG'5-S-17	130.00	226.87	420.87	0.00815	431.83	0.00942	435.55	0.00197
CRG'8-C-17	199.00	294.00	356.04	0.00783	378.17	0.00923	400.76	0.00193
CRG'8-S-17	144.50	215.00	428.07	0.00786	437.03	0.00813	440.51	0.00187

Figure 12 presents the shear force relationship with the shear deformation derived from Equation (1). Practically, there is shear deformation until the yield load is negligible and after this point, cracks develop on the concrete. Once the beams enter the strain hardening stage, the shear reinforcement yields, and eventually plastic deformation occurs until failure. It can be seen from Figure 12 that the shear strength of all strengthened beams was much larger than the reference beam at yield as well as the plastic stage. Similar to the load-axial deformation curves, the shear force-shear deformation curve of the CRG8-S-17 beam outperformed other strengthened beams. In addition, the beams strengthened with smaller CFRP grid sizes (CRG8-S-17 and CRG5-S-17) had higher shear capacity than the beams with bigger CFRP grid sizes (CRG'8-S-17 and CRG'5-S-17).

Table 9 summarizes the shear capacity and load contribution of each strengthening method for all the beams. In general, the improvement of the shear capacity of the simply supported beams strengthened with different schemes was more profound than the continuous beams. This was mainly because the simply supported beams were subjected to a pure shear load while the continuous beams were subjected to a shear-flexural load. The maximum increase in the shear force of the beam strengthened with CFRP sheets (CR-S-17) was about 31%. However, the maximum improvement of the shear for the beams strengthened with CFRP grids and ECC (CRG8-S-17) was 49%. As this table indicates, the effect of CFRP grid size was more profound in improving the shear capacity of the beam than the cross-sectional area of CFRP grids.

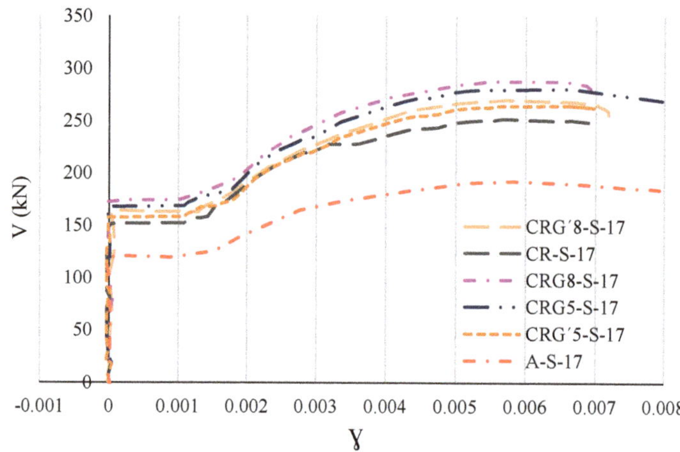

Figure 12. Shear force-shear deformation curves of reference and strengthened beams [37].

Table 9. Summary of shear capacity of beams.

Beam	V_{Total} (kN)	Shear Span Ratio, (*a/d*)	Increase in Shear Strength (%)	$V_{FRP\,grid\,+\,ECC}$ (kN)	V_{FRP} (kN)
A-C-17	183.57	3	-	-	-
A-S-17	192.74	3	-	-	-
CR-C-17	231.50	3	26.12	-	47.93
CR-S-17	253.02	3	31.28	-	60.28
CRG5-C-17	240.54	3	31.04	56.97	-
CRG5-S-17	280.93	3	45.75	88.19	-
CRG8-C-17	245.71	3	33.86	62.15	-
CRG8-S-17	287.93	3	49.38	95.19	-
CRG′5-C-17	238.54	3	29.95	54.98	-
CRG′5-S-17	265.35	3	37.8	72.61	-
CRG′8-C-17	240.20	3	30.85	56.64	-
CRG′8-S-17	268.71	3	39.42	75.97	-

3.3. Analytical Model

Generally, the total shear capacity of concrete structures externally strengthened is simply the summation of the shear contribution of the concrete (V_{CON}), the available shear of the steel shear reinforcement (V_{ST}), the shear contribution of externally bonded CFRP grid (V_g), and shear carried by the ECC (V_{ECC}):

$$V = V_{con} + V_{st} + V_{ECC} + V_g \tag{2}$$

The shear carrying capacity of concrete (V_{con}) is [44,45]:

$$V_{con}(V_{ECC}) = \frac{\beta_d \beta_p \beta_n f_{vcd} b_w d}{\gamma_b} \tag{3}$$

where, γ_b is the safety factor ($\gamma_b = 1.3$), and d is the effective height of the beam. The value for coefficient β_n relies on the bending moment and the stress caused by the axial forces and since no axial compressive force is applied in this research, $\beta_n = 1$, and f_{vcd}, β_d, and β_p can be obtained from the following equations:

$$f_{vcd} = 0.20 \sqrt[3]{f_c'} \tag{4}$$

$$\beta_d = \sqrt[4]{\frac{1000}{d}} \tag{5}$$

$$\beta_p = \sqrt[3]{100 P_w} \tag{6}$$

$$P_w = \left. A_f \middle/ (b_w d) \right. \tag{7}$$

where, f_c' is the compressive strength of the concrete, P_w is the longitudinal reinforcement ratio, b_w is the width of the section, and A_f is the cross-sectional area of the tension reinforcement.

The shear capacity of ECC-reinforced structural members can be calculated as [46,47]:

$$V_{ecc} = V_{c,ecc} + V_{f,ecc} \tag{8}$$

where, $V_{c,ecc}$ is the shear carried by the member, and $V_{f,ecc}$ is the shear carried by fibers. The shear capacity equation of the ECC-reinforced members is similar to the shear capacity equation of concrete; however, with a smaller coefficient, which is 0.7. This is mainly due to the absence of coarse aggregates in ECC that weaken the interlocking of the aggregates. Accordingly, $V_{c,ecc}$ and $V_{f,ecc}$ are calculated as:

$$V_{c,ecc} = 0.7 \times 0.2 \sqrt[3]{f_{ecc}'} \times \sqrt[4]{1/d} \times \sqrt[3]{100 \rho_w} \times t \times d \tag{9}$$

$$V_{f,ecc} = \left(\left. f_{t,ecc} \middle/ \tan\beta_u \right. \right) \times t \times z \tag{10}$$

where, f_{ecc}' and $f_{t,ecc}$ are the compressive and tensile yield strength of ECC, respectively, t is the ECC thickness, β_u is the angle between the diagonal crack on the surface with the member axis and it is generally taken as 45 degrees. z is the distance between the location of the resultant compressive stress and the centroid of the tensile steel, which is generally equal to $d/1.15$.

The shear capacity of the stirrups is

$$V_{st} = \frac{A_w f_y (\sin\alpha + \cos\alpha) z}{S \times \gamma_b} \tag{11}$$

where, f_y is the yield strength of the shear rebar, A_w is the total cross-sectional area of the stirrups, α is the angle between the shear rebar with member axis, S is the stirrup spacing, and generally the γ_b coefficient is 1.15.

Moreover, an analytical model for the shear capacity of FRP grids can be calculated by an analytical model given in the references [29,30]. The analytical model accounts for the shear contribution of the horizontal and vertical CFRP grids as well as the shear span ratio of the RC beams. The equation of for shear capacity of CFRP is:

$$V_g = \frac{2 \cdot E_W \cdot [\rho_v \cdot \varepsilon_{uv} \cdot (\sin\alpha_v + \cos\alpha_v) + K \cdot \varepsilon_{uh}(\sin\alpha_h + \cos\alpha_h)] \cdot Z}{\gamma_b} \tag{12}$$

where, α_v and α_h are angles between vertical and horizontal grids with axial orientation, respectively, ρ_v and ρ_h are the cross-sectional areas in vertical and horizontal grids per unit of length, respectively, k is the ratio between the shear contribution of vertical grids and horizontal grids, E_W is the Young's modulus of CFRP grids, d is the effective height of the beam, the value of the coefficient γ_b is 1.15. ε_{uv} and ε_{uh} are the effective strains of vertical and horizontal grids, respectively, which are:

$$\varepsilon_{uv} = \left(\frac{1}{0.0352\rho_v + 0.0079}\right)^2 \times 10^{-6} \tag{13}$$

$$\varepsilon_{uh} = k \cdot \varepsilon_{uv} \tag{14}$$

where, k is the comparing coefficient related to the contribution of the shear between vertical and horizontal grids and it is:

$$k = e^{-0.612 \, (a/d) + 0.365} \geq 0.23 \tag{15}$$

where, a is the shear span of the beam.

A comparison of the shear force of the strengthened beams obtained by the numerical analysis and the theoretical models [29,46] is made in Table 10. The shear capacity of the strengthened beams from the numerical analysis was relatively close to those obtained from the theoretical models. The average ratio of shear capacity of the beams obtained by the numerical analysis and the above analytical model was 1.04, which was close to unity. Further, the average ratio of shear capacity of the beams obtained by the numerical analysis and above analytical model by the Japan Society of Civil Engineers (JSCE) [46] was 1.07, which was also close to unity. Given that the analytical models slightly overestimate the shear capacity of the strengthened beams, it is safe to use them for design calculations.

Table 10. Comparison of the shear capacity of beams obtained by the theoretical model and numerical analysis.

Beam	V_{FEM} (kN)	V [29] (kN)	V [46] (kN)	V_{FEM}/V [29]	V_{FEM}/V [46]
CRG5-C-17	240.54	230.28	221.93	1.05	1.08
CRG5-S-17	280.93	266.88	259.86	1.05	1.08
CRG8-C-17	245.71	240.41	229.40	1.02	1.05
CRG8-S-17	287.93	269.21	269.21	1.07	1.10
CRG'5-C-17	238.54	229.54	220.86	1.04	1.08
CRG'5-S-17	265.35	260.12	252.49	1.04	1.05
CRG'8-C-17	240.20	233.99	228.42	1.03	1.05
CRG'8-S-17	268.71	265.80	257.53	1.01	1.04

4. Conclusions

This paper studies the shear capacity of RC beams strengthened with CFRP grids within externally bonded ECC layers through numerical analysis. For this purpose, two RC beams were selected from the literature [37] and modeled in Abaqus FE software [38]. Thereafter, the application of the proposed strengthening method for shear strengthening of the beams was investigated in terms of failure mechanism and load-deformation responses. Lastly, the analytical model for calculating the shear capacity of beams strengthened by this method was given. Based on the obtained results, the following conclusions are drawn:

1. The CFRP grids and ECC were effective in delaying and reducing the diagonal shear crack as well as the flexural cracks in the RC beam. Comparatively, the CFRP grids and ECC controlled the diagonal shear cracks in the beams better than the CFRP sheets. Stress analysis showed that the CFRP grid was the primary strengthening member for improving shear performance, while ECC layers mainly acted as bonding agents.
2. The effect of the shear strengthening of the simply supported beam with the proposed method was more profound as compared with the continuous beam. It was primarily due to the failure mechanism of the beams that the simply supported beam was designed to fail due to shear while the mode of the continuous beam was a shear-

flexural failure. Nevertheless, the proposed shear strengthening also reduced the flexural cracks in the continuous beam.

3. The load-deformation responses of beams strengthened with CFRP grids and ECC showed significant improvement compared to reference beams and those strengthened with CFRP sheets. The shear capacity of the RC beams was greatly improved from 30% to 50% after strengthening with the CFRP grids and ECC.

4. The main governing factor for increasing the shear capacity of the beam with the proposed method was the CFRP grid size. The shear capacity of simply supported beams strengthened with smaller CFRP grid sizes was 8% and 10% larger than the simply supported beams strengthened with larger CFRP grid sizes, respectively. Therefore, as the CFRP grid size increased, the ultimate shear strength of the beam reduced.

5. The shear capacity of the beam strengthened by the CFRP grids and ECC was calculated using the analytical model and the Japan Society of Civil Engineers (JSCE). The average shear capacity ratios of the numerical results with those calculated based on the analytical model and JSCE were 1.04 and 1.07, respectively. While the analytical models slightly overestimated the shear capacity (compared to numerical results), they were deemed safe for design purposes.

6. In summary, this numerical study demonstrated the effectiveness of the proposed strengthening method in improving the shear behavior of RC beams. Future research should explore the application of this method in experimental studies.

Author Contributions: Software, M.S.G.; investigation, A.J., M.I., N.U. and E.K.; writing—original draft, M.S.G.; writing—review and editing, A.J.; project administration, A.J. All authors have read and agreed to the published version of the manuscript.

Funding: The support of the startup fund of Fuzhou University (XRC-22046) is also greatly acknowledged.

Data Availability Statement: Not applicable.

Conflicts of Interest: The authors declare no conflict of interest.

References

1. Costa, A.; Appleton, J. Case studies of concrete deterioration in a marine environment in Portugal. *Cem. Concr. Compos.* **2002**, *24*, 169–179. [CrossRef]
2. Safehian, M.; Ramezanianpour, A.A. Assessment of service life models for determination of chloride penetration into silica fume concrete in the severe marine environmental condition. *Constr. Build. Mater.* **2013**, *48*, 287–294. [CrossRef]
3. Hanif, M.U.; Ibrahim, Z.; Ghaedi, K.; Javanmardi, A.; Rehman, S. Finite Element Simulation of Damage in RC Beams. *J. Civ. Eng.* **2018**, *9*, 50–57. [CrossRef]
4. Mulk, S.; Usman, M.; Khan, A.; Usman, M.; Javanmardi, A.; Ahmad, A. Damage assessment of reinforced concrete beams using cost-effective MEMS accelerometers. *Structures* **2022**, *41*, 602–618. [CrossRef]
5. Hanif, M.U.; Ibrahim, Z.; Ghaedi, K.; Hashim, H.; Javanmardi, A. Damage Assessment of Reinforced Concrete Structures using a Model-based Nonlinear Approach—A Comprehensive Review. *Constr. Build. Mater.* **2018**, *192*, 847–865. [CrossRef]
6. He, W.; Wang, X.; Monier, A.; Wu, Z. Shear Behavior of RC Beams Strengthened with Side-Bonded BFRP Grids. *J. Compos. Constr.* **2020**, *24*, 04020051. [CrossRef]
7. Wakjira, T.G.; Ebead, U. Hybrid NSE/EB technique for shear strengthening of reinforced concrete beams using FRCM: Experimental study. *Constr. Build. Mater.* **2018**, *164*, 164–177. [CrossRef]
8. Ghaedi, K.; Ibrahim, Z.; Javanmardi, A.; Rupakhety, R. Experimental study of a new bar damper device for vibration control of structures subjected to earthquake loads. *J. Earthq. Eng.* **2021**, *25*, 300–318. [CrossRef]
9. Sajan, K.C.; Bhusal, A.; Gautam, D.; Rupakhety, R. Earthquake damage and rehabilitation intervention prediction using machine learning. *Eng. Fail. Anal.* **2023**, *144*, 106949. [CrossRef]
10. Khatibi, H.; Wilkinson, S.; Dianat, H.; Baghersad, M.; Ghaedi, K.; Javanmardi, A. Indicators bank for smart and resilient cities: Design of excellence cities. *Built Environ. Proj. Asset Manag.* **2021**, *12*, 5–19. [CrossRef]
11. Marcinczak, D.; Trapko, T.; Musiał, M. Shear strengthening of reinforced concrete beams with PBO-FRCM composites with anchorage. *Compos. Part B Eng.* **2019**, *158*, 149–161. [CrossRef]
12. Grace, N.F.; Sayed, G.A.; Soliman, A.K.; Saleh, K.R. Strengthening reinforced concrete beams using fiber reinforced polymer (FRP) laminates. *ACI Struct. J.* **1999**, *96*, 865–874. [CrossRef]
13. Colalillo, M.A.; Sheikh, S.A. Behavior of shear-critical reinforced concrete beams strengthened with fiber-reinforced polymer-Experimentation. *ACI Struct. J.* **2014**, *111*, 1373–1384. [CrossRef]

14. Rodrigues, H.; Arêde, A.; Furtado, A.; Rocha, P. Seismic behavior of strengthened RC columns under biaxial loading: An experimental characterization. *Constr. Build. Mater.* **2015**, *95*, 393–405. [CrossRef]

15. Rodrigues, H.; Arêde, A.; Furtado, A.; Rocha, P. Seismic Rehabilitation of RC Columns Under Biaxial Loading: An Experimental Characterization. *Structures* **2015**, *3*, 43–56. [CrossRef]

16. Rodrigues, H.; Furtado, A.; Arêde, A. Experimental evaluation of energy dissipation and viscous damping of repaired and strengthened RC columns with CFRP jacketing under biaxial load. *Eng. Struct.* **2017**, *145*, 162–175. [CrossRef]

17. Ghaedi, K.; Ibrahim, Z.; Javanmardi, A.; Jameel, M.; Hanif, U.; Rehman, S.; Gordan, M. Finite Element Analysis of a Strengthened Beam Deliberating Elastically Isotropic and Orthotropic Cfrp Material. *J. Civ. Eng. Sci. Technol.* **2018**, *9*, 117–126. [CrossRef]

18. Chaallal, O.; Nollet, M.-J.; Perraton, D. Shear Strengthening of RC Beams by Externally Bonded Side CFRP Strips. *J. Compos. Constr.* **1998**, *2*, 111–113. [CrossRef]

19. Hadi, M.N.S. Retrofitting of shear failed reinforced concrete beams. *Compos. Struct.* **2003**, *62*, 1–6. [CrossRef]

20. Adhikary, B.B.; Mutsuyoshi, H. Behavior of Concrete Beams Strengthened in Shear with Carbon-Fiber Sheets. *J. Compos. Constr.* **2004**, *8*, 258–264. [CrossRef]

21. Al-Tersawy, S.H. Effect of fiber parameters and concrete strength on shear behavior of strengthened RC beams. *Constr. Build. Mater.* **2013**, *44*, 15–24. [CrossRef]

22. Karzad, A.S.; Leblouba, M.; Al Toubat, S.; Maalej, M. Repair and strengthening of shear-deficient reinforced concrete beams using Carbon Fiber Reinforced Polymer. *Compos. Struct.* **2019**, *223*, 110963. [CrossRef]

23. Yang, X.; Gao, W.Y.; Dai, J.G.; Lu, Z.D.; Yu, K.Q. Flexural strengthening of RC beams with CFRP grid-reinforced ECC matrix. *Compos. Struct.* **2018**, *189*, 9–26. [CrossRef]

24. Ye, Y.Y.; Smith, S.T.; Zeng, J.J.; Zhuge, Y.; Quach, W.M. Novel ultra-high-performance concrete composite plates reinforced with FRP grid: Development and mechanical behaviour. *Compos. Struct.* **2021**, *269*, 114033. [CrossRef]

25. Zeng, J.-J.; Zeng, W.-B.; Ye, Y.-Y.; Liao, J.; Zhuge, Y.; Fan, T.-H. Flexural behavior of FRP grid reinforced ultra-high-performance concrete composite plates with different types of fibers. *Eng. Struct.* **2022**, *272*, 115020. [CrossRef]

26. Zeng, J.-J.; Pan, B.-Z.; Fan, T.-H.; Zhuge, Y.; Liu, F.; Li, L.-J. Shear behavior of FRP-UHPC tubular beams. *Compos. Struct.* **2023**, *307*, 116576. [CrossRef]

27. Azam, R.; Soudki, K.; West, J.S.; Noël, M. Strengthening of shear-critical RC beams: Alternatives to externally bonded CFRP sheets. *Constr. Build. Mater.* **2017**, *151*, 494–503. [CrossRef]

28. Blanksvärd, T.; Täljsten, B.; Carolin, A. Shear Strengthening of Concrete Structures with the Use of Mineral-Based Composites. *J. Compos. Constr.* **2009**, *13*, 25–34. [CrossRef]

29. Guo, R.; Pan, Y.; Cai, L.; Hino, S. Study on design formula of shear capacity of RC beams reinforced by CFRP grid with PCM shotcrete method. *Eng. Struct.* **2018**, *166*, 427–440. [CrossRef]

30. Guo, R.; Cai, L.; Hino, S.; Wang, B. Experimental study on shear strengthening of RC beams with an FRP grid-PCM reinforcement layer. *Appl. Sci.* **2019**, *9*, 2984. [CrossRef]

31. Cai, L.; Liu, Q.; Guo, R. Study on the shear behavior of RC beams strengthened by CFRP grid with epoxy mortar. *Compos. Struct.* **2021**, *275*, 114419. [CrossRef]

32. Chen, C.; Cai, H.; Cheng, L. Shear Strengthening of Corroded RC Beams Using UHPC–FRP Composites. *J. Bridg. Eng.* **2021**, *26*, 1–14. [CrossRef]

33. Zhang, R.; Meng, Q.; Shui, Q.; He, W.; Chen, K.; Liang, M.; Sun, Z. Cyclic response of RC composite bridge columns with precast PP-ECC jackets in the region of plastic hinges. *Compos. Struct.* **2019**, *221*, 110844. [CrossRef]

34. Liu, H.; Zhang, Q.; Gu, C.; Su, H.; Li, V. Self-healing of microcracks in Engineered Cementitious Composites under sulfate and chloride environment. *Constr. Build. Mater.* **2017**, *153*, 948–956. [CrossRef]

35. Pan, J.; Lu, B.; Gu, D.; Xia, Z.; Xia, T. Mechanical behavior of rectangular steel-reinforced ECC/concrete composite column under eccentric compression. *Trans. Tianjin Univ.* **2015**, *21*, 269–277. [CrossRef]

36. Ge, W.; Ashour, A.; Cao, D.; Lu, W.; Gao, P.; Yu, J.; Ji, X.; Cai, C. Experimental study on flexural behavior of ECC-concrete composite beams reinforced with FRP bars. *Compos. Struct.* **2019**, *208*, 454–465. [CrossRef]

37. Pellegrino, C.; Modena, C. Fiber-reinforced polymer shear strengthening of reinforced concrete beams: Experimental study and analytical modeling. *ACI Struct. J.* **2006**, *103*, 720–728. [CrossRef]

38. Simulia, D. *ABAQUS 6.14 Analysis User's Manual. Abaqus 6.14 Documentation*; ABAQUS M, Hibbitt, Karlsson and Sorensen, Inc.: Providence, RI, USA, 2016.

39. Fink, J.; Petraschek, T.; Ondris, L. Push-out test parametric simulation study of a new sheet-type shear connector. In *Projeket an Zentralen Applikationsselenrvern, Berichte*; Zentraler Informatikdienst der Technischen University. Wien: Wien, Austria, 2006; pp. 131–153.

40. Zheng, Y.Z.; Wang, W.W.; Brigham, J.C. Flexural behaviour of reinforced concrete beams strengthened with a composite reinforcement layer: BFRP grid and ECC. *Constr. Build. Mater.* **2016**, *115*, 424–437. [CrossRef]

41. Singh, S.B.; Patil, R.; Munjal, P. Study of flexural response of engineered cementitious composite faced masonry structures. *Eng. Struct.* **2017**, *150*, 786–802. [CrossRef]

42. Nippon Steel Chemical & Material. Available online: https://www.nscm.nipponsteel.com (accessed on 10 April 2023).

43. Kameda, Y.; Nemoto, M.; Mitani, K.; Kawase, Y.; Ashino, T. Pier Strengthening Technique Using Carbon Fiber Grid and under Water Curing Type Resin. *Nippon. Steel Sumitomo Met. Tech. Rep.* **2017**, *115*, 99–108.

44. *GB 50208-2011*; Code for Acceptance of Construction Quality of Underground Waterproof. Standardization Administration of the People's Republic of China: Beijing, China, 2011.
45. *GB/T 1499.2-2007*; Steel for the Rein-Forcement of Concrete-Part 2: Hot Rolled Ribbed Bars. Standards Press of China: Beijing, China, 2007.
46. Yokota, H.; Rokugo, K.; Sakata, N. *Japan Society of Civil Engineers (JSCE): Recommendations for Design and Construction of High-Performance Fiber Reinforced Cement Composites with Multiple Fine Cracks (HPFRCC)*; Springer: Tokyo, Japan, 2008.
47. Li, H.; Leung, C.K.Y.; Xu, S.; Cao, Q. Potential use of strain hardening ECC in permanent formwork with small scale flexural beams. *J. Wuhan Univ. Technol. Mater. Sci. Ed.* **2009**, *24*, 482–487. [CrossRef]

 *buildings*

Article

Performance-Based Assessment of RC Building with Short Columns Due to the Different Design Principles

Ercan Işık [1], Hakan Ulutaş [2], Ehsan Harirchian [3,*], Fatih Avcil [1], Ceyhun Aksoylu [4] and Musa Hakan Arslan [4]

[1] Department of Civil Engineering, Bitlis Eren University, Bitlis 13100, Turkey
[2] Department of Civil Engineering, Burdur Mehmet Akif Universtiy, Burdur 15030, Turkey
[3] Institute of Structural Mechanics (ISM), Bauhaus-Universität Weimar, 99423 Weimar, Germany
[4] Department of Civil Engineering, Konya Technical University, Konya 42075, Turkey
* Correspondence: ehsan.harirchian@uni-weimar.de

Abstract: Many factors affect the earthquake vulnerability of reinforced concrete (RC) structures, constituting a large part of the existing building stock. Short column in RC structures is one of the reasons for earthquake damage. Significant damages may occur due to brittle fractures in structural elements when the shear resistances are exceeded under the effect of high shear stress in short columns formed due to architectural and topographic reasons. This study created structural models for three situations: the hill slope effect, band-type window and mezzanine floor, which may cause short column formation. The structural analyses by SAP2000 were compared with the reference building model with no short columns. Structural analyses were performed separately according to strength-based and deformation-based design approaches in the updated Türkiye Building Earthquake Code (TBEC-2018). Short column formation; the effects on soft-storey irregularity, the relative storey drifts, column shear force, plastic rotation in columns, roof displacement, base shear force and column damage levels were investigated. As a result of the analysis, it was determined that the relative drifts from the first floor of the building decreased significantly due to the band-type window and slope effect, which caused the second storey to fall into the soft-storey status. In addition, short-column formation caused a significant increase in both plastic rotation demand and shear force in short columns.

Keywords: short column; seismic behaviour; reinforced-concrete; design principles

Citation: Işık, E.; Ulutaş, H.; Harirchian, E.; Avcil, F.; Aksoylu, C.; Arslan, M.H. Performance-Based Assessment of RC Building with Short Columns Due to the Different Design Principles. *Buildings* **2023**, *13*, 750. https://doi.org/10.3390/buildings13030750

Academic Editors: Rajesh Rupakhety and Dipendra Gautam

Received: 29 January 2023
Revised: 23 February 2023
Accepted: 8 March 2023
Published: 13 March 2023

1. Introduction

The first step in protecting a settlement from an earthquake disaster is to form and possess a theoretical prediction of the consequences: structural damage and socio-economic losses that may happen after the earthquake. In fact, it is crucial to assess the effects of any potential earthquake to prepare for management during a catastrophic situation. As well as anticipating and taking appropriate measures to reduce the vulnerability and expected losses on the part of guaranteeing urban resilience [1–3].

In order to reduce the impact of an earthquake in settlement areas, existing design codes must be applied strictly. The robustness of earthquake defence mechanisms of buildings is in inverse proportion to the presence of negative parameters in the buildings. The more negative parameters mean a weaker defence mechanism for the building. Therefore, it is essential to avoid negative parameters as much as possible. These parameters that cause building damage and total collapse are usually included in seismic design codes. The factors that would reduce buildings' performances under the impact of earthquakes are also available in rapid visual safety evaluation and damage classification of present structures. These negative criteria include short columns, the hill-slope effect, the year of construction, a weak/soft storey, a lateral or vertical discontinuity and a visible building's quality. Filling RC frames with half-height infill walls, creating band windows and using inter-mediate beams on stairwells are the primary causes of short-column formation. If the building is clearly on a hill or a slope with a high slope (more than 30°), it increases the effects of earthquakes to a certain extent. The differences in stiffness and strength between stories and the variation in storey height within a building are considered in the soft storey/weak storey parameter. Vertical irregularity can be considered as the case of a substantial change in the stiffness, mass and dimensions or interruption of vertical structural elements, such as columns or lateral resisting systems. The performance of workmanship and material quality during and after the construction process reveals the quality of the building. These parameters are included in the Turkish Rapid Assessment Method, the Canadian Seismic Screening Method, the Japanese Seismic Index Method and similar rapid assessment methods [4–10]. This study investigated the short-column effect in detail according to two different design principles in the updated Türkiye Building Earthquake Code.

Short column damage is one of the common types of damage observed after earthquakes [11,12]. In particular, the decrease in the effective height of the column due to many effects, especially architectural reasons, causes the shear force on the column to increase too much, and the column reaches its bending capacity [13]. Suppose no prediction was made for this decrease in the free height of the column during the design phase. In that case, sudden and brittle shear damage occurs, especially in the column with insufficient reserve shear capacity. Particularly in the case of brittle damage to the short columns in the basement and lowest floor, they may cause a total collapse of the structure [14,15]. Therefore, in cases where it is not possible to eliminate the factors causing the short column, the effect of these columns on the structural performance should be evaluated in detail, and precautions should be taken if necessary.

In order to determine the structural performance, member-based damage conditions must be evaluated first. The earthquake performance of a building can be defined as "the building safety condition determined in connection with the level and distribution of the damages that may occur in that building under earthquake effect". In recent earthquakes, it has been reported that structures with columns that suffered shear force damage could not dissipate enough energy during the earthquake and collapsed altogether [16–18]. The number, distribution and location of columns likely to exhibit brittle behaviour in buildings are also very important for global performance.

Experimental and analytical studies in the literature emphasised that short columns are a very important parameter for structural performance. Meral [19] determined the effects of short column behaviour on reinforced concrete structures with non-linear inelastic time history analysis, and the base shear force demands were higher in short column models compared to the reference buildings, while the displacement demands were calculated as low. Şeker and Bedirhanoğlu [20] investigated the behaviour of short columns with insufficient shear strength with representative samples produced in the laboratory. As expected, the samples produced brittle collapse with shear damage, and the experiments explained the behaviour of such short columns under lateral loads. Balık and Bahadır [21] investigated the effects of window spacing on short-column behaviour using different reinforcement methods. They discussed the results of producing seven reinforced concrete frames with two floors, 1/5 scale and single span. Işık et al. [22] examined the formation of short columns in reinforced concrete structures built on sloping soils and revealed that plastic hinges formed earlier in short columns. Haji et al. [23] conducted experimental studies using three different methods for reinforcing the reinforced concrete circular short columns. They observed a good correlation between analytical values and experimental results. Chen et al. [24] investigated the behaviour of reinforced concrete short columns under constant compression loading experimentally and theoretically. Moretti and Tassios [25] presented experimental results on eight full-scale reinforced concrete short columns. The relative performance of the tested samples was evaluated according to various ductility criteria. As a result of these studies, it was observed that the stiffness of the columns increased with decreasing effective length. Therefore, the initial stiffness and load-carrying capacity of the structure increased slightly in the global behaviour of the structure. However, it has been reported that the structure could not achieve the desired ductile behaviour due to shear damage in the columns. In addition, short-column damages come to the fore in studies on damage caused by destructive earthquakes in different parts of the world [26–29]. Moreover, short-column effects are clearly identified in studies using rapid assessment methods used for reinforced concrete structures [30–32].

Within the scope of this study, first of all, short column formation is explained with theoretical details. In other words, the effects of short column formation, which negatively affects the earthquake performance of the building, are examined in detail. Then, in addition to three different structural models that may cause short column formation, including the hill slope effect, the band-type window, and the mezzanine floor, a total of four different models were created, including a structural model without any short columns, in order to make comparisons. Structural analyses were performed using SAP2000 software v20. The band-type window is placed on the outer axes in the x calculation direction in the SAP2000 model, the hill slope effect is placed on the two adjacent axes in the x direction, and the mezzanine floor is placed on three storeys. The band-type window, slope effect and mezzanine floor are modelled as bar members in SAP2000 models with short column lengths of 0.7 m, 1.5 and 2 m, respectively, to be closest to the actual formation. In the structural model, only the short-column effect is considered a variable. The results were compared among themselves and those obtained for the reference building model without short columns. In the study, most possible scenarios that may cause the formation of short columns are remarked on. Depending on these scenarios, changes in storey drifts (especially in soft-storey formation) and column shear force distributions in the structural model, changes in plastic rotation demands and capacities in columns, damage distribution and global changes in building performance were examined, different structural parameter results that were not included in previous studies were taken into account. The main novelty of the study is that the short column effect is revealed according to both strength-based and deformation-based design approaches specified in the Turkish Building Earthquake Code (TBEC-2018), which is currently in place in Türkiye.

2. Short Column in RC Structures

The two main components of the works performed to estimate the effect of earthquakes are detecting earthquake hazards and determining buildings' vulnerability. Determining the vulnerability of building systems is generally possible by examining and classifying the current building stock and other structures and obtaining the damage potential curves. These potential damage curves are used in the preparation of earthquake scenarios [33]. While obtaining these curves, it is necessary to consider the aspects of the building stock that are peculiar to the region.

Knowing buildings' characteristics that may negatively affect their earthquake behaviour in areas where strong seismic motions can be expected will enable setting forth more serious approaches to mitigate hazard risk levels. Being aware of the negative parameters both in terms of the designs of new buildings to be constructed and of examining the current building stock constitutes a significant part of reducing vulnerability.

Vertical load-bearing elements in framed system structures consist of columns. Columns have two essential roles in such systems. The first is safely transferring all the vertical and lateral loads that act on the structure to the foundation system. The second role is to keep the relative storey drifts caused by lateral loads within the allowed limits [34]. In brief, while designing columns that serve as vertical load-bearing elements in building systems, it has to be taken care that the set forth limits are not exceeded in terms of load and deformation [35].

Columns as vertical load-bearing elements may involve some negative parameters that would weaken the defence mechanism of the building. One of these parameters is short columns that may form in time. Short columns have formed in different ways. Short columns may occur either because the original column length-to-depth ratio is small or due to an obstacle along a certain height of the column (e.g., low brick masonry wall), in which case the effective length of the column is reduced. The sudden and explosive nature of shear failure in a short column was first pointed out by Yamada and Furui in 1968 [36]. Although considerable experimental and theoretical research has been conducted on this subject, especially in the 1970s and '80s, neither a generally accepted way of predicting the behaviour of short columns exists nor special provisions concerning the design of such elements are included in the majority of codes [37].

The term short column refers to columns shorter than the other columns in the same storey and expected to be subjected to brittle shear fractures [38]. Short column effects appear during earthquakes, particularly on the exterior walls of buildings where the walls do not extend up to the top floor and on the parts of the columns that are not covered by the wall. Typically, in industrial buildings, some gaps are left on the outer walls of the buildings, and some are left on the outer walls to leave space for windows and take light in. If these walls do not extend along the column height, particularly if they are not separated from the frame, the columns may suffer heavy damage during the earthquake. These effects, known as short-column behaviour, can also be seen in Türkiye as they occur worldwide. Many parameters have a role in the formation of a short column. Some of these are wall height, the gaps left on the frame wall and inadequate shear reinforcements. However, one of the primary parameters that cause this behaviour is the vertical gaps left on the frame wall. The shear force generated on the short column is inversely proportional to the short column height [39,40]. Due to the ribbon windows placed on both sides of the columns without any extra measures taken, the calculated height of the column shortens, and the shear force is received at a much higher level than the calculated strength.

Having both long and short columns in the same storey of the building causes these columns to be subjected to different shear forces during the earthquake due to the height difference. Earthquake loads primarily affect long and flexible columns and then affect short columns, where they cause energy accumulation. Due to this accumulation, X-shaped shear cracks occur at both the lower and upper ends of the column. Long and short columns with the same cross-sectional area suffer the same displacement (Δ) under earthquake load (Figure 1). However, short columns are more rigid than long ones, so they are subjected to more earthquake load. This is referred to as the short-column effect and is a point that needs to be considered in the early phases of architectural design [41]. Short column formations in the buildings are given in Figure 2.

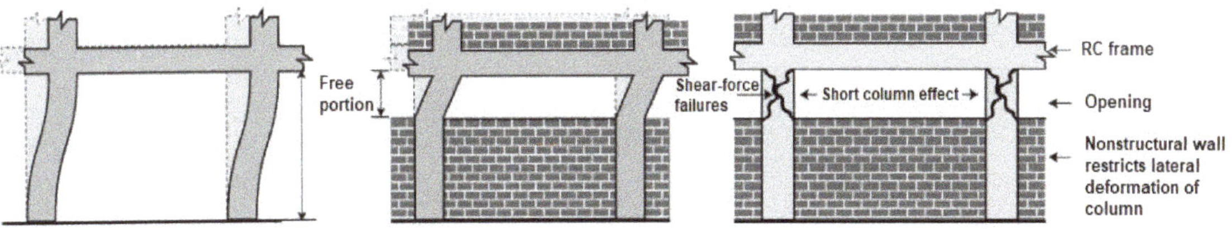

Figure 1. Behaviour of short and regular columns [42].

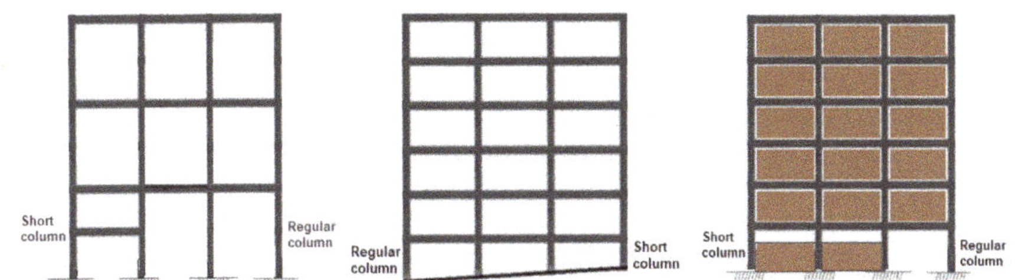

Figure 2. Short column formations.

One of the most significant causes of damage observed as a consequence of earthquakes is short-column formation. Short columns are subjected to high shear stress, and when the shear resistance is exceeded, brittle collapses take place on these elements. The early strength loss that takes place in these elements also limits the displacement capacity of the building. In turn, brittle collapses become inevitable in buildings with short columns that cannot meet the displacements forced by the earthquake. In addition, since the lateral load capacities of these elements are limited by shear capacities, their bending capacities also cannot be utilized fully [43,44]. Different levels of damage occur due to short columns in earthquakes occurring in different countries of the world [45–47]. Figure 3 presents some short-column damages observed in earthquakes.

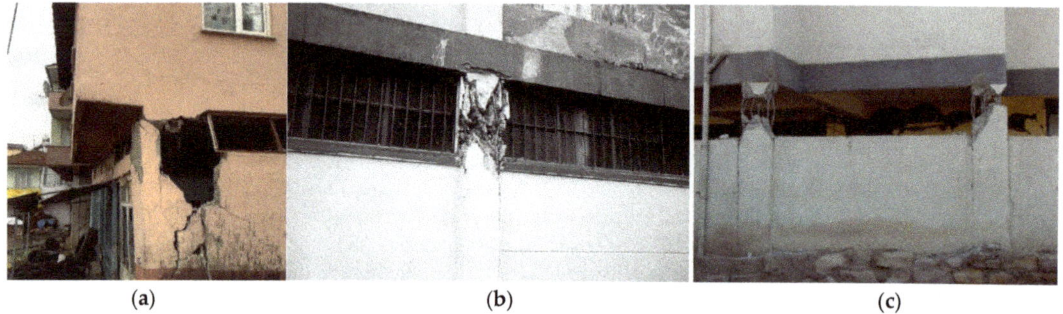

(a)   (b)   (c)

Figure 3. Short column damages: (**a**)-24 January 2020 Elazığ earthquake [48], (**b**)-25 January 1999 Armenia Colombia earthquake [49], (**c**)-26 January 2001 (Bhuj) India earthquake [50].

3. Performance-Based Assessment and Design Principles

Following numerous destructive earthquakes worldwide, the significance of research, studies and earthquake prevention has recently increased. Due to the vulnerability of both rural and urban building stock, earthquake damages grow. The magnitude of the earthquakes and the poor structural characteristics increase the size of the damage. Knowing the characteristics of buildings that will be adversely affected by their behaviour under earthquakes will reduce damage risk after earthquakes. Determining how well structures perform is the first step in minimizing earthquake damage. After earthquakes in Türkiye, particularly in the past 30 years, protecting existing buildings has become increasingly important. For RC structures already in existence, performance-based assessment procedures are frequently applied. A building's susceptibility to earthquake damage is known as its seismic vulnerability. Seismic vulnerability relationships make an effort to forecast, for a variety of building classes, the severity and scope of damage at specific seismic demand levels [51].

Using the performance-based design and evaluation technique, it is feasible to calculate the quantities of damage levels caused by design ground motion inside the structural system elements. It is examined whether this damage remains below the permitted damage thresholds for each pertinent element. Reasonable damage limits are established in a way that is consistent with the anticipated performance goals at different earthquake levels [52,53]. The evaluation process aims to calculate the required earthquake force at which the building would meet its performance goals. The highest reaction that a building gives against seismic action during an earthquake is displayed on the demand spectrum used to assess the system performance of the building [54].

Two essential performance-based design and assessment parameters are earthquake demand and capacity [55,56]. Static pushover and capacity curve serve as representations of structural capacity. This capacity curve is created by plotting the relationship between the base shear force and the displacement of the building's roof. The building system is calculated with gravity loads and lateral forces that increase proportionately up to the target point where structural capacity ends. The non-linear static method's actual goal is to calculate a building's target displacement, which is then used to do a final pushover analysis by increasing lateral loads until the target displacement is reached. In conclusion, demand values such as strains, displacements, internal forces and rotations are calculated, and the behaviour of the analysed section is measured for different cross-section performance levels by comparing the strains obtained from the total curvature of the section with the determined upper bound strains [57–59].

In this paper, structural analysis was performed by using static pushover analysis while determining the earthquake performance of the reference RC structure model. Pushover analysis is a typical method for determining seismic demand in building designs and evaluations [60]. The pushover curve is calculated using the static multiplier that was produced by applying the theorem of virtual works while considering the mechanism under study's various kinematic configurations at large displacements [61]. Pushover analysis is one of the most frequently used methods to estimate the non-linear behaviour of buildings [62]. In addition to the target displacement, stress-strain relationships for the material are also important factors in the structural analysis [63].

The building performance of the building model with a short column was determined separately according to two different methods using static pushover analysis. Design and evaluation methods were used according to the strength and deformation in the earthquake code currently used in Türkiye [64].

The strength-based design approach is one of the two main approaches for designing structural systems under the influence of earthquakes: (a) Reduced seismic loads corresponding to the ductility capacity of the structural system defined for a certain predicted performance target are determined. (b) Linear earthquake calculation of the bearing system is made under reduced seismic loads. The strength demands are obtained by combining the reduced internal forces of the element found from this calculation with the internal forces of other loads, taking the excess strength into account when necessary. (c) Member strength demands are compared with the internal force capacities (strength capacities) defined for the anticipated performance target. (d) The relative storey drifts obtained from the earthquake calculation are compared with the allowable limits. (e) The design is completed by showing that the strengths are below the strength capacities and, simultaneously, the relative storey drifts are below the permissible limits. Otherwise, the frame sections are changed, and the calculation is repeated.

The deformation-based design approach is the other main approach for structural building systems under the influence of earthquakes: (a) Internal force-deformation relations compatible with non-linear modelling approaches of existing or previously pre-designed structural systems are determined. (b) Under the earthquake ground motions selected per the predicted performance targets, the structural system's static or time history analysis is calculated with dynamic incremental methods, and deformation demands of non-linear ductile behaviour and strength demands of brittle behaviour are attained. (c) Obtained deformation and internal force demands are compared with defined deformation and strength capacities by the anticipated performance targets. (d) For existing buildings, the assessment according to deformation is completed by showing that the deformation and strength demands are below or exceed the corresponding deformation and strength capacities. (e) If the deformation and strength demand for new or existing buildings to be strengthened are below the corresponding deformation and strength capacities, the design is completed according to the deformation-based design approach. Otherwise, the frame sections are changed, the calculation is repeated and re-evaluated and the design is finalised according to the deformation.

As in countries with high earthquake risk, earthquake-resistant building design rules have been developed and updated in Türkiye over time. With the current code, design and evaluation principles have been adopted according to both strength and deformation.

This study compared the strength-based design approach, the changes in relative storey-drift, soft-storey formation and column shear force. Additionally, with the deformation-based design approach, roof displacement, capacity curves, and damage to the structural elements and plastic rotations were compared. Therefore, this study allowed a very detailed comparison of the short-column effects.

4. Structural Models and Analysis Results

4.1. Description of the Numerical Model

The chosen RC frame building comprises seven storeys, each of which is 3 m high. The blueprint of the sample RC building is shown in Figure 4. The sample RC building has been designed in accordance with TEBC-2018 principles.

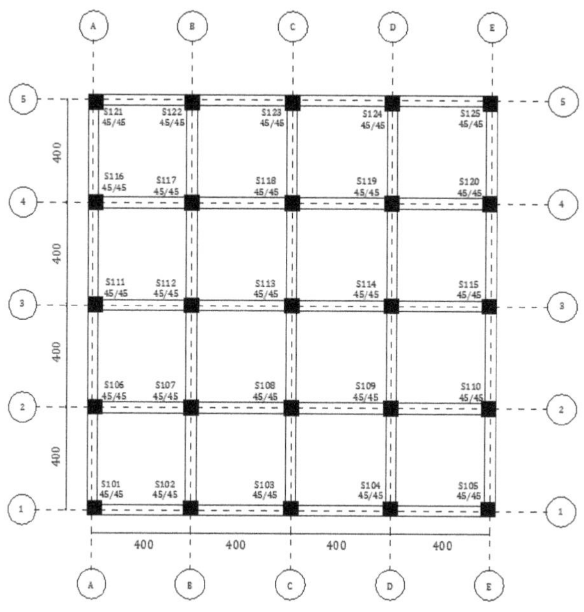

Figure 4. The blueprint of the sample RC building (all dimensions are in cm).

The structural features considered for the RC building selected as an example are shown in Table 1.

Table 1. Sample RC model features.

Number of storeys	7
Storey height	3 m
Concrete grade	C25
Reinforcement grade	B420C
Dead load (g)	4.5 kN/m^2
Live load (q)	2 kN/m^2
Location of the structure (Latitude-Longitude)	37.73534–30.305906
Infill wall load (g)	4 kN/m
Column stirrup reinforcement *	Φ8/100 (mm)
Beam stirrup reinforcement *	Φ8/100 (mm)
Local soil class	ZD
SDS	1.0721
S_{D1}	0.4786
Live load reduction coefficient (n)	0.3

* In confinement region.

In this study, since two different design approaches in the current seismic design in Türkiye are taken into account, the design spectrum used in this country has been taken into account. For this purpose, the design spectrum obtained for any geographical location in Burdur province was used. Along with the earthquake hazard map currently used in Türkiye, map spectral acceleration coefficients (S_S, S_1) have been used for the first time. By multiplying these coefficients with the local soil effect coefficients (F_s, F_1), the design spectral acceleration coefficients are obtained separately for a short period (0.2 s) (S_{DS}) and a long period (1.0 s) (S_{D1}). These values are directly obtained with the help of the Türkiye Earthquake Hazard Maps Interactive Web Application [65], which has been implemented with the current earthquake code. This study used this application was used for the location where the building will be built. The design spectrum used in structural analysis is shown in Figure 5.

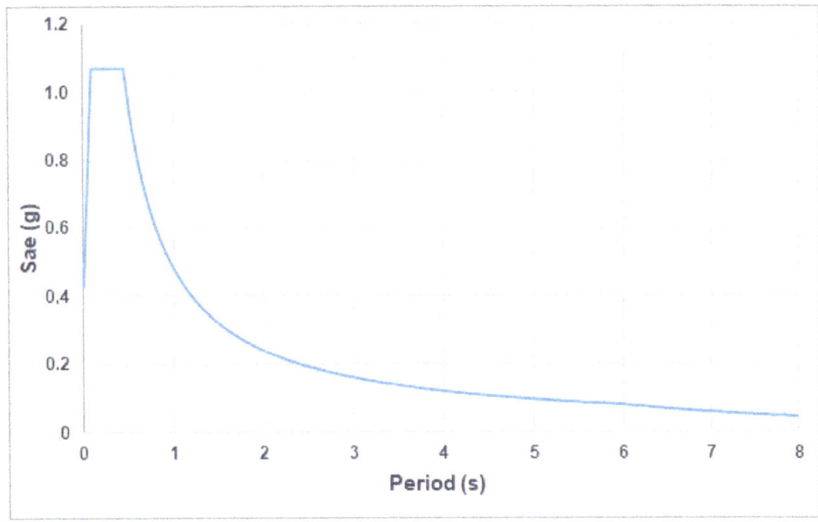

Figure 5. Design spectrum for structural analyses.

The cross-section and dimensions of the beams and columns used in the reinforced concrete structure chosen as an example are shown in Figure 6. All columns and beams have the same size and reinforcement in the sample RC building model. In the calculations, the effect of slabs on the stiffness of the beams is neglected, and slabs are considered only as loads. As can be seen in Figure 6, a reinforcement selection was made in the columns according to the minimum longitudinal reinforcement ratio (0.01) allowed for the columns of TBEC-2018. In addition, by using crossties with stirrups, the maximum spacing requirement between stirrup arms has been reduced below $25 \times \varnothing_{\text{transverse}}$. In beam supports, the tension reinforcement is above the minimum reinforcement ratio ($0.8\, f_{\text{ctd}}/f_{\text{yd}}$) and below the maximum reinforcement ratio (ρ_b, ρ_{max} etc.) given in TS-500-2000 and TBEC-2018. In addition, the compressive reinforcement/tensile reinforcement ratio of the beams in the supports was kept above 0.5and the ductility condition of the beam support section specified for $S_{DS} > 0.75$ in TBDY-2018 was also met.

The reference building model and short column formation models considered in the study are shown in Figure 7. Red-coloured columns are modelled as short columns. Only the appearance of the first 2 storeys is shown for a better understanding of the structural models.

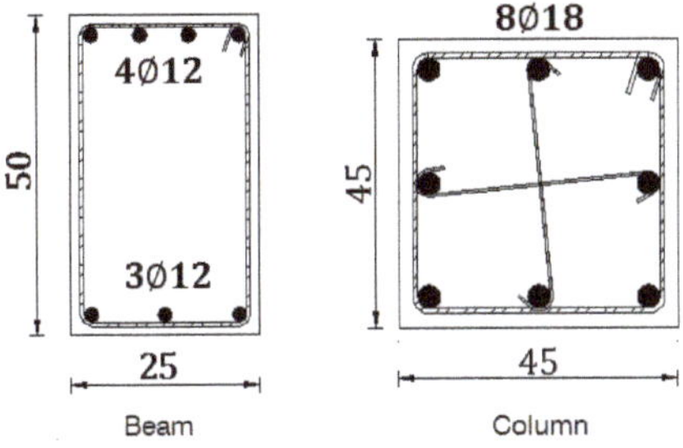

Figure 6. Cross-sections of columns and beam support.

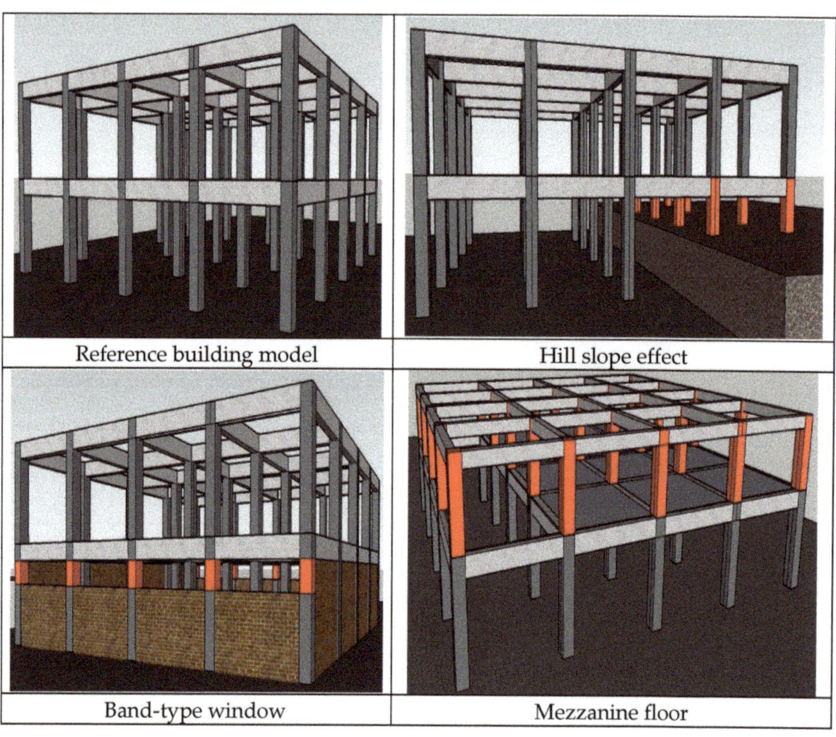

Reference building model	Hill slope effect
Band-type window	Mezzanine floor

Figure 7. Four different structural models considered in the study.

Considering the common situations that cause short column formation, a total of four different structural models, including the reference model where no short column formation is expected and three different models where potential short column effects may occur due to the hill slope effect, band-type window and mezzanine floor, were created using SAP2000 software [66]. In the structural models, the blueprint of the sample RC building has not been changed; the short column lengths and their axes are given in Table 2. All other columns not included in this table are considered non-short columns. While creating the structural models, separate short columns were created for the three cases that caused the formation of short columns and were considered in the study. Structural features were kept constant in order to reveal the short-column effect.

The fundamental natural periods of all structural models were obtained by eigenvalue analysis. The periods are given in Table 3. Since the short column heights will increase the stiffness values due to the short column formations, the period values have taken lower values in the short column structural models.

Table 2. Short columns length and their axes.

	Hill Slope Effect	**Band-Type Window**	**Mezzanine Floor**
Axes with a short column	All columns on the 4th and 5th axes of the 1st storey	All columns on the 1st and 5th axes of the 1st storey	All axes of the 2nd storey
Short column length	1.5 m	0.7 m	2 m
Number of short columns/ total columns (%)	5.71% (10/175)	5.71% (10/175)	14.29% (25/175)

Table 3. The periods of the structural models.

Model	Reference Model	Band-Type Window	Hill-Slope Effect	Mezzanine Floor
Period (s)	0.989	0.868	0.928	0.903

4.2. Investigation of Short Column Effect by the Strength-Based Design Approach

In order to take into account the possible uncertainties in the stiffness and mass distribution of the structural system of the building under the influence of earthquake ground motion, the earthquake calculation was made by considering the additional eccentricity of ±5% for each of the X and Y directions. For each storey, X direction +0.05 additional eccentricity (E_{x+}), x direction −0.05 additional eccentricity (E_{x-}), Y direction +0.05 additional eccentricity (E_{y+}), y direction −0.05 additional eccentricity (E_{y-}) earthquake calculations were made using four different earthquake directions/eccentricities, and the findings are given below.

4.2.1. Investigation of the Change in Relative Storey Drifts

Considering the highest values obtained from four different earthquake loadings (E_{x+}, E_{x-}, E_{y+}, E_{y-}) for each storey, the relative drifts in the X direction depending on the storey are shown in Figure 8.

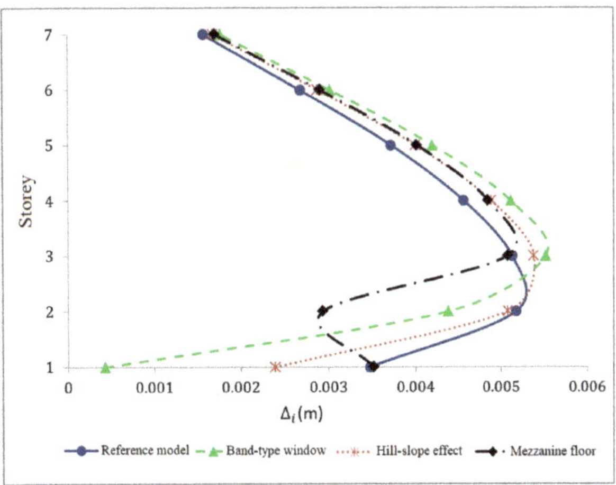

Figure 8. Comparison of the highest relative storey drifts for each storey.

Short columns formed due to the band-type window formation reduced the relative drifts of the ground storey by 88% compared to the reference model. The short columns formed due to the hill-slope effect reduced the relative drifts of the first storey by 31% compared to the reference model. The short columns formed due to the formation of the mezzanine floor reduced the relative drifts of the second storey by 43% compared to the reference model. Considering the maximum relative drifts of all floors, the highest relative drifts occurred in the band-type window, the hill-slope effect, the reference model and the mezzanine floor model building, respectively.

4.2.2. Investigation of Soft-Storey Formation

Whether there is a soft storey irregularity in a building under investigation is determined by looking at the stiffness irregularity coefficient (η_{ki}). The coefficient of stiffness irregularity (η_{ki}) is obtained by dividing the average storey drift ratio at any i-th storey by the average relative storey drift ratio at an upper or a lower storey for either of two perpendicular earthquake directions. The calculation of the stiffness irregularity coefficient (η_{ki}) for the calculation direction X is shown in Equation (1).

$$\eta_{ki,(lower)} = \frac{\Delta_{i\,(mean)}^{(X)}/h_i}{\Delta_{i-1\,(mean)}^{(X)}/h_{i-1}} \tag{1}$$

$$\eta_{ki,(upper)} = \frac{\Delta_i^{(X)}{}_{(mean)}/h_i}{\Delta_{i+1}^{(X)}{}_{(mean)}/h_{i+1}} \tag{2}$$

η_{ki} in Equation (2) is the stiffness irregularity coefficient obtained by dividing the average relative storey drift ratio of the i-th storey by the average relative storey drift ratio of a lower ((i − 1) storey; η_{ki} (upper)) expresses the stiffness irregularity coefficient obtained by dividing the average relative storey drift ratio of the i-th storey by the average relative storey drift ratio of the upper ((i + 1) storey).

Stiffness irregularity coefficients were obtained for each storey's four different earthquake types (E_{x+}, E_{x-}, E_{y+}, E_{y-}). The highest soft-storey irregularity coefficients obtained from four different earthquake types were determined as the soft-storey irregularity coefficient of the relevant storey, shown in Figure 9. According to the TBEC-2018, Inter-storey Stiffness Irregularity (Soft Storey) coefficients (n_{ki}) were obtained for each of the two orthogonal earthquake directions and four different earthquake types (E_{x+}, E_{x-}, E_{y+}, E_{y-}) of each storey by considering the effects of ±5% additional eccentricities.

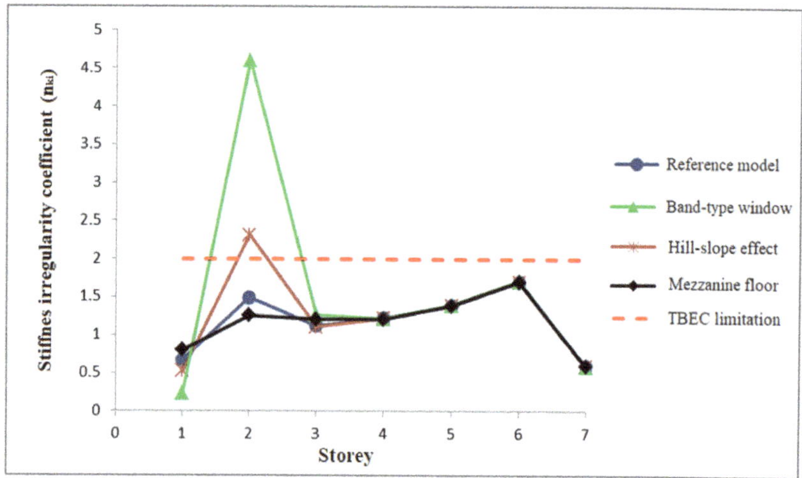

Figure 9. Stiffness irregularity coefficients (n_{ki}) of RC building models.

According to TBEC-2018, if the stiffness irregularity coefficient (n_{ki}) is more than 2 on any storey, it is stated that there is a soft storey irregularity in the building. Since the stiffness irregularity coefficients in the second storey of the band-type window and the hill-slope effect models were obtained as 4.6 and 2.3, respectively, soft storey irregularity occurred on these storeys. Therefore, due to the band-type window and the hill-slope effect, the storey where the formation of short columns occurs is considerably more rigid than the upper storey. This causes stiffness irregularity between adjacent storeys. Since the height of the second storey of the structural model, including the mezzanine storey, is lower than that of the reference model, it was a more rigid storey than the second storey of the reference model, which caused the relative drifts of the mezzanine storey to be lower than that of the reference model. Therefore, the mezzanine storey model's soft-storey irregularity coefficient is less than the reference model's second storey. Since the stiffness irregularity coefficient was less than 2 in the reference and mezzanine models, soft storey irregularity was not observed in these models.

4.2.3. Investigation of Column Shear Force Variation

In order to examine the effect of short column formation on the column shear force, the shear forces of the columns in the first and second storeys were obtained for four earthquake types (E_{x+}, E_{x-}, E_{y+}, E_{y-}), and these values are shown in Figure 10. When the shear force of the columns on the first storey is examined, the models with the highest increase in shear force compared to the reference model are the band-type window and the hill-slope effect models, respectively. Shear forces in the first storey columns of the mezzanine floor and reference models were close. When the column shear forces on the second storey are examined, the model with the highest shear force increase is the mezzanine floor model. The shear forces on the second-storey columns of the band-type window and the hill-slope effect models were close to the reference model.

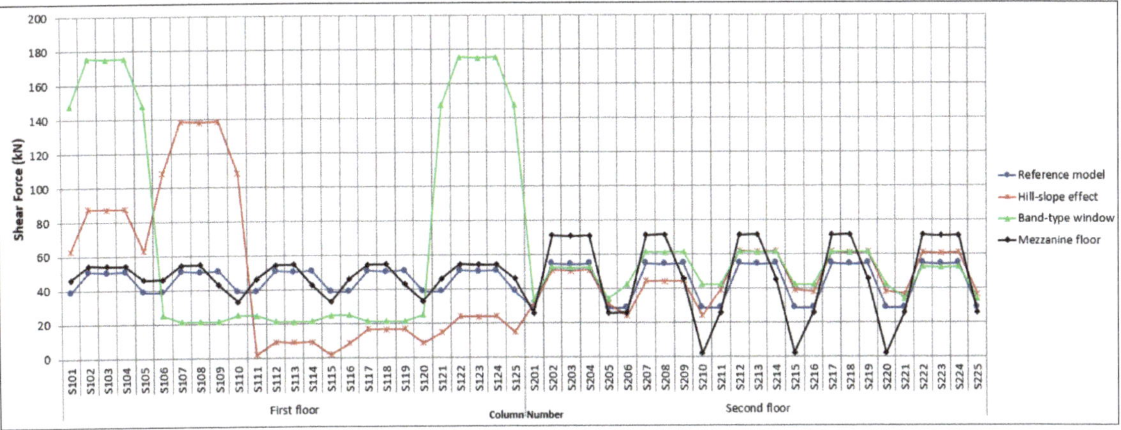

Figure 10. Column shear forces.

While there was an average increase of 261% in the short columns in the band-type window model compared to the same columns in the reference model, there was a decrease of 49% in the regular columns (non-short) in the same storey compared to the same columns in the reference model. On the other hand, while there was an average of 122% increase in the short columns in the hill-slope effect building model compared to the reference model, there was a 71% decrease in the regular columns on the same storey compared to the reference model. In other words, short columns had to carry the shear- force of non-short columns on the same storey due to the formation of short columns. In the mezzanine floor model, some columns increased by 30% compared to the same columns in the reference model, while a decrease of around 92% occurred in columns S210, S220 and S225.

4.3. Investigation of the Short Column Effect by the Deformation-Based Design Approach

In order to examine the effect of short column formation on section damages, a non-linear behaviour model of each of the reference models, the hill-slope effect, the band-type window and the mezzanine floor models were created. Plastic rotation limits were obtained for the Damage Limitation (DL), the Controlled Damage (CD) and Prevention of Collapse (PC) performance levels conditions given for the lumped plastic hinge model of columns and beams modelled as bar elements in TBEC-2018. After the plastic hinge sections of each structural element were determined, these hinges were assigned to both ends of the column and beam elements. M3 plastic hinge is used for beams, and P-M2-M3 plastic hinge is used for column sections. Moment-curvature analyses obtained plastic hinge properties of the sections. In these analyses, the unwrapped Mander stress-strain diagram was used for the shell concrete and the wrapped concrete for the core concrete. In the moment-curvature analyses, the yield unit strain of reinforcement steel is 0.0021, the tensile strain of reinforcement steel is 0.1, the yield strength of reinforcement steel is 420 MPa and the tensile strength is 550 MPa. A P-joint definition for shear power consumption has not been made, as necessary precautions have been taken regarding shearing in the elements. Second-order effects are neglected in the structural analyses. Horizontal loads were applied to the story mass centers of the model buildings according to the values obtained by multiplying the mode shape amplitude of the X-direction of the floors, and the masses and static pushover analyses were performed. The capacity curves obtained for the X direction are given in Figure 11.

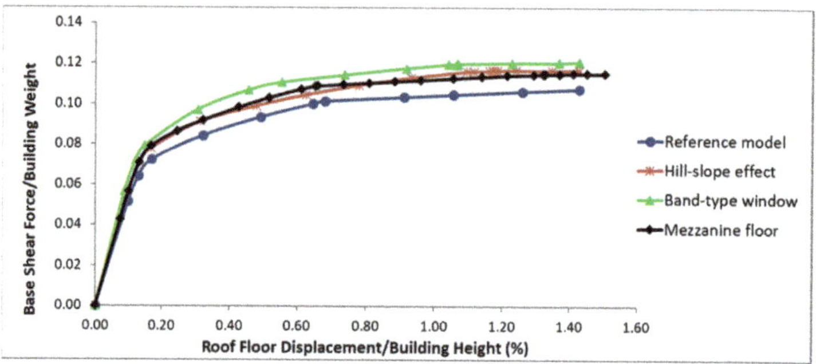

Figure 11. Comparison of capacity curves.

It is seen that the buildings with the highest ratio of base shear force to building weight are the band-type window, the hill-slope effect, the mezzanine floor and the reference model, respectively. The ratio of the highest base shear force to the weight of the building was obtained in the band-type window with the most rigid behaviour model. So, it can be said that the building with the most rigid behaviour is the band-type window, and the building with the least rigid behaviour is the reference building. In order to determine the damage to the columns and beams, the building should be subjected to a static pushover analysis until the roof displacement demand. Therefore, obtaining the roof displacement demands for each model building in the X direction is necessary. In order to obtain the roof displacement demands of the models, the modal capacity spectrum and the demand spectrum must intersect. The modal capacity spectrum is obtained by axis conversion of the capacity curve obtained as a result of the pushover analysis. The purpose of the axis conversion process is to bring together the capacity curve and the demand spectrum on the same graph. With the axis conversion process, the y-axis of the capacity curve is converted from shear force (V_t) to spectral acceleration (S_{ae}), and the x-axis is converted from roof-floor displacement (Δ) to spectral displacement (S_{de}). Obtaining the loft displacement request for the reference modal building is given in Figure 12.

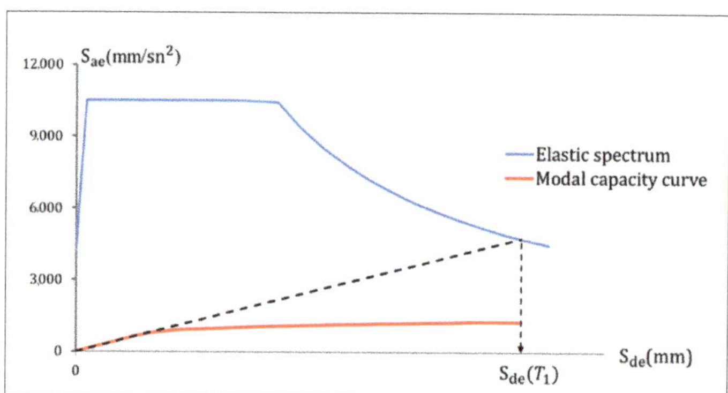

Figure 12. Obtaining the roof displacement demand for the reference model building.

The slope of the initial tangent shown in black in Figure 12 is calculated by the square of the dominant angular frequency of the reference model building (ω^2). The modal displacement demand is found by intersecting the initial tangent drawn from the modal capacity curve with the demand spectrum ($S_{de}(T_1)$). The solid roof displacement demand is obtained by multiplying the modal displacement demand with the modal participation factor. The roof displacement demands obtained for all models are given in Figure 13.

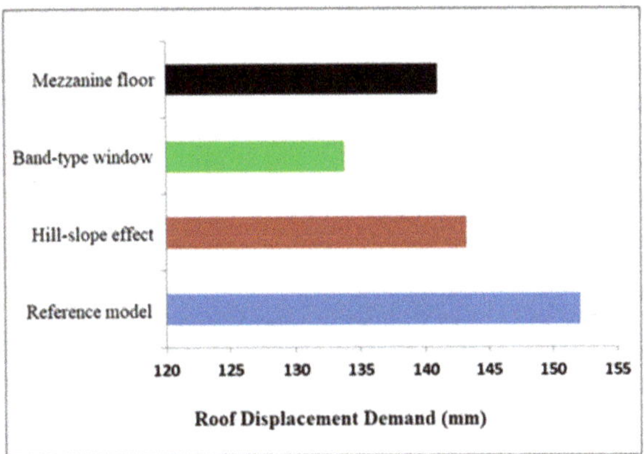

Figure 13. Comparison of roof displacement demands.

Since the structural models with short columns show a more rigid behaviour, the roof displacement demands are lower than the reference model. The band-type window model with the most rigid behaviour obtained the lowest roof displacement demand. Each of the model buildings was subjected to static pushover analysis until the target displacement demand in order to determine the damage in the structural elements. The blue-colored plastic hinge formed in the structural elements, the plastic rotation occurred in the section as a result of the static push-over analysis remains between the limits given for the Damage Limitation (DL) and the Controlled Damage (CD) in TBEC-2018, so it is in the obvious damage zone. The turquoise-coloured plastic hinge indicates that the plastic rotation in the section is between the Controlled Damage (CD) and Prevention of Collapse (PC) plastic rotation limits, that is, in the advanced damage zone. Section damage limits and specified zones are given in Figure 14. The results for damage situations are shown in Figure 15.

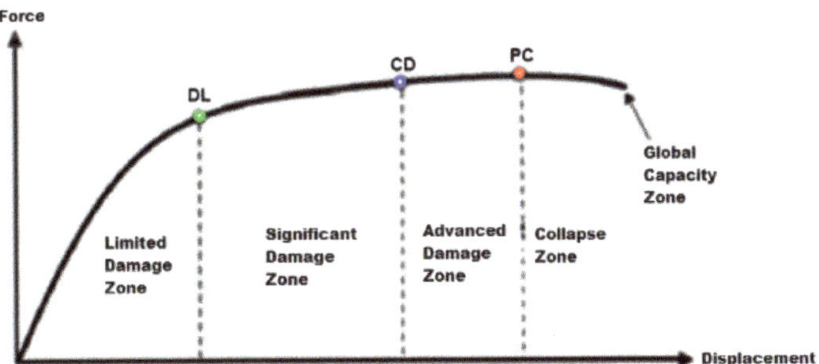

Figure 14. Section damage limits and damage zones [64].

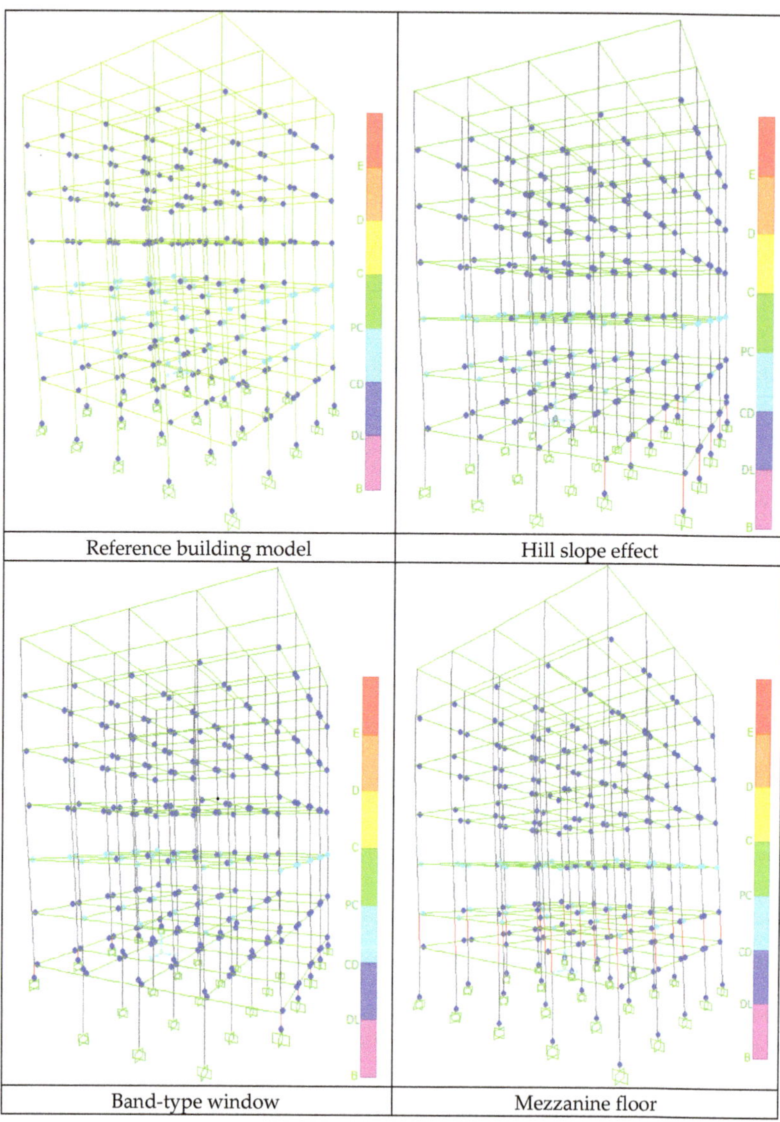

Figure 15. Damages in the structural elements.

In all models, the columns remained in the zone of either obvious damage or damage limitation zone. The number of columns remaining in the obvious and damage limitation zone for the first and second floors of each structural model is given in Figure 16.

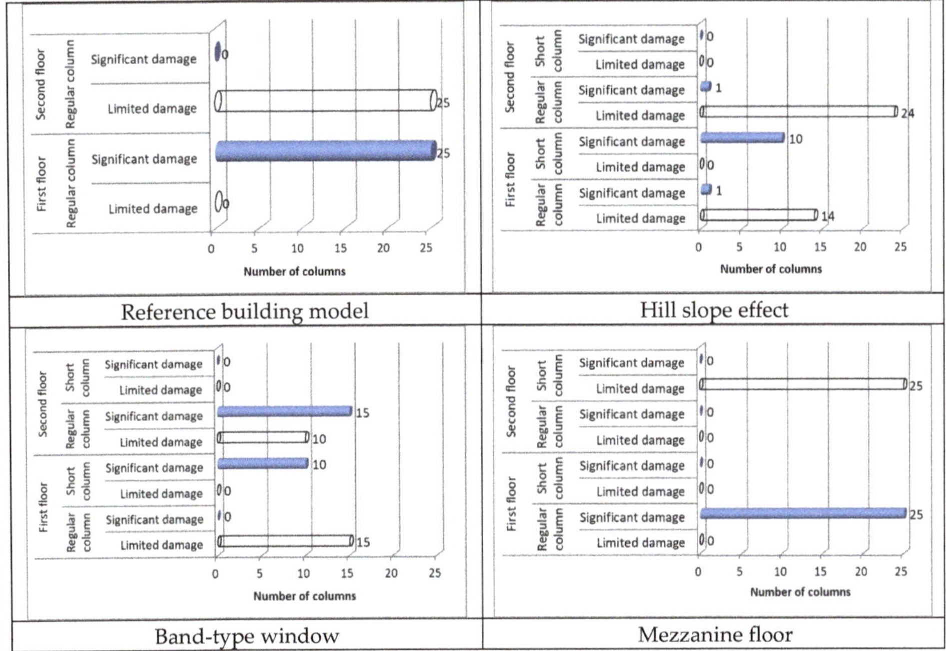

Figure 16. Column damage zones of the structural models.

The damage zones where the columns are located in the reference and mezzanine models are the same. In other words, the short column formed due to the mezzanine floor did not change the damage zone where the column was located. All the columns, which became short columns due to the band-type window and slope effect, remained in the obvious damage zone as in the reference model. However, in the hill-slope effect and band-type window models, all non-short columns (regular column) are in the damage limitation zone in these models, while they are in the obvious damage zone in the reference model with the effect of the short columns on the same storey. As it can be understood from this, since the roof displacement demands of the short column models are lower than the reference model, these models were subjected to less thrust up to the roof displacement, which contributed positively to the damaged zones of the hill-slope effect, the band-type window and the and non-short columns (regular columns).

As a result of the pushover analysis, the plastic rotations occurring in the short columns increased considerably compared to the reference model. However, since these rotations did not exceed the controlled damage plastic rotation limit in the regulation, the damaged zone in the columns remained the same. Therefore, in order to better understand the short-column effect, it would be more appropriate to compare the plastic rotation of the structural elements instead of the damaged zone where the structural elements are located. The rotations on the first-floor columns are given in Figure 17.

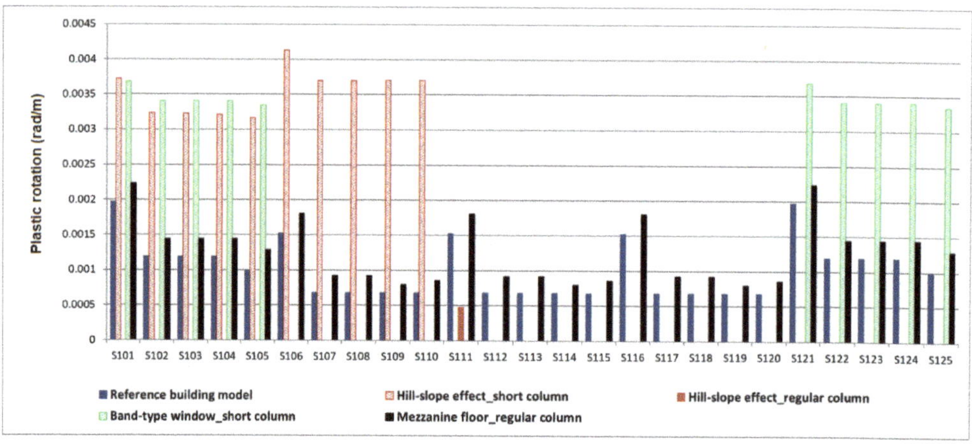

Figure 17. Comparison of plastic rotations.

Columns with zero plastic rotation are not included in Figure 17. For example, since plastic rotation does not occur in any of the non-short columns in the band-type window model, there is no data on these columns in the figure. The highest plastic rotations were obtained in band-type window models. The plastic rotations occurring in the columns in the hill-slope effect model were close to the band-type window. The rotations occurring in the mezzanine floor model were, on average, 24% higher than in the reference model. Especially in buildings with band-type windows and buildings under the effect of hill slope effect, the increase in plastic rotation demands in columns between S101 and S110 has increased the damage level.

5. Conclusions

(a) Earthquake codes in Türkiye have been updated over time, and the last earthquake code came into force in 2019. In this code, designs can be made according to strength and deformation. In this study, short column effects were tried to be revealed for four different structural models using both methods. The structural models were created for three different situations: the hill slope effect, band-type window and mezzanine floor, which may cause short column formation. The results obtained from the structural analyses using SAP2000 were compared with the reference building model with no short columns. The findings obtained from the study are summarised below;

(b) According to the Strength-based Design Approach:

(1) Short columns formed due to the band-type window formation reduced the relative drifts of the ground storey by 88% compared to the reference model. The mentioned ratio decreased to 31% in buildings with a hill-slope effect. However, the mezzanine floor reduced the second-floor relative drifts by 43% compared to the reference model. It is clear that these effects are related to the increase in rigidity with the decrease in column lengths. In addition to these, the maximum relative drifts of all floors and the highest relative drifts occurred in the band-type window. The displacements are inversely proportional to the column's stiffness (especially the length of the column). However, while the stiffness increases in short columns, the shear forces obtained by dividing the moments occurring at the column ends by the column length increase significantly;

(2) It was determined that the relative drifts from the first floor of the building decreased significantly due to the band-type window and slope effect, which caused the second floor to fall into the soft-storey status. The soft storey phenomenon is related to the average relative storey drifts of the floors. If the stiffness irregularity coefficient, which is found by dividing the average storey drift ratio in the floor in

the direction of the earthquake direction, by the average relative storey drift ratio of one upper or lower storey, is greater than 2.0, soft storey irregularity is observed;

(3) According to TBEC-2018, the stiffness irregularity coefficient (n_{ki}) can be more than 2, especially in buildings with band-type windows. This indicates that the irregularity referred to as B2 may be critical for these buildings. A similar situation exists in models with a hill-slope effect, albeit limited. In the case of this irregularity, some limitations have been introduced for the application of the Equivalent Seismic Load Method;

(4) When the changes in the shear forces in the columns are examined, it is seen that the highest increase occurs in the first and fifth axes of the Band-Type buildings and in the first and second axes of the buildings having the hill-slope effect. However, these increases were not at a level that could cause shear damage to the columns. Considering the frequency of stirrups and crossties used, concrete compressive strength and column sections, it is seen that the shear demand does not exceed the shear capacity of the columns.

(c) According to the Deformation-based Design Approach;

(1) Changing stiffness has also changed the period of the buildings and the displacement demand. Here, it is seen that the building with the least ductility requirement is the buildings with Band-Type type short columns. However, it should not be forgotten that the demand for shear force increases in this model inversely proportional to the displacement;

(2) Especially in buildings with the hill slope effect, the increase in plastic rotation demands 4–5 times compared to the reference building has significantly increased the bending damage of the columns in this building. Similarly, in buildings with band-type windows, the plastic rotation demands increase 2–3 times in the relevant axes compared to the reference building. The increase in plastic rotation demands, especially on critical floors, may cause increased column damage and insufficient global performance of the building;

(3) Since the period decreases with the increase in stiffness in short-column models, the roof displacement demands of these models are lower than the reference model. In this situation, it is incorrect to think that the short column models' performance is better than the reference model. Because of the short column formation, the plastic rotation of the short columns increases considerably, even with low roof demands. In the short-column models created within the scope of the study, the plastic rotations occurring in the short columns increased significantly, but since these values did not exceed the controlled damage limit, the performance status of the models did not change. However, in different buildings with different storey numbers, blueprints and material properties, the plastic rotations increasing in this size in short columns can change the performance status of the buildings.

The Strength-Based Design Approach and Deformation-Based Design Approach showed that short column formation is an important parameter in structural behaviour. Although there are many studies on this subject in the literature, it has been shown in this study that the most critical short column performance is in buildings with band windows, and the most innocent short columns are in buildings with mezzanine floors. The buildings considered in this study are modelled according to the strict design rules given in TBEC-2018. Therefore, none of the vulnerabilities mentioned in this article caused shear damage. However, changes in the models related to rotation and displacement demands have affected the structural performance. The authors plan future and ongoing studies to include structural defects in the models in question. For example, it is estimated that the presence of some parameters that reduce the shear capacity, such as the low concrete strength and the absence of stirrup confinement, will affect the global performance more.

Author Contributions: Conceptualization, M.H.A., C.A. and E.I.; methodology, F.A., M.H.A., E.I., F.A. and C.A.; software, H.U., M.H.A. and C.A.; validation, M.H.A., E.I. and C.A.; formal analysis, F.A., E.H. and M.H.A.; investigation, E.I., M.H.A., C.A. and H.U.; resources, E.I., M.H.A. and H.U.; data curation, E.I., M.H.A. and H.U., writing—original draft preparation, E.I., M.H.A., F.A. and E.I.; writing—review and editing, M.H.A., F.A. and E.I.; visualization, E.H.; supervision, M.H.A. and E.I.; project administration, E.I.; funding acquisition, E.H. All authors have read and agreed to the published version of the manuscript.

Funding: This research received no external funding.

Institutional Review Board Statement: Not applicable.

Informed Consent Statement: Not applicable.

Data Availability Statement: Most data are included in the manuscript.

Conflicts of Interest: The authors declare no conflict of interest.

References

1. Yel, N.S.; Arslan, M.H.; Aksoylu, C.; Erkan, İ.H.; Arslan, H.D.; Işık, E. Investigation of the Earthquake Performance Adequacy of Low-Rise RC Structures Designed According to the Simplified Design Rules in TBEC-2019. *Buildings* **2022**, *12*, 1722. [CrossRef]
2. Ademović, N.; Kalman Šipoš, T.; Hadzima-Nyarko, M. Rapid assessment of earthquake risk for Bosnia and Herzegovina. *Bull. Earthq. Eng.* **2020**, *18*, 1835–1863. [CrossRef]
3. Bilgin, H.; Shkodrani, N.; Hysenlliu, M.; Ozmen, H.B.; Isik, E.; Harirchian, E. Damage and performance evaluation of masonry buildings constructed in 1970s during the 2019 Albania earthquakes. *Eng. Fail. Anal.* **2022**, *131*, 105824. [CrossRef]
4. Kaminosono, T. Evaluation method for seismic capacity of existing reinforced concrete buildings in Japan. In *Memoria*; Centro Nacional de Prevención de Desastes (CENAPRED): Mexico City, México; Agencia de Cooperación Internacional (JICA): Tokyo, Japan; Centro para el Desarrollo Regional (UNCRD): Nagoya, Japan, 1992; pp. 44–53.
5. National Research Council of Canada (NRCC). *Manual for Screening of Buildings for Seismic Investigation*; Canadian Standard; National Research Council of Canada: Ottowa, ON, Canada, 1993.
6. Okada, T. Needs to Evaluate Real Seismic Performance of Buildings-Lessons from the 1995 Hyogoken-Nambu Earthquake. *INCEDE Rep.* **1999**, *15*, 225–231.
7. Gulay, F.G.; Kaptan, K.; Bal, E.I.; Tezcan, S.S. P25-scoring method for the collapse vulnerability assessment of R/C buildings. *Procedia Eng.* **2011**, *14*, 1219–1228. [CrossRef]
8. Bülbül, M.A.; Harirchian, E.; Işık, M.F.; Aghakouchaki Hosseini, S.E.; Işık, E. A hybrid ANN-GA model for an automated rapid vulnerability assessment of existing RC buildings. *Appl. Sci.* **2022**, *12*, 5138. [CrossRef]
9. Dogan, G.; Ecemis, A.S.; Korkmaz, S.Z.; Arslan, M.H.; Korkmaz, H.H. Buildings damages after Elazığ, Turkey earthquake on 24 January 2020. *Nat. Hazards* **2021**, *109*, 161–200. [CrossRef]
10. Bektaş, N.; Kegyes-Brassai, O. Development in fuzzy logic-based rapid visual screening method for seismic vulnerability assessment of buildings. *Geosciences* **2023**, *13*, 6. [CrossRef]
11. Ghobarah, A.; Galal, K.E. Seismic rehabilitation of short rectangular RC columns. *J. Earthq. Eng.* **2004**, *8*, 45–68. [CrossRef]
12. Moretti, M.; Tassios, T.P. Behaviour of short columns subjected to cyclic shear displacements: Experimental results. *Eng. Struct.* **2007**, *29*, 2018–2029. [CrossRef]
13. ASCE. *Seismic Evaluation and Retrofit of Existing Buildings*; ASCE/SEI, 41-17; ASCE: Reston, VA, USA, 2017.
14. Arslan, M.H.; Korkmaz, H.H. What is to be learned from damage and failure of reinforced concrete structures during recent earthquakes in Turkey? *Eng. Fail. Anal.* **2007**, *14*, 1–22. [CrossRef]
15. Tsantilis, A.V.; Triantafillou, T.C. Innovative seismic isolation of masonry infills using cellular materials at the interface with the surrounding RC frames. *Eng. Struct.* **2018**, *155*, 279–297. [CrossRef]
16. Moehle, J.P.; Mahin, S.A. Observations on the behavior of reinforced concrete buildings during earthquakes. *ACI Spec. Publ.* **1991**, *127*, 67–90.
17. Wang, Y.Y. Lessons learned from the "5.12" Wenchuan earthquake: Evaluation of earthquake performance objectives and the importance of seismic conceptual design principles. *Earthq. Eng. Eng. Vib.* **2008**, *7*, 255–262. [CrossRef]
18. Yen, W.P.; Chen, G.; Buckle, I.; Allen, T.; Alzamora, D.; Ger, J.; Arias, J.G. *Post-Earthquake Reconnaissance Report on Transportation Infrastructure: Impact of the 27 February 2010, Offshore Maule Earthquake in Chile*; FHWA-HRT-11-030; Federal Highway Administration: Richmond, VA, USA, 2011.
19. Meral, E. Investigation of short column effects in reinforcement concrete buildings. *Int. J. Eng. Res. Dev.* **2019**, *11*, 515–527.
20. Şeker, M.; Bedirhanoglu, İ. Investigation of shear behaviour of reinforced concrete captive column with low strength concrete. *Dicle Univ. J. Eng. (DUJE)* **2019**, *10*, 385–395.
21. Balik, F.S.; Bahadir, F. Investigation of short column behaviors at 1/5 scaled reinforced concrete frames using different strengthening methods. *J. Inst. Sci. Technol.* **2019**, *9*, 433–445. [CrossRef]

22. Işık, E.; Karasin, İ.B.; Ulu, A.E. Investigation of earthquake behavior of reinforced-concrete buildings built on soil slope. *Eur. J. Sci. Eng.* **2020**, *20*, 162–170.

23. Haji, M.; Naderpour, H.; Kheyroddin, A. Experimental study on influence of proposed FRP-strengthening techniques on RC circular short columns considering different types of damage index. *Compos. Struct.* **2019**, *209*, 112–128. [CrossRef]

24. Chen, C.Y.; Liu, K.C.; Liu, Y.W.; Huang, W.J. A case study of reinforced concrete short column under earthquake using experimental and theoretical investigations. *Struct. Eng. Mech.* **2010**, *36*, 197–206. [CrossRef]

25. Moretti, M.; Tassios, T.P. Behavior and ductility of reinforced concrete short columns using global truss model. *ACI Struct. J.* **2006**, *103*, 319–327.

26. Çelebi, E.; Aktas, M.; Çağlar, N.; Özocak, A.; Kutanis, M.; Mert, N.; Özcan, Z. 23 October 2011 Turkey/Van–Ercis earthquake: Structural damages in the residential buildings. *Nat. Hazards* **2013**, *65*, 2287–2310. [CrossRef]

27. Alih, S.C.; Vafaei, M. Performance of reinforced concrete buildings and wooden structures during the 2015 Mw 6.0 Sabah earthquake in Malaysia. *Eng. Fail. Anal.* **2019**, *102*, 351–368. [CrossRef]

28. Pozos-Estrada, A.; Chávez, M.M.; Jaimes, M.Á.; Arnau, O.; Guerrero, H. Damages observed in locations of Oaxaca due to the Tehuantepec Mw8. 2 earthquake, Mexico. *Nat. Hazards* **2019**, *97*, 623–641. [CrossRef]

29. Liu, C.; Fang, D.; Zhao, L. Reflection on earthquake damage of buildings in 2015 Nepal earthquake and seismic measures for post-earthquake reconstruction. *Structures* **2021**, *30*, 647–658. [CrossRef]

30. Arslan, M.H. An evaluation of effective design parameters on earthquake performance of RC buildings using neural networks. *Eng. Struct.* **2010**, *32*, 1888–1898. [CrossRef]

31. Işık, E.; Ulu, A.E.; Aydin, M.C. A case study on the updates of Turkish rapid visual screening methods for reinforced-concrete buildings. *Bitlis Eren Univ. J. Sci. Technol.* **2021**, *11*, 97–103.

32. Bektaş, N.; Kegyes-Brassai, O. Conventional RVS methods for seismic risk assessment for estimating the current situation of existing buildings: A state-of-the-art review. *Sustainability* **2022**, *14*, 2583. [CrossRef]

33. Ay, B.Ö.; Erberik, M.A. Seismic risk assessment of low-rise and mid-rise reinforced concrete structures in Turkey. In Proceedings of the Sixth National Conference on Earthquake Engineering, Istanbul, Turkey, 16–20 October 2007; pp. 37–48.

34. Doğangün, A. *Betonarme Yapıların Hesap ve Tasarımı*; Birsen Yayınevi: Istanbul, Turkey, 2013.

35. Işık, E. Size effects of columns on buckling. In Proceedings of the 5th International Science Technology and Engineering Conference, St. Petersburg, Russia, 6–8 April 2015.

36. Yamada, M.; Furui, S. Shear resistance and explosive cleavage failure of reinforced concrete members subjected to axial load. In Proceedings of the 8th International Congress IABSE, New York, NY, USA, 9–14 September 1968; pp. 1091–1102.

37. Moretti, M.L.; Tassios, T.P. Design in shear of reinforced concrete short columns. *Earthq. Struct.* **2013**, *4*, 265–283. [CrossRef]

38. Bal, İ.E.; Tezcan, S.S.; Gülay, G.F. P25 rapid screening method to determine the collapse vulnerability of r/c buildings. In Proceedings of the Sixth National Conference on Earthquake Engineering, Istanbul, Turkey, 16–20 October 2007; pp. 661–674.

39. Doğan, T.P.; Kızılkula, T.; Mohammadi, M.; Erkan, İ.H.; Kabaş, H.T.; Arslan, M.H. A comparative study on the rapid seismic evaluation methods of reinforced concrete buildings. *Int. J. Dis. Risk Reduct.* **2021**, *56*, 102143. [CrossRef]

40. Çağatay, İ.H. Investigation of parameters affecting short column of buildings. In Proceedings of the Sixth National Conference on Earthquake Engineering, Istanbul, Turkey, 16–20 October 2007; pp. 29–236.

41. İnan, T.; Korkmaz, K. Investigation of vertical structural irregularities. *Erciyes Üniversitesi Fen Bilim. Enstitüsü Derg.* **2012**, *28*, 240–248.

42. Guevara, L.T.; Garcia, L.E. The captive-and short-column effects. *Earthq. Spectra* **2005**, *21*, 141–160. [CrossRef]

43. Dowrick, D.J. *Earthquake Resistant Design for Engineers and Architects*; John Wiley & Sons: Chichester, NH, USA, 1987.

44. Colomb, F.; Tobbi, H.; Ferrier, E.; Hamelin, P. Seismic retrofit of reinforced concrete short columns by CFRP materials. *Compos. Struct.* **2008**, *82*, 475–487. [CrossRef]

45. Işık, E. The effects of 23 October 2011 Van earthquake on near-field and damaged on structures. *Int. Anatolia Acad. Online J. Sci. Sci.* **2014**, *2*, 10–25.

46. Karakostas, C.; Lekidis, V.; Makarios, T.; Salonikios, T.; Sous, I.; Demosthenous, M. Seismic response of structures and infrastructure facilities during the Lefkada, Greece earthquake of 14 August 2003. *Eng. Struct.* **2005**, *27*, 213–227. [CrossRef]

47. Aliaari, M.; Memari, A.M. Analysis of masonry infilled steel frames with seismic isolator subframes. *Eng. Struct.* **2005**, *27*, 487–500. [CrossRef]

48. Caglar, N.; Vural, I.; Kirtel, O.; Saribiyik, A.; Sumer, Y. Structural damages observed in buildings after the 24 January 2020 Elazığ-Sivrice earthquake in Türkiye. *Case Stud. Constr. Mater.* **2023**, *18*, e01886. [CrossRef]

49. Education and Job. Available online: http://educationandjob1.blogspot.com.tr/2014_05_01_archive.html (accessed on 6 May 2022).

50. Jayaguru, C.; Subramanian, K. Seismic behaviour of partially infilled RC frames retrofitted using GFRP laminates. *Exp. Tech.* **2012**, *36*, 82–91. [CrossRef]

51. Eleftheriadou, A.K.; Karabinis, A.I. Seismic vulnerability assessment of buildings based on damage data after a near field earthquake (7 September 1999 Athens-Greece). *Earthq. Struct.* **2012**, *3*, 117–140. [CrossRef]

52. Aydınoğlu, M.N. A response spectrum-based nonlinear assessment tool for practice: Incremental response spectrum analysis (IRSA), ISET. *J. Earthq. Technol.* **2007**, *44*, 169–192.

53. Kutanis, M.; Boru, O.E. The need for upgrading the seismic performance objectives. *Earthq. Struct.* **2014**, *7*, 401–414.

54. İlki, A.; Celep, Z. Betonarme yapıların deprem güvenliği. In Proceedings of the 1. Türkiye Deprem Mühendisliği ve Sismoloji Konferansı, Ankara, Turkey, 1–14 October 2011.

55. Fajfar, P. Capacity spectrum method based on inelastic demand spectra. *Earthq. Eng. Struct. Dyn.* **1999**, *28*, 979–993. [CrossRef]
56. Gholizadeh, S. Performance-based optimum seismic design of steel structures by a modified firefly algorithm and a new neural network. *Adv. Eng. Softw.* **2015**, *81*, 50–65. [CrossRef]
57. Işık, E.; Kutanis, M. Performance based assessment for existing residential buildings in Lake Van basin and seismicity of the region. *Earthq. Struct.* **2015**, *9*, 893–910. [CrossRef]
58. Chopra, A.K.; Goel, R.K. A modal pushover analysis procedure for estimating seismic demands for buildings. *Earthq. Eng. Struct. Dyn.* **2002**, *31*, 561–582. [CrossRef]
59. Foti, D. A new experimental approach to the pushover analysis of masonry buildings. *Comput. Struct.* **2015**, *147*, 165–171. [CrossRef]
60. Hsiao, F.P.; Oktavianus, Y.; Ou, Y.C. A pushover seismic analysis method for asymmetric and tall buildings. *J. Chin. Inst. Eng.* **2015**, *38*, 991–1001. [CrossRef]
61. Casapulla, C.; Argiento, L.U. The comparative role of friction in local out-of-plane mechanisms of masonry buildings. Pushover analysis and experimental investigation. *Eng. Struct.* **2016**, *126*, 158–173. [CrossRef]
62. Gholipour, M.; Alinia, M.M. Considerations on the pushover analysis of multi-story steel plate shear wall structures. *Period. Polytech. Civ. Eng.* **2016**, *60*, 113–126. [CrossRef]
63. Wang, C.; Xiao, J.; Liu, W.; Ma, Z. Unloading and reloading stress-strain relationship of recycled aggregate concrete reinforced with steel/polypropylene fibers under uniaxial low-cycle loadings. *Cem. Concr. Compos.* **2022**, *131*, 104597. [CrossRef]
64. *TBEC-2018*; Turkish Building Earthquake Code. T.C. Resmi Gazete: Ankara, Turkey, 2018.
65. Available online: https://tdth.afad.gov.tr (accessed on 2 February 2022).
66. *SAP2000*; Integrated Software for Structural Analysis & Design. Computers & Structures, Inc.: Berkeley, CA, USA, 2011.

 *buildings*

Article

Experimental Investigations and Seismic Assessment of a Historical Stone Minaret in Mostar

Faris Trešnjo [1], Mustafa Humo [2], Filippo Casarin [3] and Naida Ademović [4],*

1 Faculty of Civil Engineering, University "Džemal Bijedić" in Mostar, 88000 Mostar, Bosnia and Herzegovina
2 Freelance Structural Engineer, 88000 Mostar, Bosnia and Herzegovina
3 Expin srl, 35138 Padova, Italy
4 University of Sarajevo-Faculty of Civil Engineering, 71000 Sarajevo, Bosnia and Herzegovina
* Correspondence: naida.ademovic@gf.unsa.ba or naidadem@yahoo.com

Abstract: Minarets, tall structures, connected or not to the mosque attract attention due to their specific architectural features. Vulnerability to seismic damage has been witnessed throughout history on tall and slender structures after earthquake ground motions. In that respect, it is of the utmost importance to investigate the dynamic characteristics and resilience of historical stone minarets. This paper aims to provide the results of an on-site dynamic investigation of a stone minaret in Mostar and deliver its seismic assessment. The minaret is part of the Tabačica mosque built at the turn of the 16th and 17th century in the City of Mostar, Bosnia and Herzegovina. The on-site investigation comprised dynamic identification of the minaret by ambient vibration testing and qualitative estimation of the masonry wall by sonic pulse velocity testing. Besides the modal analysis a time-history analysis was performed by using the Applied Element Method (AEM), considered an appropriate tool for assessing the behavior of historic masonry structures. A good match is found between the first natural frequency obtained by the on-site investigation and the modal analysis which is a solid basis for further seismic assessment of the minaret as a slender tower-like structure. The concentration of stresses is observed at the transition zones.

Keywords: minaret; seismic assessment; modal analysis; stone masonry; historic structures; applied element method; dynamic testing

Citation: Trešnjo, F.; Humo, M.; Casarin, F.; Ademović, N. Experimental Investigations and Seismic Assessment of a Historical Stone Minaret in Mostar. *Buildings* **2023**, *13*, 536. https://doi.org/10.3390/buildings13020536

Academic Editors: Rajesh Rupakhety and Dipendra Gautam

Received: 14 December 2022
Revised: 16 January 2023
Accepted: 22 January 2023
Published: 15 February 2023

1. Introduction

Preserving cultural heritage structures as well as passing knowledge to future generations is of the utmost importance for the history of a certain region regarding the social life of the communities of a given period and the construction technologies utilized at that time. Many of these structures have been exposed to various extreme natural hazards (wind, earthquakes, etc.), inadequate maintenance, and sometimes they have even been neglected. Regardless of all of these negative impacts and influences, they have survived which shows the magnificent engineering work conducted by famous masons centuries ago without the application of any high-level technology [1].

Assessment of cultural heritage buildings has gained a lot of attention in the last decades and numerous researchers have dedicated their investigations in this direction. Analysis of these structures is a rather complex and multidisciplinary task as it requires the involvement of various engineers, architects, conservators, historians, etc. Once collected, the data has to be analyzed by a team of experts from different fields to make a correct decision regarding the potential preservation measures and strengthening techniques. Most of these structures, constructed several centuries ago are made of traditional materials such as stone, brick, and adobe (masonry buildings), and are vulnerable to seismic effects. It is important to state that at the time of construction there were no seismic codes, but structures were constructed based on the knowledge of individual masons. Earthquake destruction triggered the development of seismic codes and their enforcement in different countries. With the development of different software packages, numerical modeling of historic masonry buildings became possible. Each modeling technique has its advantages and disadvantages, as well as field of application. According to D'Altri et al. [2], there are four main classes of numerical strategies for masonry structures. The first class is when the structure is modeled utilizing a block-by-block definition. Masonry is conceived as an assembly of blocks linked together by mortar joints. Page [3] was the first researcher who applied nonlinear analysis to masonry structures. The advantage lies in the ability to model the actual masonry texture and the structural detail; however, in historical structures this is commonly an unknown parameter. This method is typically used for small-scale experimental tests. However, it has been applied for several historical structures [4,5], but this is not common due to its high computational time. This method has been further developed leading to five sub-classes: Interface element-based approaches; Contact-based approaches; Textured continuum-based approaches; Block-based limit analysis approaches; and Extended finite element (FE) approaches. For more details, please see D'Altri et al. [2]. The second class represents the continuum models with the application of adequate homogeneous constitutive laws. As the mesh is not connected to the size of masonry elements the computational time is greatly reduced. Description of the masonry's constitutive laws could be undertaken by direct approaches or homogenization procedures and multiscale approaches. The direct approach was used for modeling the historical church by Pantò et al. [6] and complex masonry structures by Abbati et al. [7]. The smeared crack, damage, and plastic-damage models found a massive application in the assessment of masonry structures due to their implementation in various software, and the fact that only a few mechanical characteristics of masonry are needed as input, significantly reducing computational time. In this way anisotropy of masonry is neglected but masonry is modeled as an isotropic material. Several researchers applied this modeling procedure to model historic towers [8–12]. Homogenization procedures and multiscale approaches are mainly used for small-scale elements. The shortcoming of models based on the finite element method is that it is not possible to simulate the mechanical interaction between multiple bodies, which is important in the analysis of structures exposed to impact loads as well as in the analysis of the progressive collapse of structures. With macroelement models, the structure is composed of piers and spandrels, and rigid elements. There are generally two approaches—equivalent beam-based and spring-based approaches. Finally, in the geometry-based models, the masonry structure is determined as a rigid body. Additionally, a discrete element method (DEM) as a micro modeling approach is able to provide adequate masonry representation on the level of masonry elements and joints. In that respect, only limited research is available on the application of this methodology for masonry structures, such as the work done by de Felice [13], Pulatsu et al. [14], and Lemos [15]. The mechanical behavior of block-based masonry walls is well represented by DEM as presented in [16–18]. The application of DEM was used as well by [19] for modeling the infilled masonry frames. Dimitri and Zavarise [20] explored the application of DEM in Gothic buttresses of trapezoidal and stepped shapes. It has been concluded that this procedure is useful in the prediction process of the first-order seismic behavior of unreinforced masonry (URM) structures in terms of collapse load and collapse mechanism. The model was able to predict in an accurate manner the dynamic response of the structure. Alexakis and Makris [21] elaborated on the application to masonry arches. An interesting work was presented by [22] for an aqueduct in Portugal and a multi-leaves arch-tympana of a church in Italy. Models based on the DEM were developed for the analysis of problems in which there is a mechanical interaction between multiple bodies that can have large displacements and rotations. The main characteristic of the DEM is the modeling of the structure as a set of discrete elements interconnected by contact elements, which enabled its application in the analysis of structures with complex heterogeneous compositions (masonry structures). This approach makes it possible to simulate the collapse of the structure due to rotation, sliding, and impact loading. A major disadvantage of this method is its inability to describe the state of stress and deformation within discrete elements, which is especially important when analyzing the appearance, development and propagation of cracks.

A new modeling philosophy has emerged which combines both the finite element method (FEM) and the discrete element method (DEM), and is known as the applied element method (AEM). It was originally developed for the simulation of structural demolition of reinforced concrete buildings in a controlled manner [23–25]. This new method uses the advantages of both approaches, meaning the accuracy of the FEM until elements separate and DEM when the elements become detached. The advantage of this method is that it is able to automatically simulate the separation of the elements and follow up all stages until the collapse of the structure and can predict the falling debris. With more than two decades of continuous research and development, AEM has been proven to be the only method that can track structural collapse behavior passing through all stages of loading; elastic, crack initiation and propagation in tension-weak materials, element separation, element collision (contact), etc. The advantage is seen in its implicit time-integration frameworks as opposed to DEM which uses explicit solvers and direct integration. Malomo et al. [26] utilized AEM for the assessment of a full-scale URM prototype representative of a typical Dutch detached house which was experimentally investigated on a shaking table. A good relation was obtained between the experimental data and the numerical model. The general positive impression of the behavior of the numerical model was additionally supported by the damage evolution. The model was capable of capturing the fractures' propagation. The global behavior of the structure was correctly captured in the model. Hadhoud et al. [27] modeled a Margherita Palace with the application of the AEM modeling procedure. The heritage masonry Margherita Palace was built two centuries ago and was heavily damaged by the strong 2009 L'Aquila earthquake. The model was able to capture the damage to the structure caused by the earthquake and the damage pattern was very similar to the one observed on the structure after the earthquake. The weakest point of the structure was detected leading to the need for remedial and strengthening measures. Diana et al. [28] used the AEM model to validate the results obtained from the simplified approach of the LV1-church form (the mechanism related to the belfry) and by a kinematic approach. It was noted that the non-linear verification was reasonable and true with the damage pattern obtained by the AEM model and then by the simplified procedures. Evidently, the AEM is capable of simulating both the in-plane and out-of-plane failure modes in masonry structures and their individual components exposed to either static or dynamic actions [29]. As many buildings throughout the world are constructed as masonry-infilled reinforced concrete structures, their seismic performance with the application of the AEM models was explored [30]. A structure with eight storeys was exposed to time history analysis taking into account four different configurations: one including a soft storey, then this soft storey was retrofitted, an RC bare frame, and RC with masonry infill in all floors. The AEM modeling was used as well to evaluate strengthening measures for a URM building [31].

After the 2011 Van earthquake, an extensive campaign was conducted in order to determine the dynamic characteristics of eleven minarets in Turkey [32]. Three techniques and three excitation levels were used for the determination of the dynamic characteristics. Seven frequencies were determined with the application of FDD, EFDD, and SSI techniques. In this specific case, seven minarets were modeled using a FEM without the mosque. Research on minarets, but this time those beside historic, contemporary, and combined minarets was conducted by Çaktı et al. [33]. Additionally, the Mihrimah minaret was modeled utilizing a discrete element method and ten various time history accelerations were used for non-linear dynamic analysis [33]. The Grand Mosque in Bursa was modeled with the application of FEM and seismic performance analyses were conducted on four different models. Model 1 took into account most details while the most simplified model was the fourth one. It has been noted that the stairs have to be taken into account as their influence cannot be neglected in the numerical analysis, although other elements could be neglected [34]. How geometrical characteristics affect the dynamic behavior of the minarets was investigated by Livaoğlu et al. [35]. The dynamic characteristics were determined by the ambient vibration tests and with the application of Enhanced Frequency Domain Decomposition (EFDD) and then this data was used for validation of the numerical model. In total, seven minarets in Bursa were investigated. A simple equation was proposed for the determination of the natural period which is a function of the cross-sectional area, outer diameter, inner diameter, height, and material density of the minaret as a result of OMA investigations and numerical models. However, additional investigations have to be done in order to take into account the stiffness of the boot, the ratio between different segments of the minaret, and the boundary conditions, as their influence cannot be neglected and have to be taken into account in the calculations [35].

Döven et al. [36] investigated the Yeşil Mosque (Green Mosque) in Kütahya and the changes in the dynamic behavior of the structure in case the balcony of the minaret is open or covered. A comprehensive study of the Mustafa Pasha Mosque located in the central part of the city of Skopje (Macedonia) was performed in 2011 by Portioli et al. [37]. The model of the mosque was scaled to 1:6 with respect to the length and a shaking table test was performed. A numerical model of the mosque was created. The structure was exposed to various seismic loads and the results of the numerical models were calibrated and validated with the results obtained on the shaking table test. Details of the experimental investigation can be found in Krstevska et al. [38] and Tashkov et al. [39]. Additionally, a strengthening procedure was proposed to avoid the propagation of cracks in the directions of the maximum principal stresses, out-of-plane failure, and other types of damage that were observed as results of the numerical modeling; for details, please see Krstevska et al. [38]. A vulnerability assessment of the Al-Askari Shrine exposed to gravity loading as well as seismic loads with the application of the finite element method was conducted by Yekrangnia and Mobarake [40]. The investigation of the gravity loads was needed to check the capacity of the supporting piers to take over the load of the newly constructed dome, as the original dome was destroyed by blast loading. In view of possible earthquake damage to the Al-Askari Shrine, it was required to conduct this analysis as well. Due to the complex geometry and material characteristics, macromodeling was carefully chosen as the modeling method. The whole mosque and the minaret were elaborated and again one of the most critical parts to be identified was the lower part of the minaret; thus, cracks might propagate through the base of the minaret. Calculations due to seismic actions revealed the vulnerability of the structure in the form of crack widening in the piers, leading to asymmetrical movement and finally to the formation of diagonally rising cracks in the dome. Additionally, it was the new minarets that were identified as the most vulnerable elements requiring strengthening (Yekrangnia and Mobarake [40]).

The most remarkable and glorious historic fired brick 12th-century Islamic structure, Gonbad-e Qābus in Iran, was assessed by Ebrahimiyan et al. [41]. There was very little data regarding the material properties; only the average compressive strengths of three masonry bricks were available. Behavior in compression and tension was chosen to be represented by the Concrete-damaged plasticity model. As in the case of Al-Askari Shrine, macromodeling was selected as an adequate modeling procedure. In this specific case, the equivalent static force method was used for obtaining the nonlinear behavior of the structure exposed to gravity and seismic loads. The results of the calculation showed that the collapse-prevention level is satisfied meaning that there is no fear of collapse and structure overturning. Nevertheless, it was noted that adequate measures should be considered regarding the strengthening of the structure (Ebrahimiyan et al. [41]). Five timber minarets in the city of Sakarya (Turkey), located in one of the most seismically prone zones (North Anatolian Fault Zone) were researched by Bağbancı et al. [42]. The two most devasting earthquakes which caused significant damage to the structures or even collapse were the 1999 Düzce (Ms = 7) and Kocaeli earthquakes (Ms = 7.8), the 1967 Sakarya-Akyazı earthquake (Ms = 7.8), and the 1943 Sakarya-Hendek earthquake (Ms = 6.6) [43]. As a result of experimental tests and measurements conducted on the site, Bağbancı et al. [42] proposed a new equation for obtaining the natural frequency; besides being a function of height, cross-sectional area, and material, this became a function of construction techniques of the outer walls of the minarets. The application of this proposed new equation led to an error of less than 10% which may be considered acceptable.

Işık et al. [44] determined the material characteristics of the local Bitlis stone that was used for the construction of the Five minarets (Ulu Mosque Minaret, Şerefiye Mosque Minaret, Meydan Mosque Minaret, Kalealtı Mosque Minaret, and Gökmeydan Mosque Minaret); these represent one of the most famous treasures of Bitlis (Turkey), referenced in the famous folk lyrics "Five Minarets in Bitlis". Once the material parameters were determined, modal analysis and the minarets' seismic performance were obtained with the application of the Turkish Building Earthquake Code. Further work and investigation of the Ulu Mosque's minaret in Bitlis (Turkey) was performed by Işık et al. [45]. The assessment of the minaret was conducted by applying the current and previous seismic code in Turkey, enabling the comparisons of the minaret's behavior. As well, the calculation was performed for five different geographical locations in the same earthquake hazard zone and four different earthquake ground motion levels (DD-1, DD-2, DD-3, and DD-4), with the probability of exceedance in 50 years. As in previous cases, the macromodelling technique was selected.

A comprehensive investigation and seismic assessment of a reconstructed mosque, which was originally built in the early 1800s within the Bigali castle was done by Gunes et al. [46]. After the on-site (visual inspection) and laboratory tests (mechanical, chemical, and mineralogical analyses) to obtain as much as possible information regarding the structure and its ruins, it was necessary to propose a reconstruction procedure and perform adequate numerical analysis. Due to the specific features of this case study and its importance three different levels of structural analysis were conducted (linear, kinematic, and pushover analysis). In this case also, the minaret was modeled separately from the mosque. The proposed methodology can be applied to the evaluation of similar cultural heritage buildings. Devastation and damage to the minarets after the earthquake which hit the Sivrice district of Elazig city on 24 January 2020 was investigated by Yetkin et al. [47]. The sections where the damage occurred in the minarets were determined, and the reasons for this damage were evaluated. Commonly, the soil effects in the calculations are neglected, leading to potentially incorrect results in soft and loose soils, although this approach may provide adequate results in stiff soil types [48]. Haciefendioğlu et al. [49] concluded that the soil-structure interaction affects the dynamic characteristic of the scaled reinforced concrete minaret. As the soil type changed from sand to clay-gravel the frequency of the minaret increased. Altiok and Demir [50] investigated the effect of the soil structure interaction on the seismic behavior of the Lala Mehmet Pasha minaret, located in Manisa (Turkey). They used the OMA method to calibrate the model, leading to a significant reduction of the initial Young's modulus, and the difference in the first mode frequency of only 0.03%. Linear type history (LTH) and nonlinear type history (NLTH) analyses were conducted to obtain a reliable formulation of the damage pattern on the structure exposed to significant earthquake motion. The sixth-largest mosque in the world, Faisal Mosque, located in Islamabad (Pakistan) was studied by Akhlaq et al. [51]. The four minarets are completely separated from the mosque. The total length of the reinforced concrete minaret is 90 m. Application of the OMA was utilized and the three identification techniques were used to determine the modal properties. This data was used for calibration of the FEM model by manually correcting the updating of the modulus of elasticity and correcting the weight of the nonstructural elements. Işık et al. [52] proposed a new empirical formula for determination of the minaret's periods as a function of heights. Additionally, estimation of the ten natural periods as a function of various materials (thirteen taken into account) were estimated with the application of the Artificial Neural Network (ANN) model. There was excellent agreement between the values obtained by experiment and the values estimated by ANN (less parameters were used).

There are several aims of the paper. For the first, the time horizontal to vertical spectral ratio (HVSR) method was utilized to determine the dynamic characteristics of the minaret. These results were compared with the frequently used operational modal analysis

(OMA) and a certain consistency was noted. Additionally, the quality of the three-leaf stone masonry used for the mosque building and the minaret base was determined using the sonic velocity test revealing its solid quality. The goal was to determine the seismic behavior and performance of one of the significant minarets in Bosnia and Herzegovina. For this aim, for the first time, the applied element method was used for the nonlinear seismic assessment of the minaret located in Mostar, Bosnia and Herzegovina.

The paper is organized into several sections. Section 2 is devoted to the historical background of the Tabačica mosque as one of the significant and unique monuments of Islamic sacred architecture in Mostar. Geometrical and architectural features are presented in detail together with experimental results for the "tenelija" stone. Section 3 is dedicated to the new experimental study that was conducted in 2022. Utilizing non-destructive techniques, the experimental work aimed to determine the natural frequencies and damping ratios. This was done with the application of two methods, the horizontal-to-vertical spectral ratio (HVSR) method and the operational modal analysis method (OMA). The obtained parameters are used later for the calibration of the numerical model. Secondly, to determine the quality of the masonry, the sonic velocity test (SPVT) was performed. Section 4 is devoted to the modeling of the minaret. For the first time, a nonlinear model that is based on AEM, a derivative of the FEM and DEM, is utilized. The connection between the elements in the AEM is provided by springs, meaning that the plastic hinges come as a result of the analysis and not as input as in FEM. The first modal analysis was conducted and a comparison was made with the experimental results. As previously mentioned, this was used for the validation of the model. Nonlinear time history analysis of the minaret exposed to artificial earthquakes is described in the second part of Section 4. Maximum lateral displacement in X and Y directions, maximum and minimum principal stress contours under the earthquake ground motion for the three sets are presented. The article concludes that there is a good consistency between the experimentally obtained frequency and the numerical model. The sonic velocity test revealed that the masonry is of solid quality. Once again, the transitional regions were identified as the most vulnerable points once the minaret is exposed to seismic loading.

2. Historical Background of Tabačica Mosque

Traditional minarets in Bosnia and Herzegovina dating from the period of Ottoman rule from the 15th to 19th century can be categorized into three groups or types. Type I are stone masonry minarets of predominantly polygonal outer shape; type II are stone masonry minarets of predominantly squared outer shape and type III are timber structured minarets of polygonal outer shape [53]. The most representative and common type is the type I which is the subject of the research presented in this paper. Minarets of this type in Bosnia and Herzegovina are usually constructed with one balcony, as for the minaret in Tabačica mosque (Figure 1a). The cross-section of this minaret is presented in Figure 1b. Similar-looking minarets with more than one balcony also made of stone masonry may be found in other historic areas of the Ottoman empire [54].

As far as it is known, 36 mosques and masjids were built in Mostar, and they typically bear the names of their founders, except Tabačnica. It was built on the river Radobolja, vaulted with two stone vaults, which is why it is known as "the mosque where the imam is on dry land and the congregation is on water". This mosque was built at the turn of the 16th and 17th centuries at the request of Hadži-Kurt, representative of one of the oldest families in Mostar. It is located on the right bank of the Neretva, about a hundred meters from the Old Bridge, next to "Tabhana", a neighborhood where leather was tanned, processed, and sold.

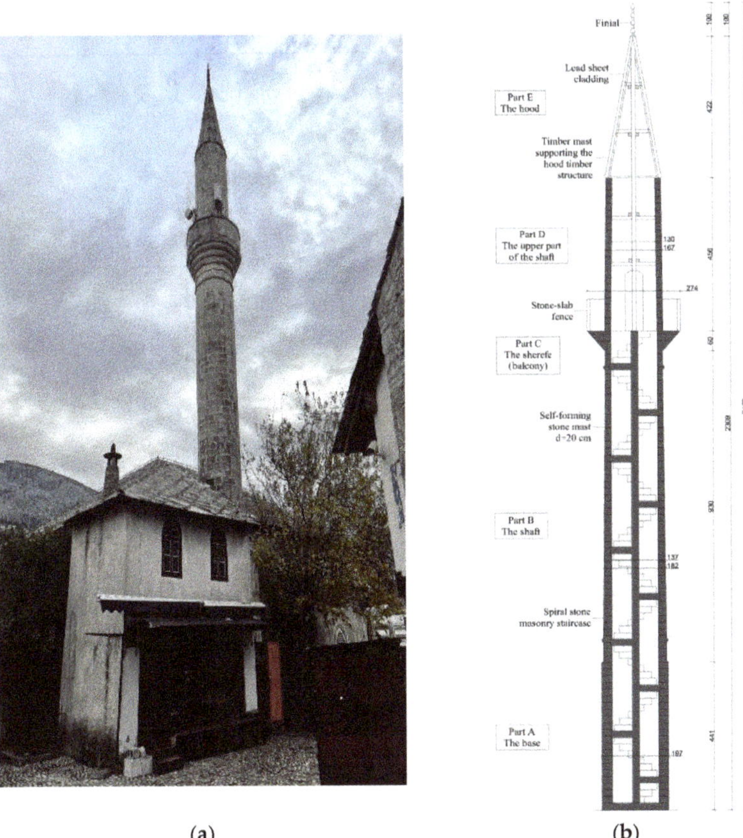

Figure 1. Tabačica mosque in Mostar (**a**) Photo of the mosque from 2022 (author's photo); (**b**) cross-section of Tabačica minaret.

The Tabačica mosque is one of the most important monuments of Islamic sacred architecture in Mostar. Its architecture differs in some details from the architecture of other Mostar mosques. The architectural specificity was mostly influenced by the place and environment in which it was built (Figure 1a). It is located directly in front of the "Tabhana", a leather processing complex, and was mostly visited by members belonging to that society. In Bosnia and Herzegovina, leather tanners built mosques in front of "tabhanas" in five cities (Sarajevo, Mostar, Banja Luka, Visoko, and Tešanj). The name Tabačica was derived from "Tabhana", and became so familiar to the people of Mostar that over time the name of its founder was completely forgotten. It is said that eighty leather tanners used to worship in it, which would speak of the number of the leather guild in Mostar. Towards the end of Ottoman rule, the Tabačica mosque was repaired and painted inside by Hadji Arif Effendi Kajtaz with his resources. The next conservation and restoration works were carried out in 1954. During the aggression against Bosnia and Herzegovina, the Tabačica mosque suffered massive damage (Figures 2 and 3). The minaret was completely destroyed. The Tabačica Mosque was reconstructed and re-equipped for prayer in 2000 [55].

Figure 2. Demolished roof structure after a war [56]. Reproduced with permission from CIDOM, 2022.

Figure 3. Demolished minaret in a war [57]. Reproduced with permission from Mufti Unit in Mostar, Islamic Community in Bosnia and Herzegovina, 2022.

The minaret was reconstructed using the "Tenelija" stone extracted from the nearby quarry. The present condition of the reconstructed minaret after 2000 is shown in Figure 1a.

2.1. Climate and Seismicity Investigation

In the period from 1858 to 2018, the average annual temperature in Mostar was 16 °C. The warmest month of the year is July, with an average temperature of 26 °C, and the coldest is January, with an average temperature of 5 °C. The highest precipitation is in November and the lowest precipitation is in July, with a difference of 47 mm. The month with the highest relative humidity is February (85%) and the lowest relative humidity is August (62%). Mostar typically receives about 93.95 cm of precipitation and annually has 141.13 rainy days. The average hourly wind speed in Mostar experiences significant seasonal variation during the year. The direction of the wind varies over the year. The wind blows for the majority of the year from the east, for two months from the south, and less than two months from the north. In the period from 1990 to 2022, several earthquakes occurred near Mostar. During past years in Mostar, moderate earthquakes with magnitude 5.0–5.9 on the Richter scale were observed. Several earthquakes had a magnitude in the range of 4.0–4.9 on the Richter scale. The city of Mostar is located in a seismically active zone with a peak ground acceleration (PGA) equal to 0.26 g for the return period of 475 years [58], and according to the calculations conducted by Ademović et al. [59], vulnerability of the buildings in Mostar is very high.

During the visual inspection of the minaret, damage due to ascending and descending moisture has been noted as illustrated in Figure 4a,b respectively. Several dead pigeons were noted in the staircase of the minaret as once they entered the staircase, they were not able to exit. The birds did not cause any damage to the structure and no threats by insects were noted.

(a)   (b)

Figure 4. (**a**) Ascending moisture and biological attack (author's photo, 2023). (**b**) Descending moisture (author's photo, 2023).

There are no signs of damage on the minaret itself after the reconstruction except for a minor biological attack at the bottom of the base (Figure 5). However, plaster spalling on the side of the mosque building is visible along the vertical joint between the minaret and the mosque building (Figure 6), which was not present shortly after the reconstruction (Figure 7). The metal grounding rod runs vertically along the joint, which might have been generating this damage over a period of 20 years.

Figure 5. Biological attack at the base of the minaret (author's photo, 2023).

(a)   (b)

Figure 6. Plaster spalling along the vertical joint between the minaret and the mosque building (**a**) view (author's photo, 2023). (**b**) detail, (author's photo, 2023).

Figure 7. Vertical joint between the minaret and the mosque building, no plaster spalling visible shortly after reconstruction (author's photo, 2020).

There are smaller damage impacts visible on the mosque building due to moisture (Figure 4a) and biological attack (Figure 4b). There is damage to the stone roof slabs at the eaves (Figure 8), and a vertical crack can be seen on the western façade of the attached shop building (Figure 9).

Figure 8. Damaged stone roof tiles at the southern eaves (author's photo, 2023).

(a)   (b)

Figure 9. Vertical crack at the western façade of the attached shop building: (**a**) view (author's photo, 2023); (**b**) detail (author's photo, 2023).

2.2. Geometry and Materials

A minaret is typically positioned at the right corner of the front façade. The base of the minaret can be separated from the mosque and in this way have its own foundation, or it can be built on the same foundations as the mosques. Regarding the minaret's height, several typical parts may be noted, moving upwards (Figure 1b) [60]:

- Part A: The base starts from the masonry footing which may be of polygonal or squared shape. The inner space of the base may be either hollow or filled with rubble masonry which depends on the starting height of the inner spiral staircase. This staircase is supported by a self-forming central column and by the minaret's outer wall.

- Part B is the longest part of the minaret. It is the central shaft where the outer shape is commonly a 12, 14, or 16-sided polygon and the inside is cylindrical.
- Part C is the balcony (in Turkish "sherefe") located at the end of part B. At this location, the inner spiral staircase ends.
- Part D is the upper shaft. It is above the balcony with a door opening providing the exit to the balcony and is typically narrower than the central shaft.

The minaret ends with Part E which is a roofed part. It usually ends with a conical hood made of copper, lead, or confined stone slabs supported by the timber structure.

The materials which were selected for the construction of minarets were usually abundant in the local areas. Limestone ashlar was used for the construction of the mosque, the minaret, and the spiral stairs. The base wall is typically constructed as a three-leaf wall consisting of outer leaves made of ashlars and the inner core made of rubble stone with hydrated lime. For the connection of the ashlars, hydrated lime with thin-layer joints was used, while hydrated lime mortar with normal joints was used for the rubble stone. Horizontal connection was done by iron cramps, which were fixed by the lead poured at the upper side of blocks in previously prepared holes. Timber was used for the construction of the hood structure and it was covered with copper or lead sheets, or with shaped stone slabs, additionally confined with iron hoops.

Numerous structures in the Herzegovina region, are constructed of stone known as "tenelija" or "miljevina". A significant number of experiments were conducted in order to determine the physical and mechanical properties of the "tenelija" stone due to its extensive applicability in the past and in the reconstruction of the famous Stari Most (Old Bridge) in Mostar [61]. Tenelija has been comprehensively investigated during the reconstruction of Stari Most and Karagöz Bey Mosque in Mostar [61,62]. The minaret of the Tabačica mosque was built from the same "tenelija" stone, from a quarry which is located in Mukoša near Mostar. The "tenelija" stone is very easy to shape, which is why more elegant, delicate, and refined buildings were built from it. However, due to its insufficient resistance to frost, the use of "tenelija" in the exterior is limited to the Mediterranean climate [63]. Masonry with "tenelija" is a tradition of the Mostar climate. During construction, "tenelija" can be found in various forms: in blocks weighing several tons (arch of the Old Bridge—Stari Most in Mostar), in smaller cubes and corners for building walls, cut slabs of various formats for interior and exterior cladding, in broken pieces as filling of walls, in various forms of tombstones, for columns and decorative masonry fences, and as a model stone for making sculptures [64].

Microscopic investigation of "tenelija" and "mukoša" blocks from the Stari Most has been conducted [65]. Besides the difference in texture and structure, these two domestic stone varieties differ in their mineralogical, petrographic, and chemical composition [61]. The most significant feature of "tenelija" limestone is its oolite structure [66]. Utilizing a scanning electron microscope (SEM) showed that "tenelija" stone is made of grain size up to 0.2–0.5 mm, classifying it into the oosparit stone class. In "tenelija" ooids are somewhat condensed and, in some places, there is a space between them [61].

For civil engineering applicability, the most important characteristics are physical and mechanical. Physical parameters of interest include, but are not limited to, density, open porosity, and water absorption, while mechanical properties of interest include, but not limited to, uniaxial compressive strength (UCS), tensile strength, Poisson's ratio, and modulus of elasticity. Hahamović and Tufegdžić were the first researchers who tested the "tenelija" stone from the Mostar's Stari Most. The density that was obtained was of a rather wide range, from 1700–2270 kg/m^3, and 2070–2320 kg/m^3, respectively [64]. After that, the most important in situ and laboratory tests were carried out during the renovation of the Stari Most in Mostar in the period from 1998 to 2002. Additional tests were performed in various periods and for various purposes. Some of the results are presented in Table 1. It should be noted that some tests were conducted in line with the ex-Yugoslavian standards and others in line with the European standards. Additionally, some researchers used cube samples and others cylinders, and the number of samples varied; thus, direct comparisons cannot be done. As can be seen, the results have a wide range. Some of the results are presented here for an informative purpose so that the readers can get an idea about the characteristics of this stone type.

Table 1. Test values of physical and mechanical characteristics of "tenelija" stone.

	Tufegdžić (1956) [64]	Hahamović (1960) [64]	"LGA" (2000) [62]	"IGH" (2002) [67]	RGGF—Tuzla [68]	"IGH" (2016) [62]
Density [g/cm^3]	2.56–2.63	2.27–2.57	2.64–2.67	2.62	2.56	2.669–2.674
Unit weight [kN/m^3]	20.3–22.75	16.67–22.26	19.4–19.8	19.39	17.76–20.09	17.7–20.3
Compressive strength [MPa]	30–48.8	12.9–57.1	11.0–45.1	32.86–45.0	21.5–23.2	18.5–27.2
Bending strength [MPa]	-	-	-	-	4.82	4.1–6.27
Shear strength	-	-	-	-	angle of internal friction $\upsilon = 34°$ cohesion $c = 0$ MPa	-
Modulus of elasticity [MPa]	-	-	16,539–17,976	-	1636–5563	-
Poisson's ratio	-	-	0.25	-	-	-

3. On-Site Experimental Investigations

Geophysical tests, as part of the site investigation at the location of the minaret of the Tabačica mosque in Mostar, were carried out in July 2022 by Terra Compacta d.o.o. from Zagreb. The main goal of the research was to determine the fundamental vibration frequencies of the minaret and the surrounding soil using non-destructive methods and to determine the qualitative parameters of the mosque wall (speed of propagation of P-waves through the wall).

3.1. Determination of the Dynamic Characteristics

For the determination of the dynamic characteristics of the elaborated structure, two methods were applied, the horizontal to vertical spectral ratio (HVSR) method and the operational modal analysis (OMA). These parameters are necessary in order to calibrate the numerical model. Besides being non-destructive methods, the data is collected under the operational conditions of the structure.

The HVSR method was widely spread by Nakamura [69]; however, it was first developed by Nogoshi and Igarashi [70]. It is commonly used to estimate the resonant frequency of sedimentary layers on top of bedrock, by means of gathering the subsurface data from single station measurements by comparing the Fourier spectrum of horizontal components to the vertical components. This ratio is a function of the frequency that will produce the H/V curve. The biggest advantage of this method is the efficient estimation of the frequency at which the effect of seismic amplification will occur, being independent of input parameters. The method is almost independent of the source excitation and gives results for both earthquakes and microseismic tremors. This method was initially used by [69,71] to investigate the risk of seismicity in Japan. The usage of this method is limited to structures.

According to Nakamura [69], the microseismic tremors measured at a location are amplified by soft layers on a solid half-space. It is assumed that the horizontal components of microtremors are amplified by multiple reflections of transverse (S) waves in the upper soil layers, while the vertical component is amplified by multiple reflections of longitudinal

(P) waves. It is assumed that P-wave amplification is not significant in the frequency range of interest; this assumption is satisfied in most cases.

The transfer function S_T of surface layers is usually defined by the Equation (1):

$$S_T = \frac{S_{HS}}{S_{HB}} \tag{1}$$

where S_{HS} and S_{HB} are the horizontal tremor spectrum on the surface of the soil and the horizontal tremor spectrum incident from the substrate to surface layers (S—surface, B—bedrock). According to Nakamura [69], the S_{HS} spectrum can be affected by Rayleigh waves propagating through the surface layers of the soil. The influence of Rayleigh waves should also be visible in the vertical component of a microseismic tremor on the surface, S_{VS}, but it is not expected in the vertical component of a microseismic tremor at the level of the bedrock (because Rayleigh waves exist in the layer above the half-space—bedrock), S_{VB}.

Assuming that the vertical component of motion is not amplified by surface layers, Nakamura [69] suggests that the influence of Rayleigh waves can be written in the form:

$$E_S = \frac{S_{VS}}{S_{VB}}. \tag{2}$$

If there is no Rayleigh wave, then $E_S = 1$. Assuming that the influence of Rayleigh waves is equal for the vertical and horizontal components, Nakamura [69] suggests using a more reliable transfer function, S_{TT}:

$$S_{TT} = \frac{S_{HS}/S_{VS}}{S_{HB}/S_{VB}} = \frac{R_S}{R_B}.S_T = \frac{S_{HS}}{S_{HB}} \tag{3}$$

R_S and R_B are the spectral ratios of the horizontal and vertical components of the microseismic tremor at the surface (R_S) and the bedrock level (R_B).

It has been experimentally shown that $R_B \cong 1$ is valid for wide frequency ranges. Therefore it follows that $S_{TT} = R_S$. This means that the transfer function can be estimated using only the tremor from the surface measurements; the vertical tremor on the surface maintains the horizontal tremor's parameters.

Empirical research has shown that the spectral spike that appears on the HVSR curves is a very good estimate of the fundamental frequency of the location (soil or object) where the microseismic tremor is measured. The amplitude of the curve, in contrast, cannot be used as a measure of the seismic amplification of a site.

In the work conducted by Triwulan et al. [72], it has been stated that the HVSR method is able to describe the geological parameters of the elaborated area, the building's structural characteristics and the correlation between them if the recording is done outside and inside the building. By this kind of application, they obtained the values of the natural frequency, the index of vulnerability, and the amplification factor. This means that the method may provide information regarding two very important things: to determine how far apart are the frequencies of the structure and the surface layer, and to determine the value of the amplitude of this frequency.

Luo et al. [73] studied how subway trains impact the ambient noise in reinforced concrete buildings. During the study, they were able to obtain information on the building's eigenfrequency, the frequency that resulted from the train traffic near the elaborated building, and the fundamental frequency of the area where the structure is built.

Pentaris [74] used the HVSR method to obtain the resonance frequency of movement for two concrete buildings constructed of different materials and constructed in different years. It was noted that with the age of the building and crack visibility there is an increase in the values obtained by HVSR. As well, HVSR also indicated higher differential acceleration from floor to floor and higher structural vulnerability. Results indicate that the HVSR process using acceleration data proves to be an easy, cost-effective, and fast method for assessment of the fundamental frequency of structures; it also provides estimates on the building's vulnerability, as an assessment method for building vulnerability estimation.

On the other hand, operational modal analysis (OMA) is frequently used for the determination of the natural frequencies, mode shapes, and damping of structures. In the literature, several names are allocated to this method, from output-only modal analysis, and ambient modal identification, to in-operation modal analysis [75]. Applying the enhanced frequency domain decomposition (EFDD) method provides higher accuracy in relation to the frequency domain decomposition (FDD). The EFDD is capable of estimating within a high degree of accuracy the closely spaced modes [75]. OMA has been used by numerous researchers for the determination of the dynamic characteristics of mosques and minarets. For the determination of the dynamic features of the Küçük Fatih mosque in Trabzon before and after the rehabilitation, the OMA was applied. After the strengthening, the structure became more rigid, with an increase of the frequencies [76]. Nohutcu et al. [77] with the application of OMA determined the frequency, damping ratios, and mode shapes of the Hafsa Sultan mosque in Manisa. The difference between the OMA and numerical results was up to 10% which was reduced in a stepwise manner by the reduction of the modulus of elasticity in the FEM model and in this way the model was calibrated. Demir et al. [78] improved the initial elasticity module of the historical Hafsa Sultan mosque by using the OMA method; once the model was calibrated, the damage to the structure was represented in a more accurate manner. Haciefendioğlu et al. [49] used OMA to determine how the type of soil impacts the natural frequencies of the scaled reinforced concrete minaret.

At the location of the Tabačica mosque, the microseismic tremors were measured at six locations within the minaret and the mosque, and the seventh was located in front of the mosque outside the influence of the building. The measurement in the mosque is marked as T-1. There were five measurements in the minaret, marked as T-2 at the base of the minaret, T-3 at the 25th stair, T-4 at the 40th stair, T-5 at the 56th stair, and T-6 at the 70th stair. Measurement T-7 was taken in the ground in front of the mosque outside the influence of the building (Figures 10 and 11).

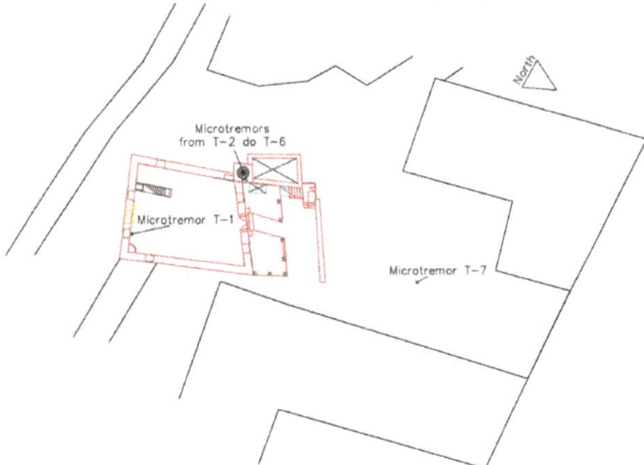

Figure 10. Locations of microtremors from T-1 to T-7.

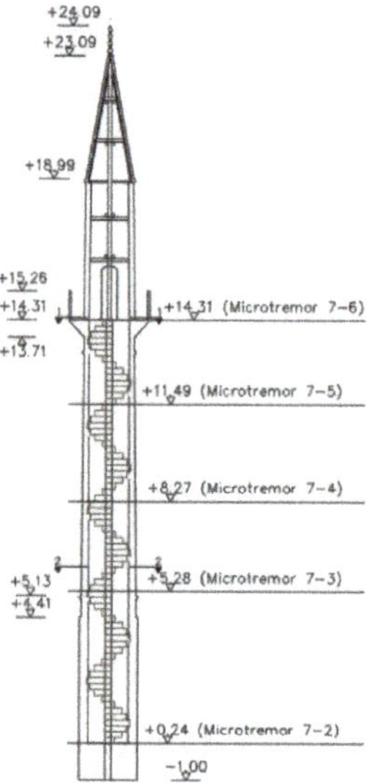

Figure 11. Locations of microtremors from T-2 to T-6 in the minaret.

The measurements were performed with the equipment shown in Figure 12. The Seismograph GEA24 16077, compact-sized with 24 channels, was used. It is suitable for all applications and is very reliable. GEA24 can acquire data using geophones with any resonance frequency (even 1 Hz). In this specific case, a three-component geophone with a resonant frequency of 2 Hz was used.

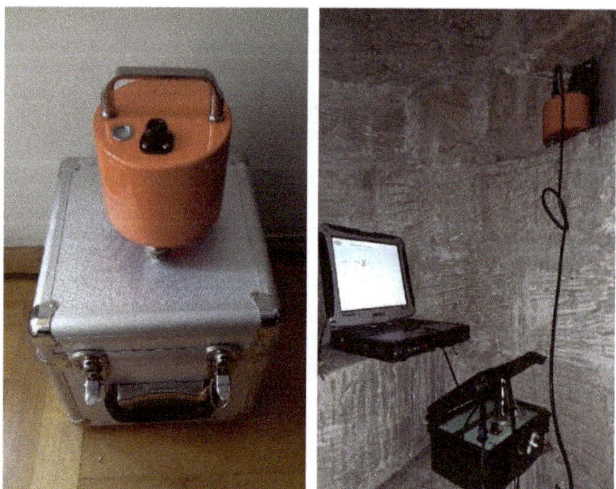

Figure 12. Geophysical equipment for seismic surveys (author's photo).

Two microtremor measurements were performed at each location. The recordings were filtered with a low-pass filter that passes frequencies lower than 30 Hz. All three components of a single record are divided into time windows of 60 s with a five percent overlap, and the mean of the horizontal components is determined for each window; the spectrum of the mean of the horizontal components is divided by the spectrum of the vertical component. The results of individual windows are averaged. The resulting curve was then smoothed using the Konnoi–Ohmachi function [79] in the software Geopsy, developed in the frame of the SESAME program (Site Effects assessment using Ambient Excitations). The results are shown together with the standard deviations (dashed line in Figure 13).

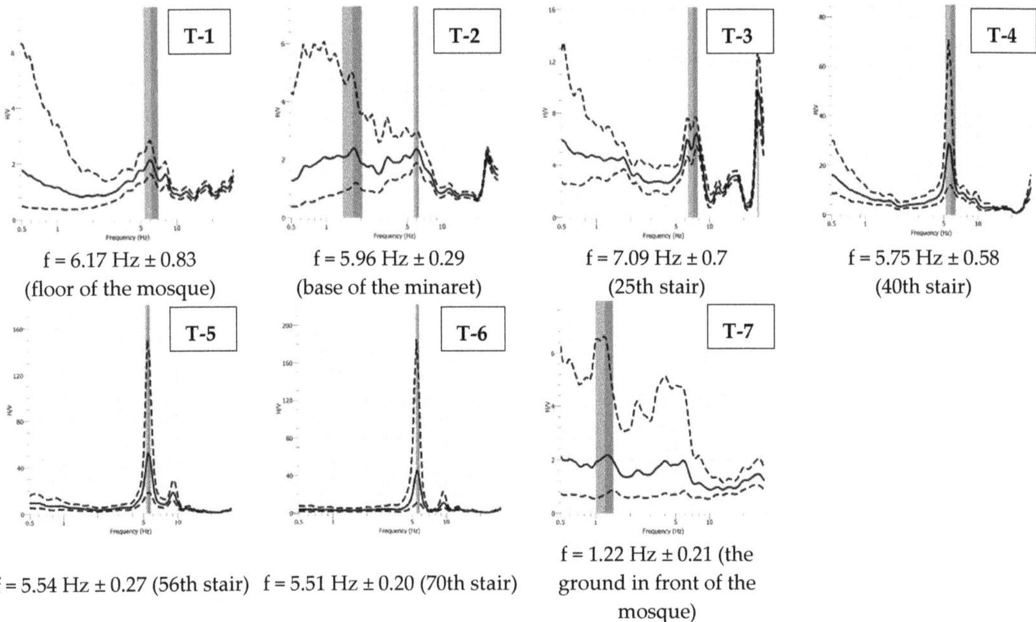

f = 6.17 Hz ± 0.83
(floor of the mosque)

f = 5.96 Hz ± 0.29
(base of the minaret)

f = 7.09 Hz ± 0.7
(25th stair)

f = 5.75 Hz ± 0.58
(40th stair)

f = 5.54 Hz ± 0.27 (56th stair)   f = 5.51 Hz ± 0.20 (70th stair)

f = 1.22 Hz ± 0.21 (the ground in front of the mosque)

Figure 13. First frequency values using the HVSR method [80]. Reproduced with permission from Terra Compacta, 2022.

Data were also elaborated by employing an OMA-specific analysis software [81], relying both on time domain—UPC SSI—and frequency—sEFDD—based identification methods. From the analyses carried out, there emerges a strict coupling of the bending modes of the minaret. The first frequency of the minaret using the OMA method test results was 1.85 Hz in all test setups (bending in the Y direction), very close to the second identified frequency, equal to 1.91 Hz (bending in the X direction), while the second bendings were found, respectively, at about 5.53 and 5.67 Hz. A supposed torsional mode was located at 7.40 Hz, and the successive coupled bending modes were identified at 9.20 (Y) and 9.40 (X) Hz, and again at 12.90 (Y) and 13.40 (X) Hz. Since no reference sensors were employed, it is not possible to present the graphical depiction of the identified modes. Results are reported as a graph in the frequency domain (Singular Value Decomposition—Figure 14) and numerically in Table 2 The average damping values for F1 to F4 are consistent with the level of vibration and are within the reported ranges under operational conditions as defined in manuals for civil engineering and seismic design [82,83]. A significant value of damping has been detected for the torsional mode indicating that it may be affected by biases, as reported by Karatzetzou et al. [84].

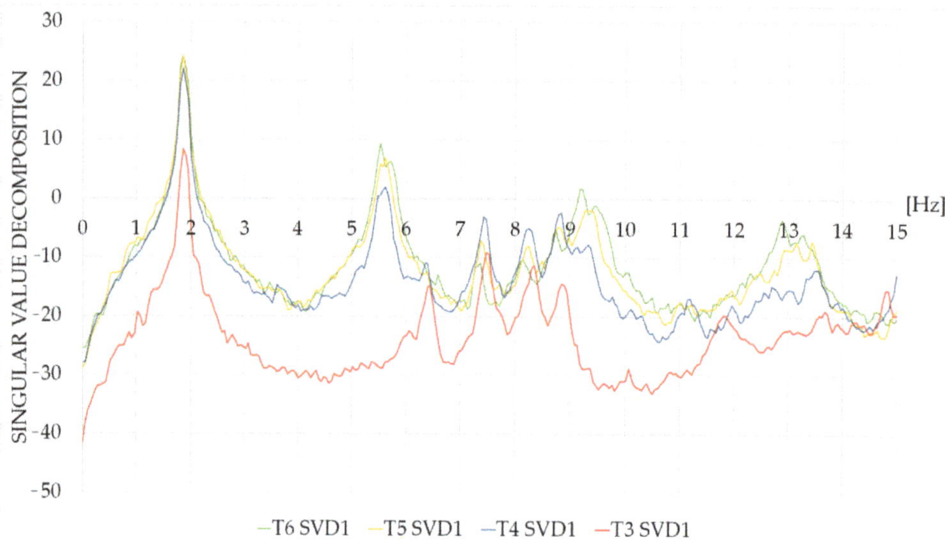

Figure 14. Singular value decomposition of the acquired data in positions T3 to T6.

Table 2. Eigenvalues and related damping derived from the acquisition configurations in the minaret (T6 to T4).

	T6		T5		T4		T3	
	EFDD	**UPC**	**EFDD**	**UPC**	**EFDD**	**UPC**	**EFDD**	**UPC**
F1—1st bending Y	1.84	1.84	1.85	1.85	1.85	1.84	1.86	1.86
damping	1.62%	0.63%	1.72%	0.73%	1.99%	1.05%	2.03%	1.22%
F2—1st bending X	1.92	1.91	1.92	1.90	1.92	1.91	1.92	1.91
damping	1.76%	2.28%	1.64%	0.72%	1.60%	0.45%	1.65%	0.84%
F3—2nd bending Y	5.52	5.51	5.53	5.55	5.50	5.53	- - -	- - -
damping	1.04%	0.73%	0.70%	1.47%	0.50%	3.49%		
F4—2nd bending X	5.71	5.68	5.61	5.61	5.66	5.62	- - -	- - -
damping	0.72%	1.11%	0.47%	1.16%	0.67%	2.16%		
F5—torsion	7.33	- - -	7.41	- - -	7.44	- - -	7.49	7.54
damping	0.92%		1.31%		1.00%		0.94%	8.22%
F6—3rd bending Y	9.21	9.17	9.28	9.11	8.80	- - -	- - -	- - -
damping	1.23%	1.44%	- - -	2.68%				
F7—3rd bending X	9.46	9.37	9.38	9.31	9.33	- - -	- - -	- - -
damping	1.73%	2.49%	0.45%	2.30%	0.45%			
F8—4th bending Y	12.89	12.83	13.14	- - -	- - -	- - -	- - -	- - -
damping	0.71%	3.06%	0.31%					
F9—4th bending X	13.27	- - -	13.44	- - -	13.46	- - -	- - -	13.31
damping	0.71%		0.46%		1.98%			6.93%

In addition to the values indicated in Figure 13, the resonance at approximately 1.93 Hz is visible also in the HVSR curves at T-2, T-3, and T4. The first and second frequencies according to the OMA method are 1.85 Hz and 1.91 Hz, respectively.

3.2. Sonic Pulse Velocity Test (SPVT)

The second test that was conducted was the sonic velocity test (SPVT). The SPVT is based on measuring the travel time of elastic longitudinal P-waves (acoustic waves, "sonic waves") through a given material. The speed of acoustic waves is determined according to the length of the path and the calculated propagation time of the wave. The obtained results provide useful information about the quality and consistency of the investigated element. The speed is affected by the composition of the material as well as the presence of substances, voids, and damage. It is a non-destructive method frequently used for diagnosing existing masonry and evaluating the effectiveness of interventions.

As per Binda et al. [85], there are two aims for the implementation of the sonic tests in masonry structures. Based on the morphology of the wall section it is possible to quantify masonry, as well as locate cracks and determine the damage pattern. This method enables the detection of voids and eventual flaws in the masonry. Once a structure is repaired, the sonic tests provide information that can be used to check the effectiveness of the remedial injection measures. The pulse sonic velocity is individual for each masonry typology and no generalization is possible. These tests provide qualitative information and are combined with other non-destructive tests and minor-destructive tests.

Sonic testing could not be performed on the walls of the minaret due to inaccessibility, so the testing was performed on the wall of the mosque, which is built from the same stone as the minaret and built in the same way as the base of the minaret. The testing was carried out in a grid of 0.9 × 0.9 m, with 16 test micro-locations spaced at 30 cm along the horizontal and vertical axis (Figure 15).

Figure 15. Gird of test micro-locations (sonic test).

The emission of acoustic waves was done with a rubber hammer. The hammer applies an impulse on the surface of the wall and generates a sonic wave. The arrival times of the P-waves were recorded on an accelerometer which is connected to the instrument on the opposite side of the wall (Figures 16 and 17). This enables the transformation of the mechanical energy into acoustic-vibration energy and enables the wave to propagate through the material. A receiving sensor (monoaxial accelerometer) is responsible for recording and transmitting to the acquisition system the response of the material to the propagation of the signal in the section.

Evaluation of the characteristics of the walls of historic masonry buildings by using a sonic pulse velocity method has been performed by numerous researchers, see, e.g., [86–90].

The results of the SPVT test are shown in Table 3 and Figure 18. The table contains the serial numbers of the recording points with the associated digital record number; the wall thickness in cm; X and Y coordinates of the position of initiating the elastic wave; and recording with the accelerometer, propagation time, and calculated speed of P-wave propagation through the wall.

Figure 16. The emission of acoustic waves with a rubber hammer (outside) (author's photo, 2022).

Figure 17. Receiving waves using an accelerometer (inside) (author's photo, 2022).

Table 3. Results of the Sonic Pulse Velocity Test (SPVT) [80].

Point No.	Record	Wall Thickness (cm)	x (cm)	y (cm)	dt (s)	v (m/s)
1	164406	80	0	90	0.0016	5000.0
2	165409	80	30	90	0.0015	5333.3
3	165952	80	60	90	0.0017	4848.5
4	170605	80	90	90	0.0017	4705.9
5	171209	80	0	60	0.0017	4848.5
6	171921	80	30	60	0.0017	4705.9

Table 3. *Cont.*

Point No.	Record	Wall Thickness (cm)	x (cm)	y (cm)	dt (s)	v (m/s)
7	172410	80	60	60	0.0016	4848.5
8	173031	80	90	60	0.0015	5333.3
9	173656	80	0	30	0.0017	4705.9
10	174133	80	30	30	0.0019	4102.6
11	174701	80	60	30	0.0017	4705.9
12	175107	80	90	30	0.0021	3809.5
13	175610	80	0	0	0.0017	4571.4
14	180020	80	30	0	0.0019	4210.5
15	180456	80	60	0	0.0018	4444.4
16	180859	80	90	0	0.0019	4210.5

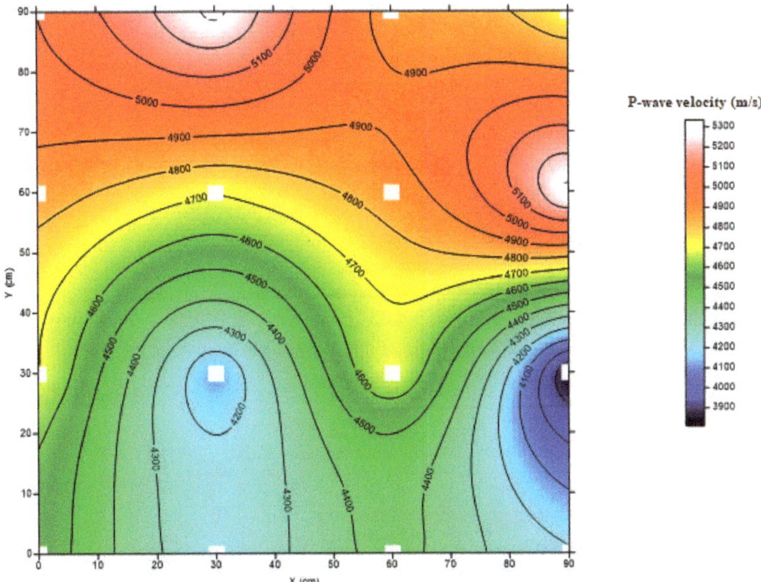

Figure 18. Distribution of P-wave propagation velocities through the tested wall of the minaret in the Tabačica mosque [80]. Reproduced with permission from Terra Compacta, 2022.

Figure 18 shows the distribution of P-wave propagation speeds through the examined wall. Velocity values are approximately in the range of 3800 to 5300 m/s, which indicates the high compactness and quality of the wall. Slightly lower velocities (3800 to 4500 m/s) were found in the first half of the wall (from the height Y = 0 to about 50 cm), which indicates that sporadic, localized discontinuities are possible in that part of the wall, or that the lower part of the wall is made of material with different characteristics.

4. Nonlinear Dynamic Structural Analysis

In this specific case, it was decided to use the nonlinear model that is based on AEM, a derivative of the FEM and DEM which is utilized in the Extreme Loading for Structures (ELS) software. With this software, it is possible to study the behavior of structures through the continuum phase and the discrete phase of loading, which is of the utmost importance when the structure collapses. Seismic assessment of existing structures using ELS software was done by several researchers [27,91–93].

In AEM, along their connecting surfaces, the elements are connected with a set of normal and shear springs that are responsible for the transfer of normal and shear stresses among adjacent elements. Each spring denotes stresses and deformations of a certain volume of material. Two elements are separated from each other if the springs connecting them are broken (failed). In AEM, elements may automatically be separated, re-contacted, or contacted with other elements. When elements are connected with the contact springs at contact points, different types of element contact can be formulated as shown in Figure 19a,b [27].

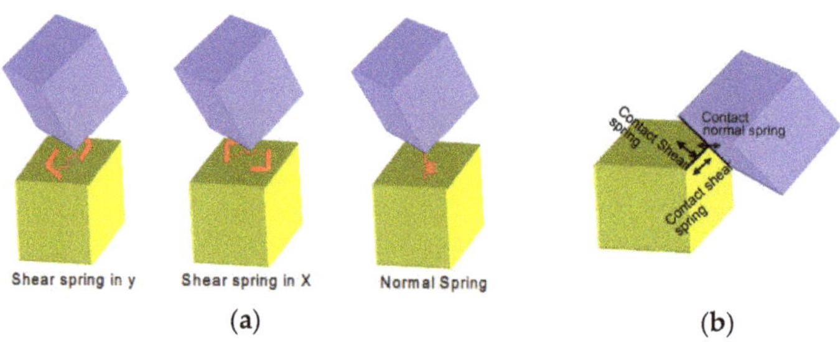

Figure 19. (**a**) Corner-to-face or corner-to-ground contact [27]; (**b**) Edge-to-edge contact [27]. Reproduced with permission from Salem, H. 2022.

The AEM is found to be much more user-friendly for modeling solid elements for non-linear structural analysis than the FEM. First of all, nodal completability is not needed which makes meshing much easier. AEM connects elements through springs which are located between the solid elements. There is no need for an a priori definition of hinge locations. This means that the AEM gives the plastic hinges as results and not as input data. Modeling of masonry units is very easy and all types of masonry patterns and mortar joints can be implemented. This technology allows modeling of the existing structures with true detailing of connections. This includes applying weakening at sections with existing cracks (crack width and custom shapes of cracks). Due to all the above, it was decided that the AEM is the most representative numerical tool for this research.

4.1. Description of the Structural Model

The minaret of Tabačica mosque was modeled as a stand-alone structure. According to earlier research by Karatzetzou et al. [84] analyzing the dynamic properties of the Suleiman Mosque in the city of Rhodes in Greece, it was concluded that there were no significant changes in the frequency values between the model of the minaret only and the model of the minaret together with the mosque.

The minaret is made of stone elements which are disseized by the number of small elements connected with mortar springs at the interface between the elements. For the units, the principal stress failure criterion is implemented while Mohr–Coulomb's friction model with tension cut-off is used to represent the interface behavior with mortar. The shear behavior in the tensile regime is represented through the softening regime once the cohesion is lost and debonding takes place. The elements are connected by the strings and the constitutive properties of the area material in the particular area of the representations are formulated in these springs. Frequencies were extracted from a linear model, while the material nonlinearity was taken into account in the time-history analysis.

The minaret of Tabačica was modeled in ELS software [94] and it has a total of 23,302 quadrilateral elements and 1,856,800 springs. The exact dimensions were determined by on-site measurements of the minaret. The 3D models developed in the ELS software are shown in Figure 20. The stairs were taken into account in the model and drawn together with the outer wall to avoid possible problems in the mesh creation and intersection. The mechanical properties of the minaret are shown in Table 4.

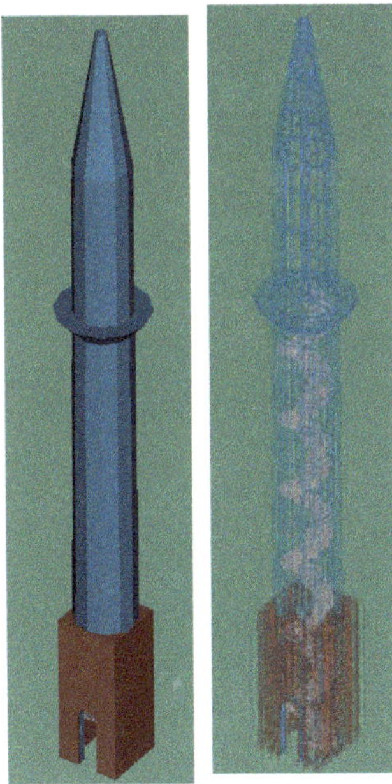

Figure 20. 3D minaret models in ELS software.

Table 4. Mechanical properties of the minaret.

Mechanical Properties	Value [MPa]
Compressive strength	3.66
Tensile strength	0.366
Modulus of elasticity	3660
Poisson's ratio	0.25

The natural frequencies of a structure are obtained as the result of modal analysis and are presented in Table 5.

Table 5. Modal analysis results of the minaret model.

Mode	Frequency (Hz)	Period (s)	Mass Participation (X)%	Mass Participation (Y)%	Total Mass Participation (X)%	Total Mass Participation (Y)%
1	1.96	0.509	0.02	45.91	0.02	45.91
2	2.12	0.498	45.66	0.02	45.68	45.93
3	4.45	0.225	0	25.87	45.68	71.80
4	4.55	0.220	26.86	0	72.54	71.80
5	6.58	0.152	0.05	0	72.59	71.80

The first period obtained for the minaret of the Tabačica mosque is compared with the values obtained by four empirical formulas as suggested in the literature [95–100]. The empirical formulas are shown in Table 6 and it can be concluded that the first period of the minaret is in good agreement with the empirical formulas provided in the literature.

Table 6. Empirical formulas for calculating the period of vibration.

Empirical Formula	Period of Oscillation Minaret	Description	References
$T_1 = C_t \cdot H^{0.75}$	$T_1 = 0.544$	$C_t = 0.05$; H: total height	[95–97]
$f_1 = Y \cdot (H/B)^{-z}$	$T_1 = 0.932$	$Y = 8.03$; $z = 0.86$	[98]
$T_1 = 0.0187 \cdot H$	$T_1 = 0.450$	H: total height	[99]
$T_1 = 0.01137 \cdot H^{1.138}$	$T_1 = 0.425$	H: total height	[100]

The largest difference between the experimental and numerical frequencies is about 9.54% for the first five modes (Table 7), which can be considered acceptable.

Table 7. Comparison between experimental frequencies and frequencies of the AEM model.

Frequency [Hz]	Vibration m Ode				
	1	2	3	4	5
Experimental (OMA method)	1.85	1.91	5.53	5.67	7.40
AEM model	1.93	2.04	5.32	5.40	8.18
ε [%]	4.14	6.37	3.95	5	9.54

The mode shapes obtained from the analysis in the software ELS of the minaret are shown in Figure 21. The first four mode shapes are longitudinal, but in the fifth mode, torsion has occurred, as can be seen in Figure 21. In the work done by Işık et al. [44], Işık et al. [45], and Mutlu and Sahin [34], the fifth mode was torsional as well. Ural and Çelik [101] investigated the seismic behavior of seven single balcony minarets. In this case as well, the fifth mode was one of torsion. Interestingly, the analysis of five timber minarets investigated by Bağbancı et al. [42] revealed the torsional mode as the fifth mode. In the work of Altiok and Demir [50] the torsional mode was identified as the sixth. For the reinforced concrete minaret, located in Islamabad, the dominant-torsional was also identified in the fifth position [51].

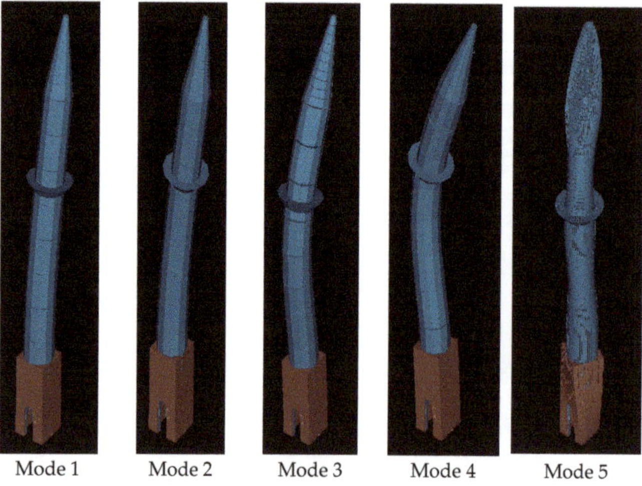

Figure 21. Mode shapes for the first five modes.

4.2. Non-Linear Time History Analysis

In the nonlinear analysis of the earthquake records, two main approaches for the generation of an accelerogram can be used; real earthquake records or synthetic (artificial accelerograms). Very often artificial accelerograms are used for numerical simulations and the same will be applied in this case. According to Eurocode 8 [102], several requirements have to be fulfilled for seismic analysis. It is necessary to use at least three accelerograms and the same accelerogram cannot be implemented at the same time in both horizontal directions. In the case when site-specific data are not available, the minimum duration of the stationary part of the accelerograms (T_s) should be 10 s, which is the case here as well.

The artificial acceleration records were used to determine the seismic resistance of the Tabačica mosque minaret. According to the National Annex of Bosnia and Herzegovina [58] and the accompanying interactive seismic map of Bosnia and Herzegovina, the peak ground acceleration (PGA) for Mostar is 0.26 g for the return period of 475 years. Taking this data into account, together with the soil type B and 5% damping, three sets of three artificial accelerograms were obtained as presented in Figure 22. An example of the artificial accelerograms are applied for X, Y, and Z directions.

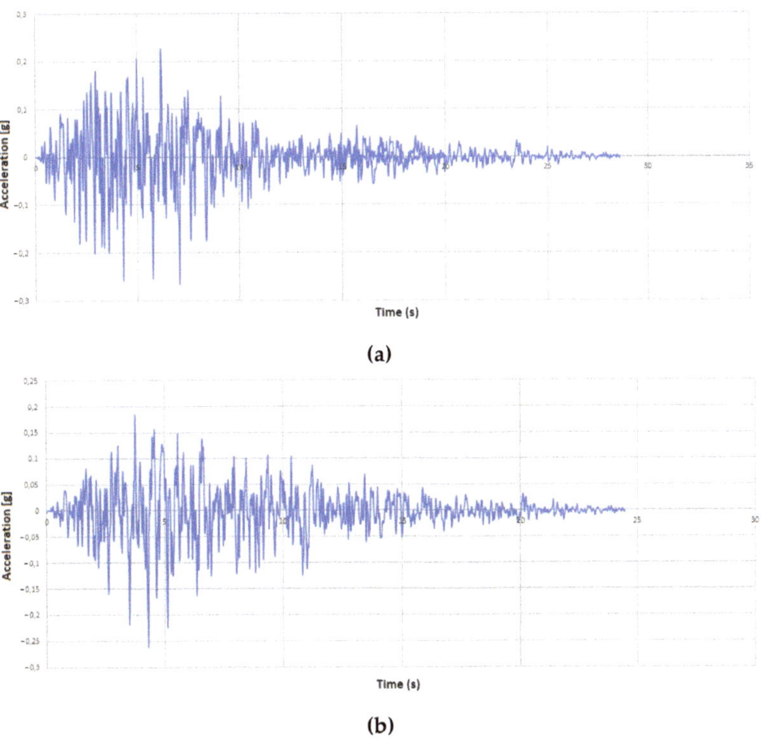

Figure 22. *Cont.*

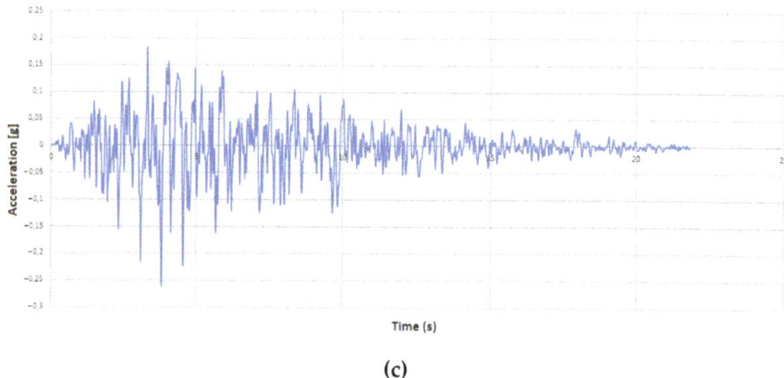

(c)

Figure 22. (**a**) Accelerogram in the X direction. (**b**) Accelerogram in the Y direction. (**c**) Accelerogram in the Z direction.

Nonlinear analysis of the minaret under the first set of accelerograms shows that the maximum lateral displacement values are 16.2 cm and 15.3 cm in the X and Y directions, respectively (Figure 23 a and b). The total response of the structure is the highest in the range from around 2.5 s to 6 s. It is interesting to note that in the range from eight to 15 s, there is a reduction of the displacement, and then at 15 s the displacement reaches the value of 9.02 cm.

The critical maximum and minimum principal stress contours under the earthquake ground motion are presented in Figure 24 a and b respectively.

Analyzing the distribution of the principal stresses, maximum and minimum principal stresses occurred in the same location, above the transition from the base to the shaft, and were obtained as 2.186 MPa and 1.706 MPa, respectively (Figure 24a,b). The concentration of stresses is located at the contact of the base and the shaft (transitional region) as indicated in [95], which is usually the location where the failure of the minaret is to be expected [54,103–105].

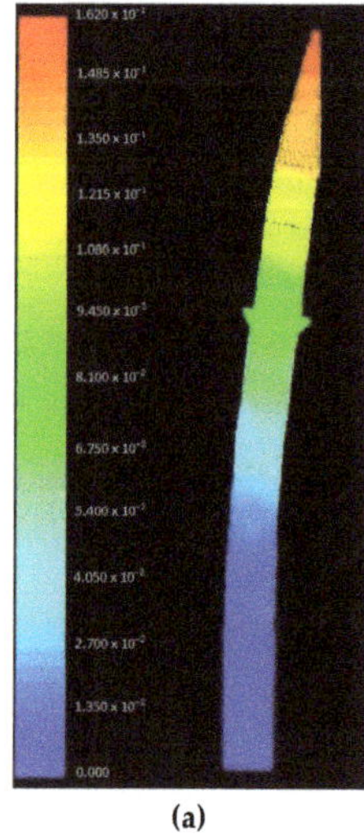

(a)

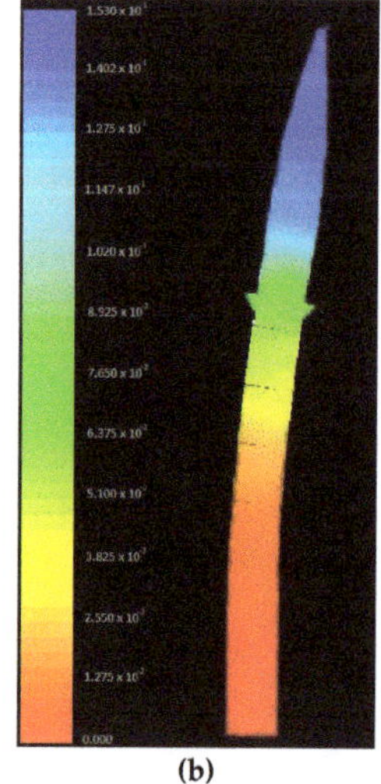

(b)

Figure 23. *Cont.*

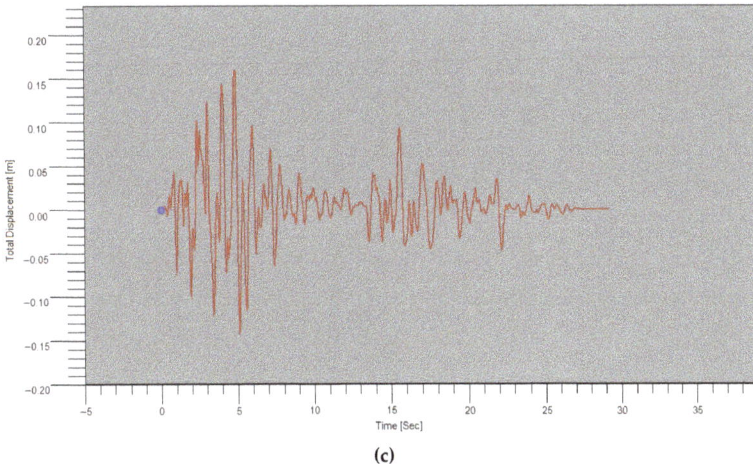

(c)

Figure 23. (**a**) Maximum displacement in the X direction [m]. (**b**) Maximum displacement in the Y direction [m]. (**c**) Total displacement of the top minaret element during the earthquake.

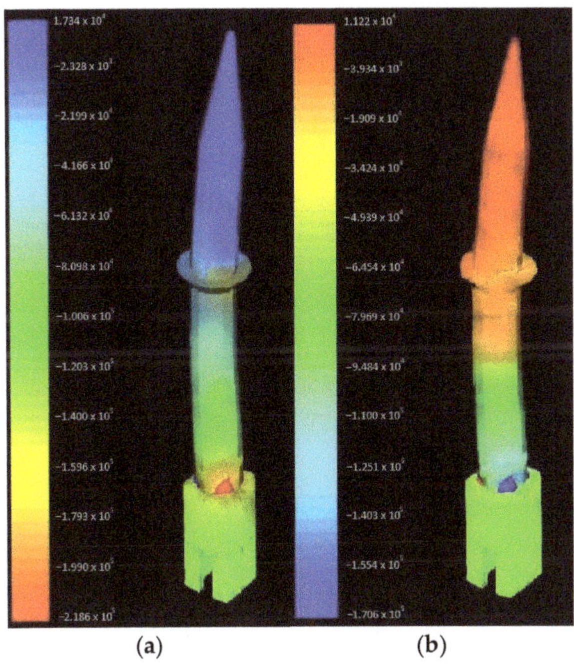

(a)   (b)

Figure 24. (**a**) Maximum principal stresses [kgf/m^2]. (**b**) Minimum principal stresses [kgf/m^2].

The time histories of total displacements above the transition from the base to the shaft (Figure 25a) are presented in Figure 25b. The largest value of the displacement is 0.0164 cm at 4.81 s. The same response pattern is observed in the case of the top displacement of the minaret. Formation of the horizontal crack is observed at the transition from the base to the shaft (Figure 25c).

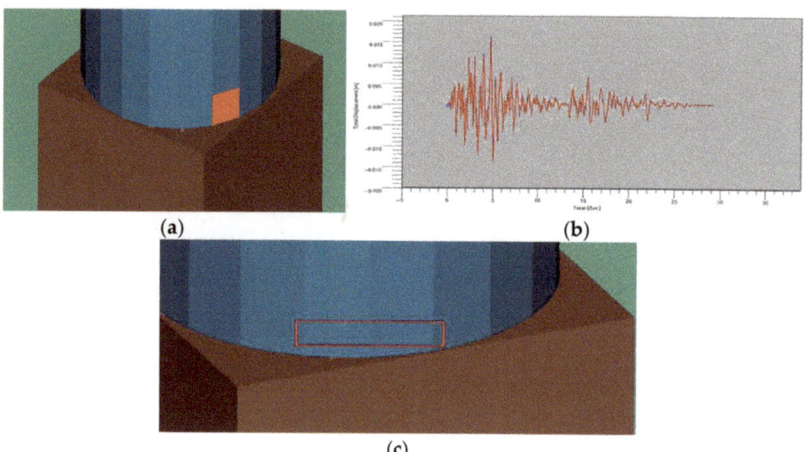

Figure 25. (**a**) Location at the transition from the base to the shaft. (**b**) Total displacements above the transition from the base to the shaft. (**c**) Horizontal crack.

The total displacement of the top minaret element during the two other artificial accelerograms is shown in Figure 26a,b.

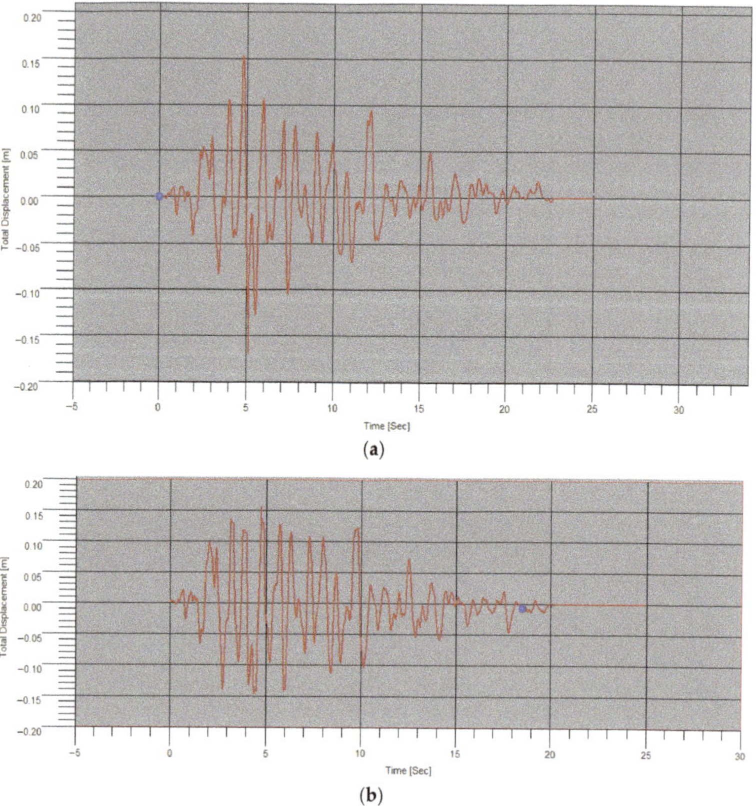

Figure 26. Total displacement of the top minaret element during (**a**) the second artificial accelerogram earthquake; (**b**) the third artificial accelerogram earthquake.

For the second set of accelerograms, slightly lower values are obtained. The maximum lateral displacement value for the X direction is 15.1 cm and 15.2 cm for the Y direction. For the third set of accelerograms, similar values are obtained as for the first set, except that the displacement in the Y direction is higher than the displacement in the X direction. The maximum lateral displacement value for the X direction is 15.7 cm and 16.1 cm for the Y direction. Maximum X and Y displacements for all three sets of accelerograms are shown in Table 8.

Table 8. Comparison of the maximum displacements for all three sets of the accelerograms.

Maximum Displacement	Set 1	Set 2	Set 3
X direction [cm]	16.2	15.1	15.7
Y direction [cm]	15.3	15.2	16.1

The values of principal stresses for the second and third sets are shown in Table 9. The concentration of stresses is located at the contact of the base and the shaft (transitional region).

Table 9. Comparison of the maximum and minimum principal stresses for all three sets of the accelerograms.

Principal Stresses	Set 1	Set 2	Set 3
Maximum [MPa]	2.186	1.911	2.311
Minimum [MPa]	1.706	1.476	1.888

In the aftermath of the 1999 Duzce/Bolu earthquake, Doğangün et al. [54] reported the collapse of two more than 500-years old minarets at the location of the transition segments due to extreme values of tensile stresses at these locations. Çaktı et al. [33] identified three zones with potential excessive deformation and stresses. Two locations were identified in the transition segments and the third location was identified at the mid-height of the body. The concentration of the tensile stresses was noted in the transition zone between the square base and the circular section. The maximum tensile stress values were in the range of 2.17 MPa to 10.46 MPa, which was significantly above the tensile strength of the minarets (around 0.4 MPa) [101]. The same location of the damage was determined by Döven et al. [36]. In the aftermath of the 2020 earthquake that hit the Sivrice district of Elazig city, damage was noted on the transition section in several minarets [47], as well as propagation of cracks and in some cases even collapse in the cone section and cracks in the pulpit section were detected. Transition zones, as one of the most vulnerable regions, were also identified by Altiok and Demir [50] in their study of one of the most famous minarets in Turkey, the minaret of the Lala Mehmet Pasha Mosque. During the detailed elaboration and assessment of the Ulu Mosque's minaret, Işık et al. [45] identified the concentration of stresses at the location of the balcony and the connection of the squared base and circular body of the minaret. It was interesting to note that the obtained stresses once the structure was exposed to earthquake actions were, as per new Turkish regulations, lower than the allowable Bilitis stone's strength.

This may be connected to the change in the cross-section and the difference in the stiffness. Additionally, the second concentration is located at the second transition zone where the balcony is located, as found by [95]. It is likely that under much stronger earthquakes, the minaret will be significantly damaged in the transitional region.

5. Conclusions

The first frequency of the minaret obtained by the ambient vibration test is 1.85 Hz, and the one obtained by the modal analysis is 1.96 Hz. There are also four empirical formulas for calculating the first period of vibration for slender structures. It may be concluded that there is a good match between the results of the on-site investigation and the modal analysis. For the first time, two methods were used for the determination of the eigen frequency, the horizontal-to-vertical spectral ratio (HVSR) method and the operational modal analysis (OMA). However, there is a need for fine calibration of the model in order to get the results even closer to those obtained by on-site investigation. In this connection, it is necessary to perform the parametric analysis to determine the influence of certain factors. These factors may include material properties, connection joints, attachment of the minaret to the mosque building, foundation soil, etc. It has been indicated that there is a good agreement between the different methods utilized for the determination of the dynamic characteristics.

The on-site investigation included a qualitative assessment of the three-leaf masonry wall of the mosque building by sonic pulse velocity test. Obtained velocities between 3800 and 5300 m/s indicate a solid quality of the masonry wall which is similar to the one at the base of the minaret.

The nonlinear analysis of three sets of accelerograms revealed and confirmed once again the weak point of the minaret being the transitional regions, the location where geometry changes from a rectangular to a circular shape, and the location of the balcony. Above the transition from the base to the shaft, for all three sets of accelerograms, a horizontal crack appears after the earthquake. Under stronger earthquakes, this crack would widen and other cracks would appear, which potentially would lead to the collapse of the minaret. Tabačica minaret is vulnerable to earthquakes of moderate intensity due to its slenderness and material characteristics.

The elegant pencil-shaped minarets built in classical Ottoman style are susceptible to damage by moderate earthquakes due to their particular geometry and height, in addition to the deterioration of their construction materials as a result of the aging process and environmental conditions.

In addition to the aforementioned parametric analysis, further work on the subject should comprise the structural analysis for stronger seismic excitation with vanishing tensile strength of masonry. Special care regarding the possible change of the structural system should be taken in case the strengthening of the minaret is considered.

Author Contributions: Conceptualization, N.A.; methodology, N.A.; software, F.T. and N.A.; validation, N.A., F.T. and F.C.; formal analysis, N.A., F.T. and F.C.; investigation, M.H. and F.T.; resources, F.T., M.H. and N.A.; writing—original draft preparation, N.A. and F.T.; writing—review and editing, N.A., F.C. and M.H.; visualization, N.A.; supervision, N.A., M.H. and F.C.; project administration, N.A. and F.T.; funding acquisition, N.A. and F.T. All authors have read and agreed to the published version of the manuscript.

Funding: This research was funded by the Federal Ministry of Education and Science, Federation of Bosnia and Herzegovina, Bosnia and Herzegovina in 2021 (No.05-35-1935-1/21), Project "Analysis of the bearing capacity of masonry minarets under seismic action" acronym ANZMSD.

Institutional Review Board Statement: Not applicable.

Informed Consent Statement: Not applicable.

Data Availability Statement: The data are not publicly available due to privacy reasons.

Acknowledgments: The authors give special thanks to the Federal Ministry of Education and Science, Federation of Bosnia and Herzegovina, Bosnia and Herzegovina for its financial support for this project.

Conflicts of Interest: The authors declare no conflict of interest.

References

1. Hadzima-Nyarko, M.; Ademović, N.; Pavić, G.; Sipos, T.K. Strengthening techniques for masonry structures of cultural heritage according to recent Croatian provisions. *Earthq. Struct.* **2018**, *15*, 473–485. [CrossRef]
2. D'Altri, A.M.; Sarhosis, V.; Milani, G.; Rots, J.; Cattari, S.; Lagomarsino, E.; Sacco, A.; Tralli, G.; Castellazi, G.; de Miranda, S. Chapter 1—A review of numerical models for masonry structures. In *Woodhead Publishing Series in Civil and Structural Engineering, Numerical Modeling of Masonry and Historical Structures*; Woodhead Publishing: Kidlington, UK, 2019; pp. 3–53. [CrossRef]
3. Page, A. Finite element model for masonry. *J. Struct. Div.* **1978**, *104*, 1267–1285. Available online: https://ascelibrary.org/doi/10.1061/JSDEAG.0005406 (accessed on 3 November 2022). [CrossRef]
4. Çaktı, E.; Saygili, Ö.; Lemos, J.V.; Oliveira, C.S. Discrete element modeling of a scaled masonry structure and its validation. *Eng. Struct.* **2016**, *126*, 224–236. [CrossRef]
5. Rafiee, A.; Vinches, M. Mechanical behaviour of a stone masonry bridge assessed using an implicit discrete element method. *Eng. Struct.* **2013**, *48*, 739–749. [CrossRef]
6. Pantò, B.; Cannizzaro, F.; Caddemi, S.; Caliò, I. 3D macro-element modelling approach for seismic assessment of historical masonry churches. *Adv. Eng. Softw.* **2016**, *97*, 40–59. [CrossRef]
7. Abbati, S.D.; D'Altri, A.M.; Ottonelli, D.; Castellazzi, G. Seismic assessment of interacting structural units in complex historic masonry constructions by nonlinear static analyses. *Comput. Struct.* **2019**, *213*, 51–71. [CrossRef]
8. Bartoli, G.; Betti, M.; Vignoli, A. A numerical study on seismic risk assessment of historic masonry towers: A case study in San Gimignano. *Bull. Earthq. Eng.* **2016**, *14*, 1475–1518. [CrossRef]
9. Castellazzi, M.; D'Altri, A.M.; de Miranda, S.; Chiozzi, A.; Tralli, A. Numerical insights on the seismic behavior of a non-isolated historical masonry tower. *Bull. Earthq. Eng.* **2018**, *16*, 933–961. [CrossRef]
10. Milani, G.; Shehu, R.; Valente, M. Seismic Assessment of Masonry Towers by means of Nonlinear Static Procedures. In Proceedings of the X International Conference on Structural Dynamics (EURODYN 2017), Rome, Italy, 10–13 September 2017; Volume 199, pp. 266–271. [CrossRef]
11. Micelli, F.; Cascardi, A.; Aiello, M.A. Seismic Capacity Estimation of a Masonry Bell-Tower with Verticality Imperfection Detected by a Drone-Assisted Survey. *Infrastructures* **2020**, *5*, 72. [CrossRef]
12. Torelli, G.; D'Ayala, D.; Betti, M.; Bartoli, G. Analytical and numerical seismic assessment of heritage masonry towers. *Bull. Earthq. Eng.* **2020**, *18*, 969–1008. [CrossRef]
13. de Felice, G. Out-of-Plane Seismic Capacity of Masonry Depending on Wall Section Morphology. *Int. J. Archit. Herit.* **2011**, *5*, 466–482. [CrossRef]
14. Pulatsu, B.; Bretas, E.M.; Lourenco, P.B. Discrete element modeling of masonry structures: Validation and application. *Earthq. Struct.* **2016**, *2016*, 563–582. [CrossRef]
15. Lemos, J.V. Discrete Element Modeling of the Seismic Behavior of Masonry Construction. *Buildings* **2019**, *9*, 43. [CrossRef]
16. Sarhosis, V.; Sheng, Y. Identification of material parameters for low bond strength masonry. *Eng. Struct.* **2014**, *60*, 100–110. [CrossRef]
17. Sarhosis, V.; Garrity, S.W.; Sheng, Y. Influence of brick–mortar interface on the mechanical behaviour of low bond strength masonry brickwork lintels. *Eng. Struct.* **2015**, *88*, 1–11. [CrossRef]
18. Giamundo, V.; Sarhosis, V.; Lignola, G.P.; Sheng, Y.; Manfredi, G. Evaluation of different computational modelling strategies for the analysis of low strength masonry structures. *Eng. Struct.* **2014**, *73*, 160–169. [CrossRef]
19. Mohebkhah, A.; Sarhosis, V. Discrete Element Modeling of Masonry-Infilled Frames. In *Computational Modeling of Masonry Structures Using the Discrete Element Method*; IGI Global: Hershey, PA, USA, 2016; pp. 200–235. [CrossRef]
20. Dimitri, R.; Zavarise, G. Numerical Study of Discrete Masonry Structures under Static and Dynamic Loading. In *Computational Modeling of Masonry Structures Using the Discrete Element Method*; IGI Global: Hershey, PA, USA, 2016; pp. 254–291. [CrossRef]
21. Alexakis, H.; Makris, N. Validation of the Discrete Element Method for the Limit Stability Analysis of Masonry Arches. In *Computational Modeling of Masonry Structures Using the Discrete Element Method*; IGI Global: Hershey, PA, USA, 2016; pp. 292–325. [CrossRef]
22. Drei, A.; Milani, G.; Sincraian, G. Application of DEM to Historic Masonries, Two Case-Studies in Portugal and Italy: Aguas Livres Aqueduct and Arch-Tympana of a Church. In *Computational Modeling of Masonry Structures Using the Discrete Element Method*; IGI Global: Hershey, PA, USA, 2016; p. 41. [CrossRef]
23. Meguro, K.; Tagel-Din, H. Applied Element Method For Structural Analysis: Theory and application for linear materials. *Struct. Eng. Earthq. Eng. JSCE* **2000**, *17*, 21–35. [CrossRef]
24. Meguro, K.; Tagel-Din, H. Applied Element Simulation of RC Structures under Cyclic Loading. *J. Struct. Eng.* **2001**, *127*, 1295–1305. [CrossRef]
25. Meguro, K.; Tagel-Din, H. Applied Element Method Used for Large Displacement Structural Analysis. *J. Nat. Disaster Sci.* **2002**, *24*, 25–34.
26. Malomo, D.; Pinho, R.; Penna, A. Applied Element Modelling of the Dynamic Response of a Full-Scale Clay Brick Masonry Building Specimen with Flexible Diaphragms. *Int. J. Archit. Herit.* **2020**, *14*, 1484–1501. [CrossRef]
27. Salem, H.; Gregori, A.; Helmy, H.; Fassieh, K.; Tagel-Din, H. Seismic Assessment of the Damaged Margherita Palace. In Proceedings of the 16th World Congress on Earthquake Engineering (16WCEE), Santiago, Chile, 9–13 January 2017. Available online: https://www.wcee.nicee.org/wcee/article/16WCEE/WCEE2017-2399.pdf (accessed on 1 November 2022).

28. Diana, L.; Reuland, Y.; Lestuzzi, P. Seismic Vulnerability Assessment of "Sion Cathedral" (Switzerland): An Integrated Approach to Detect and Evaluate Local Collapse Mechanisms in Heritage Buildings. In Proceedings of the 3rd International Conference on Protection of Historical Constructions, PROHITECH'17, Lisbon, Portugal, 12–15 July 2017.
29. Karbassi, A.; Nollet, M.J. Performance-Based Seismic Vulnerability Evaluation of Masonry Buildings Using Applied Element Method in a Nonlinear Dynamic-Based Analytical Procedure. *Earthq. Spectra* **2013**, *29*, 399–426. [CrossRef]
30. Zerin, A.I.; Hosoda, A.; Salem, H.; Amanat, K.M. Seismic Performance Evaluation of Masonry Infilled Reinforced Concrete Buildings Utilizing Verified Masonry Properties in Applied Element Method. *J. Adv. Concr. Technol.* **2017**, *15*, 227–243. [CrossRef]
31. Clotaire, M.; Karbassi, A.; Lestuzzi, P. Evaluation of the seismic retrofitting of an unreinforced masonry building using numerical modeling and ambient vibration measurements. *Eng. Struct.* **2018**, *158*, 124–135. [CrossRef]
32. Oliveira, C.S.; Çaktı, E.; Stengel, D.; Branco, M. Minaret Behaviour under Earthquake Loading: The Case of Historical Istanbul. *Earthq. Eng. Struct. Dyn.* **2012**, *41*, 19–39. [CrossRef]
33. Çaktı, E.; Oliveira, S.; Carlos, S.; Lemos, J.V.; Özden, S.; Serkan, G.; Esra, Z. Earthquake behavior of historical minarets in Istanbul. In Proceedings of the 4th ECCOMAS Thematic Conference on Computational Methods in Structural Dynamics and Earthquake Engineering, Kos Island, Greece, 12–14 June 2013.
34. Mutlu, Ö.; Sahin, A. Investigating the effect of modeling approaches on earthquake behavior of historical masonry Minarets-Bursa, Grand Mosque case study. *Sigma* **2016**, *7*, 123–136.
35. Livaoğlu, R.; Baştürk, M.H.; Doğangün, A.; Serhatoğlu, C. Effect of geometric properties on dynamic behavior of historic masonry minaret. *KSCE J. Civ. Eng.* **2016**, *20*, 2392–2402. [CrossRef]
36. Döven, M.S.; Serhatoğlu, C.; Kaplan, O.; Livaoğlu, R. Dynamic behaviour change of Kütahya Yeşil minaret with covered and open balcony architecture. *Eskişehir Tech. Univ. J. Sci. Technol. B Theor. Sci.* **2018**, *6*, 192–203.
37. Portioli, F.; Mammana, O.; Landolfo, R.; Mazzolani, F.M.; Krstevska, L.; Tashkov, L.; Gramatikov, K. Seismic Retrofitting of Mustafa Pasha Mosque in Skopje: Finite Element Analysis. *J. Earthq. Eng.* **2011**, *15*, 620–639. [CrossRef]
38. Krstevska, L.; Tashkov, L.; Gramatikov, K.; Landolfo, R.; Mammana, O.; Portioli, F.; Mazzolani, F.M. Large-scale experimental investigation on Mustafa-Pasha mosque. *J. Earthq. Eng.* **2010**, *14*, 842–873. [CrossRef]
39. Tashkov, L.A.; Krstevska, L.S.; Safak, E.; Çakti, E.; Edincliler, A.; Erdik, M. Comparative study of large and medium scale mosque models tested on seismic shaking table. In Proceedings of the 15th World Conference on Earthquake Engineering 2012, Lisbon, Portugal, 24–28 September 2012.
40. Yekrangnia, M.; Mobarake, A.A. Restoration of Historical Al-Askari Shrine. II: Vulnerability Assessment by Numerical Simulation. *J. Perform. Constr. Facil.* **2016**, *30*, 04015031. [CrossRef]
41. Ebrahimiyan, M.; Golabchi, M.; Yekrangnia, M. Field Observation and Vulnerability Assessment of Gonbad-e Qābus. *J. Archit. Eng.* **2017**, *23*, 05017008. [CrossRef]
42. Bağbancı, M.B.; Bağbancı, K.Ö. The Effects of Construction Techniques and Geometrical Properties on the Dynamic Behavior of Historic Timber Minarets in Sakarya, Turkey. *Shock. Vib.* **2018**, *2018*, 9853896. [CrossRef]
43. Boğaziçi University Kandilli Observatory and Earthquake, Research Institute. 15 OCAK 2023 ML = 4.9 Kavakköy Sivrice Elaziğ Depremi. 2017. Available online: http://www.koeri.boun.edu.tr/sismo/2/deprem-bilgileri/buyuk-depremler/ (accessed on 5 December 2022).
44. Işık, E.; Harirchian, E.; Arkan, E.; Avcil, F.; Günay, M. Structural Analysis of Five Historical Minarets in Bitlis (Turkey). *Buildings* **2022**, *12*, 159. [CrossRef]
45. Işık, E.; Avcil, F.; Harirchian, E.; Arkan, E.; Bilgin, H.; Özmen, H.B. Architectural Characteristics and Seismic Vulnerability Assessment of a Historical Masonry Minaret under Different Seismic Risks and Probabilities of Exceedance. *Buildings* **2022**, *12*, 1200. [CrossRef]
46. Gunes, B.; Cosgun, T.; Sayin, B.; Ceylan, O.; Mangir, A.; Gumusdag, G. Seismic assessment of a reconstructed historic masonry structure: A case study on the ruins of Bigali castle mosque built in the early 1800s. *J. Build. Eng.* **2021**, *39*, 102240. [CrossRef]
47. Yetkin, M.; Dedeoğlu, İ.Ö.; Calayır, Y. Investigation and assessment of damages in the minarets existing at Elazig after 24 January 2020 Sivrice earthquake. *Fırat Univ. J. Eng. Sci.* **2021**, *33*, 379–389.
48. Casolo, S.; Diana, V.; Uva, G. Influence of soil deformability on the seismic response of a masonry tower. *Bull. Earthq. Eng.* **2017**, *15*, 1991–2014. [CrossRef]
49. Haciefendioğlu, K.; Alpaslan, E.; Demir, G.; Dinc, B.; Birinci, F. Experimental modal investigation of scaled minaret embedded in different soil types. *Gradevinar* **2018**, *70*, 201–212. [CrossRef]
50. Altiok, T.Y.; Demir, A. Collapse mechanism estimation of a historical masonry minaret considered soil-structure interaction. *Earthq. Struct.* **2021**, *21*, 161–172. [CrossRef]
51. Akhlaq, H.; Butt, F.; Alwetaishi, M.; Riaz, M.; Benjeddou, O.; Hussein, E.E. Structural Identification of a 90 m High Minaret of a Landmark Structure under Ambient Vibrations. *Buildings* **2022**, *12*, 252. [CrossRef]
52. Işık, E.; Ademović, N.; Harirchian, E.; Avcil, F.; Büyüksaraç, A.; Hadzima-Nyarko, M.; Akif Bülbül, M.; Işık, M.F.; Antep, B. Determination of Natural Fundamental Period of Minarets by Using Artificial Neural Network and of the Impact of Different Materials on Their Seismic Vulnerability. *Appl. Sci.* **2023**, *13*, 809. [CrossRef]
53. Casarin, F.; Humo, M.; Lorenzoni, F.; da Porto, F.; Girardello, P.; Modena, C.; Cantini, L.; Kulukčija, S. Experimental Investigations On Two Traditional Minarets in Bosnia and Herzegovina. *Struct. Anal. Hist. Constr.* **2012**, *3*, 2349–2357.

54. Doğangün, A.; Sezen, H.; Tuluk, Ö.İ.; Livaoğlu, R.; Acar, R. Traditional Turkish masonry monumental structures and their earthquake response. *Int. J. Archit. Herit.* **2007**, *1*, 251–271. [CrossRef]

55. Hasandedić, H. *Mostarski Vakifi I Njihovi Vakufi*; Medžlis Islamske Zajednice Mostar: Mostar, Bosnia and Herzegovina, 2000; pp. 115–120.

56. Center for information and documentation Mostar. Džamija Tabačica Razorena. 2015. Available online: https://www.cidom.org/?attachment_id=4972 (accessed on 9 November 2022).

57. Islamic Community in Bosnia and Herzegovina, Mufti unit of Mostar. HVO Je U Mostaru, Stocu I Čapljini Srušio 40 Džamija. 2017. Available online: http://www.muftijstvo-mostarsko.ba/arhiva/index.php/glamoc/item/1554-hvo-je-u-mostaru-stocu-i-capljini-srusio-40-dzamija (accessed on 9 November 2022).

58. Institute for Standardization of Bosnia and Herzegovina. *National Annex BAS EN 1998-1:2018—Design of Structures for Earthquake Resistance—Part 1: General Rules. Seismic Actions and Rules for Buildings*; BAS: Istocno Sarajevo, Bosnia and Herzegovina, 2018.

59. Ademović, N.; Kalman Šipoš, T.; Hadzima-Nyarko, M. Rapid assessment of earthquake risk for Bosnia and Herzegovina. *Bull. Earthq. Eng.* **2020**, *18*, 1835–1863. [CrossRef]

60. Uluengin, F.; Uluengin, B.; Uluengin, M.B. *Classic Construction Details of Ottoman Monumental Architecture*; Yapı-Endüstri Merkezi Yayinlari: Istanbul, Turkey, 2007.

61. Ademović, N.; Kurtović, A. Influence of planes of anisotropy on physical and mechanical properties of freshwater limestone (Mudstone). *Constr. Build. Mater.* **2021**, *268*, 121174. [CrossRef]

62. Šaravanja, K.; Popić, D.; Marić, J.; Radić-Kustura, J. Analysis of the available test results of the "Tenelija" stone. *Electron. Collect. Work. Fac. Civ. Eng. Univ. Most.* **2018**, 245–273. Available online: https://hrcak.srce.hr/file/303671 (accessed on 13 December 2022).

63. Zvonić, Z. Tenelija—The stone of Mostar. *Most. Mag. Educ. Sci. Cult.* **2000**, 130. Available online: http://www.most.ba/041/067.htm. (accessed on 8 November 2022).

64. Elaborate on the classification, categorization and calculation of stocks of architectural and building stone of oolitic limestone (Tenelija and Miljevina) in the research area "Mukoša" in Mostarsko polje. In *"Dent" Research, Exploitation and Processing of Stone*; Company Dent: Mostar, Bosnia and Herzegovina, 1999.

65. Marić, L. *O Kamenu Od Koga Je Sadjelan Stari Most U Mostaru*; Prosvjeta: Zagreb, Croatia, 1972.

66. Landesgewerbanstalt Bayern-Historical Bridges Group. *Masonry Tests and Monitoring System*; Landesgewerbanstalt Bayern-Historical Bridges Group: Nürnberg, Germany, 2003.

67. Bilopavlović, V.; Šaravanja, K.; Pekić, S. Testing of petrographic, physical and mechanical properties of "Tenelija" and "Miljevina" stones. *Electron. Collect. Work. Fac. Civ. Eng. Univ. Most.* **2013**, 104–111. Available online: https://fgag.sum.ba/e-zbornik/e_zbornik_06_08.pdf. (accessed on 2 November 2022).

68. Madžić, K. Geotehničke Karakteristike Tenelije. Inžinjersko i Geotehnička Svojstva Kamena Tenelije. Master's Thesis, Univerzitet u Tuzli, Tuzla, Bosnia and Herzegovina, 2006; pp. 31–67.

69. Nakamura, Y. A Method for Dynamic Characteristics Estimation of Subsurface using Microtremor on the Ground Surface. *Q. Rep. Railw. Tech. Res. Inst.* **1989**, *30*, 25–33.

70. Nogoshi, M.; Igarashi, T. On the amplitude characteristics of microtremor (part 2). *J. Seismol. Soc. Jpn.* **1971**, *24*, 26–40. Available online: https://www.jstage.jst.go.jp/article/zisin1948/24/1/24_1_26/_article/-char/en (accessed on 2 November 2022).

71. Okada, H. *The Microtremor Survey Method*; Society of Exploration Geophysicists: Houston, TX, USA, 2003; Volume 12. [CrossRef]

72. Triwulan, W.; Utama, D.; Warnana, D.; Sungkono, S. Vulnerability index estimation for building and ground using microtremor. *Second. Int. Semin. Appl. Technol. Sci. Arts* **2010**, 1194–1997. Available online: https://www.academia.edu/3520308/vulnerability_index_estimation_for_building_and_ground_using_microtremor (accessed on 2 November 2022).

73. Luo, G.C.; Liu, L.B.; Cheng, Q.; Chen, Q.F.; Chen, Y.P. Structural response analysis of a reinforced concrete building based on excitation of microtremors and passing subway trains. *Chin. J. Geophys.* **2011**, *54*, 2708–2715.

74. Pentaris, F.P. A novel horizontal to vertical spectral ratio approach in a wired structural health monitoring system. *J. Sens. Sens. Syst.* **2014**, *3*, 145–165. [CrossRef]

75. Zahid, F.B.; Ong, Z.C.; Khoo, S.Y. A review of operational modal analysis techniques for in-service modal identification. *J. Braz. Soc. Mech. Sci. Eng.* **2020**, *42*, 398. [CrossRef]

76. Calik, I.; Bayraktar, A.; Türker, T.; Karadeniz, H. Structural dynamic identification of a damaged and restored masonry vault using Ambient Vibrations. *Measurement* **2014**, *55*, 462–472. [CrossRef]

77. Nohutcu, H.; Demir, A.; Ercan, E.; Altıntaş, G.; Hökelekli, E. Investigation of a historic masonry structure by numerical and operational modal analyses. *Struct. Des. Tall Spec. Build.* **2015**, *24*, 821–834. [CrossRef]

78. Demir, A.; Nohutcu, H.; Ercan, E.; Hokelekli, E.; Altintas, G. Effect of model calibration on seismic behaviour of a historical mosque. *Struct. Eng. Mech.* **2016**, *60*, 749–760. [CrossRef]

79. Konno, K.; Ohmachi, T. Ground-motion characteristics estimated from spectral ratio between horizontal and vertical components of microtremor. *Bull. Seismol. Soc. Am.* **1998**, *88*, 228–241. [CrossRef]

80. *Report about the Performed Geophysical Investigations at the Location of Tabačica Mosque in Mostar*; Terra Compacta d.o.o: Zagreb, Croatia, 2022.

81. *ARTeMIS Modal 7.2.0.2 ×64*; Structural Vibrations Solution A/S; NOVI Science Park, DK 9220: Aalborg, Denmark. Available online: https://www.energy-xprt.com/companies/structural-vibration-solutions-a-s-8720 (accessed on 13 December 2022).

82. Rainieri, C.; Fabbrocino, G.; Cosenza, E. Some remarks on experimental estimation of damping for seismic design of civil constructions. *Shock Vib.* **2010**, *17*, 383–395. [CrossRef]

83. Elnashai, A.S.; Disarno, L. *Fundamentals of Earthquake Engineering*; Wiley: Hoboken, NJ, USA, 2008.

84. Karatzetzou, A.; Pitilakis, D.; Karafagka, S. System Identification of Mosques Resting on Soft Soil. The Case. *Geosciences* **2021**, *11*, 275. [CrossRef]

85. Binda, L.; Saisi, A.; Tiraboschi, C. Application of sonic test to the diagnosis of damage and repaired structures. *Non-Destr. Test. Eval. Int.* **2001**, *34*, 123–138. [CrossRef]

86. Casarin, F.; Valluzi, M.R.; da Porto, F.; Modena, C. Evaluation of the structural behaviour of historic masonry buildings by using a sonic pulse velocity method. In *Structural Studies, Repairs and Maintenance of Heritage Architecture X. WIT Transactions on The Built Environment*; WIT Press: Billerica, MA, USA, 2007; Volume 95, pp. 227–236. [CrossRef]

87. Bindemitteln, F. MS.D.1. Measurement of mechanical pulse velocity for masonry. *Mater. Struct.* **1996**, *29*, 463–466.

88. Schuller, M.P. Nondestructive testing and damage assessment of masonry structures. In Proceedings of the NSF/RILEM Workshop, In-Situ Evaluation of Historic Wood and Masonry Structures, Prague, Czech, 10–14 July 2006; pp. 67–86.

89. Grazzini, A. Sonic and impact test for structural assessment of historical masonry. *Appl. Sci.* **2019**, *9*, 5148. [CrossRef]

90. Casarin, F.; Modena, C. Seismic Assessment of Complex Historical Buildings: Application to Reggio Emilia Cathedral, Italy. *Int. J. Archit. Herit.* **2008**, *2*, 304–327. [CrossRef]

91. Reuland, Y.; Jaoude, A.; Lestuzzi, P.; Smith, I. Usefulness of ambient-vibration measurements for seismic assessment of existing structures. In Proceedings of the Fourth International Conference on Smart Monitoring, Assessment and Rehabilitation of Civil Structures (SMAR), Zurich, Switzerland, 13–15 September 2017; p. 162. Available online: https://www.researchgate.net/publication/319998448 (accessed on 13 December 2022).

92. Belec, G. Seismic Assessment of Unreinforced Masonry Buildings in Canada. Master's Thesis, University of Ottawa, Ottawa, ON, Canada, January 2016. Available online: https://ruor.uottawa.ca/handle/10393/34301 (accessed on 5 November 2022).

93. Helmy, H.; Salem, H.; Tageldin, H. Numerical Simulation of Charlotte Coliseum Demolition Using the Applied Element Method. In Proceedings of the 6th International Engineering and Construction Conference (IECC'6), Cairo, Egypt, 28–30 June 2010; pp. 528–537.

94. Extreme Loading®for Structures Software (ELS Software). 2021. Available online: https://www.appliedscienceinteurope.com/extreme-loading-for-structures/ (accessed on 6 November 2022).

95. Usta, P. Assessment of seismic behavior of historic masonry minarets in Antalya, Turkey. *Case Stud. Constr. Mater.* **2021**, *15*, e00665. [CrossRef]

96. Norme Tecniche per le Costruzioni (NTC). Ministero delle Infrastrutture e dei Trasporti, Norme Tecniche per le Costruzio-ni, NTC 2008, (English: Technical Standards for Construction), D.M. del Ministero delle Infrastrutture e dei Trasporti del 14/01/2. *Case Stud. Constr. Mater.* **2020**, *13*, 235. Available online: https://www.studiopetrillo.com/files/ntc2008-completa.pdf. (accessed on 8 November 2022).

97. Adam, M.A.; El-Salakawy, T.S.; Salama, M.A.; Mohamed, A.A. Assessment of structural condition of a historic masonry minaret in Egypt. *Case Stud. Constr. Mater.* **2020**, *13*, e00409. [CrossRef]

98. Shakya, M.; Varum, H.; Vicente, R.; Costa, A. Empirical formulation for estimating the fundamental frequency of slender masonry structures. *Int. J. Archit. Herit.* **2016**, *10*, 55–66. [CrossRef]

99. Faccio, S.; Podestà, A.; Saetta, A. Venezia, il campanile della chiesa di Sant'Antonin, in MIBAC. In *Linee Guida per la Valutazione e Riduzione del Rischio Sismico del Patrimonio Culturale*; Gangemi: Rome, Italy, 2010; pp. 292–323; ISBN 9788849220292.

100. Rainieri, C.; Fabbrocino, G. Estimating the Elastic Period of Masonry Towers. In *Topics in Modal Analysis I*; Springer: New York, NY, USA, 2012; Volume 5, pp. 243–248. [CrossRef]

101. Ural, A.; Çelik, T. Dynamic analyses and seismic behavior of masonry minarets with single balcony. *Aksaray J. Sci. Eng.* **2018**, *2*, 13–27. [CrossRef]

102. EN 1998-1:2004; Eurocode 8: Design of Structures for Earthquake Resistance—Part 1: General Rules, Seismic Actions and Rules for Buildings. *European Union: Brussels, Belgium*. 2004. Available online: https://www.confinedmasonry.org/wp-content/uploads/2009/09/Eurocode-8-1-Earthquakes-general.pdf (accessed on 20 November 2022).

103. Turk, A.M.; Cosgun, C. Seismic Behaviour and Retrofit of Historic Masonry Minaret. *Građevinar* **2012**, *64*, 39–45. [CrossRef]

104. Dogangun, A.; Acar, R.; Sezen, H.; Livaoglu, R. Investigation of dynamic response of masonry minaret structures. *Bull. Earthq. Eng.* **2008**, *6*, 505–517. [CrossRef]

105. Dogangun, A.; Acar, R.; Livaoglu, R.; Tuluk, I. Performance of Masonry Minarets Against Earthquakes and Winds in Turkey. In Proceedings of the First International Conference on Restoration of Heritage Masonry Structures, Cairo, Egypt, 24–27 April 2006.

Article

Post-Earthquake Assessment and Strengthening of a Cultural-Heritage Residential Masonry Building after the 2020 Zagreb Earthquake

Naida Ademović [1,*], Mirko Toholj [2], Dalibor Radonić [3,†], Filippo Casarin [4], Sanda Komesar [3] and Karlo Ugarković [3]

1   University of Sarajevo—Faculty of Civil Engineering, 71000 Sarajevo, Bosnia and Herzegovina
2   T&E Sarajevo, 71000 Sarajevo, Bosnia and Herzegovina
3   Projekt na kvadrat d.o.o., 10000 Zagreb, Croatia
4   Expin srl, 35138 Padovana, Italy
*   Correspondence: naida.ademovic@gf.unsa.ba or naidadem@yahoo.com
†   This author passed away.

Abstract: After a long period of no excessive ground shaking in Croatia and the region of ex-Yugoslavia, an earthquake that woke up the entire region was the one that shook Croatia on 22 March 2020. More than 25,000 buildings were severely damaged. A process of reconstruction and strengthening of existing damaged buildings is underway. This paper presents proposed strengthening measures to be conducted on a cultural-historical building located in the city of Zagreb, which is under protection and located in zone A. After a detailed visual inspection and on-site experimental investigations, modeling of the existing and strengthened structure was performed in 3Muri. It is an old unreinforced masonry building typical not only for this region but for relevant parts of Europe (north, central, and east). The aim was to strengthen the building to Level 3 while respecting the ICOMOS recommendations and Venice Charter. Some non-completely conservative concessions had to be made, to fully retrofit the building as requested. The structural strengthening consisted of a series of organic interventions relying on—in the weakest direction—a new steel frame, new steel-ring frames, and FRCM materials, besides fillings the cracks. Such intervention resulted in increasing the ultimate load in the X and Y directions, respectively, more than 650 and 175% with reference to the unstrengthened structure. Good consistency was obtained between the numerical modeling, visual inspection, and on-site testing.

Keywords: masonry cultural heritage building; strengthening; pushover analysis; FRCM; 3MURI; earthquake; Venice Charter; ICOMOS recommendations; ambient testing; numerical modeling

Citation: Ademović, N.; Toholj, M.; Radonić, D.; Casarin, F.; Komesar, S.; Ugarković, K. Post-Earthquake Assessment and Strengthening of a Cultural-Heritage Residential Masonry Building after the 2020 Zagreb Earthquake. *Buildings* **2022**, *12*, 2024. https://doi.org/10.3390/buildings12112024

Academic Editors: Rajesh Rupakhety and Dipendra Gautam

Received: 14 October 2022
Accepted: 15 November 2022
Published: 18 November 2022

Publisher's Note: MDPI stays neutral with regard to jurisdictional claims in published maps and institutional affiliations.

1. Introduction

Earthquakes are one of the natural hazards which can have a dreadful influence on existing buildings, population, economy, and community as a whole. Multiple consequences of the ground shaking can lead to the devastation of complete districts and even some small villages can be completely erased by destructive earthquakes, such as the case of the village of Onna in Italy in 2009. On 22 March 2020, around half past seven in the morning, an earthquake woke up the citizens of Zagreb. The earthquake was of magnitude $M_L = 5.5$, $Mw = 5.3$, and the intensity in the Zagreb Metropolitan area was measured as VII-VIII according to the Medvedev–Sponheuer–Karnik scale [1]. Additionally, the fact that this was a shallow earthquake having a depth of only 10 km influenced the degree of the building's damage. The measured peak ground acceleration (PGA) for the leading North-South direction was 0.22 g [2]. This main shock was followed by numerous aftershocks and the ground continued to shake for several months. It was 142 years before that Zagreb was shaken by an earthquake, known as the Great Zagreb earthquake, with an estimated magnitude (M_L = 6.3) [3], according to macroseismic observations, which had caused tremendous damage. In the building's assessment report, it was stated that 1758 residential buildings were damaged and out of that 27.6% were heavily damaged (Figure 1a). Numerous buildings of various usage (residential buildings, public buildings, medical facilities, historical monuments, etc.) were damaged by the 2020 Zagreb earthquake, having different levels of damage from minor damage to severe damage, and even partial collapse of buildings (Figure 1b). The most affected area was in the vicinity of the epicenter, the Lower Town of the city of Zagreb, constituted of cultural heritage buildings. Additionally, buildings constructed in the late 19th and 20th centuries, mainly unreinforced masonry structures were damaged to a great extent. It is estimated that up to 25,000 buildings were affected by this earthquake. Luckily, not many human fatalities were registered (one human life was lost) and this can be connected to the COVID-19 restrictions at that time.

(a) **(b)**

Figure 1. Building devastation after: (**a**) 1880 Zagreb earthquake [4] Reproduced with permission from Zagreb City Museum, 2022.; (**b**) 2020 Zagreb earthquake (author's figures).

Based on the available statistical census data [5], the percentage of dwellings constructed before 1945 was 13.3%, from 1946 to 1970 was 29.1%, from 1971 to 1980 was 22.1%, from 1981 to 1990 was 16.8%, from 1991 to 2005 was 13.6%, and after 2006 was 5.1%. The first group can be subcategorized into a period before 1919 (7.63%) and from 1919 to 1945 (5.78%). The categorization of building according to construction age (six groups) is connected with the introduction, revision, and upgrading of seismic codes in ex-Yugoslavia which were applied to Bosnia and Herzegovina [6], Croatia, and other ex-Yugoslavian Republics. For the city of Zagreb, a very similar trend is observed for certain construction periods (13.3%; 30.4%; 17.8%; 14.7%; 15.2; and 8.8%, respectively) [7]. Unreinforced masonry structures (URM) mainly constructed between 1860 and 1920 make up the historical cores of Croatian cities, so does the city of Zagreb as well, where this percentage is 3.9. It was all until 1920 that the floors were made with timber flexible floors which were not well connected to the walls and the connection of the wall was poor and inadequate. Buildings constructed until 1948 were built during the period when there were no standards regarding seismic actions in this region. The first official document which had minimum seismic requirements was the PTP2 (1948) [8], having no limitation on the height of the various masonry building, which would be modified in 1964 regulations PTP-GuSP64 [9].

During the 19th century, the material which was mainly used for construction was adobe, stone, brick, timber, and steel profiles to a minimum extent. Family buildings (an example of the considered case study) in this period (until 1920) were constructed as URM structures with wooden floors, except for the basement ceilings which were very often constructed with iron beams and segmental brick vaults. This is the case as well for residential multistorey buildings. These structures usually have an adequate load-bearing capacity for standard loads except for seismic loads. It was in 1930 that the timber floors were starting to be replaced by the semi-prefabricated ribbed reinforced concrete floor, having a concrete layer of around 6.5 cm which could be considered as a rigid diaphragm [10]. Buildings with rigid floors had a smaller level of damage compared to the ones with wooden floors. After II World War, until 1964, monolithic reinforced concrete floors started to be introduced in the building's construction. After the 1963 Skopje earthquake, confined masonry was massively implemented in this region. This was as well an era of construction of residential buildings made of reinforced concrete load-bearing systems in line with the regulations for 1964 and then later stricter regulation regarding seismic actions enforced in 1981 after the 1979 Montenegro earthquake. Eurocodes were introduced in a step-by-step sequence firstly having a pre-standard status for the duration of 6 years (1992–1998) and then with a full EN label from 1998. The final implementation of Eurocode standards started in 2005 [7].

After the 2020 Zagreb earthquake, it was necessary to collect information about the type and level of damage. As there were no post-earthquake forms for Croatia, it was decided to use the Italian forms and adjust them to Croatia [11]. After conducting the preliminary inspection, each inspected building was marked with the appropriate color and label. Buildings were categorized into three usability categories marked with three colors. The green color indicated that the building is usable either with limitations (U1) or with recommendations (U2). Buildings marked with a yellow sticker indicated that they are temporarily unusable: either that a detailed inspection is required (PN1) or that short-term countermeasures are required (PN2). Unusable buildings either due to external risk (N1) or due to damage (N2) were marked with red color.

Several papers have been published regarding the rehabilitation of damaged buildings after the 2020 Zagreb earthquake. One building of the infantry barrack of Prince Rudolf (built from 1887 to 1889) located in the Lower Town of the city of Zagreb was researched by [12]. It is listed as one of the Protected Cultural Heritage buildings being located within the A protection zone of the Historical and Urban Entity of the city of Zagreb. The building is representative of a typical URM structure, where the load-bearing walls are made of Austro-Hungarian solid bricks of the old standard (14 $\times$ 6.5 $\times$ 29 cm). The thickness of the walls reduces from the basement to the upper floors. The ceiling floor is made of a brick vault while the other floors are made of wooden and steel beams. In order to increase the capacity of the brick load-bearing walls, fibre reinforced cementitious matrix (FRCM) strengthening system or concrete jacketing was proposed. Additionally, the removal of brick partition walls was proposed and replacement with a drywall system. In order to increase the stiffness of the wooden floors, a thin reinforced concrete compression slab was proposed [12]. Moreover, two additional strengthening ideas proposed a new steel equivalent system and seismic isolation. A step further was conducted in [13], where three possible ways of strengthening a URM structure used for education purposes built in the 19th century were elaborated. Strengthening was proposed solely by FRCM, shotcrete, and then their combination. The analyses were implemented through the application of the static nonlinear method in 3Muri software. Besides the increase of the load-bearing capacity, the estimation of the expected cost and environmental impact of each proposed remedial measure was investigated. The lowest cost and the highest capacity increase were obtained for strengthening the structure with shotcrete; however, its application was proven to be too invasive for cultural heritage structures [13]. The FRCM system had a much smaller emission of CO_2 than the shotcrete application.

A remarkable building located at Matice Hrvatske 2 Street in Zagreb, identified as one of the most significant structures in the whole of Croatia [14], was investigated in the paper [15]. The building was constructed in 1887 and during its life, several adaptations were conducted. It is an L-shaped corner building within an old masonry building block. The vertical loadbearing system is composed of Austro-Hungarian solid bricks connected with lime mortar. This structure is as well under heritage protection. The building was modeled as a stand-alone structure due to a lack of information regarding the adjacent buildings. Good consistency was obtained between the modeling and the noted damage after the 2020 Zagreb earthquake.

Our paper is structured in the way that in Section 2 a description of the methodology which has been applied is briefly explained and discussed. The intention was to emphasize the importance of the application of the ICOMOS recommendations [16] and the principles of the Venice Charter when reconstructing buildings of cultural heritage. Section 3 is dedicated to the selected case study building Ribnjak 44, a building located within the protection zone "A". The section opens with the historical information which was obtained about the buildings, and the defined classification of the building after the 2020 Zagreb earthquake according to the EMS-98 classification. In Section 3.3, a detailed explanation is provided about the conducted visual inspection of the buildings and crack pattern survey. Once the visual inspection was conducted, it was necessary to make on-site investigations and perform tests required for the determination of the in-situ compressive stress level as well as deformability characteristics of masonry (modulus of elasticity) and dynamic characteristics of the structure (Section 3.4). Section 4 is dedicated to the modeling of the existing structure, taking into account all the data provided in the previous sections. The nonlinear static analysis (pushover analysis) was performed in the 3Muri software. Once the results were obtained, it was concluded that the structure does not possess adequate capacity, requiring it to be strengthened. This leads to Section 5 which elaborates on various strengthening procedures and their modeling in the 3Muri software. An adequate choice of various strengthening procedures had to be selected in order to follow and respect the recommendations regarding material compatibility with existing materials, application of reversible methods as much as possible, and methods that are least invasive and most compatible with heritage values, matching the need for safety and durability. Section 6 compares the results of the unstrengthened and strengthened structure and indicates the effectiveness of the strengthening techniques. The paper closes with a conclusion and provides several recommendations and explains the contributions of the research to knowledge in the European context. It indicates some encountered challenges and provides suggestions on how to overcome these issues.

2. Applied Methodology

In order to propose and design a suitable intervention and strengthen the existing buildings, proper investigation and diagnosis are the key factors. The importance of investigations for the determination of structural diagnosis (including historical aspects, materials, and structures) is emphasized in the Venice Charter (1964). Investigations must take into account a variety of different aspects including the typology of the building, the type of masonry elements, connections between elements, characteristics of materials, etc. Each structure has unique characteristics and the investigations must be planned and executed to ensure that an adequate understanding of the structure is obtained [17]. The methodology can be divided into two phases. In the first phase, it is necessary to gather as much information as possible regarding the structure (historical information, description of the building, survey and description of the damage pattern, and in-situ and laboratory experimental tests). The second phase is devoted to the numerical analysis where it is necessary to select what is the most effective modeling type (block-based models, continuum homogeneous models, geometry-based models, or equivalent frame models) and type of analysis (linear or nonlinear). Once numerical modeling is conducted, based on the obtained results it is necessary to determine appropriate interventions for strengthening the structure. The use of this phased multidisciplinary procedure is essential for an in-depth understanding of structures. This kind of analysis allows the performance of knowledge-based structural analysis and thus defines with more confidence the strengthening interventions. It can as well prevent the execution of intrusive repair works. Its application gains special interest when structures are located in areas with high seismic risk [18].

ICOMOS [16] has published and approved the Recommendations for the Analysis, Conservation and Structural Restoration of Architectural Heritage. To make a decision regarding which remedial measures will be selected and applied, it is required to apply a methodology (Figure 2) which is an iterative process between the tasks of data acquisition, structural behavior, and diagnosis and safety [19]. The correct interpretation of the diagnosis and safety from a qualitative and quantitate perspective is of the utmost importance as it will lead to the type and degree of remedial measures.

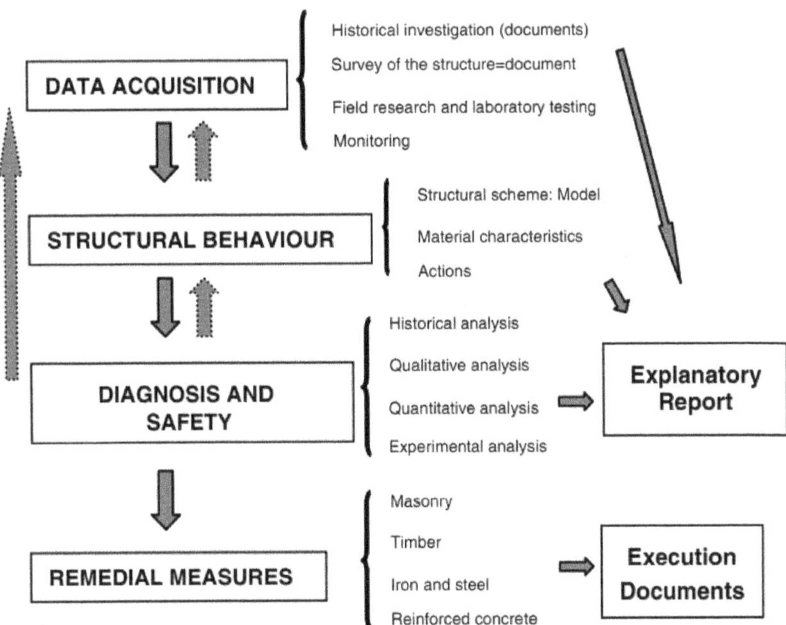

Figure 2. Flowchart with the methodology for structural interventions proposed by [16,19]. Reproduced with permission from Lourenço, P.B., 2022.

Comprehensive and in-depth knowledge of the structure and material characteristics, state of damage, and crack pattern with the causes is a prerequisite for adequate and correct rehabilitation of the structure. It is only once an accurate diagnosis of the structure's damage stated has been determined that adequate prevention and rehabilitation measures can be effectively conducted. Incorrect diagnosis may lead to inadequate strengthening recommendations, which may be more harmful [20]. This methodology has been used for the assessment of several buildings [17,19,21–26]. The research conducted by Ademović [17] and Ademović et al. [21] applied this methodology in the seismic assessment of a typical residential building constructed in Sarajevo (Bosnia and Herzegovina) in 1957. Iterative procedure in the view of various strengthening procedures was investigated in [22]. In this work, three strengthening proposals were given: the addition of new walls, the addition of new walls and a tie, and the addition of new wall and fiber reinforced polymers (FRP) on the ground floor. It was noted that the inclusion of ties provides a higher capacity of the global structure in relation to the strengthening by FRP. However, this did not solve the issue of localized damage on the ground floor, which was fixed with the inclusion of FRP. The mode of failure transformed from shear failure to bending damage, being a more tolerable failure mode for masonry structures. The recommended iterative process proposed in the ICOMOS methodology was implemented for the assessment of the historical structures of the Monastery of Jerónimos in Lisbon [19]. From the gathered results, it was proposed to conduct additional on-site experimental investigations and additional monitoring. This is all with the goal to obtain as much information as possible that leads to the application of the least invasive and most adequate strengthening measures. It has been emphasized that a correct and adequate choice of strengthening methodology and techniques in cultural heritage buildings requires an iterative approach. In [23], the Cathedral of Porto dating back to the middle of the 12th century was investigated. The absence of a suitable preliminary diagnosis was recouped with excessive multidisciplinary activity during the execution period. The work conducted by Lourenço et al. [24] presents all the above-mentioned steps for a building from the 1930s. The case study highlights the importance of suitable methodologies that allow a clear understanding of the behavior of complex structures; correct assessment of the need for structural strengthening; and analysis of the economic implications of different strengthening interventions in the specific case. The instrumental case of Julianos Church in Umm el-Jimal [25] elaborates on the influence of the local construction material and techniques that have been used for construction on the arrangement of complex roofing structures. The data obtained from comprehensive historical research is combined with structural engineering reconstruction leading to the generation of 2D and 3D digital reconstruction models. The archaeological complex of "Chokepukio" in Cusco and the church "San Pedro Apostol de Andahuaylillas" were assessed in the same city [26] with the application of a modern scientific approach organized in an iterative manner and phases. Combining aerial photogrammetry with in-situ operational modal analysis and IR thermography is beneficial as more precise information is obtained regarding geometry, material, and state-of-damage. This information is then used as input data for numerical modeling. In this specific case, equivalent static nonlinear analyses were conducted. The obtained failure patterns interpreted correctly can lead to less invasive strengthening interventions [26].

3. Case Study-Building in the Street Ribnjak

The building considered as a case study in this paper is located on the eastern side of Ribnjak street in Zagreb, in a building row. It was constructed at the end of 1904. In 1904, the cluster construction method was prescribed, together with semi-open and open-mode construction [27]. This residential building (Ribnjak 44) is located in the area of the protection zone of the cultural property of the Republic of Croatia, within the area of the cultural-historical complex of the city of Zagreb, Protection Zone—A (Figure 3a). This zone represents an area of extremely well-preserved and particularly valuable historical structures. By valorization, the protection zone 'A' has been established for urban settlements or their parts with pronounced urban-architectural, cultural-historical, landscape, or ambient values of emphasized significance for the narrower and wider picture of the city, with a preserved architectural structure of high monumental value. It is interesting to mention that the famous Villa Peroš is located at Ribnjak 46 which belongs to the built structure "in the group", which together with the buildings at house number 42 and 44 forms a series of one-story houses in accordance with the so-called cluster construction method. This represents a transitional form from the closed (Lower Town block) to the open way of building in the northern areas. More details about the history of Villa Peroš and damage after the 2020 Zagreb earthquake can be found in [28].

A complete preservation procedure of the historic urban structure, spatial and landscape features, and individual buildings are applied in this zone [29]. The register of all cultural assets that have been damaged in the Zagreb 2020 earthquake in the protection zone area A is shown in Figure 3b.

(**a**)

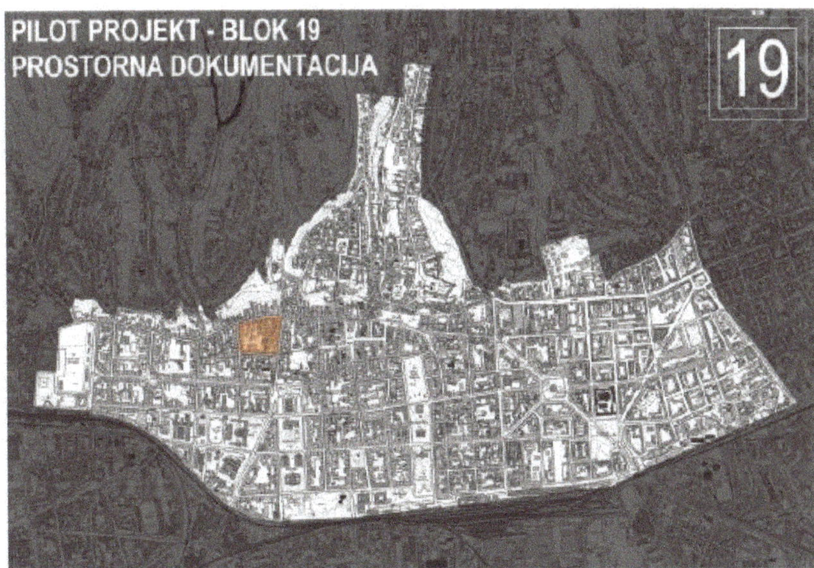

(**b**)

Figure 3. (**a**) Location of the elaborated building—Zone A; (**b**) register of cultural assets-damages in a plan. Data source: Geoportal of cultural assets, Ministry of culture and media, 2020. Prepared for the needs of the complete reconstruction of the historical urban area of the city of Zagreb—authors of the study, Urban model of the reconstruction, Faculty of Architecture—Department of Urban Planning, Spatial Planning and Landscape Architecture [30]. Reproduced with permission from Institute for Spatial Planning of the City of Zagreb, Program cjelovite obnove povijesne urbane cjeline grada Zagreba, prijedlog za javnu raspravu 22_ožujak 2022, 2022.

3.1. Historic Information

As the building under investigation is a building of cultural-historical importance, it is of the utmost importance to follow the procedure explained previously. Historic buildings, regardless of their importance (famous monuments or so-called "minor"), characterize a vital part of our cultural heritage [31] that have to be preserved as testimonies of our past for future generations.

First of all, it was necessary to obtain the archive drawings from the State Archives in Zagreb (Figure 4a–c). As can be seen from the drawings, the design was created in 1901. Since its construction, the building has undergone several reconstructions and upgrades. The first one was conducted in 1937 when several walls were demolished and a new lodge in the courtyard was constructed (Figure 4b,c). Demolition of the façade load-bearing walls looking at the courtyard additionally decreased the loadbearing capacity of the structure in the X direction. To make a wider open space, the construction of the new walls did not follow the position of the existing walls on the ground floor causing the change in load transfer. The reconstruction of the apartments was not synchronized causing a change in the global behavior of the structure. In 1998, several reconstructions were conducted on the first floor, while in 2015, in the apartments, several steel girders have been constructed. The demolishment of the longitudinal walls, as will be proved later on by the calculations, caused a lot of difficulties in the reconstruction process. A new lodge was constructed (see Figures 4c and 5b) founded on new steel columns. The constructed lodge is not adequately connected to the existing structure, the steel columns do not have adequate foundations, and this additionally modified the behavior of the structure.

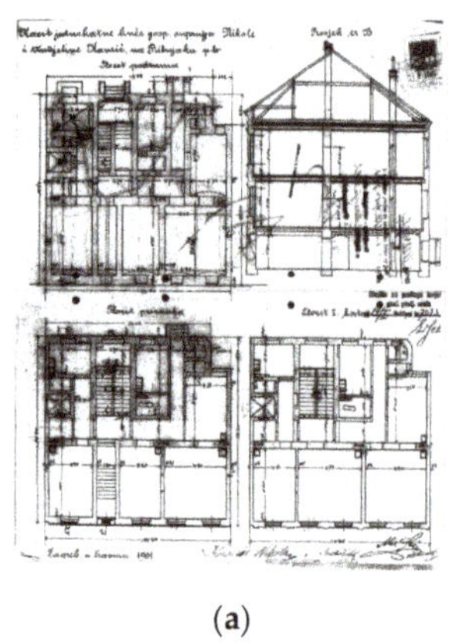

(**a**)

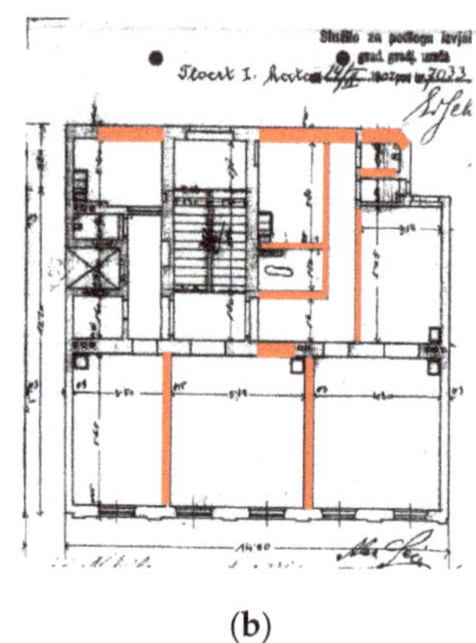

(**b**)

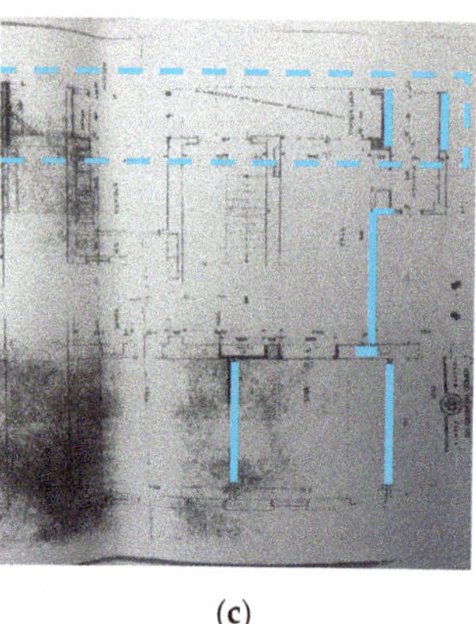

(**c**)

Figure 4. Investigated buildings Ribnjak 44: (**a**) original drawings from the State Archives in Zagreb; (**b**) reconstruction-demolished walls shown in red color State Archives in Zagreb; (**c**) construction of new walls and a lodge [32]. Reproduced with permission from State Archives in Zagreb, 2022.

(**a**)

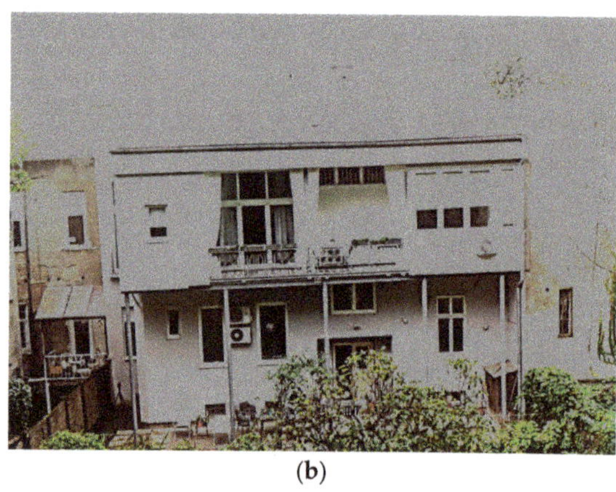

(**b**)

Figure 5. Investigated buildings Ribnjak 44: (**a**) street view (building in the middle); (**b**) courtyard view [32]. Reproduced with permission from Projekt na kvadrat d.o.o., 2022.

The present state of the unreinforced masonry (URM) buildings is presented in Figure 5a,b. It is a row building (the yellow one) located between two buildings of the same height (on the north and the south side). The building has a regular shape with dimensions 14.5 × 17.20 m in plan, the floor plan of the building is 217 m^2, and the gross construction area is equal to 799.07 m^2. The structure consists of a basement, two storeys (ground floor and first floor), and an attic.

Due to the undergone changes, and the fact that not all these changes and upgrades have been documented, it was of the utmost importance to make a point cloud laser scanning of the building in order to obtain the correct geometry of all the elements of the buildings. Laser scanning imaging was used to form a point cloud of the exterior and the interior of the entire building. The 3D point cloud and 3D vector digital model of the existing building were constructed as presented in Figure 6a,b. The height of the basement floor is 3.48 m, the height of the ground floor is 4.10 m, and the height of the first floor is 4 m. The height of the attic varies and the one looking at the street is 1.1 m while the one towards the courtyard is 0.5 m (Figure 6b). The load-bearing walls are built with solid "old format" brick (29 × 14 × 6.5 cm). Load-bearing walls (longitudinal and transverse) have different thicknesses and range from 30 to 75 cm. The largest thickness is in the basement and the smallest is in the attic.

(a)

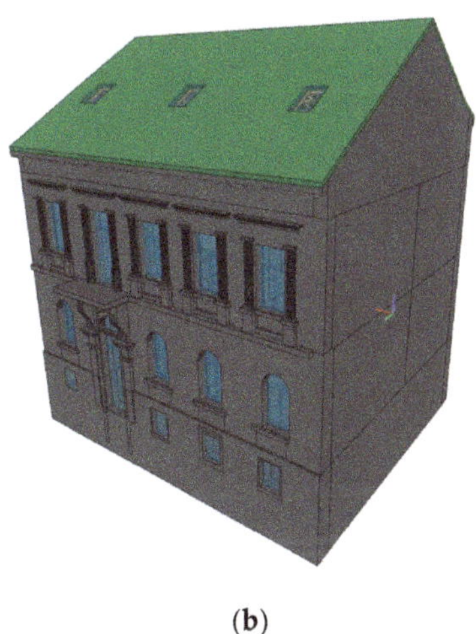

(b)

Figure 6. (**a**) 3D point cloud; (**b**) 3D vector digital model of the existing building. Reproduced with permission from GO2BIM d.o.o. Zagreb, 2022.

The ceilings of the basement are made of brick vaults, supported on load-bearing walls and brick arches, while the ceilings of the ground floor and first floor are made of wooden beams. One concrete beam with dimensions 45/50 cm was identified on the first floor above the newly constructed lodge.

3.2. Description of the Current State of the Building

Based on the preliminary (quick) assessment of the damage to the building, an assessment of the carrying capacity and stability was conducted along with the first risk assessments and the need to evacuate the building in the event of an emergency repeated earthquake. Inspection and assessment of damage to the building after the earthquake was carried out according to the EMS-98 classification for brick buildings. It was marked with a yellow label and PN1 (temporarily unusable) meaning that a detailed inspection is required. PN1 means that the building has moderate damage with no danger of collapse. The bearing capacity of the building is partially violated. It is not recommended to stay in the building, that is, citizens stay at their own risk in such a building. A shorter stay in the building is possible, with the advice of the structural experts related to the necessary measures and the restriction of stay (depending on the danger). A structural expert makes recommendations to eliminate hazards.

3.3. Visual Inspection of the Buildings and Crack Pattern Survey

A detailed visual inspection of the building was performed on 28 June 2021. During the inspection, minor to moderate damage was noted. There was no damage observed in the basement. The most pronounced damage to the structure was observed at a gable wall, as it partially collapsed (Figure 7a). On the roof, the rotation of the purlins towards the outer part of the wall was observed. This is visible in several purlins where the rotation of the purlin located at the corner is minimal, whereas the maximum rotation is observed on the seventh purlin (Figure 7b). The chimney collapsed and immediately after the earthquake the roofing has been rehabilitated.

Figure 7. (**a**) Partial collapse of a gable wall; (**b**) oration of purlin's; (**c**) damage in the spandrels; (**d**) cracks at the contact of the ceiling; (**e**) vertical crack at the contact of two walls; (**f**) vertical crack along the wall [32]. Reproduced with permission from Projekt na kvadrat d.o.o., 2022.

The interpretation of the crack pattern can be of substantial assistance in understanding the structure's state of damage, its possible causes, and the type of survey to be performed [33]. Cracks were observed on the largest number of lintels, both vertical and diagonal in one direction and cross (diagonal) cracks (Figure 7c). Moreover, minor diagonal and X-pattern cracks are visible in the walls at the lintels (above the doors) and parapets between the windows on the façade. These are minor cracks that indicate that there has been an exceedance of the shear strength, but since the width of the cracks is small, it enables the transfer of forces through friction. Additionally, vertical cracks were observed on the lintels near the openings. These are mostly minor cracks that are located at the top and bottom of the walls. The appearance of these smaller cracks in most cases is caused by arch action and indicates that the tensile strength has been exceeded. This represents characteristic damage to such structures. Some of the cracks were repaired after the earthquake. In any case, it is necessary to determine the exact width of the cracks and, depending on the degree of damage, the decision will be made regarding an appropriate rehabilitation method (injection of crack or installation of spiral reinforcement bars). The observed horizontal cracks, which were recorded at the bottom and top of the walls between the openings, occur mainly due to stress concentration along the corners of the opening. These are mostly minor cracks that can be rehabilitated with the injection process.

The floors of the ground and first floor of this URM building are made of light and flexible timber so the appearance of cracks in the ceiling structure was expected, as most probably there is no good connection between the floors and the wall and between the walls (Figure 7d,e). This is an additional parameter that makes these structures vulnerable to earthquakes. More severe damage was noted in wooden ceilings, load-bearing walls, and connection between the load-bearing walls and the ceiling. The cracks extend throughout the entire height of the wall at the connection of the two walls Figure 7e and along the height of the wall Figure 7f.

In this case, as the floor is flexible, the loads on the walls are transferred in relation to their surface, and not as per their stiffness as in the case of the structures with rigid floors which exhibit a "box" behavior [21]. The appearance of horizontal cracks located at the level of the floors and the gable wall was also observed. Generally, when it comes to very small movements (which is not the case here) sliding occurs on the surfaces between the wall and the floor structures, and in this case, it is a damage of a minor degree. In case the cracks are followed by dislocations in the value of several mm, it is a serious slip between the floor structure and the wall. In this case, it is necessary to carry out appropriate rehabilitation strengthening of the floor structure, and appropriate connection of the floor and the walls.

During the preparation for the flat-jack testing, the type of masonry and its units were determined, as well as the quality of mortar. As is usually the case in these types of structures, bricks are found to be in a good shape and of good quality, while the mortar is of very bad quality.

Damage was observed to the steel columns located on the western façade (courtyard side). One of the columns was displaced from its original position, and under the second column damage to the concrete part was observed. All the damages were photographed and described in detail which became a part of the report on the assessment of the existing state of the building structure which was conducted by this team [32].

3.4. Onsite Testing

Accurate diagnosis results from comprehensive on-site and laboratory experimental tests. In cultural-heritage buildings, the on-site tests should be non-destructive as much as possible, or at least minor destructive [31,33]. Identification of geometry, details, and materials of the structural elements (type of masonry units, types of floors, geometry, connections between elements, joints, etc.) was carried out in accordance with the provisions defined in points C.1 and C.2 of the standard EN 1998-3 [34].

To define the in-situ compressive stress level as well as deformability characteristics of masonry (modulus of elasticity), the flat-jack technique [35–37] was employed. Tests were carried out in the basement walls, with one wall in the longitudinal and the other in the transversal direction. A test layout is presented in Figure 8a and the measurement of the stress-strain relationship is shown in Figure 8b. It is seen that the load-bearing walls are built with "old format" solid bricks (29 × 14 × 6.5 cm), assembled with lime mortar which was a common material used in the late 19th and at the beginning of the 20th century.

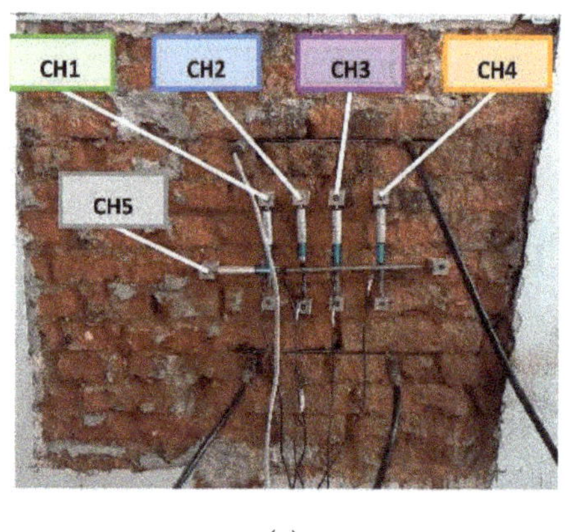

(**a**)

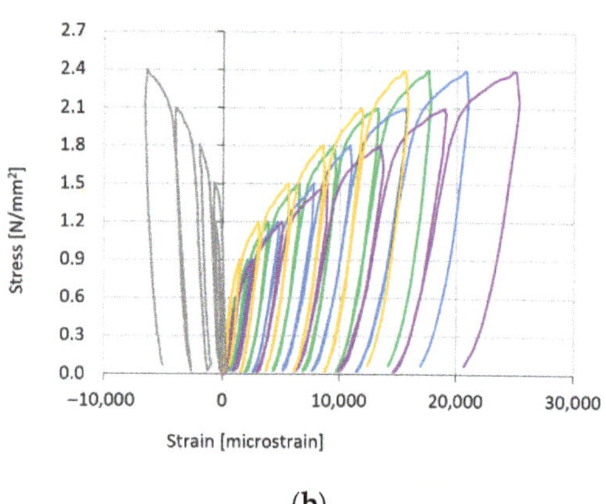

(**b**)

Figure 8. (**a**) 3D Sensors' layout; (**b**) Stress-strain relationship measured at different sensors.

The tests were made in accordance with the ASTM C1197-14a [38]. The maximum attained stress values were in the range of 2.4 to 2.7 N/mm^2. Maximum values were limited also for the great deformability encountered in masonry. The modulus of elasticity at the first and second sampling positions were in fact 554 and 470 N/mm^2 at the beginning of the loading paths (between 0.00 and 0.30 N/mm^2), and tests were carried out almost until material plasticization or strong Young's modulus reduction of masonry. The obtained values are rather low compared to the suggested values found in the literature, and this may be explained by the relatively thick joints and by a very deformable mortar. It has also to be stressed that maximum tests values do not correspond to the material strength, since the tests induced "compaction" in the material (seen in the hardening effect at the end of the second test) and that almost no cracks—indicating masonry crushing—were noticed throughout the tests. This once again emphasizes the importance of on-site testing and the fact that each building represents a case study of its own and has to be separately examined. The values that were selected as input in the numerical modeling were cautiously defined as f_m = 2.2 N/mm^2 for the compressive strength of masonry and the modulus of elasticity E = 500 N/mm^2.

As the floors are made of wooden beams, it was necessary to determine their dimensions and direction. The direction of the beams was from east to west. Dimensions of the wooden beams on the first floor and the attic were 14 × 24 cm placed 0.80 m and 0.85 apart, respectively, with a 2.4 cm thick wooden plank over them. The built-in timber is of good quality. The load is carried in one direction by these types of floors. Several steel beams were noted at the ceiling of the first floor where the removal of the walls took place during the 2015 reconstruction. The dimensions of all openings (doors and windows) were accurately determined by the laser scanner technique.

Operational modal analysis (OMA) was used to define the dynamic features of the investigated structure (natural frequencies, mode shapes, and damping) as a means of verification of the structural behavior and integrity of a building. The obtained information was used for the calibration of the numerical model. OMA determines the modal properties of a structure based on vibration data collected when the structure is under its operating conditions. It is as well-known as output-only modal analysis, ambient modal identification, and in-operation modal analysis [39]. Data are elaborated with ARTEMIS

MODAL 7.2.0.0/2022 licensed software, using enhanced frequency domain decomposition (EFDD) method. This method enables the determination of the damping ratios which is not the case with the frequency domain decomposition (FDD). Additionally, the accuracy of dynamic identification is higher compared to FDD. It is able to estimate closely spaced modes with good accuracy [39]. The vibrations were measured on two different setups (1–2). Sensors are placed to identify global modes acquiring the vibrations mainly at the attic level, at the building corners, and on the long side. Reference sensors no.1 and 2 (in blue) are in the S–W corner of the building—ch. 1 dir. N-S (Y), ch. 2 dir. E–W (X) sensors no.3–6 moved from the South to the North side (Figure 9a). Sensor no.7 was used only in setup 2, at a lower level (Figure 9b). The singular value decomposition of the acquired data is shown in (Figure 9c).

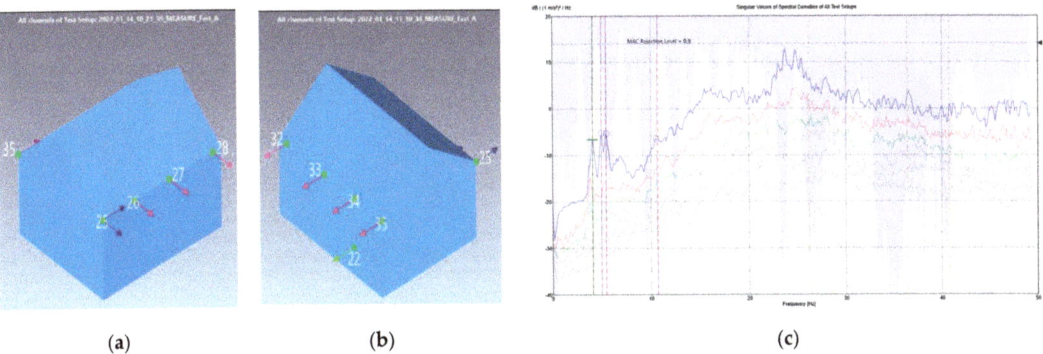

(a)	(b)	(c)

Figure 9. (**a**) Set up no. 1; (**b**) Set up no. 2; (**c**) Singular value decomposition of the acquired data.

The obtained measured data for the first two modes is given in Table 1.

Table 1. Identifies measured dynamic properties.

Mode	Frequency [Hz]	Damping [%]
1	4.09	3.23
2	4.99	2.52

The first two modes were out-of-plane modes, the first one in bending and the second one translational (Figure 10a,b) both in the Y direction.

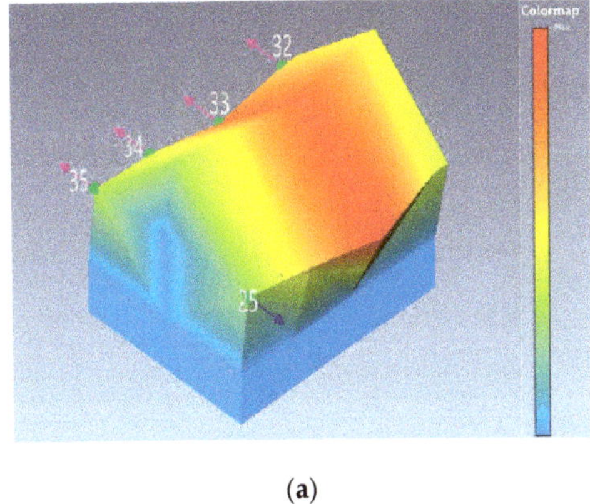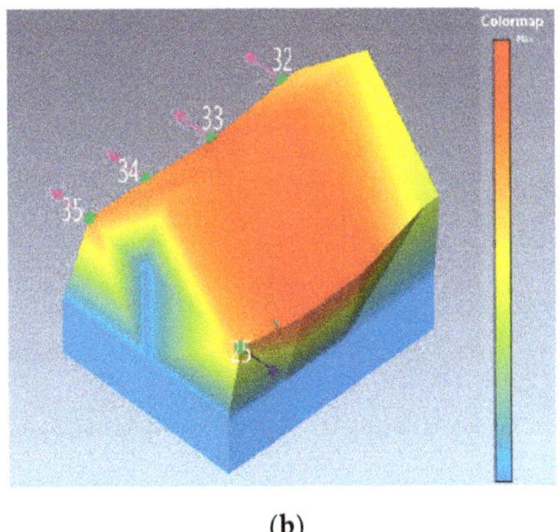

(**a**)	(**b**)

Figure 10. (**a**) 1st mode; (**b**) 2nd mode.

4. Numerical 3D Model of the Existing Buildings

It was decided to construct a 3D model in 3Muri software [40] based on a macro-element approach. This software has been used by numerous researchers [17,21,41–51] and many others. The benefits of this approach are its simplicity, adequate precision, low computational complexity, and computational unpretentiousness [17,52,53]. The non-linear response is determined in masonry macro-elements, composed of piers and spandrels and rigid elements that connect the piers and the spandrels; for more details, see [40]. The 3D model of the building is presented in Figure 11a,b. It was decided to model the building as a single building because there was no information regarding the characteristics of the adjacent buildings and the location of their load-bearing walls. It is clear that this is a conservative approach, and if the structure was to be looked at as being a part of the aggregate there would be stiffness and resistance increase [54].

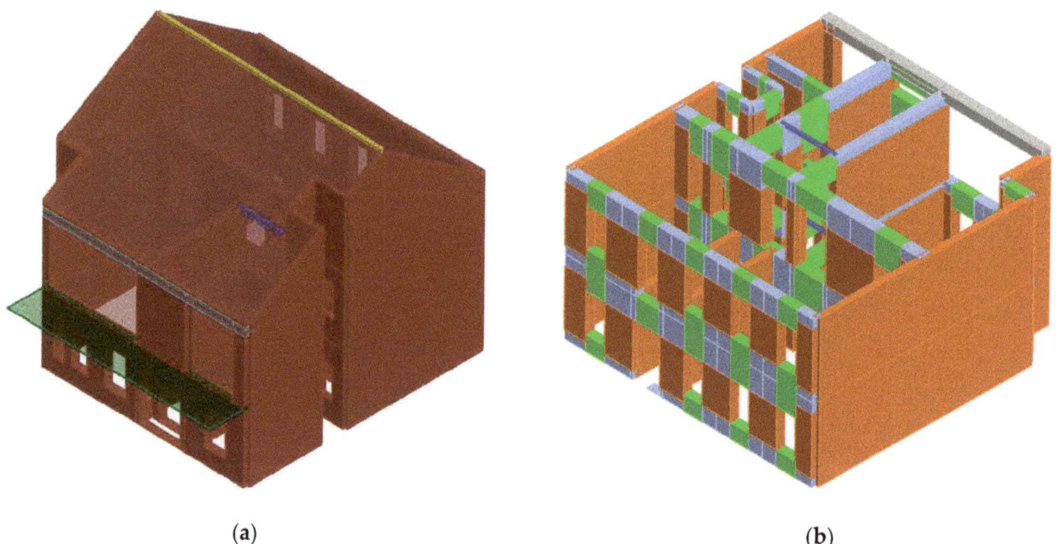

(**a**) (**b**)

Figure 11. (**a**) The 3D model; (**b**) 3D equivalent frame of the elaborated building.

The non-linear static pushover analysis became a popular tool for the seismic assessment of existing and new structures. It provides adequate information on seismic demands imposed by the design ground motion on the structural system and its components. It is a performance-based methodology, which is based on constant gravity loads and an incremental increase of the horizontal force distribution on a structure. An envelope of all the responses derived from the non-linear dynamic analysis is obtained as a result that represents the structural behavior [21]. Both load distributions which are available in 3MURI (uniform pattern and modal pattern) were elaborated. It was seen that the modal pattern distribution (horizontal loads proportional to the first vibration mode shape) prevails. For the existing structures and the type of cracks, it was appropriate to choose the Turnsek–Cacovic law.

4.1. Geometry and Materials

The exact geometrical data was obtained through laser scanning and this was implemented in the model. Regarding the material characteristic, the input data used was the one obtained from the flat-jack tests, while the other data required for the input was calculated according to the suggested formulas provided in the literature and based on engineering judgment. Masonry mechanical properties that were taken in the analysis are given in Table 2. The damage conditions were treated as existing taking into account stiffness reduction.

Table 2. Masonry mechanical properties.

Mechanical Property	Abbreviation	Value
Compressive strength of masonry	f_m	2.2 N/mm^2
Specific weight of masonry	γ_m	18 kN/m^3
Modulus of elasticity	E	500 N/mm^2
Shear modulus	G	200 N/mm^2

Concrete class C20/25 was chosen for the concrete element with the mean values as proposed in Eurocode 2 [55]. Reinforcement for the tie beams was taken to be equal to B420 with a characteristic yield strength of 420 N/mm^2. During the visual inspection, it was noted that the timber was of good quality and the strength class was determined based on engineering judgment. The one-way timber floor with a single wood plank of 2.4 cm was selected for the floor structure having a strength class of C24 according to EN 338:2016 [56] meaning that the type of timber is softwood having a bending strength of 24 N/mm^2 in the major axis. This material is considered new. The newly constructed lodge is supported by steel columns made of structural steel S235 with a modulus of elasticity of E = 210,000 N/mm^2 and yield strength of 235 N/mm^2.

4.2. Conducted Analysis and Their Results

A new Law on Reconstruction of Earthquake-Damaged Buildings in the city of Zagreb, Krapina-Zagorje County, and Zagreb County [57] was enforced after the 2020 Zagreb earthquake. The law clearly stated which return periods have to be taken into account for certain locations and levels of strengthening. In this specific case, in order to check the state of the structure, the return period of 475 years representing the limit state of significant damage (SD) and the 95 years return period representing the limit state of damage limitation (DL) was examined. Regarding the strengthening procedure, Level 3 which envisaged strengthening the building to the return period of 225 years which corresponds to a probability of exceedance of 20% in 50 years was selected. So, in Table 3, PGA values for different limit states are presented.

Table 3. Peak ground accelerations for various return periods for the selected location.

Return Period [Years]	PGA
95	0.128 g
225	0.183 g
475	0.255 g

According to the latest geological research of the city of Zagreb, the soil is classified as soil type C. The seismic analysis was conducted in line with EN 1998-1 [58] and EN 1998-3 [34]. The building belongs to the importance class III (buildings whose seismic resistance is of importance in view of the consequences associated with a collapse) [58] and the recommended value of the importance factor is 1.2. According to the available data from the original drawings, conducted laser scanning, and on-site experimental tests, Knowledge Level 2 was assumed with the confidence factors (CF) equal to 1.2.

As in any calculations, first of all, static analyses are conducted. Due to very low masonry characteristics, many of the elements had already bending damage and did not possess adequate capacity to take over the gravity loads as indicated in Figure 12. According to the calculation, 32 out of 75 elements did not possess adequate vertical load resistance. This had a direct effect on the seismic capacity of the walls.

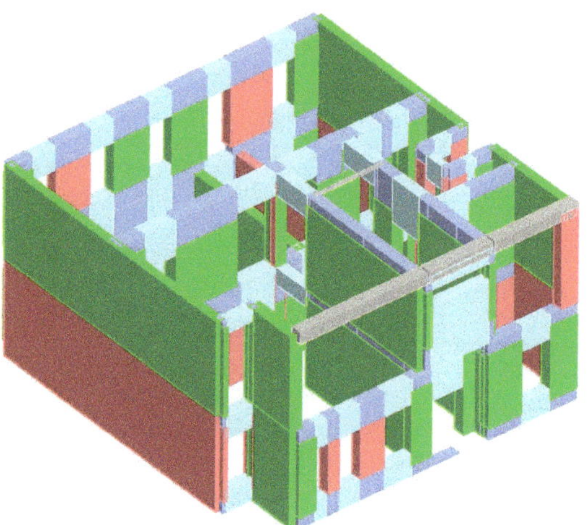

Figure 12. Damage stated due to static loads.

The second calculation which was conducted was the modal analysis which provides information regarding the eigenfrequencies and eigenmodes of the structure. Here, only the dominant global modes in the Y (mode 1) and X (mode 2) directions are given in Table 4.

Table 4. Modal analysis information.

Mode	Frequency [Hz]	Period [s]	Mx [%]	My [%]	Mz [%]
1	3.41	0.29317	0.08	74.58	0
2	2.49	0.40236	24.83	0	0

A rather good consistency is obtained between the measured and modeled eigenfrequency in the Y direction by which the valuation of the model was performed.

In order to correctly implement the nonlinear static pushover analysis, several parameters have to be correctly chosen. First of all, the correct choice of the seismic load pattern is required; secondly, the selection of the correct control node is required for the optimization of the numerical convergence. Finally, representative displacement is to be considered in the pushover curve [59]. This is of great importance when the structures are irregular and with flexible diaphragms, which is the case in this building. In this case, both uniform and modal load distribution were elaborated together with various levels of eccentricities leading to 12 pushover analyses in the X and 12 pushover analyses in the Y direction. The correct choice of the control node in-plane was a very sensitive task as the results are a function of different stiffnesses and strengths of masonry walls. The control node was selected in the wall that had the largest displacement, which would first collapse [59]. An average displacement of all nodes at the same level, weighted by the seismic nodal mass, was chosen instead of the displacement of the control node as proposed by [59]. This provides a generalized explanation of the structure behavior, independent from the different stiffnesses and strengths of the walls, identifying a single failure point.

As a result, capacity curves are obtained for all the cases as presented in Figure 13. The structure has rather inferior resistance in the X direction, the appearance of the first crack was observed already at 95.70 kN, while the maximum reached load was 204.17 kN. The structure has rather superior resistance in the Y direction, with the first crack occurring at the force 13.30 times larger, and the maximum load reached was equal to 1468.35 kN, being 7.19 times larger than in the X direction. This can be connected to a large number of openings and a much lower percentage of load-bearing walls in the X direction.

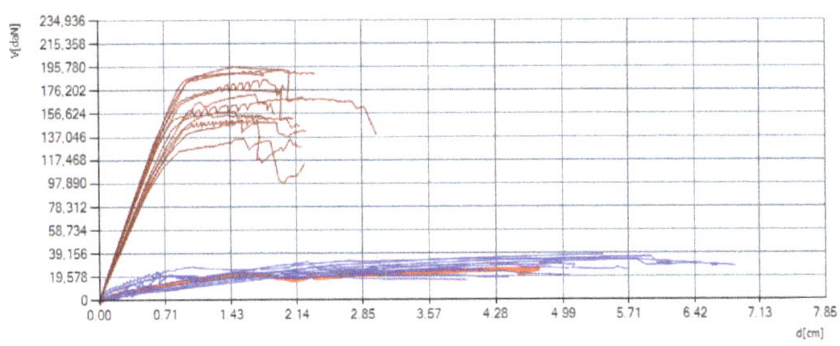

Figure 13. Capacity curves in X and Y directions.

The type of damage at the maximum displacement capacity for X and Y directions is presented in Figures 14 and 15a. The predominant damage in the walls is caused by bending (pink). A rather similar situation is observed in the Y direction; however, additionally, one of the walls experienced shear failures (orange color) and bending failure (red color) (Figure 15b). The bending damage above the windows in the lintels is clearly observed on the façade wall (Figure 15c) and the visible cracks on the walls due to bending have been presented in Figure 7f.

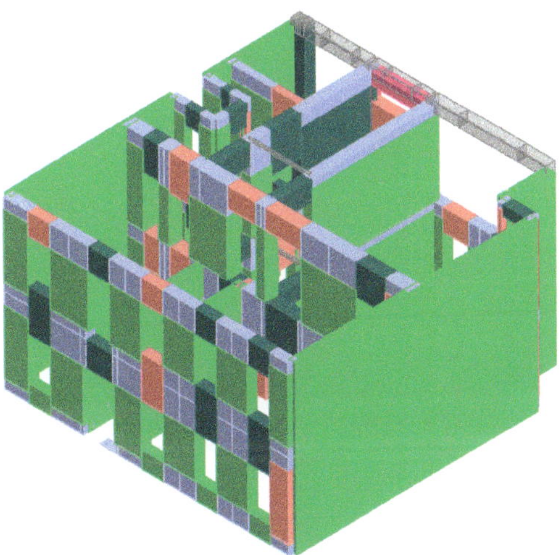

Figure 14. Damage at maximum displacement capacity for pushover in the X direction.

According to the obtained results, not even one of the twenty-four conducted analyses fulfilled the requirements. The parameter (α) is used for the seismic vulnerability evaluation. It is defined by Equation (1).

$$\alpha_{PGA} = PGA/a_{gR} \tag{1}$$

where: PGA represents the peak ground acceleration and a_{gR} limit capacity acceleration of the given building, which represents the ratio between the limit capacity acceleration of the given building and the reference PGA. For the representative analysis in the X and Y directions, these values were $\alpha(SD) = 0.373$, $\alpha(DL) = 0.311$ and $\alpha(SD) = 0.521$, $\alpha(DL) = 0.771$, respectively. This means that the limit state of significant damage (SD) for the return period of 475 years and the limit state of limited damage (DL) for the return of 95 years was exceeded requiring the strengthening of the existing structure.

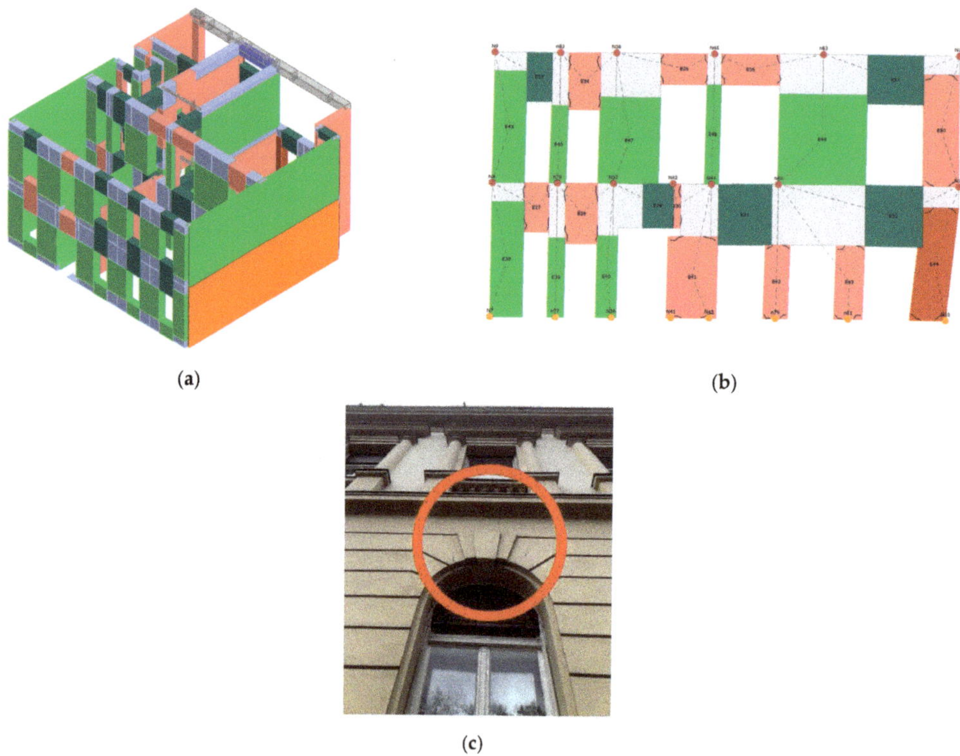

Figure 15. Damage at maximum displacement capacity for pushover in the Y direction; (**a**) shear failure of the wall: (**b**) bending failure of the wall; (**c**) bending damage above the windows in the lintels.

Besides checking the global in-plane behavior of masonry structures, it is as well necessary to check the out-of-plane behavior which is basically the first mode of failure in traditional URM structures, especially the ones with flexible diaphragms. The timber floors are usually not well connected to the walls and the connection of the walls is not constructed in a proper way, not providing a "box" behavior of the structure and leading to out-of-plane failures. As expected, the façade wall looking at the courtyard did not pass the out-of-plane verification (Figure 16). The failure mode is consistent with the 1st eigenfrequency mode.

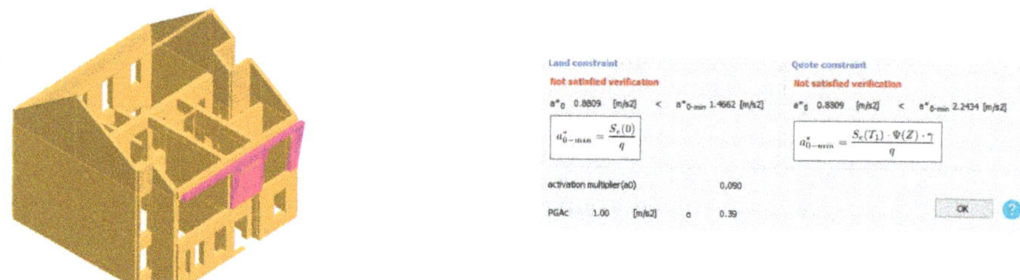

Figure 16. Local mechanism out-of-plane failure of the façade wall.

5. Numerical 3D Model of the Strengthened Buildings

As already mentioned in Subsection 4.2, Level 3 as defined in [57] was chosen as the adequate level for strengthening this masonry building. Modeling was conducted in

3Muri with the inclusion of the strengthening elements. Due to the very low mechanical characteristics of the masonry, several strengthening measures had to be taken in order to improve the seismic resistance of the existing building. First of all, it was necessary to add a new steel moment frame structure inside the existing masonry building with the goal to increase the horizontal resistance, as presented in Figure 17a. This has been identified as one of the most popular methods due to its several advantages such as its reversibility, which is one of the requirements provided in the Charter of Venice [60], fast and easy construction, strength and stiffness increase, and minimal spatial disruption. Additionally, with this kind of strengthening there is no significant increase in the building self-weight [61]. In their work, Karantoni and Sarantitis [62] investigated the influence of constructing steel and concrete frames every 2 m at the floor level on masonry walls. It was noted that this kind of strengthening procedure increases the shear strength and decreases the bending deflection. Steel frames can be connected or not connected to the existing masonry load-bearing structure. Papalou [63] investigated how the masonry structure behaves once the steel frames are connected to the masonry walls. Special care in the modeling procedure was devoted to the connections as they have been considered as hinges not permitting the transfer of the bending moment. Improvement of the seismic behavior is obtained once a close space connection of the steel frame with the masonry wall is created opposed to the situation if a gap is left between the steel structure and the load-bearing wall. Besides the bare frames in several locations, braced frames were envisaged as more rigid elements. In this way, the load-bearing of the whole structure has been increased as well as its ductility. This is foreseen in the location of full masonry walls without openings not affecting the physical and visual access Figure 17b.

During renovations, very often new perforations are created for new windows or doors, affecting the strength and stiffness of the wall and the seismic behavior of the whole structure. Recently, only several researchers conducted studies in this domain. Proença et al. [64] and Billi et al. [65] focused on experimental and numerical modeling of steel ring-frame around the openings. The inclusion of the steel frame affects the mode of failure. Once a steel ring-frame has been installed, the mechanism of solid masonry changes from a shear mechanism to a mixed shear-rocking mechanism with a spread cracking pattern [64]. In order to reestablish the wall's stiffness, it is necessary to use a steel profile with a very large moment of inertia. In-plane strength and ductility will be considerably increased only if a perfect connection between the wall and the steel frame is established [65]. A very comprehensive and in-depth experimental and numerical analysis in line with the work conducted by [64,65] was conducted by Oña Vera et al. [66]. It was shown that it is quite acceptable to select a steel profile with a smaller moment of inertia and their choice was (HEA140) which was capable of re-establishing the in-plane lateral load and the displacement capacity of the elaborated solid wall. Maximal wall stiffness that could be restored amounts to 50% of the original wall regardless of the steel profile size [66]. This was taken into account in the steel profile selection for strengthening the openings (Figure 17c).

The second measure consists of strengthening the walls with FRCM, a material that has been introduced recently for strengthening masonry and concrete structures. This material has pushed aside FRP as the usage of polymeric resins showed to be mechanically, physically, and heritage preservation incompatible with the existing masonry structures [67]. Contrary to FRP, the FRCM uses a thick layer of inorganic plaster as opposed to polymer resins which are used in FRP. The chemical and physical properties of FRCM composites are very similar to masonry, especially in the case of Basalt and AR-glass fibers. This material in the literature is known by other names such as fibre reinforced mortar (FRM) and textile reinforced mortar (TRM). The experimental investigation and numerical modeling of FRCM are still ongoing [68–74] with several applications in practice [75,76]. In our case, some walls were strengthened on both sides and some only on one side. The number of strengthened layers varied depending on the level of damage and capacity of the walls Figure 17d.

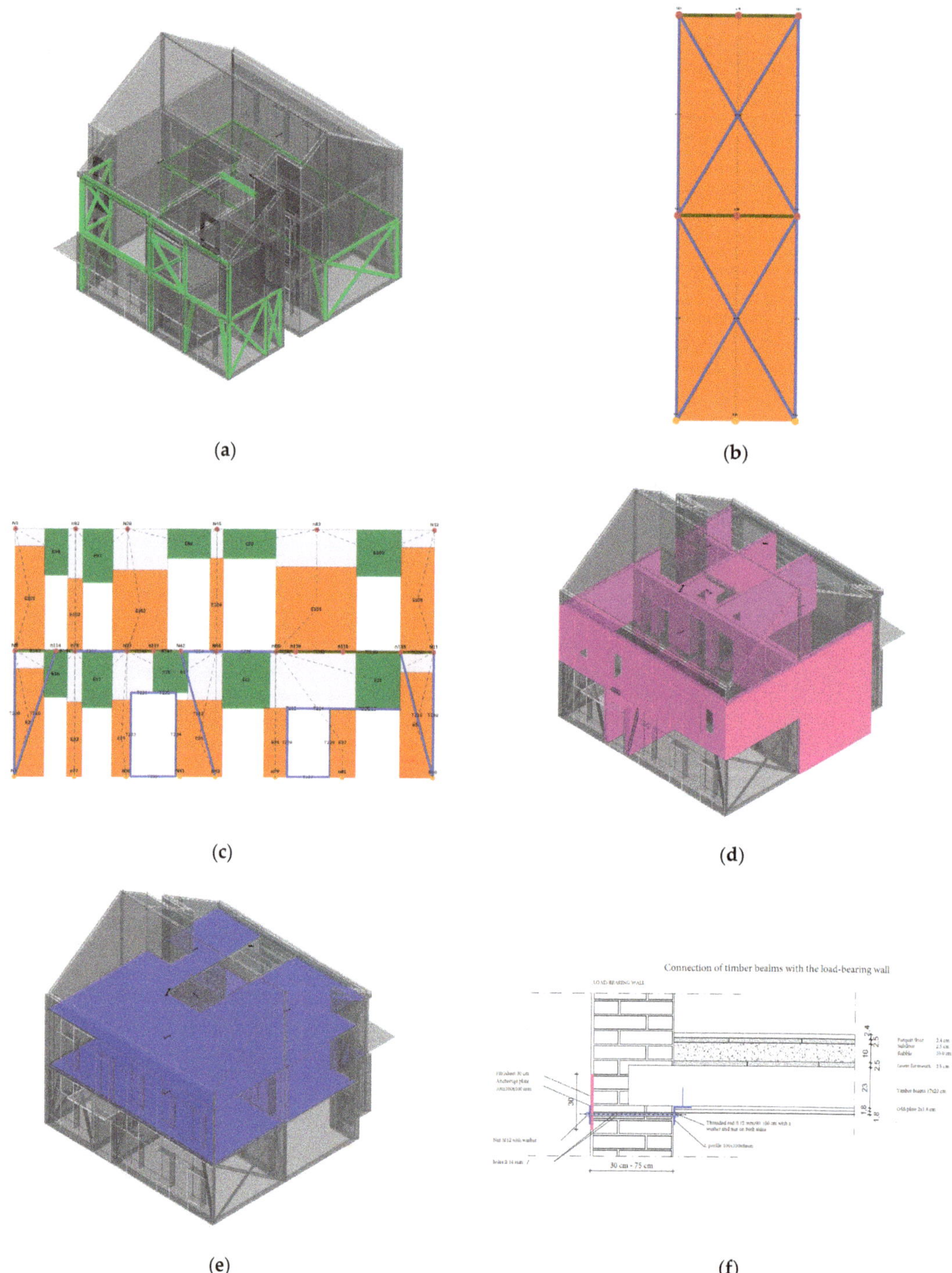

Figure 17. 3Muri 3D strengthened model (**a**) steel structure; (**b**) braced frame; (**c**) steel ring-frame around the doors; (**d**) Strengthening of the walls with FRCM; (**e**) Strengthening of the walls with OSB; (**f**) Connection of the timber beams with load-bearing walls.

One of the main flaws of traditional timber floors is their low in-plane stiffness and lack of effective connections to the load-bearing masonry walls leading to out-of-plane local failures [77]. The decision on which kind of strengthening technique will be selected was guided by the fact that many researchers indicated the important decisive role of the in-plane stiffness of timber diaphragms, and emphasized that disproportionate stiffening of the floors could even have a negative effect on the seismic behavior of the existing building [78–81].

Strengthening and replacement of timber slabs with reinforced concrete slabs have been proven to be unproductive and even worsen the behavior of buildings made of low masonry quality as reported by on-site aftermath earthquake investigations in Italy [82–84]. Using a thin concrete layer would increase the in-plane stiffness if properly connected; however, additional weight would be added consequently increasing the seismic loads on the existing structure.

Gubana and Melotto [82] conducted comprehensive experimental testing of oriented strand board (OSB) panels and it was noted that there was an increase of the initial secant stiffness associated with the top displacement of about six times regardless of the panel orientation in relation to the joists. The initial stiffness of the floor increased seven times with the application of screws for fastening the OSB panels. The increase of strength was from six to nine times with the application of OSB panels which were connected with nails, while with the application of screws, the strength increased 15 times [82], both connection types being reversible and minimally invasive as requested per [60]. This all gives a good basis for the proposal of this kind of strengthening of the URM buildings in seismic-prone areas. In that respect, to improve the in-plane stiffness of the slabs, but still keep the same behavior of the structure, was the purpose to strengthen the timber slabs with OSB having a thickness of 18 mm and laid in two layers (Figure 17e). In order not to open the slab from the upper side as the apartments are all occupied and in usage, it was proposed to construct the strengthening from the lower side of the slab. This strengthening procedure has several benefits, from its low mass and thin thickness to irreversibility and minimum invasiveness, which is in agreement with the requirements for cultural-heritage buildings [60].

Connection of the existing timber slab with the walls will be conducted with the application of the L steel profiles as presented in Figure 17f. It is well known that increasing the in-plane stiffness of the floor has the largest impact on the improvement of the seismic behavior of the existing traditional masonry building. This enables the "box" behavior of the structure (to some degree) and the transfer of the loads to the walls according to their stiffness.

If the crack width is relatively small (e.g., less than 10 mm) and the wall thickness is relatively small, cracks that occur may be sealed with injection material. This was selected as a strengthening measure for walls with small cracks. Around the crack, it is necessary to remove the plaster (preferably traditional, not using electric tools, in order to avoid vibrations and their negative impact on the walls), the area around the crack and inside the cracks has to be cleaned, the cracks will be sealed with grouting material, and a new layer of plaster will be applied.

After several iterations, the final strengthening procedures were adopted.

Examples of the proposed strengthening measures are shown in Figures 18–20.

The capacity curves in the X and Y directions of the strengthened existing URM structure are presented in Figure 21. The first crack is noted at the value of 1152 kN in the X direction while the ultimate load reached 1345 kN. In the Y direction, the structure has a much larger capacity, and the occurrence of the first crack is noted at 2423 kN and the ultimate load reached 2549 kN.

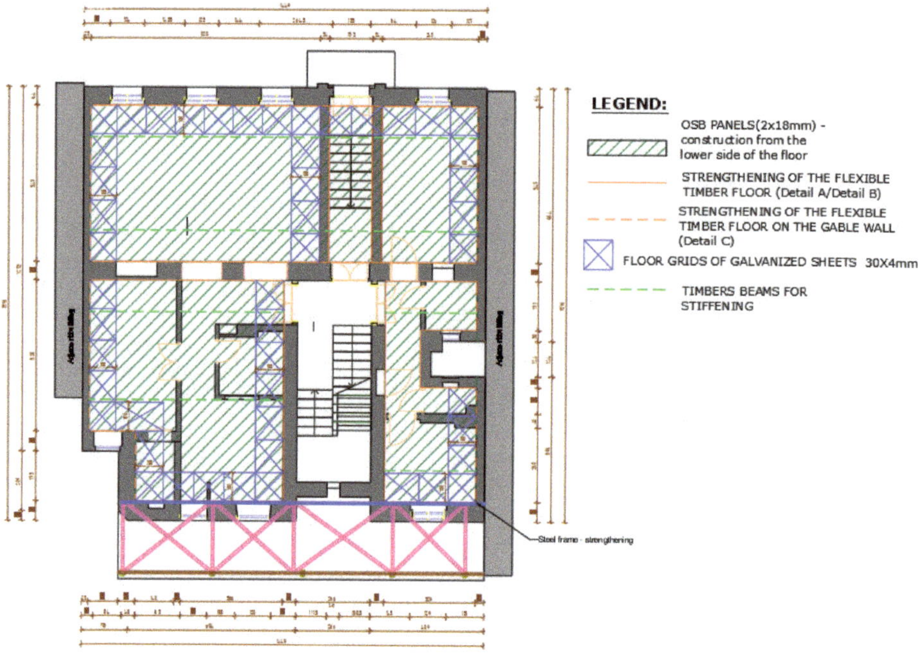

Figure 18. Strengthening measures of the floor at the ground floor level.

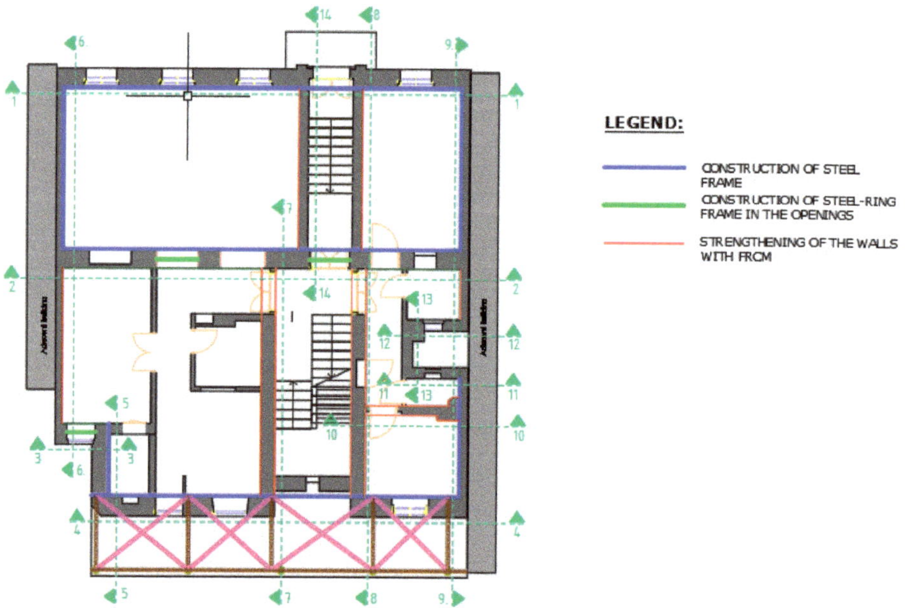

Figure 19. Strengthening measures of the walls at the ground floor level.

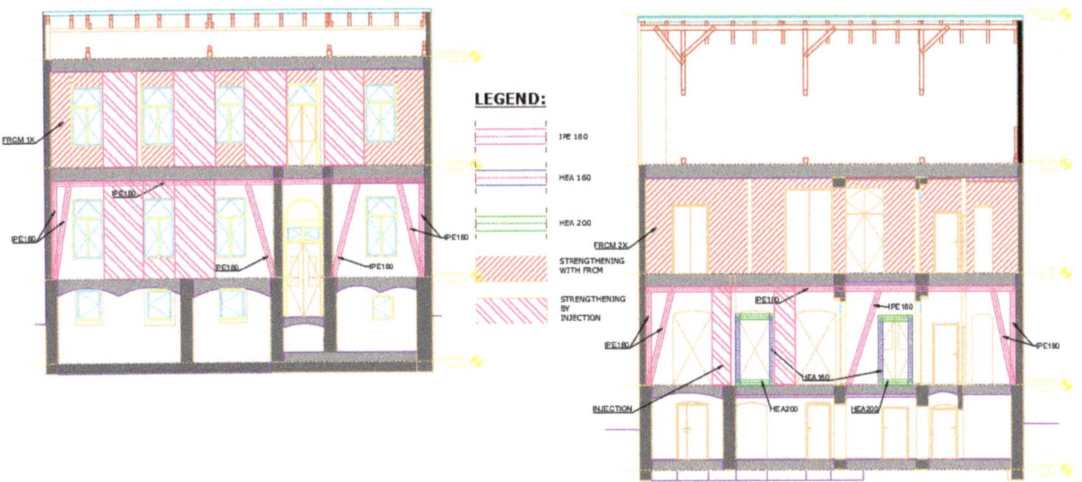

Figure 20. Cross-sections 1-1 and 2-2, Strengthening measures.

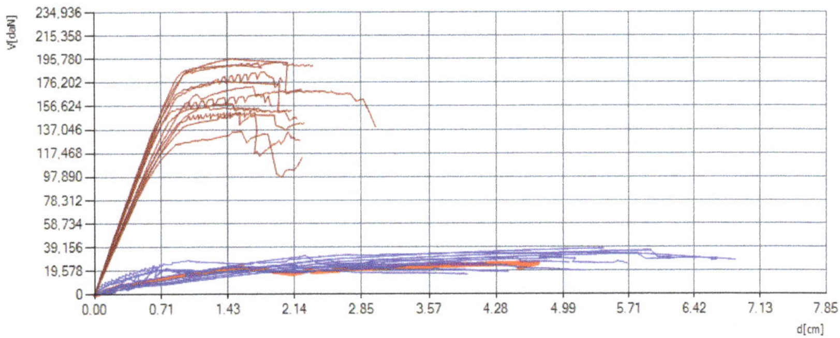

Figure 21. Capacity curves in the X and Y directions of the strengthened structure.

The modal parameters of the strengthened structure are presented in Table 5.

Table 5. Modal analysis information—strengthened structure.

Mode	Frequency [Hz]	Period [s]	Mx [%]	My [%]	Mz [%]
1	4.06	0.24648	0	82.47	0
2	2.87	0.34845	92.06	0	0

For the representative analysis in the X and Y directions, these values were $\alpha(SD) = 1.188$, and $\alpha(SD) = 1.068$, respectively. This means that the limit state of significant damage (SD) for the return period of 225 years was satisfied and that the strengthening methodologies are effective.

6. Comparison of Results

In order to see the effectiveness of the proposed strengthening methods, it is feasible to compare the capacity curves of the unstrengthened and strengthened existing URM structure under consideration. With the application of all given strengthening procedures, it was possible to increase the seismic capacity of the building in both directions. The inferior X direction was now upgraded so that the occurrence of the first crack happened at a load that is 12.04 times larger than for the unstrengthened structure. The increase in initial stiffness in the X direction was 6.4 times, while in the Y direction it was 1.4 times. Additionally, the ultimate reached load in the X direction is more than 6.5 times larger compared to the unstrengthened structure. The ultimate load reached in the Y direction increased by 75% in relation to the unstrengthened structure. It is important to state that the global eigenmodes did not change significantly and this is beneficial, as there should not be a major difference in this aspect. It was observed that now the mode shape in the X direction is translational, while in the unstrengthened structure there were unequal movements of the nodes. In this way, better structural behavior was obtained. The strengthened structure passed all the seismic verification. The value of α was greater than 1 for all the conducted analyses and limit states. The effect of the strengthening is shown in Figure 22 for the two load distribution patterns.

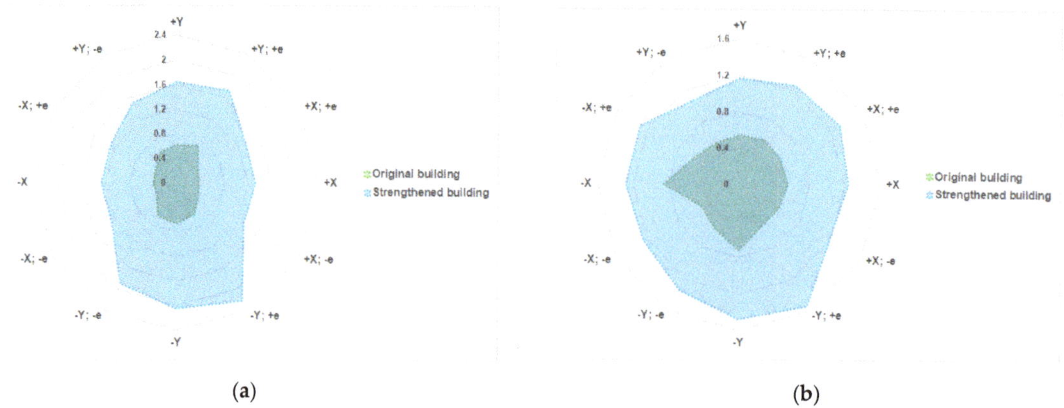

(a)   (b)

Figure 22. Comparison of α values for the original building and strengthened building for (**a**) uniform load distribution; (**b**) modal load distribution.

7. Conclusions

Earthquakes are one of the most devastating natural forces that cause damage to structures, human fatalities, and economic loss. Several variables affect the level of damage on different buildings, such as structural type, type of soil, maintenance, buildings' age, and many others. The aftermath of the 2020 Zagreb earthquake was more than 25,000 damaged buildings mainly located in the historic part of the city of Zagreb. An example of the strengthening procedures which follow the principles of ICOMOS and the Venice Charter have been suggested for the damaged building in Ribnjak 44 street, in order not to jeopardize the cultural character of the building, which would be done if invasive strengthening techniques were proposed. In this study, the aim was to strengthen the building to Level 3 (225 years return period) as per the Law on Reconstruction by Earthquake-Damaged Buildings in the city of Zagreb, Krapina-Zagorje County, and Zagreb County. The recommended strengthening procedures were the inclusion of a new steel frame, steel ring-frames, strengthening with FRCM, and injection of the cracks. The behavior of the buildings being located in a building row (clustered construction method) or as aggregates is affected by numerous parameters which leads to rather complex behavior. Due to the lack of information regarding the adjacent buildings, it was decided to model the building as a single building. In that respect, a perfect match between modeling and experimental data was not obtained; however, the obtained results were more than adequate and acceptable.

The study intended to show the behavior of a cultural-historical building exposed to the 2020 Zagreb earthquake. After a detailed visual inspection, a crack pattern was created, identification of damage was performed, and material characteristics and global behavior of the structure were determined by Ambiental investigations, leading to numerical modeling and proposed strengthening methods respecting the principles of ICOMOS and the Venice Charter.

During the modeling of the building, it was necessary to overcome several issues. First of all, it was necessary to select an adequate modeling strategy, which in this case was the equivalent frame model, substantially the one used by engineers in practice. The benefits are seen in its modeling simplicity, ease of application, and limited computational effort. The next challenge was how should the Ribnjak 44 building be modeled, as a single building or with the entire cluster taken into account. Due to a lack of information about the adjacent buildings, it was decided to model the structure as a single building, taking special care regarding the interpretation of the results and their comparison with the in-situ investigations. In order to overcome this issue, it would be feasible to conduct a simplified procedure for seismic vulnerability assessment of masonry building aggregates as proposed by [56] and model the entire cluster if time and cost permit such activity to be conducted.

Masonry construction is one of the most common construction typologies not only in Europe but in the world and even in zones of significant seismic hazard. Brickwork masonry structures with timber floors are widely geographically distributed throughout most urban and rural settlements in Europe with some specific features for each region. The paper provides a comprehensive example of the selected strengthening methods of a cultural-historic URM building in a seismically active area, avoiding overly invasive interventions that do not jeopardize the historical value of the architecture, which can find its application in different European countries.

Author Contributions: Conceptualization, N.A.; methodology, N.A.; software, M.T. and N.A.; validation, N.A., M.T. and F.C.; formal analysis, N.A., M.T. and F.C.; investigation, D.R., F.C., K.U. and S.K.; resources, D.R., K.U. and S.K.; data curation, D.R. and K.U.; writing—original draft preparation, N.A.; writing—review and editing, N.A., F.C. and S.K.; visualization, N.A.; supervision, N.A and F.C.; project administration, D.R. and K.U. All authors have read and agreed to the published version of the manuscript.

Funding: With the offer issued on 8 March 2021, three items were foreseen: an architectural survey of the building's existing state, an evaluation report on the existing state of the building's construction structure, and a construction renovation project. According to the accepted offer, the contract was signed on 8 October 2021, and the project was registered in the technical journal under number 33-21. The company projekt na kvadrat d.o.o. on its own initiative additionally organized experimental testing of the masonry and dynamic tests of the building structure due to the extremely poor structural condition of the walls.

Institutional Review Board Statement: Not applicable.

Informed Consent Statement: Not applicable.

Data Availability Statement: The data are not publicly available due to privacy reasons.

Acknowledgments: The authors give special thanks to Dalibor Radonić†, B.Sc.Eng. for his multidisciplinary approach to design and comprehensive engagement. Radonić† formed an international team of experts around him who, even after his departure, continue to successfully collaborate on projects of reconstruction and constructive renovation of buildings.

Conflicts of Interest: The authors declare no conflict of interest.

References

1. Stepinac, M.; Lourenço, P.B.; Atalić, J.; Kišiček, T.; Uroš, M.; Baniček, M.; Šavor Novak, M. Damage classification of residential buildings in historical downtown after the ML5.5 earthquake in Zagreb, Croatia in 2020. *Int. J. Disaster Risk Reduct.* **2021**, *56*, 102140. [CrossRef]
2. Uroš, M.; Todorić, M.; Crnogorac, M.; Atalić, J.; Šavor Novak, M.; Lakušić, S. *Potresno Inženjerstvo—Obnova Zidanih Zgrada*, 1st ed.; Građevinski fakultet Sveučilišta u Zagrebu: Zagreb, Croatia, 2021; pp. 1–594.

3. Torbar, J. *Report on the Zagreb Earthquake of November 9, 1880*; Book I; Jugoslovenska Akademija Znanosti i Umjetnosti: Zagreb, Croatia, 1882; pp. 1–144. (In Croatian)

4. Zagreb City Museum. Available online: https://www.mgz.hr/hr/dogadanja/tematska-vodstva-stalnim-postavom---potres-u-zagrebu-1880-godine,3324.html (accessed on 15 September 2022).

5. Croatian Bureau of Statistics. 2011. Available online: https://web.dzs.hr/default_e.htm (accessed on 28 October 2022).

6. Ademović, N.; Kalman Šipoš, T.; Hadzima-Nyarko, M. Rapid assessment of earthquake risk for Bosnia and Herzegovina. *Bull. Earthq. Eng.* **2020**, *18*, 1835–1863. [CrossRef]

7. Kalman Šipoš, T.; Hadzima-Nyarko, M. Seismic Risk of Croatian Cities Based on Building's Vulnerability. *Technical Gazette* **2018**, *25*, 1088–1094. [CrossRef]

8. PTP2 (1948) Provisional technical regulations (PTP) for loading of structures, part 2, no. 11730, 12 July 1948—PTP2. *Official Gazette of FNRY*, 17 June 1948; No. 61/48.

9. PTP-GuSP64 (1964) Provisional technical regulations for construction in seismic regions. *Official Gazette of SFRY*, 19 September 1964; No. 39/64.

10. Drysdale, R.G.; Hamid, A.A.; Baker, L.R. *Masonry Structures: Behaviour and Design*, 2nd ed.; The Masonry Society: Boulder, CO, USA, 1999; pp. 1–809.

11. Uroš, M.; Šavor Novak, M.; Atalić, J.; Sigmund, Z.; Baniček, M.; Demšić, M.; Hak, S. Post-earthquake damage assessment of buildings-Procedure for conducting building inspections. *Građevinar* **2021**, *56*, 1089–1115. [CrossRef]

12. Milić, M.; Stepinac, M.; Lulić, L.; Ivanišević, N.; Matorić, I.; Čačić Šipoš, B.; Endo, Y. Assessment and Rehabilitation of Culturally Protected Prince Rudolf Infantry Barracks in Zagreb after Major Earthquake. *Buildings* **2021**, *11*, 508. [CrossRef]

13. Moretić, A.; Stepica, M.; Lourenço, P.B. Seismic upgrading of cultural heritage-A case study using an educational building in Croatia from the historicism style. *Case Stud. Constr. Mater.* **2022**, *17*, e01183. [CrossRef]

14. Official Site of Matica Hrvatska. Available online: https://www.matica.hr (accessed on 30 October 2022).

15. Hafner, I.; Lazarević, D.; Kišiček, T.; Stepinac, M. Post- Earthquake Assessment of a Historical Masonry Building after the Zagreb Earthquake—Case Study. *Buildings* **2022**, *12*, 323. [CrossRef]

16. ICOMOS International Council on Monuments and Sites. Recommendations for the Analysis, Conservation and Structural Restoration of Architectural Heritage. Available online: https://www.icomos.org/en/about-the-centre/179-articles-en-francais/ressources/charters-and-standards/165-icomos-charter-principles-for-the-analysis-conservation-and-structural-restoration-of-architectural-heritage (accessed on 16 September 2022).

17. Ademović, N. Structural and Seismic Behavior of Typical Masonry Buildings from Bosnia and Herzegovina. Master's Thesis, University of Minho, Guimarães, Portugal, 2011.

18. Casarin, F. Structural Assessment and Seismic Vulnerability Analysis of a Complex Historical Building. Ph.D. Thesis, University of Padova, Padova, Italy, 2006.

19. Lourenço, P.B.; Krakowiak, K.J.; Fernandes, F.M.; Ramos, L.F. Failure analysis of Monastery of Jerónimos, Lisbon: How to learn from sophisticated numerical models. *Eng. Fail. Anal.* **2017**, *14*, 280–300. [CrossRef]

20. Binda, L.; Modena, C. *Lectures in Advanced Masters in Structural Analysis of Monuments and Historical Constructions*; Slides 2012; University of Padova: Padova, Italy, 2012; pp. 1–150.

21. Ademovic, N.; Hrasnica, M.; Oliveira, D.V. Pushover analysis and failure pattern of a typical masonry residential building in Bosnia and Herzegovina. *Eng. Struct.* **2013**, *50*, 13–29. [CrossRef]

22. Ademović, N.; Hrasnica, M. Seismic Assessment and Strengthening of an Existing Multi-Storey Masonry Building in Sarajevo, Bosnia and Herzegovina. In Proceedings of the 2nd European Conference on Earthquake Engineering and Seismology, Istanbul, Turkey, 25–29 August 2014.

23. Lourenço, P.B.; Melo, A.; Carneiro, M. Remedial measures for Cathedral of Porto: Apost-modern conservation approach. In *Structural Analysis of Historical Constructions IV*, 1st ed.; Modena, C., Lourenço, P.B., Roca, P., Eds.; A.A. Balkema Publishers: Leiden, The Netherlands, 2005; Volume 1, pp. 51–62.

24. Lourenço, P.B.; Lourenço, J.B.; Oliveira, D.V. Inspecção e reabilitação doSeminário Conciliar de Braga. In Proceedings of the BE 2004—Encontro Nacional de Betão Estrutural, Universidadedo Porto, Porto, Portugal, 17–19 November 2004; pp. 815–822.

25. Al Rabady, R.; Rababeh, S. Engineering the reconstruction of Hawrān's Ecclesiae during late antiquity: Case of Julianos church in Umm el-Jimal, Jordan. *Herit. Sci.* **2022**, *10*, 81. [CrossRef]

26. Aguilar, R. Applications of modern technologies for the seismic assessment of heritage constructions in Peru. In Proceedings of the 6th World Conference on Earthquake Engineering, Santiago, Chile, 9–13 January 2017; Paper No. 2445. pp. 1–11.

27. Kahle, D. Building Code for the City of Zagreb between 1850 and 1918. *Prostor* **2004**, *12*, 203–216.

28. Gazivoda, N.; Sironić, M. "Osobni dosjei" vile Frangeš i vile Peroš prije i nakon potresa 2020. godine u Zagrebu. *Rad. Inst. Povij. Umjet.* **2020**, *44*, 103–124. [CrossRef]

29. Odluka o izmjenama i dopunama Odluke o donošenju Prostornog plana Grada Zagreba. 2014. Available online: https://www1.zagreb.hr/zagreb/slglasnik.nsf/rest-akt/467ebf4469afb49ac1257d96004f65dd?Open (accessed on 3 July 2022).

30. Program Cjelovite Obnove Povijesne Urbane Cjeline Grada Zagreba, Prijedlog za Javnu Raspravu 22_ožujak 2022, Zavod za Prostorno Uređenje Grada Zagreba, in Croatian. Available online: https://www.zzpugz.hr/wp-content/uploads/2022/03/Program_obnove_prijedlog.pdf (accessed on 5 August 2022).

31. Binda, L.; Saisi, A. *Knowledge of the Building, on Site Investigation and Connected Problems*; Cosenza, E., Ed.; Eurocode 8 Perspectives from the Italian Standpoint Workshop; Doppiavoce: Napoli, Italy, 2009; pp. 213–224.

32. Čizmar, D.; Ademović, N.; Toholj, M.; Radović, D. *Elaborat ocjene postojećeg stanja građevinske konstrukcije*; Report on the state of the structure; Projekt nakvadrat d.o.o.: Zagreb, Croatia, February 2022. (In Croatian)

33. Binda, L.; Cardani, G.; Cantini, L.; Tiraboschi, C. On site and laboratory detection of the quality of masonry in historic buildings. In Proceedings of the International Symposium on Studies on Historical Heritage, Antalya, Turkey, 17–21 September 2007; pp. 667–674, ISBN 978-975-461-433-6.

34. *EN 1998-3:2005+AC:2010*; Eurocode 8: Design of Structures for Earthquake Resistance—Part 3: Assessment and Retrofitting of Buildings. European Union: Brussels, Belgium, 2005. Available online: https://www.phd.eng.br/wp-content/uploads/2014/07/en.1998.3.2005.pdf (accessed on 10 September 2022).

35. Cescatti, E.; Dalla Benetta, M.; Modena, C.; Casarin, F. Analysis and evaluations of flat jack test on a wide existing masonry buildings sample. In *Brick and Block Masonry—Trends, Innovations and Challenges*, 1st ed.; Modena, C., da Porto, F., Valluzzi, M., Eds.; Taylor and Francis Group: London, UK, 2016; pp. 1485–1491.

36. Simões, A.; Bento, R.; Gago, A.; Lopes, M. Mechanical Characterization of Masonry Walls With Flat-Jack Tests. *Exp. Tech.* **2016**, *40*, 1163–1178. [CrossRef]

37. Gregorczyk, P.; Lourenço, P.B. A Review on Flat-Jack Testing. *Engenharia Civil.* **2000**, *9*, 39–50.

38. *ASTM C1197-14a*; Standard Test Method for in situ Measurement of Masonry Deformability Properties using the Flatjack Method. American Society for Testing and Materials: West Conshohocken, PA, USA, 2020.

39. Zahid, F.B.; Ong, Z.C.; Khoo, S.Y. A review of operational modal analysis techniques for in-service modal identification. *J. Braz. Soc. Mech. Sci. Eng.* **2020**, *42*, 398. [CrossRef]

40. S.T.A. DATA. 3 Muri Program 12.5.0. Available online: http://www.stadata.com (accessed on 15 September 2021).

41. Ademović, N.; Oliveira, D.V.; Lourenço, P.B. Seismic Evaluation and Strengthening of an Existing Masonry Building in Sarajevo, BiH. *Buildings* **2019**, *9*, 30. [CrossRef]

42. Formisano, A.; Ademovic, N. An overview on seismic analysis of masonry building aggregates. *Front. Built Environ.* **2022**, *8*, 966281. [CrossRef]

43. Ademović, N.; Oliveira, D.V. Seismic Assessment of a Typical Masonry Residential Building in Bosnia and Herzegovina. In Proceedings of the 15th World Conference on Earthquake Engineering, Lisboa, Portugal, 24–28 September 2012.

44. Simões, A.; Bento, R.; Cattari, S.; Lagomarsino, S. Seismic performance-based assessment of "Gaioleiro" buildings. *Eng. Struct.* **2014**, *80*, 486–500. [CrossRef]

45. Simões, A.; Milosevic, J.; Meireles, H.; Bento, R.; Cattari, S.; Lagomarsino, S. Fragility curves for old masonry building types in Lisbon. *Bull. Earthq. Eng.* **2015**, *13*, 3083–3105. [CrossRef]

46. Formisano, A.; Marzo, A. Simplified and refined methods for seismic vulnerability assessment and retrofitting of an Italian cultural heritage masonry building. *Comput Struct.* **2017**, *180*, 13–26. [CrossRef]

47. Chieffo, N.; Fasan, M.; Romanelli, F.; Formisano, A.; Mochi, G. Physics-Based Ground Motion Simulations for the Prediction of the Seismic Vulnerability of Masonry Building Compounds in Mirandola (Italy). *Buildings* **2021**, *11*, 667. [CrossRef]

48. Valluzzi, M.R.; Sbrogiò, L.; Saretta, Y. Intervention Strategies for the Seismic Improvement of Masonry Buildings Based on FME Validation: The Case of a Terraced Building Struck by the 2016 Central Italy Earthquake. *Buildings* **2021**, *11*, 404. [CrossRef]

49. Lagomarsino, S.; Degli Abbati, S.; Ottonelli, D.; Cattari, S. Integration of Modelling Approaches for the Seismic Assessment of Complex URM Buildings: The Podestà Palace in Mantua, Italy. *Buildings* **2021**, *11*, 269. [CrossRef]

50. Chieffo, N.; Formisano, A.; Mochi, G.; Mosoarca, M. Seismic vulnerability assessment and simplified empirical formulation for predicting the vibration periods of structural units in aggregate configuration. *Geosciences* **2021**, *11*, 287. [CrossRef]

51. Lamego, P.; Lourenço, P.B.; Sousa, M.L.; Marques, R. Seismic vulnerability and risk analysis of the old building stock at urban scale: Application to a neighbourhood in Lisbon. *Bull. Earthq. Eng.* **2017**, *15*, 2901–2937. [CrossRef]

52. Mouyiannou, A.; Rota, M.; Penna, A.; Magenes, G. Identification of suitable limit states from nonlinear dynamic analyses of masonry structures. *J. Earthq. Eng.* **2014**, *18*, 231–263. [CrossRef]

53. Aşıkoğlu, A.; Vasconcelos, G.; Lourenço, P.B.; Pantò, B. Pushover analysis of unreinforced irregular masonry buildings: Lessons from different modeling approaches. *Eng. Struct.* **2020**, *218*, 110830. [CrossRef]

54. Formisano, A.; Florio, G.; Landolfo, R.; Mazzolani, F.M. Numerical calibration of an easy method for seismic behaviour assessment on large scale of masonry building aggregates. *Adv. Eng. Softw.* **2015**, *80*, 116–138. [CrossRef]

55. *EN 1992-1-1*; 2015 Eurocode 2: Design of Concrete Structures, Buildings. European Union: Brussels, Belgium, 2004. Available online: https://www.phd.eng.br/wp-content/uploads/2015/12/en.1992.1.1.2004.pdf (accessed on 10 August 2022).

56. *EN 338:2016*; Structural Timber—Strength Classes. European Standard: Brussels, Belgium, 2016.

57. Law on the Reconstruction of Earthquake-Damaged Buildings in the City of Zagreb, Krapina-Zagorje County and Zagreb County (NN 102/2020). Available online: https://www.zakon.hr/z/2656/Zakon-o-obnovi-zgrada-o%C5%A1te%C4%87enih-potresom-na-podru%C4%8Dju-Grada-Zagreba,-Krapinsko-zagorske-%C5%BEupanije,-Zagreba%C4%8Dke-%C5%BEupanije,-Sisa%C4%8Dko-moslava%C4%8Dke-%C5%BEupanije-i-Karlova%C4%8Dke-%C5%BEupanije (accessed on 10 August 2022).

58. *EN 1998-1*; Eurocode 8: Design of Structures for Earthquake Resistance—Part 1: General Rules, Seismic Actions and Rules for Buildings. European Union: Brussels, Belgium, 2004. Available online: https://www.phd.eng.br/wp-content/uploads/2015/02/en.1998.1.2004.pdf (accessed on 10 August 2022).

59. Lagomarsino, S.; Cattari, S. *Seismic Performance of Historical Masonry Structures Through Pushover and Nonlinear Dynamic Analyses In Perspectives on European Earthquake Engineering and Seismology, Geotechnical, Geological and Earthquake Engineering*, 1st ed.; Ansal, A., Ed.; Springer: Cham, Switzerland, 2015; Volume 39. [CrossRef]

60. Charter of Venice. Decisions and resolutions. In Proceedings of the 2nd International Congress of Architects and Technicians of Historical Monuments, Venezia, Italy, 25–31 May 1964; Volume 5, pp. 25–31.

61. Giuriani, E. Consolidamento degli edifici storici [in Italian]. In *UTET Scienze Technique*; UTET: Torino, Italy, 2012; ISBN 978-8859807636.

62. Karantoni, F.; Sarantitis, D. Seismic behaviour of masonry buildings after alterations of the load bearing system. In Proceedings of the 3rd National Conference in Seismic Mechanics and Technical Seismology, Athens, Greece, 24 November 2008. (In Greek).

63. Papalou, A. Strengthening of masonry structures using steel frames. *Int. J. Eng. Technol.* **2013**, *2*, 50–56. [CrossRef]

64. Proença, J.; Gago, A.; Vilas Boas, A. Structural window frame for in-plane seismic strengthening of masonry wall buildings. *Int. J. Archit. Herit.* **2018**, *13*, 98–113. [CrossRef]

65. Billi, L.; Laudicina, F.; Salvatori, L.; Orlando, M.; Spinelli, P. Forming new steel-framed openings in load-bearing masonry walls: Design methods and nonlinear finite element simulations. *Bull. Earthq. Eng.* **2019**, *5*, 2647–2670. [CrossRef]

66. Oña, V.; Mónica, Y.; Metelli, G.; Barros, J.A.O.; Plizzari, G.A. Effectiveness of a steel ring-frame for the seismic strengthening of masonry walls with new openings. *Eng. Struct.* **2021**, *226*, 111341. [CrossRef]

67. Di Tommaso, A.; Focacci, F. *Strengthening Historical Monuments with FRP: A Design Criteria Review, Composites in Construction: A Reality*; ASCE: Reston, VA, USA, 2001; ISBN 0-7844-0596-4. [CrossRef]

68. Pantò, B.; Malena, M.; de Felice, G. Non-Linear Modeling of Masonry Arches Strengthened with FRCM. *Key Eng. Mater.* **2017**, *747*, 93–100. [CrossRef]

69. Scacco, J.; Ghiassi, B.; Milani, G.; Lourenço, P. A fast modeling approach for numerical analysis of unreinforced and FRCM reinforced masonry walls under out-of-plane loading. *Compos. B Eng.* **2020**, *180*, 107553. [CrossRef]

70. Aiello, M.A.; Cascardi, A.; Ombres, L.; Verre, S. Confinement of Masonry Columns with the FRCM-System: Theoretical and Experimental Investigation. *Infrastructures* **2020**, *5*, 101. [CrossRef]

71. Murgo, F.S.; Ferretti, F.; Mazzotti, C.A. discrete-cracking numerical model for the in-plane behavior of FRCM strengthened masonry panels. *Bull. Earthq. Eng.* **2021**, *19*, 4471–4502. [CrossRef]

72. Angiolilli, M.; Gregori, A.; Pathirage, M.; Cusatis, G. Fiber Reinforced Cementitious Matrix (FRCM) for strengthening historical stone masonry structures: Experiments and computations. *Eng. Struct.* **2020**, *224*, 111102. [CrossRef]

73. Angiolilli, M.; Gregori, A.; Cusatis, G. Simulating the Nonlinear Mechanical Behavior of FRCM-strengthened Irregular Stone Masonry Walls. *Int. J. Archit. Herit.* **2021**. [CrossRef]

74. Türkmen, Ö.S.; De Vries, B.T.; Wijte, S.N.M. Mechanical characterization and out-of-plane behaviour of Fabric Reinforced Cementitious Matrix (FRCM) overlay on clay brick masonry. *Civ. Eng. Des.* **2019**, *1*, 131–147. [CrossRef]

75. D'Ambrisi, A.; Focacci, F.; Luciano, R.; Alecci, V.; De Stefano, M. Carbon-FRCM materials for structural upgrade of masonry arch road bridges. *Compos. Part B.* **2015**, *75*, 355–366. [CrossRef]

76. Berardi, F.; Focacci, F.; Mantegazza, G.; Miceli, G. Rinforzo di un viadotto ferroviario con PBO-FRCM. In Proceedings of the 1 Convegno Nazionale Assocompositi, Milan, Italy, 25–26 May 2011. (In Italian).

77. Lourenço, P.B.; Mendes, N.; Ramos, L.F.; Oliveira, D.V. Analysis of masonry structures without box behavior. *Int. J. Archit. Herit.* **2011**, *5*, 369–382. [CrossRef]

78. Trutalli, D.; Marchi, L.; Scotta, R.; Pozza, L. Dynamic simulation of an irregular masonry building with different rehabilitation methods applied to timber floors. In Proceedings of the 6th International Conference Computational Methods in Structural Dynamics and Earthquake Engineering, National Technical University of Athens, Athens, Greece, 15–17 June 2017.

79. Borri, A.; Corradi, M. Structural engineers vs. conservators safety vs. preservation: Problems, doubts and proposals. In Proceedings of the 1st International Conference (TMM_CH) Transdisciplinary Multispectral Modelling and Cooperation for the Preservation of Cultural Heritage, Athens, Greece, 10–13 October 2018.

80. Gubana, A.; Melotto, M. Discrete-element analysis of floor influence on seismic response of masonry structures. *Proc. Inst. Civ. Eng. Struct. Build.* **2021**, *174*, 459–472. [CrossRef]

81. Trutalli, D.; Marchi, L.; Scotta, R.; Pozza, L. Seismic capacity of irregular unreinforced masonry buildings with timber floors. *Proc. Inst. Civ. Eng. Struct. Build.* **2021**, *174*, 473–490. [CrossRef]

82. Gubana, A.; Melotto, M. Experimental tests on wood-based in-plane strengthening solutions for the seismic retrofit of traditional timber floors. *Constr. Build. Mater.* **2018**, *191*, 290–299. [CrossRef]

83. Binda, L.; Gambarotta, L.; Lagomarsino, S.; Modena, C. A multilevel approach to the damage assessment and seismic improvement of masonry buildings in Italy. In *Seismic Damage to Masonry Buildings*, 1st ed.; Bernardini, A., Ed.; Balkema: Amsterdam, The Netherlands, 1999; pp. 170–195.

84. Valluzzi, M.R. On the vulnerability of historical masonry structures: Analysis and mitigation. *Mater. Struct.* **2007**, *40*, 723–743. [CrossRef]

Article

Nonlinear Modeling of RC Substandard Beam–Column Joints for Building Response Analysis in Support of Seismic Risk Assessment and Loss Estimation

Naveed Ahmad [1],*, Muhammad Rizwan [2], Babar Ilyas [3], Sida Hussain [4], Muhammad Usman Khan [5], Hamna Shakeel [3] and Muhammad Ejaz Ahmad [3]

[1] Department of Civil and Environmental Engineering, Stanford University, Stanford, CA 94305, USA
[2] Department of Civil Engineering, Sarhad University of Science & Information Technology, Peshawar 25000, Pakistan
[3] Department of Civil Engineering, Abasyn University, Peshawar 25000, Pakistan
[4] Department of Civil Engineering, University of Engineering & Technology, Peshawar 25000, Pakistan
[5] Military College of Engineering, National University of Sciences and Technology (NUST), Islamabad 24080, Pakistan
* Correspondence: drnaveed@stanford.edu

Citation: Ahmad, N.; Rizwan, M.; Ilyas, B.; Hussain, S.; Khan, M.U.; Shakeel, H.; Ahmad, M.E. Nonlinear Modeling of RC Substandard Beam–Column Joints for Building Response Analysis in Support of Seismic Risk Assessment and Loss Estimation. *Buildings* 2022, *12*, 1758. https://doi.org/10.3390/buildings12101758

Academic Editors: Rajesh Rupakhety and Dipendra Gautam

Received: 2 October 2022
Accepted: 18 October 2022
Published: 20 October 2022

Abstract: The paper discusses how joint damage and deterioration affect the seismic response of existing reinforced concrete frames with sub-standard beam–column joints. The available simplified modeling techniques are critically reviewed to propose a robust, yet computationally efficient, technique for simulating the nonlinear behavior of substandard beam–column joints. Improvements over the existing models include the simulation of the cyclic deterioration of joint stiffness and strength, as well as pinching in the hysteretic response, implemented considering a deteriorating hysteretic rule. A fiber-section forced-based inelastic beam–column element is developed, considering improved material models and fixed-end rotation due to bond failure, rebars-slip, and inelastic extension, to simulate the deteriorating cyclic behavior of existing pre-cracked beam–column members. For the assessment of frames with substandard exterior beam–column joints, a nonlinear model for the exterior joint is developed and validated through a full-scale quasi-static cyclic test performed on a substandard T-joint connection. The proposed model allows considering structural performance in risk assessment while accounting for true inelastic mechanisms at the joints. An assessment of a five-story RC frame revealed that the activation of the joint shear mechanism increases the chord rotation demand on the connecting beam members by up to 85%, with increases of up to 62% (mean drift) and 89% (mean + 1.std.) on the lower floors when determining the inter-story drift demand, and the collapse probability of structures subjected to design base ground motions increased from 4.20% to 29.20%.

Keywords: beam–column joint; fiber-based section modeling; joint shear hinge; substandard beam–column joints; stiffness and strength deterioration; reinforced concrete; seismic vulnerability; risk

1. Introduction

A beam–column joint in a reinforced concrete moment-resisting frame is a most critical component; it experiences high axial and vertical/horizontal shear stresses during earthquakes, and its behavior has a significant influence on the building seismic response. If not adequate, joint shear failure can result. However, if a joint is assumed to be stiff throughout the analysis during the assessment process, this may be overlooked. The shear failure of the beam–column joint often has a brittle character, which does not provide an adequate level of structural performance, particularly under extreme seismic actions [1,2]. There have been several reports of catastrophic building collapses during strong earthquakes (Figure 1), which have been linked to beam–column joint failure, including those that occurred in the 1999 earthquakes in Chi-Chi, Taiwan, and Izmit, Turkey, and in the 1994 Northridge earthquake in California, USA [3]. Therefore, it is crucial to comprehend joint behavior in nonlinear building response analysis in order to make an informed judgment regarding the evaluation of building damages.

Figure 1. Partial collapse of buildings documented during large damaging earthquakes: from left to right: 1999 Izmit earthquake in Turkey, 1999 Chi-Chi earthquake in Taiwan, and 1994 Northridge earthquake in the USA.

Over the past several decades, a number of studies have focused on understanding how beam–column joints respond to seismic actions [4–11]. Moreover, several international seismic codes of practice have undergone repeated improvements to put the findings into practice. The most recent ACI joint design recommendations [12] account for inelastic deformation in the joint panel. The present joint design procedures and detailing provisions offer resistance to the gravity loads, seismic actions, and the interaction of multidirectional forces applied to the joint from surrounding frame elements. To ensure an improved cycle behavior and a minimum plastic deformation capacity, sufficient development length and confinement in the joint panel are provided.

Recent research on the quantification of the seismic risk of existing reinforced concrete structures has focused on frames with the substandard beam–column joints that are prevailing in most seismically active countries. This paper focuses on comprehending the primary mechanism influencing the seismic nonlinear response of beam–column connections and presents a simplified nonlinear numerical modeling technique, which is crucial for assessing the seismic building's response. With particular reference to bond failure and rebar slip (resulting in member fixed-end rotation) and joint shear strength deterioration, this research studied the effects of exterior joint nonlinear behavior and underlines the crucial factors that influence structural performance, which is fundamental for risk assessment and repair cost estimation.

The experimental reference specimen of the exterior beam–column joint under consideration was designed in accordance with the previous building code of Pakistan but lacked seismic details and incorporated deficiencies prevalent in the typical beam–column joint in existing RC structures [13]. The test specimen was initially subjected to force-controlled cyclic loading in order to induce cracks in the specimen and simulate the pre-cracked condition of the current RC frame. The main finding is presented in terms of damage observations and the connection's force–displacement hysteretic response to guide numerical modeling. The study is unique in that it takes into account pre-cracking in the substandard exterior beam–column joints that exhibit deterioration in the stiffness and strength of the connection.

2. Critical Review of Simplified Nonlinear Modeling Techniques for RC Beam–Column Joints

Modeling frames with weaker joints for nonlinear seismic analysis has been attempted using different simplified techniques (Figure 2). These are briefly reviewed here to serve as a basis for the development of a more robust, yet simplified and computationally efficient, analytical model of reinforced concrete frames that incorporates both shear and bond mechanisms behavior of beam–column joints and fixed-end rotation of connecting members due to pre-cracked beam–column members common to existing buildings and that can be used for both local and global damage evaluation to help guide retrofitting efforts.

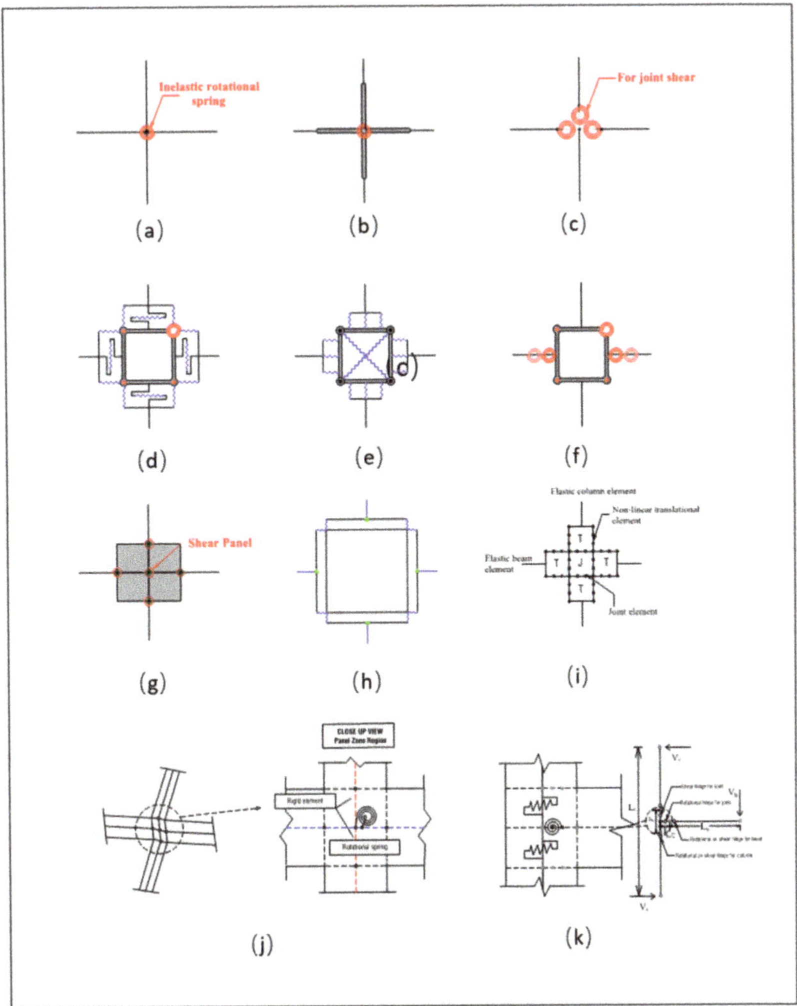

Figure 2. Existing simplified analytical models for the nonlinear modeling of reinforced concrete beam–column joint. (**a**) EI-Metawally and Chen (1988) [14]. (**b**) Alath and Kunnath (1995) [15]. (**c**) Biddah and Ghobarah (1999) [16]. (**d**) Lowes and Altoontash (2003) [17]. (**e**) Youssef and Ghobarah (2001) [18]. (**f**) Shin and LaFave (2004) [19]. (**g**) Altoontash (2004) [20]. (**h**) Ning et al. (2016) [21]. (**i**) Elmorsi et al. (2000) [22]. (**j**) Pampanin et al. (2003) [23]. (**k**) Sharma et al. (2011) [24].

The flexural strengths of the beams and columns framing the joint were decreased by Kunnath et al. [25] to take into consideration the insufficient joint shear strength capacity. However, such implicit models are incapable of accounting for the additional deformation resulting from the joint bond mechanism and the pinching effects on the frame hysteretic response due to joint shear stiffness and strength deterioration. Neverthless, the model is promising for the assessment of global performance. El-Metwally et al. [14] used a zero-length spring with nonlinear behavior to model the shear deformation of the joint panel. Through the use of rotational springs, Biddah and Ghobarah [16] modeled joint shear and bond-slip deformations. Later, Ghobarah and Biddah [26] utilized the model to demonstrate that joint deformations led to greater flexibility and drifts under seismic actions. The models were promising for frames with joints with stable behavior; however, these models were unable to accurately simulate the joint shear stiffness and strength deterioration essential for substandard joints. Alath and Kunnath [15] used a rotational spring model to explicitly simulate the joint shear deformation, with deteriorating hysteresis being established empirically, but this has the drawback that it must repeatedly be calibrated for various beam–column joints. Pampanin et al. [23] proposed a similar simplified model with a moment–rotational spring and deteriorating hysteretic response. The spring's properties were directly deduced from the corresponding principal tensile stress vs. shear deformation curve, which, in turn, was based on a large experimental database [27], using the equilibrium consideration. This, however, falls short of accurately simulating the deterioration of joint shear strength. Sharma et al. [24] improved upon this approach by including additional shear springs in the joint panel to account for the panel's shear deformation and to take strength degradation into consideration. The model still does not adequately account for stiffness and strength deterioration, i.e., cyclic deterioration (number-of-cycles effect), which is essential for substandard beam–column joints. The same is true for the method proposed by Khan et al. [28].

Adding an extra infinitely stiff bar attached to the plastic beam sub-element with plastic hinges at the ends, Filippou et al. [29] and Filippou and Issa [30] developed an explicit model to simulate the fixed-end rotations and the sliding due to shear at the beam–column interface. Although promising in simulating the rotational and bending-moment resistance of a beam–column joint, it is unable to simulate the joint shear deterioration and the pinching that results in the hysteretic response of the frame.

Youssef and Ghobarrah [18] developed a 14-spring element assembly for a beam–column joint that included two diagonal springs to simulate joint shear deformation and twelve translational springs placed at the panel zone interface to represent the inelastic mechanisms of connecting beam/column members. It models the shear strength deterioration but implicitly considers rebar slip and rebar inelasticity. A model developed by Ning et al. [21] included one spring for the panel zone and eight springs for the rebar-slip and employed the Bouc–Wen–Baber–Noori model [31] to simulate the hysteretic behavior of the panel zone. The model, however, is unable to truly replicate the cyclic strength and stiffness deterioration that often occurs at substandard beam–column joints.

Another model suggested by Elmorsi et al. [22] involved idealizing beams and columns as elastic elements connected to the joint by the intervention of non-linear transitional elements, modeled using other elements made up of 12 parts. The model takes into account the effects of bond-slip and joint shear deformations and can simulate the deterioration of bonds and eventual pullout of reinforcing bars during extreme cyclic loads. However, the model is clumsy and computationally expensive when a large number of analyses are required for building performance assessment.

A 4-node 12-degree-of-freedom joint panel was proposed by Lowes and Altoontash [17] that constitutes a panel zone component with a zero-length rotational spring simulating the shear deformation of the joint with four additional zero-length shear springs for simulating the interface–shear deformations. A total of eight zero-length translational springs were included to simulate the bar slip. The panel zone's shear stress–strain relationships were determined using the modified compression field theory (MCFT) [32]. Later, Lowes et al. [33] attempted to simulate the interface-shear based on experimental data; this work anticipated an interface-shear response that was stiff and elastic. Moreover, specimens with at least a minimal degree of transverse reinforcement in the panel zone were included in the experimental data for validation, which is compatible with the model's intended usage. However, they did not include joints that lack transverse reinforcement; therefore, it could not be used to analyze the weaker joints of frames that lack transverse reinforcement. The model proposed by Lowes and Altoontash [17] was further simplified by Altoontash [20] by adding four rotational springs to member ends to simulate the bar-slip phenomenon and a rotational spring that was defined in the panel zone to simulate the shear distortion of the joint. However, it is still insufficient for the analysis of joints lacking transverse reinforcement and members exhibiting large fixed-end rotation due to pre-cracked beam–column members.

Shin and LaFave [19] proposed an analytical model for a beam–column joint wherein the joint panel was made up of stiff elements along the panel zone's edges and three parallel rotational springs provided in one of the four hinges connecting the parallelogram's sides. The MCFT was used to anticipate the joint shear stress–strain response envelope while the cyclic response was calibrated using experimental data. Two rotating springs (in series) were positioned at the interfaces between the beam and the joint to separately simulate the member-end rotations produced by the bond-slip behavior of the longitudinal beam reinforcement and plastic hinge rotations induced by the inelastic behavior of the beam. The model is robust in simulating joint shear deterioration, bond mechanisms, and fixed-end rotation. However, the joint rotational springs needs simplification.

The assessment of beam–column joints without transverse reinforcement presents some difficulties for most of the modeling techniques discussed above. Some of these, for example, are computationally inefficient and need simplification in order to be implemented in the available finite element-based programs for large analysis, while others are only appropriate for joints that have been seismically designed and detailed and less accurate when applied to substandard beam–column joints with pre-existing cracks in the structural members. The goal of the present study is to propose a more robust modeling technique that would still be computationally efficient and capable of modeling the deterioration of stiffness and strength common to existing pre-cracked beam–column frames with substandard beam–column joints subjected to cyclic loading.

3. Proposed Nonlinear Modeling Technique for RC Substandard Beam–Column Joints

3.1. Mechanics of Exterior Beam–Column Joints

Figure 3 illustrates the internal forces and reactions acting on a reinforced concrete exterior beam–column joint under seismic actions. If it is assumed that the points of the zero bending moment are located at the half-height of the column and the half-span of the beam, respectively, then it can be assumed that column shear $V_a = V_c$ and column moment $M_a = M_c$. The reversal of the moment across the joint necessitates that the reinforcement of the beam is in compression on one side and at a tensile yield on the other. The internal horizontal tension T_b, compression C_b, and vertical beam shear V_b forces introduced by the beam to the column are shown. Making the approximations that $C_b = T_b$, the required horizontal column shear force V_{jh} across the joint region based on the equilibrium of free body is:

$$V_{jh} = T_b - V_c \tag{1}$$

$$T_b = \frac{M_b}{j_b} \tag{2}$$

$$T_b = \frac{V_b l_b}{j_b} \tag{3}$$

where M_b is the beam moment, l_b is the length of beam, h_c is the depth of column, and j_b is the internal level arm between the tensile force and the centroid of the compressive forces. This can be determined by moment-curvature analysis of the beam or approximated as follows:

$$j_b = d_b - d'_b \tag{4}$$

where d_b is the effective depth of beam and d'_b is the effective cover to compression reinforcement. The column top load V_c can be calculated corresponding to the beam moment extended linearly to the centerline of the joint:

$$V_c = \frac{V_b(l_b + 0.5h_c)}{l_c} \tag{5}$$

where l_c is the length of column. Substituting Equations (5) and (3) in Equation (1), the following expression is obtained for horizontal joint shear force V_{jh}:

$$V_{jh} = V_b\left(\frac{l_b}{j_b} - \frac{l_b + 0.5h_c}{l_c}\right) \tag{6}$$

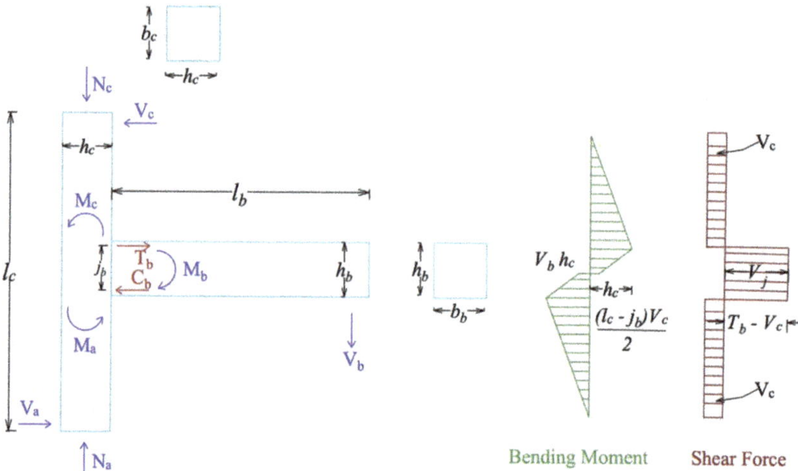

Figure 3. Internal forces and reactions in beam–column joints under lateral loads.

The resulting horizontal and vertical shear stress, v_{jh} and v_{jv}, at the mid-depth of the joint core is:

$$v_{jh} = v_{jv} = \frac{V_{jh}}{h'_c b'_c} \tag{7}$$

where h'_c and b'_c are the length and width of joint core. The joint region is subjected to horizontal and vertical shear stresses that are typically many times greater than those in the adjacent beams and columns. However, the stresses are also dependent on the joint aspect ratio α:

$$\alpha = \frac{V_{jv}}{V_{jh}} = \frac{h_b}{h_c} \tag{8}$$

Moreover, the axial compressive stress f_a at the mid-depth of the joint core due to vertical force N_c acting on the column is:

$$f_a = \frac{N_c}{h'_c b'_c} \tag{9}$$

The joint shear and axial stresses lead to the diagonal compression and tension principal stresses in the joint core. The principal compression stress f_c and tension stress f_t at mid-depth of the joint core can be found using Mohr's circle:

$$f_{t,c} = \frac{f_a}{2} \pm \sqrt{\left(\frac{f_a}{2}\right)^2 + v_{jh}^2} \tag{10}$$

Once the joint core develops diagonal tension cracks, the joint core's diagonal compression strut and truss mechanism, consisting of a concrete diagonal compression field and horizontal and vertical reinforcement, transfers the beam and column forces across the joint core (Figure 4). A beam-column joint may experience various damage or failure mechanisms as a result of the adopted structural detailing. It has been demonstrated that the use of poor-quality concrete, a lack of transverse reinforcement in the joint region, and inadequate anchoring details all constitute potential causes of a highly brittle failure mechanism. Because the shear and bond mechanisms that control the joint response have poor hysteretic properties, as a result, the joint's rotational resistance degrades rapidly [34,35].

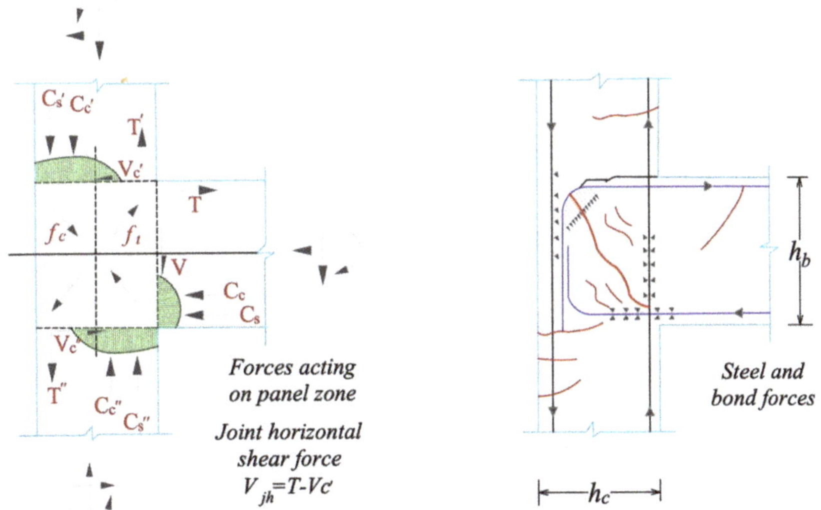

Figure 4. Principal stresses developed in the joint core, and the resulting cracking of the joint, bond, and bearing forces developed after the initiation of diagonal cracks.

3.2. Modeling of Joint Shear Panel

Figure 5 illustrates the proposed analytical model for a substandard RC external beam–column joint subjected to seismic actions. A simplified flowchart is given in Appendix A (Figure A1) to help guide the preparation of a numerical model of structural systems. The beam elements and the column elements are located at the beam mid-depth and the column mid-depth, respectively. The assembly is envisaged, as the same is tested experimentally. The joint panel distortion is modeled by four rigid link elements arranged along the edges of the joint panel and one nonlinear rotational spring is incorporated in one of the four hinges connecting the adjacent rigid elements. The failure of an exterior joint is primarily related to the principal tensile stress developed in the joint core.

The critical parameter is the principal tension stress within the joint core, and joint cracking begins at a stress of 0.29 $(f_c')^{0.5}$ MPa. In the case of exterior joints lacking transverse reinforcement, the experimental data support the highest principal tension stress of 0.42 $(f_c')^{0.5}$. Joint shear strength deterioration is governed by the gradual reduction of the effective joint principal tension stress [27,36]. The principal tensile stress vs. shear deformation envelope curve illustrated in Figure 6a is used to directly deduce the moment–rotation properties of the joint spring based on the equilibrium considerations. The nonlinear rotational spring at the joint is assigned a multilinear deteriorating hysteretic rule [37] to simulate the joint stiffness and strength deterioration and pinching in the hysteretic response (Figure 6b), which is crucial for the shear failure of substandard beam–column joints.

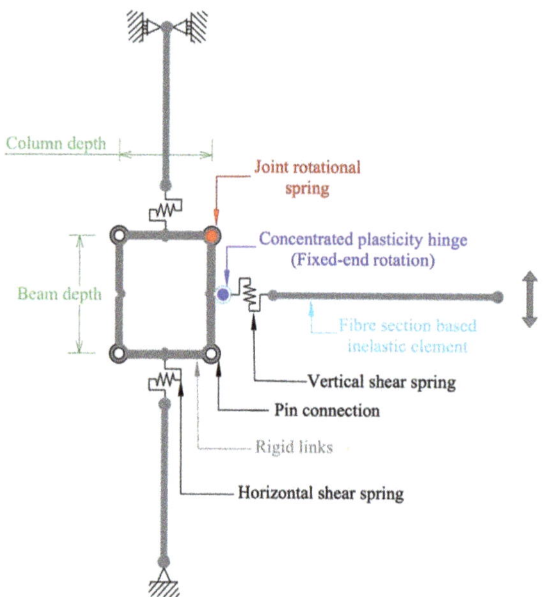

Figure 5. Nonlinear FE-based numerical model for an exterior beam–column joint.

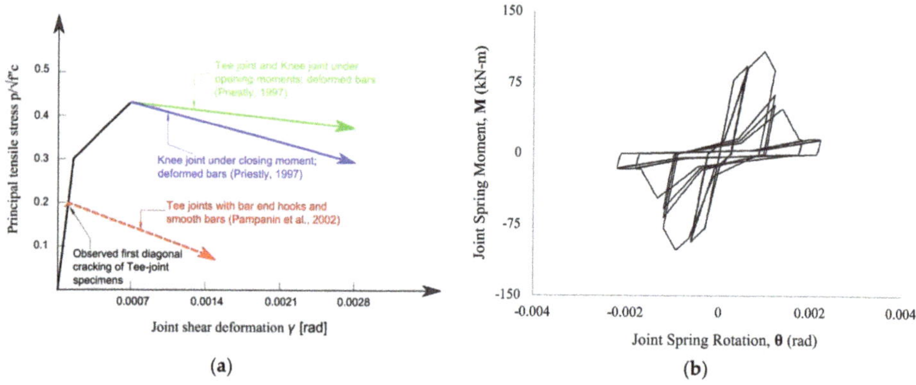

Figure 6. (**a**) Joint shear strength degradation model for exterior and corner joints [27,36]. (**b**) Hysteretic rule used to model joint stiffness and strength deterioration.

The joint shear strength in the present model is only dependent on the concrete compressive strength, with no regard for the influence of additional factors such as the joint aspect ratio or longitudinal reinforcement of connecting elements. Although more contemporary shear strength models [10,38,39] can be used, the considered model is simple, requires less input information, and provides conservative estimates, which is useful for assessing existing RC frames.

3.3. Modeling of Beam–Column Connecting Members

The flexibility-based element formulation, such as that developed by Spacone et al. [40], which is an extension of the Ciampi and Carlesimo [41]-proposed consistent flexibility-based method for formulating frame members, is established using the framework of the mixed method formulation [42]. The element formulation strictly satisfies the equilibrium of bending moments and axial force along the element using force interpolation functions rather than displacement interpolation functions. The element state determination relies on a nonlinear iterative method based on residual deformations [42] that continuously maintain equilibrium within the element and finally converge to a state that satisfies the element constitutive relation within a set tolerance.

The element stiffness matrix is obtained by inverting the element flexibility matrix. The use of improved material models to simulate the element's deteriorating hysteretic response owing to pre-existing cracks in the member, the rigid body fixed-end rotation due to rebars slip and inelastic extension, and geometric nonlinearity are improvements in the present model over the original formulation of Spacone et al. [40]. This solution technique is especially well suited for the study of the highly nonlinear deteriorating hysteretic behavior of the substandard reinforced concrete elements.

The fiber-based approach, in which each fiber is assigned a uniaxial stress–strain relationship, is used by the frame beam–column element to simulate the cross-section behavior. The nonlinear uniaxial stress–strain response of the individual fibers into which the section has been divided is then integrated to provide the sectional stress–strain state. Because the material constitutive models already specify hysteretic response, a fiber section-based element has the advantage of not requiring prior moment-curvature analysis and calibration. Moreover, it directly simulates the stiffness and strength interactions between an axial load and a bending moment, as well as the interactions between flexural strength in orthogonal directions. The discrete number of the controlling sections along the element that are used for the numerical integration is the only approximation in this formulation, and it does not actually impose any restrictions on the displacement field of the element, this formulation is always "exact": regardless of the degree of inelasticity, the force field is always exact. To prevent under-integration, a minimum of 3 Gauss–Lobatto integration sections are necessary and in general 5–7 integration points (IPs) are utilized [43,44] to adequately model the spread of inelasticity.

3.3.1. Beam–Column Element Formulation

Model assumptions:

- Beam–column element in local reference system x, y, z (Figure 7a)
- A discrete number of cross sections placed at control points of the numerical integration
- Beam–column member geometry is linear.
- Plane sections remain plane and are normal to the longitudinal axis throughout the deformation history.
- Strains and stresses act parallel to the longitudinal axis.
- Member behavior in torsion is linearly elastic and uncoupled from flexure and axial response.

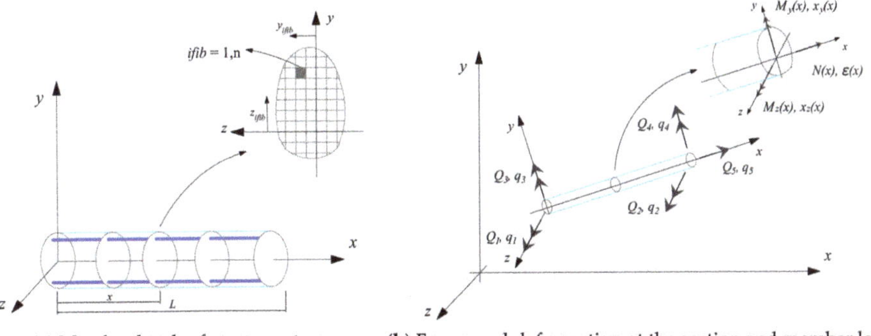

(**a**) Member local reference system   (**b**) Forces and deformation at the section and member level

Figure 7. Beam element in the local reference system: subdivision of cross-section into fibers.

Vectors of element forces and deformations and the corresponding section forces and deformations are given below:

$$Q = \{Q_1, Q_2, Q_3, Q_4\}^T \tag{11}$$

$$q = \{q_1, q_2, q_3, q_4\}^T \tag{12}$$

$$D(x) = \{M_z(x), M_y(x), N(x)\}^T \tag{13}$$

$$d(x) = \{X_z(x), X_y(x), \bar{\varepsilon}(x)\}^T \tag{14}$$

where Q is the element force vector, q is the element deformation vector, $D(x)$ is the section force vector, $d(x)$ is the section deformation vector, $X(x)$ is the section curvature about the reference axis, and $\bar{\varepsilon}(x)$ is the axial strain at the reference axis. With the use of a simple geometric transformation matrix, the fiber strains are related to section deformations. The section forces and deformations are related to element forces and deformations using the force and deformation interpolation functions:

$$\Delta d(x) = a(x)\Delta q \tag{15}$$

$$\Delta D(x) = b(x)\Delta Q \tag{16}$$

where matrix $a(x)$ is the deformation interpolation function, matrix $b(x)$ is the force interpolation function, and Δ is the increments of the corresponding quantities. The linearization of the incremental section constitutive relation is according to the Equation (17):

$$\Delta d^j(x) = f^{j-1}(x)\Delta D^j(x) + r^{j-1}(x) \tag{17}$$

where $f^{j-1}(x)$ is the section flexibility and $r^{j-1}(x)$ is the residual deformations from the previous iteration. The residual deformation $r^{j-1}(x)$ is the linear approximation to the deformation error that results from linearizing the section force-deformation relation. Equation (17) may be presented in the integral form as given:

$$\int_0^L \delta D^T(x)\left[\Delta d^j(x) - f^{j-1}(x)\Delta D^j(x) - r^{j-1}(x)\right]d_x = 0 \tag{18}$$

Substituting Equations (15) and (16) in Equation (18) gives:

$$T\Delta q^j - F^{j-1}\,\Delta Q^j - s^{j-1} = 0. \tag{19}$$

where T is the matrix dependent on the interpolation functions, F is the element flexibility matrix, and s is the element residual deformation vector given as follows:

$$T = \int_0^L b^T(x)a(x)d_x \tag{20}$$

$$F = \int_0^L b^T(x)f(x)b(x)d_x \tag{21}$$

$$s = \int_0^L b^T(x)r(x)d_x \tag{22}$$

Furthermore, the virtual displacement principle is used to obtain the integral form of the equilibrium equation:

$$\int_0^L \delta d^T(x)\left[D^{j-1}(x) + \Delta D^j(x)\right]d_x = \delta q^T Q^j \tag{23}$$

where $D^{j-1}(x) + D^j$ is the new internal force distribution and Q^j is the corresponding vector of nodal forces in equilibrium. Substituting Equations (15) and (16) in Equation (23), results in the following matrix expression, which is equivalent of the integral form of the element equilibrium equations:

$$T^T Q^{j-1} + T^T \Delta Q^j = Q^j \tag{24}$$

The combination of Equations (19) and (24) results in the following expression:

$$\begin{bmatrix} -F^{j-1} & T \\ T^T & 0 \end{bmatrix} = \begin{Bmatrix} \Delta Q^j \\ \Delta q^j \end{Bmatrix} = \begin{Bmatrix} s^{j-1} \\ Q^j - T^T Q^{j-1} \end{Bmatrix} \tag{25}$$

Solving the first equation for ΔQ^j and substituting it in the second equation results in the following expression:

$$T^T \left[F^{j-1} \right]^{-1} \left(T \Delta q^j - s^{j-1} \right) = Q^j - T^T Q^{j-1} \tag{26}$$

Assuming $T = I$, as peculiar to the proposed Bernoulli beam, where I is a 3×3 identity matrix, the above equation is simplified as:

$$\left[F^{j-1} \right]^{-1} \left(\Delta q^j - s^{j-1} \right) = \Delta Q^j \tag{27}$$

where ΔQ^j is the element force increment, $(\Delta q^j - s^{j-1})$ is the corresponding deformation increment including the residual deformation s^{j-1} that results from the linearization of the section's non-linear constitutive relations, and $[F^{j-1}]^{-1}$ is the element stiffness matrix obtained by inverting the flexibility matrix.

The element forces provide the most difficulty since they cannot be easily calculated from the section forces, even if the element stiffness matrix is obtained by inverting the element flexibility matrix. The procedure outlined in Spacone et al. [42] constitutes an iteration scheme at the element level that is similar to the Newton–Raphson method utilized for the structural level solution of the equilibrium equations. Force increments are applied to the structural degrees of freedom, and Newton–Raphson iterations are used to reduce the imbalanced forces down to acceptable levels at each load step. The structural level solution of these equations results in displacement increments at the end nodes of each element. The algorithm's iterations during the element state determination phase intend to reduce the deformation residuals to levels that are acceptable. In step i of the Newton–Raphson algorithm at the structural degrees of freedom, Figure 8 illustrates the relationship between element and section state determination. The following describes the iteration scheme used to determine the incremental element forces:

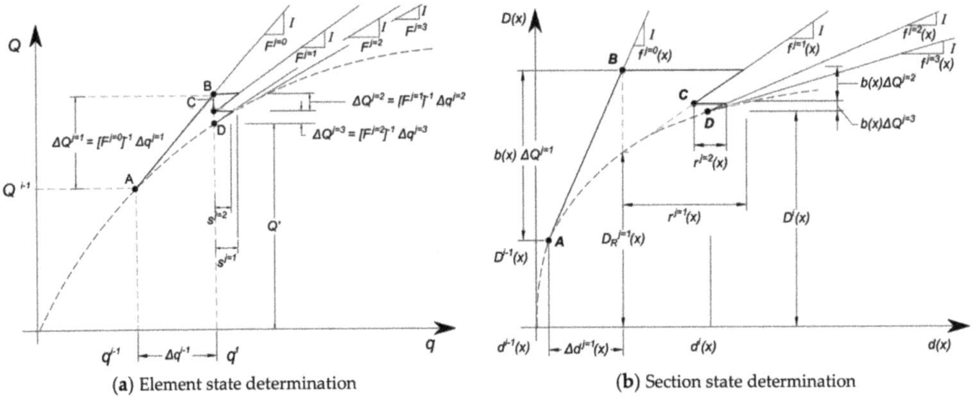

(a) Element state determination   (b) Section state determination

Figure 8. Determination of element resisting forces Q^i corresponding to element deformations q^i using the element and state determination for a flexibility-based element.

At point A, $i = 1$ and $j = 0$:

$$\text{element deformation} \quad q^i = q^{i-1} + \Delta q^i \tag{28}$$

iteration starts, $j = 1$

$$\text{element force increment} \quad \Delta Q^{j=1} = \left[F^{j=0} \right]^{-1} \Delta q^{j=1} \tag{29}$$

$$\text{initial element tangent stiffness matrix} \quad \left[F^{j=0} \right]^{-1} = \left[F^{i-1} \right]^{-1}$$

$$\text{element deformation increment} \quad \Delta q^{j=1} = \Delta q^i$$

$$\text{section deformation increment} \quad \Delta d^{j=1}(x) = f^{j=0}(x)\Delta D^{j=1}(x) \tag{30}$$

$$f^{j=0}(x) = f^{i-1}(x)$$

$$\Delta D^{j=1}(x) = b(x)\Delta Q^{j=1}$$

$$\text{section deformation} \quad d^{j=1}(x) = d^{j-1}(x) + \Delta d^{j=1}(x) \tag{31}$$

Equation (31) provides the updated section deformation that corresponds to point B in Figure 8; this updated section deformation is used as the basis to determine the section stiffness and resisting forces. Assuming that plane sections remain plane and normal to the longitudinal axis, it is straightforward to determine the strain distribution in a section. The stress and tangent modulus of the section is obtained using the constitutive equations of the steel and concrete fibers, which are integrated over the cross-sectional area to determine the section's resisting forces and tangent stiffness matrix.

$$\text{section stiffness matrix} \quad k^{j=1}(x) = \int_{A(x)} I^T(y,z)E(x,y,z)I(y,z)dA \tag{32}$$

$$\text{section resisting forces} \quad D_R^{j=1}(x) = \int_{A(x)} I^T(y,z)\sigma(x,y,z)dA \tag{33}$$

$$\text{geometric vector} \quad I(y,z) = \{-y \; z \; 1\}$$

The Gauss-Lobatto numerical integration scheme is used to solve Equations (32) and (33). The section stiffness is inverted to produce the flexibility matrix $f^{j=1}(x)$. The difference between the applied and resisting forces gives the unbalanced forces at the section.

$$\text{unbalanced forces at the section} \quad D_U^{j=1}(x) = D^{j=1}(x) - D_R^{j=1}(x) \tag{34}$$

$$\text{residual section deformations} \quad r^{j=1}(x) = f^{j=1}(x)D_U^{j=1}(x) \tag{35}$$

$$\text{residual element deformations} \quad s^{j=1} = \int_0^L b^T(x)r^{j=1}(x)dx \tag{36}$$

second iteration, $j = 2$

$$\text{updated element forces} \quad Q^{j=2} = Q^{j=1} + \Delta Q^{j=2} \tag{37}$$

$$\text{compatibility corrective element forces} \quad \Delta Q^{j=2} = -\left[F^{j=1} \right]^{-1} s^{j=1}$$

$$\text{updated section forces} \quad D^{j=2}(x) = D^{j=1}(x) + \Delta D^{j=2}(x) \tag{38}$$

$$\text{updated section deformation} \quad d^{j=2}(x) = d^{j=1}(x) + \Delta d^{j=2}(x) \tag{39}$$

$$\text{force increment applied at all IPs} \quad \Delta D^{j=2}(x) = b(x)\Delta Q^{j=2}$$

$$\text{deformation increment induced at all IPs} \quad \Delta d^{j=2}(x) = r^{j=1}(x) + f^{j=1}(x)\Delta D^{j=2}(x)$$

$$r^{j=2}(x) = f^{j=2}(x)D_U^{j=2}(x) \tag{40}$$

At the completion of the second iteration, the state of the element and the sections corresponds to point C in Figure 8. The section flexibility matrices and section residual deformation vectors are determined at each control IPs. The new element flexibility matrix is obtained by integrating the section flexibility matrices in accordance with Equation (21). The residual section deformations are then integrated to yield the residual element deformations. The third and subsequent iterations adhere to this iteration scheme. When the specified element convergence criterion is satisfied, convergence is achieved. The measures of energy are used for the purpose [45].

3.3.2. Material Models

The solution of Equations (32) and (33) necessitates the definition of suitable material models since a numerical solution algorithm should include stress–strain relations for concrete and reinforcing rebars for computing the section resisting forces and stiffness matrix.

The [46] nonlinear model, as updated by Filippou *et al.* [29] to include isotropic strain hardening, has been used in the present study to characterize the stress-strain relation of the reinforcing steel. This is the most appealing since it has the best agreement with experimental data and is computationally efficient. The stress-strain model takes on the following form to describe a curved transition from a straight-line asymptote with slope E_0 to another asymptote with slope bE_0, and allowing a good representation of the Bauschinger effect:

$$\sigma^* = b\varepsilon^* + \frac{(1-b)\varepsilon^*}{\left[1 + (\varepsilon^*)^R\right]^{\frac{1}{R}}} \tag{41}$$

$$\sigma^* = \frac{(\sigma - \sigma_r)}{(\sigma_0 - \sigma_r)}$$

$$\varepsilon^* = \frac{(\varepsilon - \varepsilon_r)}{(\varepsilon_0 - \varepsilon_r)}$$

$$R^n = R_0 - \frac{a_1 \zeta_p^n}{\left(a_2 + \zeta_p^n\right)}$$

$$\zeta_p^n = \varepsilon_r^n - \varepsilon_y^n$$

where σ^* and ε^* are the normalized stress and strain, σ_o and ε_o are the stress and strain at first yielding, b is the strain hardening, σ_r and ε_r indicate the stress and the strain at the point of the last strain reversal where the stress of equal sign took place, R is the curvature parameter, R_0 is the value of R during the first loading, a_1 and a_2 are experimentally determined parameters, and ζ_p^n is the plastic excursion at the current semicycle (Figure 9). Fragiadakis et al. [47] provide more in-depth information on the model's numerical implementation that also takes into consideration the buckling of steel bars.

In order to compute the concrete compressive forces, the cross-section can be divided into layers, and the distribution of concrete strain is determined assuming a linear distribution of concrete strain increments and strain increment compatibility between steel and concrete. For this purpose, a nonlinear constant confinement concrete model (Figure 10), which is a uniaxial nonlinear model that is based on the constitutive relationship proposed by Mander et al. [48] with modifications made by Martinez-Rueda and Elnashai [49] for the purpose of maintaining numerical stability under large deformations, has been used to determine the stress in each layer. The ratio of the compressive stress of confined concrete to unconfined concrete defines the constant confinement factor. Given the effective confined concrete core area A_e (mid-way through two consecutive stirrups) and area of concrete core A_{cc}, the effective confinement factor K_e is determined:

$$K_e = \frac{A_e}{A_{cc}} = \frac{A_e}{A_c(1 - \rho_{cc})} \tag{42}$$

$$A_e = \left(b_c d_c - \sum_{i=1}^{n} \frac{(w_i')^2}{6} \right)\left(1 - \frac{s'}{2b_c} \right)\left(1 - \frac{s'}{2d_c} \right)$$

where ρ_{cc} is the ratio of area of longitudinal reinforcement to the area of the core section, A_c is the area of the core of the section enclosed by the center lines of the perimeter tie, b_c and d_c are the core dimensions to the centerlines of the perimeter lateral tie, w_i' is the ith clear distance between adjacent longitudinal bars, s' is the clear vertical spacing between lateral ties. Figure 11 illustrates a concrete model for both confined and unconfined concrete based on the stress–strain relation proposed by Mander et al. [48], which gives the longitudinal compressive concrete stress:

$$f_c = \frac{f_{cc}' x r}{r - 1 + x^r} \tag{43}$$

$$x = \frac{\varepsilon_c}{\varepsilon_{cc}}$$

$$\varepsilon_{cc} = \varepsilon_{co}\left[1 + 5\left(\frac{f_{cc}'}{f_{co}'} - 1 \right) \right]$$

$$r = \frac{E_c}{E_c - E_{sec}}$$

$$E_{sec} = \frac{f_{cc}'}{\varepsilon_{cc}}$$

$$E_c = 5000\sqrt{f_{co}'}\ \text{MPa}$$

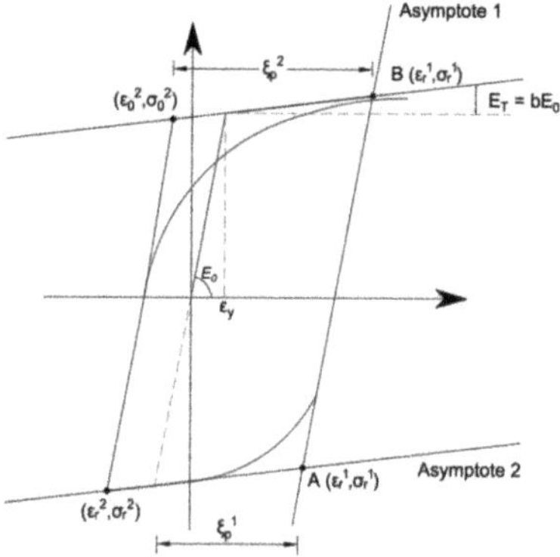

Figure 9. Stress–strain relationship of the reinforcing steel material models used in inelastic beam–column members.

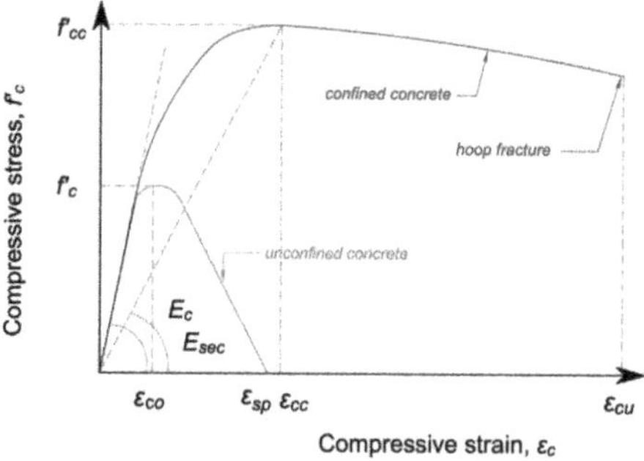

Figure 10. Stress–strain relationship of confined and unconfined concrete material models used in inelastic beam–column members.

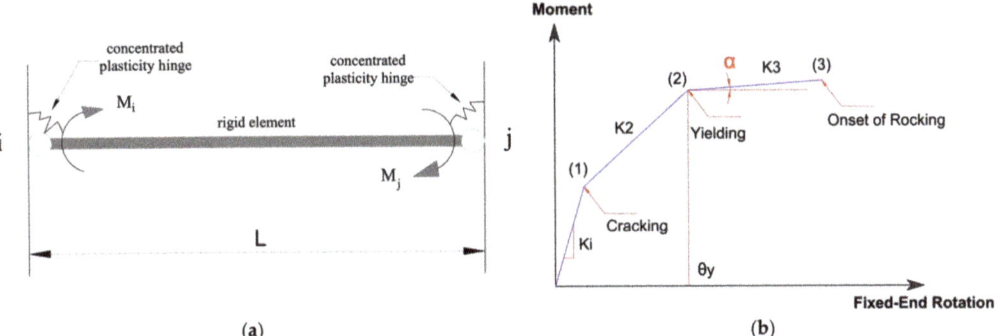

Figure 11. Analytical model for a beam–column element with fixed-end rotation: (**a**) concentrated plasticity hinges at beam ends and (**b**) a proposed moment–rotation relationship based on experimental studies performed on full-scale beams [50].

The cover concrete has been designated as unconfined concrete, and the stress–strain behavior of the segment of the falling branch in the area where $\varepsilon_c > 2\varepsilon_{co}$ is considered to be a straight line that reaches zero stress at the spalling strain ε_{sp}.

The confined compressive strength f'_{cc} is determined, assuming the confined concrete core is placed under triaxial compression with equal effective lateral confining forces f'_l from lateral ties:

$$f'_{cc} = f'_{co}\left(-1.254 + 2.254\sqrt{1 + \frac{7.94 f'_l}{f'_{co}}} - 2\frac{f'_l}{f'_{co}}\right) \tag{44}$$

$$f'_{lx} = K_e \rho_x f_{yh}$$

$$f'_{ly} = K_e \rho_y f_{yh}$$

$$\rho_x = \frac{A_{sx}}{sd_c}$$

$$\rho_y = \frac{A_{sy}}{sb_c}$$

where f_{yh} is the yield strength of the transverse reinforcement, A_{sx} and A_{sy} are the total area of transverse bars running in the x and y directions respectively, f'_{lx} and f'_{ly} the effective lateral confining stresses in the x and y directions respectively. In tension, a linear stress-strain relationship is assumed up to the tensile strength, provided the tensile strength has not been exceeded. However, it can be ignored for pre-cracked member with zero tensile strength. The cyclic loading stress-strain response is assumed to be enclosed by the monotonic loading stress-strain curve (Figure 10).

When the transverse reinforcement confining the core fractures, it is regarded as the ultimate limit to confined concrete compression strain. This may be determined by comparing the increase in the strain energy absorbed by the concrete over the value suitable for unconfined concrete to the strain-energy capacity of the confining steel. The following equation for the ultimate compression strain for confined concrete may be derived by assuming that the ultimate strain of the unconfined concrete is 0.004:

$$\varepsilon_{c,dc} = 0.004 + 1.4\frac{\rho_v f_{yh}\varepsilon_{su}}{f'_{cc}} \tag{45}$$

where ε_{su} is a strain value at fracture of lateral confining ties, and ρ_v is the volumetric confinement ratio, which is the sum of ρ_x and ρ_y.

3.3.3. Geometric Nonlinearity

Large displacements, large rotations, and large independent deformations relative to the chord of the frame element cause nonlinearities in kinematic quantities, and these sources of nonlinearity are referred to as geometric nonlinearities, also known as p-delta effects. These nonlinearities are negligible under normal load conditions, but they become significant in the presence of extreme loads and in large/slender structures. The internal forces demand is amplified under large lateral deflections, which results in a reduction in the effective lateral stiffness. The potential capacity of the structure to resist lateral loads decreases as internal forces rise, resulting in a drop in the structure's effective lateral strength. Since they may eventually result in the loss of lateral resistance, ratcheting, and dynamic instability, this warrants consideration in numerical models [51].

A total co-rotational formulation developed by Correia and Virtuoso [52] is used to take into account the large displacements/rotations and large independent deformations relative to the chord of the frame elements in the model. The total co-rotational formulation makes use of an exact description of the kinematic transformations associated with large displacements and three-dimensional rotations of the element. As a result, the independent deformations and forces of the element are correctly defined, and the effects of geometric non-linearities on the stiffness matrix are naturally defined. The use of this formulation takes into account small deformations in relation to the element's chord without losing its generality, despite the existence of large nodal rotations and displacements.

3.4. Modeling of Fixed-End Rotation

The fixed-end rotation at the interface of the beam and column caused by bond failure, longitudinal rebar-slip, and inelastic extension is simulated for each element using a single nonlinear rotational spring located at the beam–column member ends (Figure 11a), and a concentrated plasticity hinge with a nonlinear moment–rotational behavior (Figure 11b, Table 1) is used to model the rigid-body rotational deformation of the member. The formulation developed herein for fixed-end rotation is similar to the rigid-bar rotation model proposed by Filippou and Issa [30], with the improvement including a more realistic representation of the moment–rotation relationship developed recently by Ahmad et al. [50] based on quasi-static cyclic tests performed on full-scale beams exhibiting fixed-end rotation. A trilinear force–deformation envelope is proposed to model inelastic mechanisms (cracking, yielding, and rocking) with a deteriorating hysteretic rule for model stiffness and strength deterioration and pinching behavior. Due to its high initial stiffness, the hinge may only contribute marginally before the beam yields but contributes significantly after the rocking mode is initiated.

Table 1. Proposed moment–rotation values for fixed-end rotational spring to model beam–column member rigid-body rotational deformation [50].

Parameter	θ_{cr} (rad)	M_{cr}/M_{max} *	θ_y (rad)	M_y/M_{max} *	$\alpha = K_3/K_1$
Value	0.00091	0.27	0.01177	0.76	0.055

* M_{max} is the beam maximum moment capacity.

The flexibility matrix's off-diagonal components may simply be demonstrated to be zero in this situation and as a result, the flexibility matrix of the element is given:

$$F_{FER} = \begin{bmatrix} f_i & 0 \\ 0 & f_j \end{bmatrix} \tag{46}$$

where f_i is the flexibility coefficient of the rotational spring at the end i and f_j is the flexibility coefficient of the rotational spring at end j. Assuming that the point of inflection remains fixed and lies in the middle of the beam throughout the loading history, in this case, each half of the member can be viewed as a cantilever beam and the flexibility coefficients can be easily formulated. The problem can be further simplified, ignoring gravity load, this corresponds to a cantilever beam subjected to load P (Figure 12).

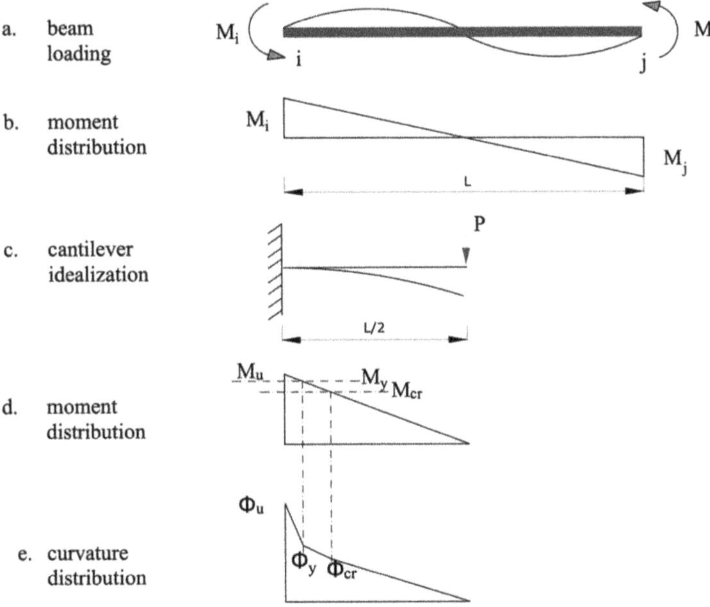

Figure 12. Mechanical parameters of fixed-end rotation beam–column member, considering only the rigid-body rotation mode of the beam.

To determine the flexibility coefficients of the concentrated hinge, the rotation at the cantilever's root (beam ends) caused by the curvature distribution is first determined for various values of the load P, and if the moment-curvature relation is known, this can be simplified. The moment-rotation relationship proposed (Figure 11b) is trilinear, due to which the approach produces a nonlinear flexibility coefficient of the corresponding concentrated spring. The stiffness of the hinge can be determined as:

$$K_i = \frac{M_{cr}}{\theta_{cr}} \quad 0 \leq M \leq M_{cr} \tag{47}$$

$$K_2 = \frac{M_y - M_{cr}}{\theta_y - \theta_{cr}} \quad M_{cr} \leq M \leq M_y \tag{48}$$

$$K_3 = \frac{M - M_y}{\theta - \theta_y} \quad M_y \leq M \tag{49}$$

where K_i is the initial stiffness, K_2 is the post-cracking stiffness, and K_3 is the post-yield stiffness of the hinge. The limit state values of moment (M) and rotation (θ) are given in Table 1.

It is typical to represent the flexibility matrix of a rotational spring stiffness K in terms of the prismatic beam element's elastic stiffness for convenience:

$$K_i = \frac{4.EI}{\gamma.L} \tag{50}$$

$$K_2 = \beta_1 \frac{4.EI}{\gamma.L} \tag{51}$$

$$K_3 = \beta_2 \frac{4.EI}{\gamma.L} \tag{52}$$

where γ, β_1, and β_2 are coefficient vary as a function of the moment-rotation demand. The experimental data suggests $\beta_1 = 0.165$ and $\beta_2 = 0.055$, and $\gamma = 0.406$ if EI is based on the nominal moment and yield curvature of the beam section [50]. The flexibility matrix can be formulated as given:

$$F_{FER,1} = \frac{L}{6EI} \begin{bmatrix} 1.5\gamma_i & 0 \\ 0 & 1.5\gamma_j \end{bmatrix} \quad 0 \leq M \leq M_{cr} \tag{53}$$

$$F_{FER,2} = \frac{L}{6EI.\beta_1} \begin{bmatrix} 1.5\gamma_i & 0 \\ 0 & 1.5\gamma_j \end{bmatrix} \quad M_{cr} \leq M \leq M_y \tag{54}$$

$$F_{FER,3} = \frac{L}{6EI.\beta_2} \begin{bmatrix} 1.5\gamma_i & 0 \\ 0 & 1.5\gamma_j \end{bmatrix} \quad M_y \leq M \tag{55}$$

This concentrated plasticity model has the advantage of being computationally simple can be readily implemented in a finite element computer program and can accurately represent the hysteretic response of RC members whose behavior is controlled by fixed-end rotation. The flexibility matrix of the element is calculated by simply adding the flexibility matrices of the constituent elements (i.e., beam-column inelastic member and concentrated plasticity hinge), since the plasticity hinge element and the inelastic beam-column member act in series.

3.5. RC Beam–Column Frame Member Shear Strength

3.5.1. Basics of RC Member Shear Strength

A key element of nonlinear modeling is including the most probable inelastic deformation mechanisms; therefore, the model should specify aspects for evaluating the performance of RC frames, especially when the shear failures of columns can anticipate the failure of the buildings. A comparison of the beam/column members' flexural and shear strengths is required to ascertain if flexural or shear failure is predicted. If individual member shear deformation is negligible, less detailed modeling is appropriate, provided that members do not develop shear forces that overwhelm the capacity of the member, as this will result in a brittle mechanism and the rapid deterioration of the member's stiffness and strength. Although shear damage was not noticed in the beam and column members for the present test specimen, since the members had sufficient shear strength and the mechanism of joint shear occurred owing to joint-region weakness, the member shear modeling is included for completeness.

It is advised that member shear strength be determined using more suitable equations that illustrate the dependence of shear strength on flexural ductility. The seismic shear strength model of columns proposed by Priestley [27] is discussed, which presents shear strength as the product of the concrete strength contribution (V_C), a truss mechanism of the transverse steel (V_S) assuming a 30° angle between the diagonal compression struts and the longitudinal axis of the column, and an arch mechanism for the axial load (V_P):

$$V_n = V_C + V_S + V_P \tag{56}$$

$$V_C = 0.8 A_g k \sqrt{f'_c} \tag{57}$$

$$V_S = \frac{A_v f_{yh}(D - c)}{s} \cot 30° \tag{58}$$

$$V_P = P \tan \alpha \tag{59}$$

$$\tan \alpha = \frac{D - c}{2a}$$

where A_g is the gross area of section, k is the degradation factor dependent on the curvature ductility, f_c' is the compressive strength of concrete, D is the depth of the section, c is the neutral axis depth, A_v is the total area of transverse reinforcement, s is the spacing of the lateral ties, f_{yh} is the yield strength of the transverse reinforcement, and a is shear span or distance from the maximum moment to the point of inflection. The value of k varies approximately from 3.50 [psi units] to 1.20 [psi units] as the member displacement ductility varies from one to three for bi-axial loading and two to four for uni-axial loading, respectively. As a result, the concrete contribution drops to as little as one-third of its initial value as the displacement ductility increases. The magnitude of the member flexural and shear strengths influence whether shear or flexural failure occurs and consequently, the ultimate section curvature ductility. The initial and residual shear strengths are represented by the two limiting values, V_{ni} and V_{nd}, respectively. Brittle shear failure happens when the shear force V_f, related to the initial flexural strength, is greater than V_{ni}. If $V_f < V_{nd}$, which ensures a ductile flexural response, the detailing of the flexural confinement determines the maximum section ductility. A shear failure is anticipated at a certain curvature ductility when $V_{nd} < V_f < V_{ni}$. Although the shear strength given by Equation (56) has been rather extensively used for column sections, with no axial load, it would appear that there should not be much conceptual difference between a beam and a column, and therefore Equation (56) should also immediately apply to beams. For simplicity, using Equation (57), with $k = 2.40$ [psi units] for curvature ductility < 3 and $k = 0.60$ [psi units] for curvature ductility > 7, determine the shear strength of a negative moment plastic hinge in beams.

3.5.2. Shear Strength Model Implementation

For implementation in nonlinear modeling, the shear capacity of an RC column is determined according to ASCE 41-17 [53], whereas the shear capacity of the beam section is determined according to ACI 318-19. For columns, the shear strength V_{Col} [psi units] shall be permitted to be calculated as:

$$V_{Col} = k_{nl} V_{Col0} \tag{60}$$

$$V_{Col0} = \alpha_{Col}\left(\frac{A_v f_{ytL/E} d}{s}\right) + \lambda\left(\frac{6\sqrt{f'_{cL/E}}}{M_{UD}/V_{UD}d}\sqrt{1 + \frac{N_{UG}}{6A_g\sqrt{f'_{cL/E}}}}\right)0.8A_g \tag{61}$$

where $k_{nl} = 1.0$ for displacement ductility demand is less than or equal to 2, 0.7 for displacement ductility greater than or equal to 6 and varies linearly for displacement ductility between 2 and 6, $\lambda = 1.0$ for normal-weight aggregate concrete, N_{UG} is the axial compression force, $M_{UD}/V_{UD}d$ is the largest ratio of moment to shear times effective depth for the column under design loadings but shall not be taken as greater than 4 or less than 2, $\alpha_{Col} = 1.0$ for $s/d \leq 0.75$, 0.0 for $s/d \geq 1.0$, and varies linearly for s/d between 0.75 and 1.0.

Equation (22.5.1.1) of ACI 318-19 is used to calculate the shear capacity of beam sections: Equation (22.5.8.5.3) of ACI 318-19 computes the shear strength provided by transverse reinforcement, and Section 22.5.5.1 of ACI 318-19 provides equations for the shear strength provided by concrete:

$$V_C = \left[2\lambda\sqrt{f'_c} + \frac{N_u}{6A_g}\right]b_w d \quad A_v \leq A_{v,min} \tag{62}$$

$$V_C = \left[8\lambda(\rho_w)^{\frac{1}{3}}\sqrt{f'_c} + \frac{N_u}{6A_g}\right]b_w d \quad A_v \leq A_{v,min} \tag{63}$$

$$V_C = \left[8\lambda_s\lambda(\rho_w)^{\frac{1}{3}}\sqrt{f'_c} + \frac{N_u}{6A_g}\right]b_w d \quad A_v < A_{v,min} \tag{64}$$

The shear strength equations (Equations (60)–(64)) were implemented in the SeismoStruct program and used to determine if the member has reached its shear capacity at each time step of the analysis. When the limit was reached, the member was assigned a residual strength (10% to 20% of maximum resistance) or remained inactive with no residual strength.

4. Validation of the Proposed Modeling Technique

4.1. Quasi-Static Cyclic Testing of Tee Beam–Column Joint

4.1.1. Description of Test Specimen

A low-rise building was considered as a prototype for choosing the geometric and reinforcement detailing of a typical substandard beam–column joint [13] in order to study the behavior of deteriorating joints and serve as a benchmark for testing and validating the proposed numerical modeling technique. Several deficiencies were discovered during the field survey of reinforced concrete building stock in Pakistan [13]. Because it was not possible to include all of the defects in the experimental models due to time and financial constraints, it was decided to analyze just those deficiencies that would have a significant influence on the seismic performance of the reinforced concrete buildings in Pakistan.

The selected beam–column joint subassembly (Figure 13) included columns that were 12 inches wide and 12 inches deep and a beam that was 12 inches wide and 18 inches deep. This model took into account all deficiencies found between design standards and actual reinforced concrete building constructions in Pakistan. These deficiencies included: spacing of ties—double of specified spacing, lack of seismic hooks, location of splice-near beam–column joint, lap length-reduced by 45%, 20% reduction in the diameter of rebars, lack of ties in the beam–column joint core (except a single tie), concrete having compressive strength of 2000 psi, and rebars having yield strength of 40,000 psi. The model's beam was reinforced with 3#5 bars at the top and bottom to represent the use of undersized reinforcing bars in constructions. A #3 stirrup was used as the lateral tie in the beam, positioned uniformly at 6 inch on centers. There were no seismic hooks, and the ties were closed at a 90-degree angle.

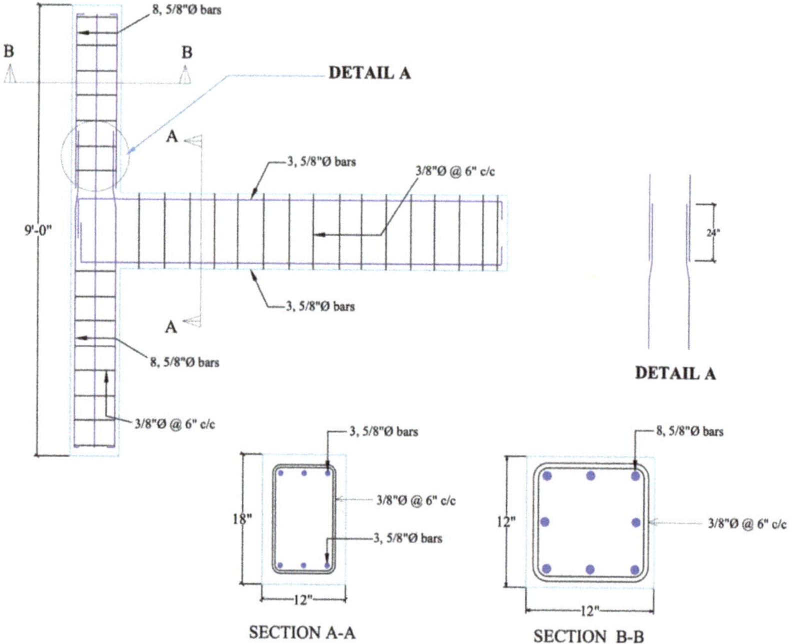

Figure 13. Geometric and reinforcement details of substandard beam–column joint.

4.1.2. Experimental Program and Specimen Behavior

Figure 14 illustrates the selected test setup and loading of the beam–column joint. The load was applied directly at the point of the contra flexure of the beam (i.e., free end of T-joint). Boundary conditions were selected to accurately reflect the end restraints of a two-dimensional frame. A hinge support was used to pin-connect the bottom of the column: pins were firmly fastened to a steel girder, and that steel girder was in turn fastened to the strong floor. The top of the column was provisioned with a roller support to enable unrestricted movement in a vertical direction, while lateral movement was restricted.

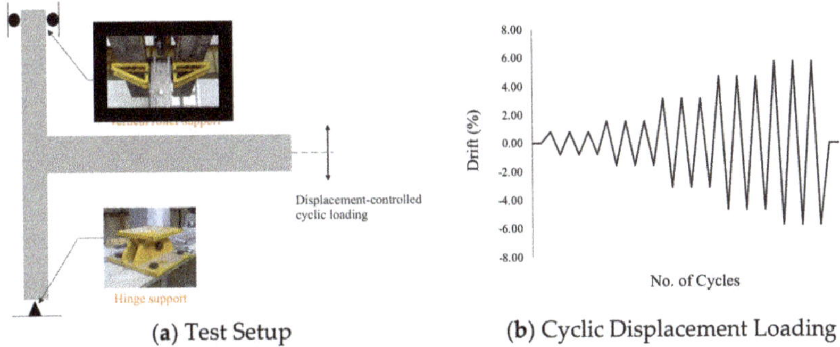

(**a**) Test Setup   (**b**) Cyclic Displacement Loading

Figure 14. Schematic representation of test specimen setup and loading.

A permanent gravity load of 26 tons (20% of the gross-section capacity) was applied at the top of the column. A reverse cyclic load was applied at the beam end of the beam–column assembly. To simulate the pre-cracked conditions, initially the assembly was tested under force-controlled loading. First, three tons of load was determined to be the theoretical yield capacity (F_y) of the system. This was used to guide load control testing; the force was applied in four equal increments: 0.25 F_y, 0.5 F_y, 0.75 F_y and F_y, and each increment was repeated three times. This was followed by displacement-controlled testing till the specimen attained extensive damages (Figure 15). Initially, the assembly was tested under force-controlled loading to simulate the pre-cracked conditions of existing buildings. The system's theoretical yield capacity, F_y, was first determined to be 2.76 tons. The force was applied in four equal increments—0.25 F_y, 0.5 F_y, 0.75 F_y, and F_y—and each increment was repeated three times during the load control tests. Displacement controlled testing was conducted afterwards until the specimen had sustained extensive damages (Figure 15).

Diagonal cracks initiated in the beam–column joint and vertical cracks appeared at the beam–column interface at a drift demand of 1.0%. The existing cracks in joints became aggravated, and additional multiple cracks appeared with increasing drift demand from 1.5% to 3.0% drift. Consequently, a concrete wedge mechanism was initiated. Upon further increasing drift demand, the width of existing cracks widened, while a concrete wedge was detached from the joint at a drift demand of 5.0%. This failure mechanism is especially brittle, and the bearing load capacity was lost as a result of the pushing out of a concrete wedge; therefore, the test was terminated.

4.2. Comparison of the Numerical and Experimental Prediction

The analytical model illustrated in Figure 5 was generated using the nonlinear finite element SeismoStruct program. The analytical model is composed of the inelastic beam and column members modeled as inelastic fiber-section-based elements that use the force-based formulation with the improved materials models discussed earlier. The material model for rebars is a uniaxial steel model that was first developed by Monti et al. [54] and can comprehend the post-elastic buckling behavior of reinforcing bars in compression. It makes use of the stress–strain relationship presented by Menegotto and Pinto [46], as well as the isotropic hardening rules of Filippou et al. [29] and the buckling rules of Monti and

Nuti [55]. For increased numerical stability/accuracy under transient seismic stress, it considers a further memory rule suggested by Fragiadakis et al. [47]. The material model for concrete is a uniaxial nonlinear constant confinement model described earlier that was first developed by Madas [56] and that adheres to the stress–strain relationship established by Mander et al. [48] and the cyclic rules proposed by Martinez-Rueda and Elnashai [49]. The Mander et al. [48] rule, which assumes continuous confining pressure over the whole stress–strain range, integrates the confinement effects offered by the lateral transverse reinforcement.

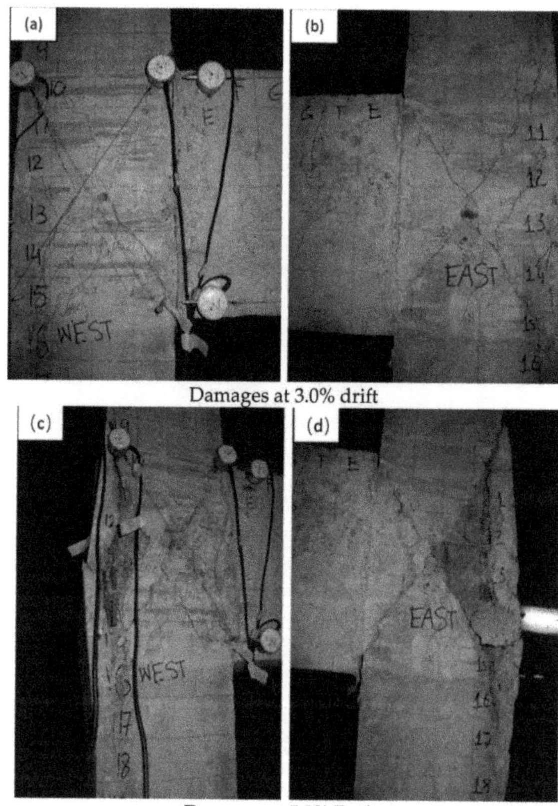

Figure 15. Damages observed in a beam–column joint with increasing drift demand. (**a**) represents the west-side face of damages at 3.0% drift (**b**) represents the east-side face of damages at 3.0% drift. (**c**) represents the west-side face of damages at 5.0% drift (**d**) represents the east-side face of damages at 5.0% drift.

To simulate fixed-end rotation, a concentrated plasticity hinge was introduced at the beam end using a zero-length link element. The link element connected two contemporaneous structural nodes and necessitated the development of a separate force–displacement (or moment–rotation) response curve for each of its local six degrees of freedom. An infinitely stiff elastic force–displacement rule was assigned to all degrees of freedom, except the in-plane rotational dof that was assigned with the multilinear moment–rotation rule (Figure 11b) featuring pinching in the hysteretic force–displacement behavior to simulate the opening/closing of a vertical crack at the beam end. The selected hysteresis loop is described in Sivaselvan and Reinhorn [57], which is capable of simulating the deterioration of strength, stiffness, and bond slip. The vertical dof of the beam element and the horizontal dof of the column element were assigned with linear force–displacement behavior to idealize the member shear deformation. However, the shear strength limit criteria described earlier were used to identify member shear failure.

Four stiff link elements were positioned along the edges of the panel to idealize the joint panel, and one nonlinear moment–rotation spring was introduced at one corner hinge. The backbone moment–rotation curve represents the behavior for monotonic loading, establishes strength and deformation bounds (Figure 6a), and uses a pinched hysteretic model from Ibarra et al. [37] to model the deteriorating hysteresis of the backbone curve (Figure 6b).

Figure 16 compares the analytical and experimental hysteretic responses of the Tee beam–column joint. This demonstrates that the initial stiffness, maximum strength, and hysteretic response of the analytical model agree well with the experimental response. The proposed joint modeling technique accurately simulates the inelastic response of the Tee beam–column joint viz. the plastic hinge flexural damage mechanism, with the slip of the reinforcing bars in the inelastic beam member and joint damage shear hinge mechanism within the panel zone, as well as the deterioration of stiffness and strength important for the considered substandard beam–column joint. Therefore, the proposed joint panel idealization using a nonlinear zero-length rotational spring with a deteriorating hysteretic rule may accurately model the nonlinear behavior of the exterior beam–column joints in existing structures.

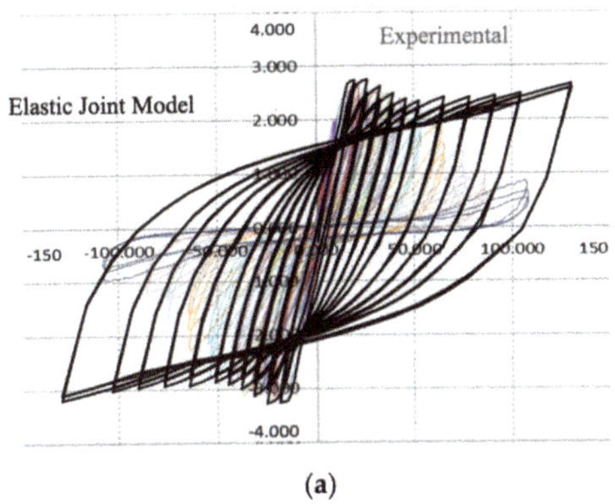

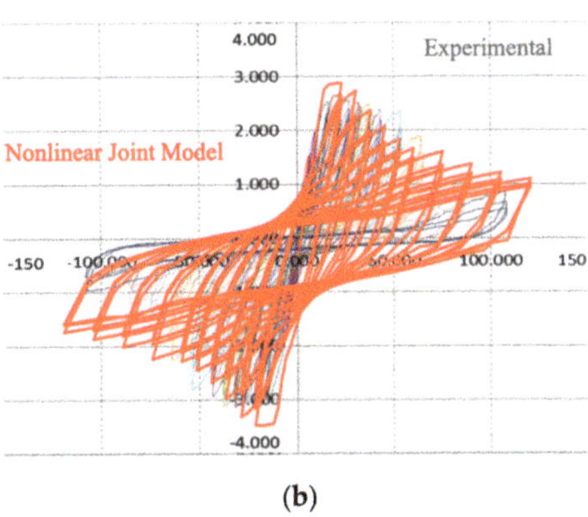

Figure 16. Comparison of the analytical and experimental force–displacement hysteretic responses. A model with elastic joints and conventional beam–column element formulation is shown in (**a**) and a proposed model with nonlinear joints and improved beam–column element formulation is shown in (**b**). Displacement is given in mm, and force is given in tons.

5. Frame Structure Nonlinear Response Analysis

5.1. Description of Frame Structure

A five-story reinforced concrete moment-resisting frame structure in the present study serves as a representative example of a Pakistani substandard frame building construction that falls short of code requirements (Figure 17). The assumed footprint of the structure is 45.72 m (150 ft) by 36.576 m. (120 ft). The structure has five bays in each direction, with the following dimensions: 9.144 m (30 ft) longitudinal bay width, 7.3152 m (24 ft) transverse bay width, and 3.6576 m story height (12 ft).

In compliance with the building regulations, a live load of 2.40 kN/m^2 was used. A 203 mm (8 in)-thick two-way floor slab and a superimposed dead load of 1.0 kN/m^2 (20 lbs/ft^2) were included in the dead load, in addition to the members' self-weight. The beams and columns were designed using regular reinforced concrete with a 28-day unconfined compressive strength of 13.80 MPa (2000 psi). The weight of the concrete was taken 23.60 kN/m^3 (150 lbs/ft^3). Reinforcement steel of grade 40 with a design yield tensile strength of 276 MPa (40 ksi) was used. Building analysis and design were performed for the short direction in accordance with BCP-SP (2007), assuming a fixed base to signify an adequate foundation.

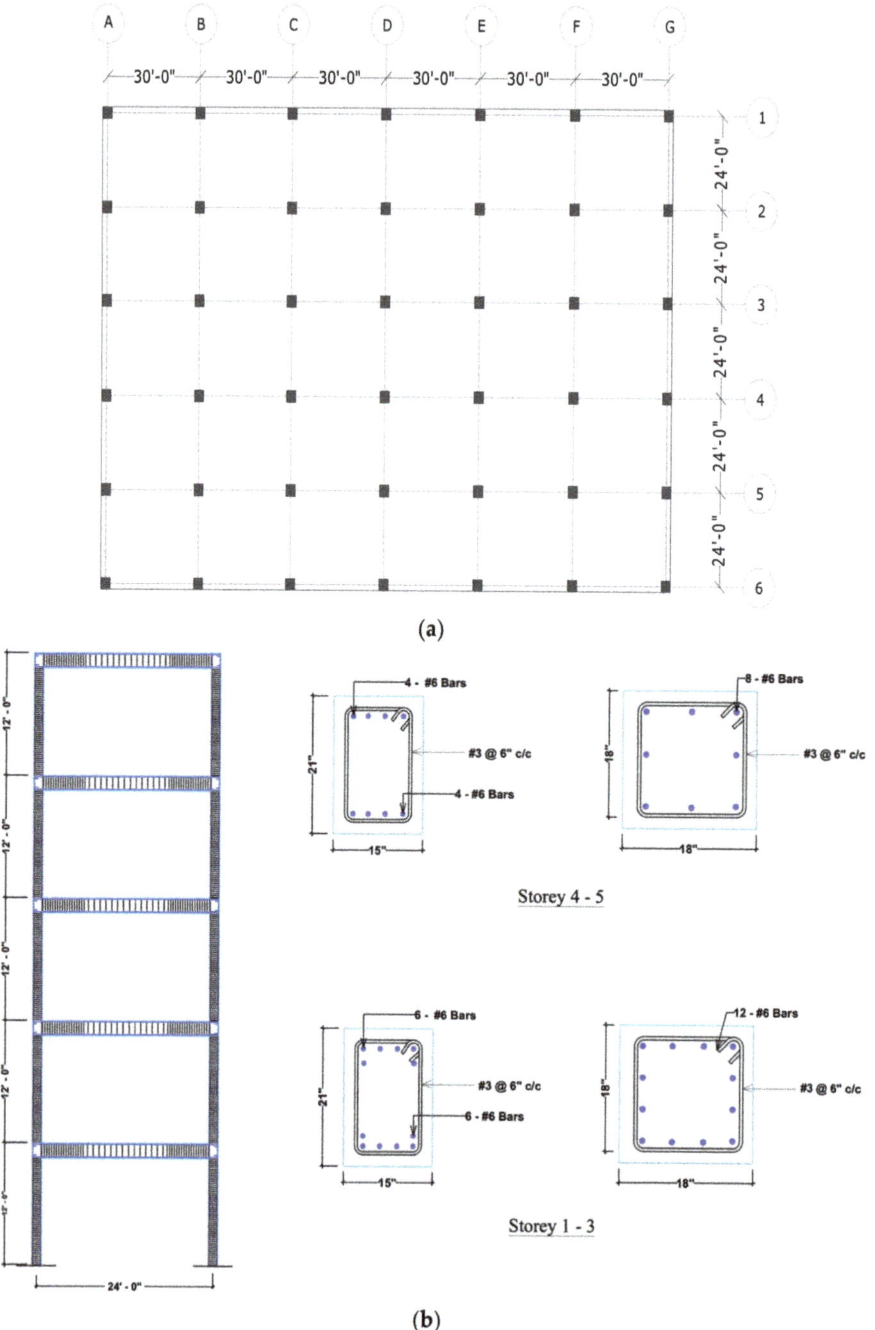

Figure 17. Geometric and reinforcement details of the selected frame. (**a**) The footprint of the considered building and (**b**) one-bay multi-story frame extracted for nonlinear response analysis.

A single-bay, two-dimensional, multi-story frame was extracted for nonlinear modeling and building response analysis. Based on the tributary area, weights were applied to the single-bay frame. Due to the same structural and dynamic properties that follow (i.e., building vibration period, frame damage mechanisms, and nonlinearities), it was intended that the behavior of the buildings resemble that of a planar frame. While other 3-dimensional effects and the slab/joists contribution can affect a structure's seismic response [58–60], the planar frame simplification allows for a significant reduction in the number of elements and degrees of freedom in the numerical model, which reduces the computation time needed for the building nonlinear response analysis and is a conservative approach to assessment.

5.2. Numerical Modeling

The nonlinear finite element SeismoStruct program was used to numerically model the considered prototype multi-story frame in a manner similar to the Tee beam–column joint discussed earlier. The floor loads and masses were lumped evenly at floor structural nodes. In order to account for the large displacements, the total corotational geometric transformation was used [52]. According to earlier research, the mass and stiffness proportional Rayleigh damping in the first two modes was set at 2% of critical for the considered frame Ahmad et al. [61]. The building frame elements were discretized using the fiber-section-based element with mixed formulation developed by Spacone et al. [40,42] that uses the improved material model formulations [47,49]. To account for the fixed-end rotation caused by bond failure, rebars yielding, and inelastic extension, the frame elements were provided with concentrated plasticity hinges that maintain the formulation of Filippou et al. [29] and uses the pinched degrading hysteretic rule as proposed by Sivaselvan and Reinhorn [57]. The frame was idealized using elastic joints (center-line model) and nonlinear joint panel models with the deterioration of stiffness and strength to evaluate the significance of the improved formulation. The joint panels were idealized as a parallelogram of stiff elements and provisioned with a corner nonlinear moment–rotation spring that uses the pinched deteriorating hysteresis of Ibarra et al. [37].

5.3. Selected Ground Motions

Twenty-four far-fault earthquake ground motions were obtained from the PEER NGA online ground motion database. Given that an existing structure may experience a variety of moderate-to-strong ground motions throughout its design life, it was preferable to subject the numerical model to a range of ground motions that represented variation in the key seismological parameters, such as magnitude (Mw: 6 to 7.62), fault mechanism (reverse, reverse-oblique, and strike-slip), source-to-site distance (Rjb: 17 km to 29 km), and significant duration (D5-95%: 11 s to 70. A single ground motion from each earthquake event was chosen in order to account for record-to-record variability. Earthquake events from numerous active tectonic zones (US, Japan, New Zealand, Iran, Taiwan, Armenia, Mexico) were taken into consideration. The median spectrum for the elastic 5% damped single degrees of freedom systems with periods ranging from 0.02 s to 4.0 s and subjected to unscaled ground movements is shown in Figure 18 and reported Table A1 in Appendix B. The selected unscaled ground motions have a peak acceleration of 0.35 g at a period of 0.26 s and a median peak ground acceleration of 0.16 g.

5.4. Frame Nonlinear Response
5.4.1. Damage Distribution

The seismic response of the selected frame was evaluated using a nonlinear response history analysis procedure, with both a conventional frame model with elastic joints and an improved frame model with nonlinear joint model and improved element formulation taken into account.

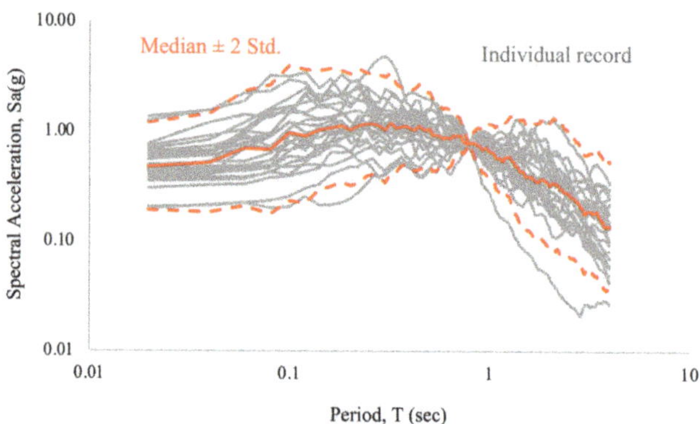

Figure 18. Acceleration response spectrum of the selected twenty-four ground motions linearly scaled and conditioned on the building period $T_1 = 0.78$ s.

The ground motion of the Kocaeli Turkey earthquake of 17 August 1999 (recording station: YARIMC, KOERI330; Source: PEER Strong Motion Database) was considered and scaled linearly by a factor of 1.75 to induce large nonlinearity in the frame for contrasting the behavior of frames. Member chord rotation indicates the extent of plasticity the member experiences during ground motion, which is an important indicator of seismic damage distribution. Damage distribution diagrams were developed (Figure 19), depicting elements' maximum chord rotation values (in percentage) to graphically compare the effects of joint modeling on the damage distribution. Shear hinge formation in the joints was also reported. The joint damage corresponded to the initiation of diagonal cracking when the principal tensile stress in the joint exceeded $0.29\ (f_c')^{0.5}$ *MPa*, moderate damage when the principal tensile stress in the joint was between $0.29\ (f_c')^{0.5}$ *MPa* and $0.42\ (f_c')^{0.5}$ *MPa*, and extensive damages in the panel zone when the principal tensile stress in the joint exceeded $0.42\ (f_c')^{0.5}$ *MPa*. Due to the flexibility that joint deformation provides, the activation of the joint shear mechanism increased the chord rotation demand on the connecting beam members. The performance of the conventional frame with elastic joints was satisfactory under the considered ground motions. The improved model, on the other hand, shows critical shear mechanism and increased member chord rotation demand sufficient to cause soft-story mechanism and consequent frame collapse.

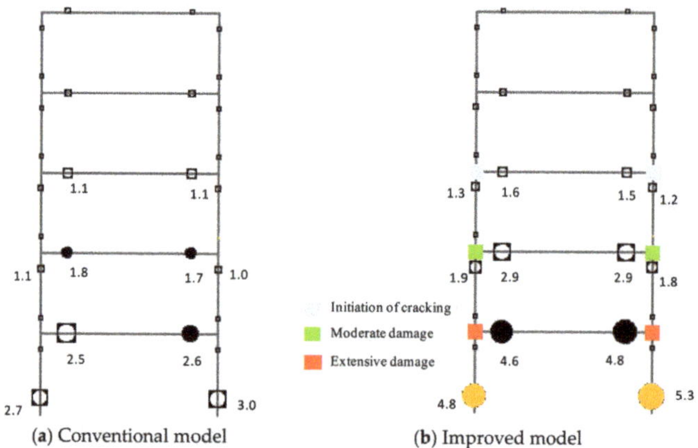

Figure 19. Response comparison for a five-story frame using a conventional model with elastic joint (**a**) and a model with a nonlinear joint model and improved frame element formulation (**b**) The member chord rotation value is given in percentage.

5.4.2. Inter-Story Drift Demand

The maximum inter-story drift ratio per record was obtained in order to evaluate the modeling technique with regard to the distribution of lateral deformation demand. Figure 20 shows the distribution of the maximum inter-story drift values while taking the average of the suite of ground motions. The inter-story drift distribution emphasizes the significance of the joint nonlinear model and the improved element formulation because, when compared to the drift demand for a conventional frame model, the drift demand for the frame model increases when joint nonlinearity is included and the improved element formulation is used. For the improved frame model with nonlinear joints, an increase of up to 62% (mean drift) and 89% (mean + 1.std.) is shown in the lower floors when assessing the inter-story drift distribution. As a result, the frame drift responses are drastically underestimated by the conventional frame model.

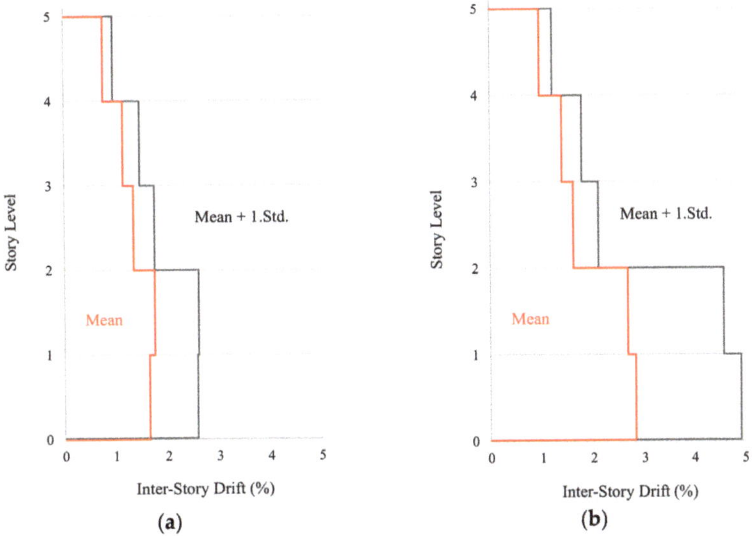

Figure 20. Inter-story drift comparison for a five-story frame using a conventional model with elastic joint (**a**) and a model with a nonlinear joint model and improved frame element formulation (**b**).

5.4.3. Collapse Risk

According to the experimental results discussed earlier, 4.50% of story drift is a realistic approximation for the beam–column connection's collapse limit since, after this drift limit was exceeded, the joint lost its ability to support the gravity loads. Figure 21 reports the maximum story drift demand for each individual ground motion. Comparing the maximum story drift demand for different ground motions can give the instances the demand exceeds the capacity. We observed seven instances of the demand exceeding the capacity in the case of an improved frame model, resulting in a collapse risk of $7/24 = 0.292$ (probability of failure = 29.20%). However, in the case of the conventional frame model, there was only one event wherein the demand exceeded the capacity, which resulted in a collapse risk of $1/24 = 0.042$ (probability of failure = 4.20%). The relevant fitting procedure for fitting fragility functions that use the maximum likelihood method, as outlined by Baker [62], was used to derive fragility functions for the considered frames. Figure 22 shows the fragility functions of both conventional and improved frame models, indicating that the conventional model grossly underestimates the collapse risk of frames.

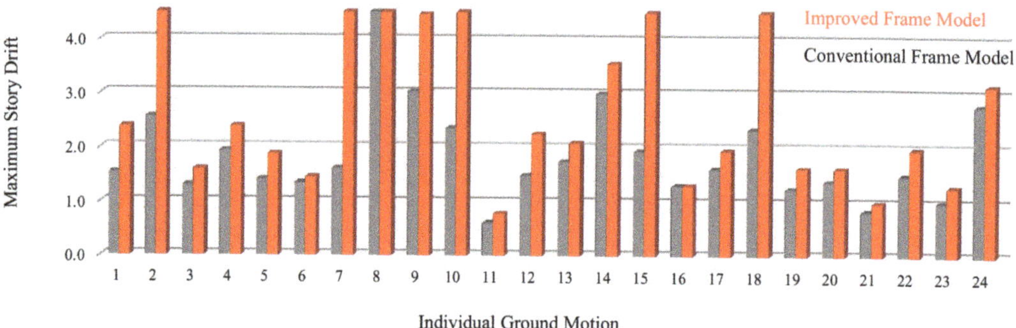

Figure 21. The maximum story drifts demand in individual ground motion for both conventional and improved frame models.

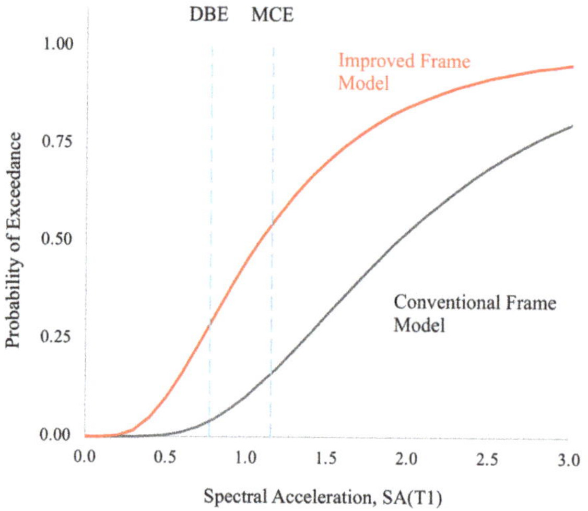

Figure 22. Collapse fragility functions for conventional and improved frame models.

6. Conclusions

The majority of existing nonlinear modeling techniques are deficient in one or more crucial characteristics (lack of fixed-end rotation deformation or stiffness and strength deterioration due to joint shear hinge), making them less appropriate for simulating the nonlinear behavior of existing substandard beam–column joints with pre-cracks that exhibit the deterioration of stiffness and strength under cyclic loading. In order to effectively model and simulate the deteriorating response of substandard beam–column joints, as demonstrated in the experimental test and supported by numerical simulation, fixed-end rotation must be taken into consideration, in addition to joint nonlinearity. The issue with pre-cracked RC members is that they exhibit fixed-end rotation under seismic actions that considerably contribute to deformation.

To accurately determine the member stiffness matrix, resisting forces, and deformation, it is therefore necessary to improve the conventional nonlinear finite element fiber-section-based element to account for the appropriate section/element force–deformation behavior. Since only the end rotational deformation needs to be additionally included, the formulation takes into account supplementing the conventional beam–column element with an additional rigid-bar with concentrated plasticity hinges at the end.

The nonlinear response history analysis of a five-story RC frame showed that activating the joint shear hinge and fixed-end rotation mechanisms increased the chord rotation demand on the connecting beam members by up to 85%. When determining the inter-story drift demand, the collapse probability of structures subjected to design base ground motions increased from 4.20% to 29.20%.

The importance of accurate nonlinear modeling becomes evident when assessing the performance of structures for strong ground motions, as the collapse risk is grossly underestimated when using conventional modeling techniques, in contrast to a more accurate nonlinear model. It is of particular importance when determining the collapse risk of structures for strong ground shaking with a large duration, and it becomes crucial in the case of reverse/oblique faults because most collapses (five out of seven) are observed in earthquakes caused by these faults.

The formulation and modeling proposed in this research are best suited for pre-cracked beam–column members that exhibit fixed-end rotation due to rebar slip/inelastic extension and the deterioration of joint stiffness/strength. Because conventional models are applicable in most situations, they must be improved when used to assess existing structures with the issues highlighted in the present research (i.e., fixed-end rotation, joint shear hinging, and deterioration). To generalize the application of the suggested formulation to structures with joints of adequate strength and without fixed-end rotation, considering the elastic response of the joint panel and discarding the fixed-end rotational springs in modeling RC beam–column members are suggested. Although the proposed detailed modeling increases analysis time in comparison to the conventional model, it is recommended for the risk assessment of individual buildings in order to obtain detailed information that can effectively guide retrofitting efforts. As the joint deformability decreases significantly under bi-axial loading, the proposed modeling may be extended for the analysis of 3D structures.

Author Contributions: Conceptualization, N.A. and M.R.; methodology, N.A., M.R., B.I. and S.H.; software, N.A., M.R., B.I., M.E.A., S.H. and H.S.; validation, N.A. and M.R.; formal analysis, N.A., M.R. and M.E.A.; investigation, N.A., M.R., M.E.A. and S.H.; resources, N.A., M.R., M.E.A., S.H., M.U.K. and H.S.; data curation, N.A. and M.E.A.; writing—original draft preparation, N.A. and M.E.A.; writing—review and editing, M.R., B.I., S.H., M.U.K. and H.S.; visualization, N.A., M.E.A. and S.H.; supervision, N.A.; project administration, N.A. and M.E.A. All authors have read and agreed to the published version of the manuscript.

Funding: This research received no external funding.

Data Availability Statement: Requests for the data presented here can be sent to drnaveed@stanford.edu.

Acknowledgments: The first author is grateful to the United States Educational Foundation in Islamabad for sponsoring his Visiting Scholar appointment at Stanford University under the Fulbright Scholar 2021 program. The authors are grateful to the reviewers for their constructive remarks, which helped to improve the manuscript's quality.

Conflicts of Interest: The authors declare no conflict of interest.

Appendix A

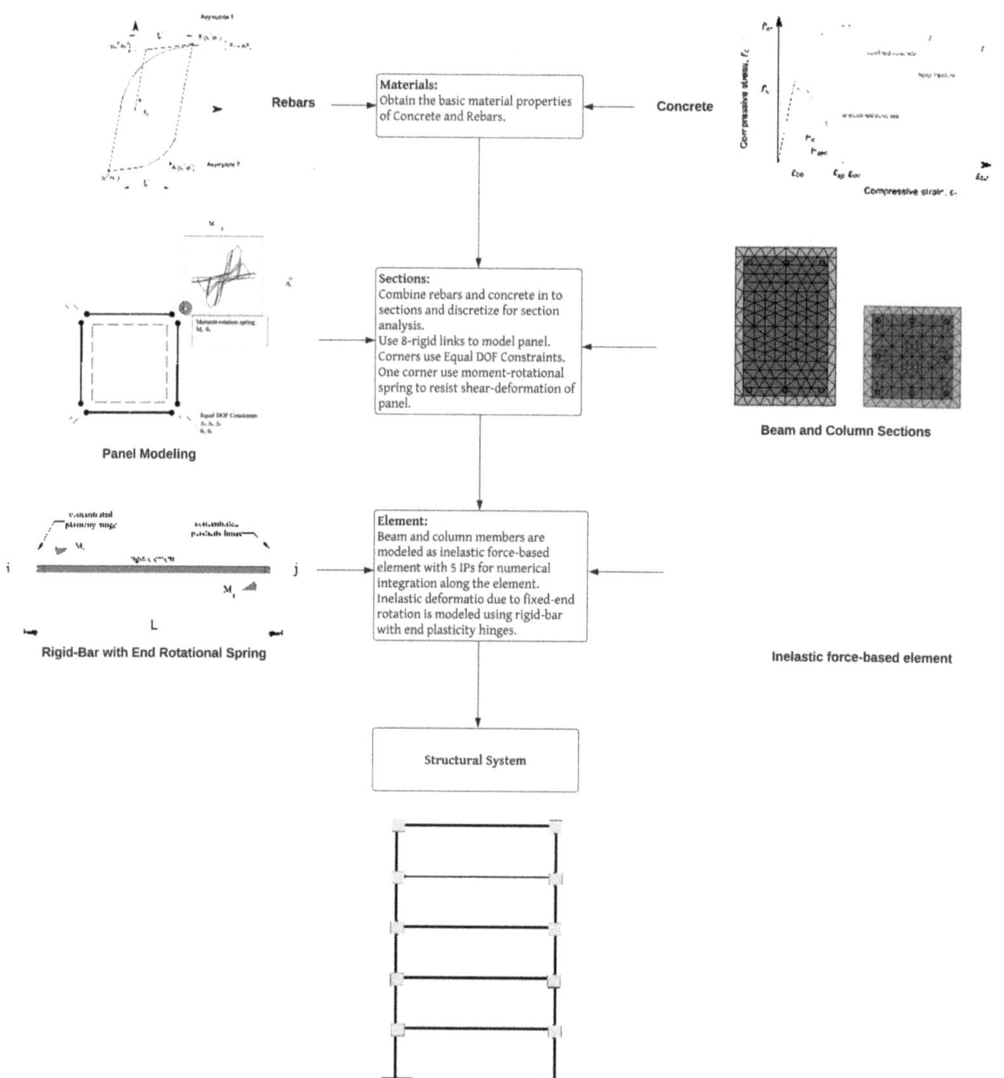

Figure A1. Simplified flowchart for preparing numerical model of structural system.

Appendix B

Table A1. Details of the ground motions used for nonlinear response history analysis. Rev. represents reverse faults; Rev. Ob. represents reverse and oblique faults, and SS represents strike-slip faults.

S. No.	Earthquake Event	Year	Recording Station	Mw	Fault	Rjb (km)	Duration (s), 5–95%
1	San Fernando	1971	LA–Hollywood Stor FF	6.61	Rev.	22.77	13.4
2	Tabas_ Iran	1978	Boshrooyeh	7.35	Rev.	24.07	19.5
3	Coalinga-01	1983	Parkfield–Fault Zone 15	6.36	Rev.	28.00	19.7

Table A1. *Cont.*

S. No.	Earthquake Event	Year	Recording Station	Mw	Fault	Rjb (km)	Duration (s), 5–95%
4	Spitak_ Armenia	1988	Gukasian	6.77	Rev. Ob.	23.99	10.5
5	Loma Prieta	1989	Hollister–South & Pine	6.93	Rev. Ob	27.67	28.8
6	Northridge-01	1994	LA–W 15th St	6.69	Rev.	25.59	20.2
7	Chi-Chi_ Taiwan	1999	CHY025	7.62	Rev. Ob.	19.07	35.3
8	St Elias_ Alaska	1979	Icy Bay	7.54	Rev.	26.46	34.6
9	Niigata_ Japan	2004	NIG018	6.63	Rev.	21.55	70.3
10	Chuetsu-oki_ Japan	2007	Joetsu Kita	6.80	Rev.	28.97	30.8
11	Iwate_ Japan	2008	IWT012	6.90	Rev.	20.47	30.8
12	Christchurch_ New Zealand	2011	LINC	6.20	Rev. Ob.	18.47	13.3
13	Northern Calif-03	1954	Ferndale City Hall	6.50	SS	26.72	19.4
14	Imperial Valley-06	1979	Delta	6.53	SS	22.03	51.4
15	Victoria_ Mexico	1980	Chihuahua	6.33	SS	18.53	19
16	Morgan Hill	1984	Agnews State Hospital	6.19	SS	24.48	40.9
17	Superstition Hills-02	1987	Brawley Airport	6.54	SS	17.03	14.3
18	Landers	1992	North Palm Springs	7.28	SS	26.84	37.9
19	Kobe_ Japan	1995	Fukushima	6.90	SS	17.85	35.7
20	Tottori_ Japan	2000	OKY005	6.61	SS	28.81	24
21	Parkfield-02_ CA	2004	Coalinga–Fire Station 39	6.00	SS	22.45	27.7
22	El Mayor-Cucapah_ Mexico	2010	Chihuahua	7.20	SS	18.21	51.2
23	Joshua Tree_ CA	1992	Thousand Palms Post Office	6.10	SS	17.15	11.1
24	Darfield_ New Zealand	2010	WSFC	7.00	SS	24.36	26.2

References

1. Moehle, J.P.; Mahin, S.A. Observations on the behavior of reinforced concrete buildings during earthquakes. *ACI Spec. Publ.* **1991**, *127*, 67–90.
2. Varum, H. Seismic assessment, strengthening and repair of existing buildings. Ph.D. Thesis, Department of Civil Engineering, University of Aveiro, Aveiro, Portugal, 2003.
3. EERI. *Northridge Earthquake January 17*; Preliminary Reconnaissance Report, Earthquake Engineering Research Institute (EERI): Oakland, CA, USA, 1994.
4. Aycardi, L.E.; Mander, J.B.; Reinhorn, A.M. Seismic resistance of reinforced concrete frame structures designed only for gravity loads–experimental performance of subassemblages. *ACI Struct. J.* **1994**, *91*, 552–563.
5. Beres, A.; Pessiki, S.P.; White, R.N.; Gergely, P. Implications of experiments on the seismic behaviour of gravity load designed RC beam-to-column connections. *Earthq. Spectra* **1996**, *12*, 185–198. [CrossRef]
6. Gautam, D.; Adhikari, R.; Rupakhety, R. Seismic fragility of structural and non-structural elements of Nepali RC buildings. *Eng. Struct.* **2021**, *232*, 111879. [CrossRef]
7. Ahmad, N.; Shahzad, A.; Rizwan, M.; Khan, A.N.; Ali, S.M.; Ashraf, M.; Naseer, A.; Ali, Q.; Alam, B. Seismic performance assessment of non-compliant SMRF reinforced concrete frame: Shake-table test study. *J. Earthq. Eng.* **2019**, *23*, 444–462. [CrossRef]
8. Rizwan, M.; Ahmad, N.; Khan, A.N. Seismic performance of compliant and noncompliant special moment-resisting reinforced concrete frames. *ACI Struct. J.* **2018**, *115*, 1063–1073. [CrossRef]
9. Ricci, P.; De Risi, M.T.; Verderame, G.M.; Manfredi, G. Experimental tests of unreinforced exterior beam-column joints with plain bars. *Eng. Struct.* **2016**, *118*, 178–194. [CrossRef]
10. De Risi, M.T.; Verderame, G.M. Experimental assessment and numerical modelling of exterior non-conforming beam-column joints with plain bars. *Eng. Struct.* **2017**, *150*, 115–134. [CrossRef]
11. Melo, J.; Varum, H.; Rossetto, T. Experimental assessment of the monotonic and cyclic behavior of exterior RC beam-column joints built with plain bars and non-seismically designed. *Eng. Struct.* **2022**, *270*, 114887. [CrossRef]
12. ACI-352-R02. In *Recommendations for the Design of Beam-Column Joints in Monolithic Reinforced Concrete Structures*; American Concrete Institute (ACI): Farmington Hills, MI, USA, 2002.
13. Badrashi, Y.I. Response modification factors for reinforced concrete buildings in Pakistan. Ph.D. Thesis, Department of Civil Engineering, UET Peshawar, Peshawar, Pakistan, 2016.
14. El-Metwally, S.E.; Chen, W.F. Moment-rotation modeling of reinforced concrete beam-column connections. *Struct. J.* **1988**, *85*, 384–394.

15. Alath, S.; Kunnath, S.K. Modeling inelastic shear deformations in RC beam-column joints. In Proceedings of 10th Engineering Mechanics Conference ASCE, Boulder, Colorado, USA, 30 April 1995; pp. 822–825.

16. Biddah, A.; Ghobarah, A. Modelling of shear deformation and bond slip in reinforced concrete joints. *Struct. Eng. Mech.* **1999**, *7*, 413–432. [CrossRef]

17. Lowes, L.N.; Altoontash, A. Modeling reinforced-concrete beam-column joints subjected to cyclic loading. *J. Struct. Eng. ASCE* **2003**, *129*, 1686–1697. [CrossRef]

18. Youssef, M.; Ghobarah, A. Modelling of RC beam-column joints and structural walls. *J. Earthq. Eng.* **2001**, *5*, 93–111. [CrossRef]

19. Shin, M.; LaFave, J.M. Testing and modelling for cyclic joint shear deformations in RC beam-column connections. In Proceedings of the 10th World Conference on Earthquake Engineering, Vancouver, BC, Canada, 1 August 2004; p. 301.

20. Altoontash, A. Simulation and damage models for performance assessment of reinforced concrete beam-column joints. Ph.D. Thesis, Department of Civil and Environmental Engineering, Stanford University, Stanford, CA, USA, 2004.

21. Ning, C.L.; Yu, B.; Li, B. Beam-column joint model for nonlinear analysis of non-seismically detailed reinforced concrete frame. *J. Earthq. Eng.* **2016**, *20*, 476–502. [CrossRef]

22. Elmorsi, M.; Kianoush, M.R.; Tso, W.K. Modeling bond-slip deformations in reinforced concrete beam-column joints. *Can. J. Civ. Eng.* **2000**, *27*, 490–505. [CrossRef]

23. Pampanin, S.; Magenes, G.; Carr, A. Modeling of shear hinge mechanism in poorly detailed RC beam-column joints. In *Concrete Structures in Seismic Regions: Fib, Symposium*; University of Canterbury, Civil Engineering: Athens, Greece, 2003; p. 171.

24. Sharma, A.; Eligehausen, R.; Reddy, G.R. A new model to simulate joint shear behavior of poorly detailed beam-column connections in RC structures under seismic loads. Part I: Exterior joints. *Eng. Struct.* **2011**, *33*, 1034–1051. [CrossRef]

25. Kunnath, S.K.; Hoffmann, G.; Reinhorn, A.M.; Mander, J.B. Gravity-load-designed reinforced concrete buildings–Part I: Seismic evaluation of existing construction and Part II: Evaluation of detailing enhancements. *ACI Struct. J.* **1995**, *92*, 343–478.

26. Ghobarah, A.; Biddah, A. Dynamic analysis of reinforced concrete frames including joint shear deformation. *Eng. Struct.* **1999**, *21*, 971–987. [CrossRef]

27. Priestley, M.J.N. Displacement-based seismic assessment of reinforced concrete buildings. *J. Earthq. Eng.* **1997**, *1*, 157–192. [CrossRef]

28. Khan, M.S.; Basit, A.; Ahmad, N. A simplified model for inelastic seismic analysis of RC frame have shear hinge in beam-column joints. *Structures* **2021**, *29*, 771–784. [CrossRef]

29. Fillippou, F.C.; Popov, E.P.; Bertero, V.V. *Effects of Bond Deterioration on Hysteretic Behaviour of Reinforced Concrete Joints*; Technical Report, Report No. UCB/EERC-83/19; EERC, University of California: Berkeley, CA, USA, 1983.

30. Filippou, F.C.; Issa, A. *Nonlinear Analysis of Reinforced Concrete Frames under Cyclic Load Reversals*; Technical Report, Report No. UCB/EERC-88/12; EERC, University of California: Berkeley, CA, USA, 1988.

31. Baber, T.T.; Noori, M.N. Random vibration of degrading pinching systems. *J. Eng. Mech. ASCE* **1985**, *111*, 1010–1026. [CrossRef]

32. Vecchio, F.J.; Collins, M.P. The modified-compression field theory for reinforced concrete elements subjected to shear. *ACI J.* **1986**, *83*, 219–231.

33. Lowes, L.N.; Mitra, N.; Altoontash, A.A. *A Beam-Column Joint Model for Simulating the Earthquake Response of Reinforced Concrete Frames*; Technical Report, Report no. PEER 2003/10; Pacific Earthquake Engineering Research Center, University of California: Berkeley, CA, USA, 2004.

34. Park, R. A summary of results of simulated seismic load tests on reinforced concrete beam-column joints, beams and columns with substandard reinforcing details. *J. Earthq. Eng.* **2002**, *6*, 147–174. [CrossRef]

35. Paulay, T.; Priestley, M.J.N. *Seismic Design of Reinforced Concrete and Masonry Buildings*; John Wiley & Sons Inc.: New York, NY, USA, 1992.

36. Calvi, G.M.; Magenes, G.; Pampanin, S. Relevance of beam-column joint damage and collapse in RC frame assessment. *J. Earthq. Eng.* **2002**, *6*, 75–100. [CrossRef]

37. Ibarra, L.F.; Medina, R.A.; Krawinkler, H. Hysteretic models that incorporate strength and stiffness deterioration. *Earthq. Eng. Struct. Dyn.* **2005**, *34*, 1489–1511. [CrossRef]

38. Metelli, G.; Messali, F.; Beschi, C.; Riva, P. A model for beam–column corner joints of existing RC frame subjected to cyclic loading. *Eng. Struct.* **2015**, *89*, 79–92. [CrossRef]

39. Hwang, S.J.; Lee, H.J. Analytical model for predicting shear strengths of exterior reinforced concrete beam-column joints for seismic resistance. *ACI Struct. J.* **1999**, *96*, 846–858.

40. Spacone, E.; Filippou, F.C.; Taucer, F. Fibre beam-column model for non-linear analysis of R/C frames. *Earthq. Eng. Struct. Dyn.* **1996**, *25*, 711–725. [CrossRef]

41. Ciampi, V.; Carlesimo, L. A nonlinear beam element for seismic analysis of structures. In Proceedings of the 8th European Conference on Earthquake Engineering, Lisbon, Portugal, 1986.

42. Spacone, E.; Ciampi, V.; Filippou, F.C. Mixed formulation of nonlinear beam finite element. *Comput. Struct.* **1996**, *58*, 71–83. [CrossRef]

43. Papadrakakis, M.; Charmpis, D.C.; Lagaros, N.D.; Tsompanakis, Y. *Computational Structural Dynamics and Earthquake Engineering: Structures and Infrastructures*; CRC Press/Balkema: Leiden, The Netherlands, 2008; Volume 2.

44. Calabrese, A.; Almeida, J.P.; Pinho, R. Numerical issues in distributed inelasticity modelling of RC frame elements for seismic analysis. *J. Earthq. Eng.* **2010**, *14* (Suppl. S1), 38–68. [CrossRef]

45. Taucer, F.F.; Spacone, E.; Filippou, F.C. *A Fiber Beam-Column Element for Seismic Response Analysis of Reinforced Concrete Structures*; EERC Report 91/17; Earthquake Engineering Research Center, University of California: Berkeley, CA, USA, 1991.

46. Menegotto, M.; Pinto, P.E. *IABSE Symposium on Resistance and Ultimate Deformability of Structures*; Symposium, International Association for Bridge and Structural Engineering: Zurich, Switzerland, 1973; pp. 15–22.

47. Fragiadakis, M.; Pinho, R.; Antoniou, S. Modeling inelastic buckling of reinforcing bars under earthquake loading. In Proceedings of the ECCOMAS Thematic Conference on Computational Methods, Crete, Greece, 13–16 June 2007.

48. Mander, J.B.; Priestley, M.J.N.; Park, R. Theoretical stress-strain model for confined concrete. *J. Struct. Eng. ASCE* **1988**, *114*, 1804–1826. [CrossRef]

49. Martinez-Rueda, J.E.; Elnashai, A.S. Confined concrete model under cyclic load. *Mater. Struct.* **1997**, *30*, 139–147. [CrossRef]

50. Ahmad, N.; Masoudi, M.; Salawdeh, S. Cyclic response and modelling of special moment resisting beams exhibiting fixed-end rotation. *Bulletin of Earthquake Engineering.* **2021**, *19*, 203–240. [CrossRef]

51. Deierlein, G.G.; Reinhorn, A.M.; Willford, M.R. *Nonlinear Structural Analysis for Seismic Design, NIST GCR 10-917-5*; National Institute of Standards and Technology: Gaithersburg, MD, USA, 2010.

52. Correia, A.A.; Virtuoso, F.B.E. *Nonlinear analysis of space frames. III European Conference on Computational Mechanics*; Springer: Dordrecht, Netherlands, 2006.

53. ASCE 41-17. In *Seismic Evaluation and Retrofit of Existing Buildings*; Technical Report 2017; American Society of Civil Engineers (ASCE): Reston, VA, USA, 2017.

54. Monti, G.; Nuti, C.; Santini, S. *CYRUS-Cyclic Response of Upgraded Sections*; Report No. 96-2; University of Chieti: Chieti, Italy, 1996.

55. Monti, G.; Nuti, C. Nonlinear cyclic behavior of reinforcing bars including buckling. *J. Struct. Eng.* **1992**, *118*, 3268–3284. [CrossRef]

56. Madas, P. Advanced Modelling of Composite Frames Subjected to Earthquake Loading. Ph.D. Thesis, Imperial College London, London, UK, 1993.

57. Sivaselvan, M.; Reinhorn, A.M. *Hysteretic Models for Cyclic Behavior of Deteriorating Inelastic Structures*; Report MCEER-99-0018; MCEER, SUNY at Buffalo: New York, NY, USA, 1999.

58. Masoudi, M.; Khajevand, S. Revisiting flexural overstrength in RC beam-and-slab floor systems for seismic design and evaluation. *Bull. Earthq. Eng.* **2020**, *18*, 5309–5534. [CrossRef]

59. Santarsiero, G.; Masi, A. Analysis of slab action on the seismic behavior of external RC beam-column joints. *J. Build. Eng.* **2020**, *32*, 101608. [CrossRef]

60. Montuori, R.; Nastri, E.; Piluso, V. Modelling of floor joists contribution to the lateral stiffness of RC buildings designed for gravity loads. *Eng. Struct.* **2016**, *121*, 85–96. [CrossRef]

61. Ahmad, N.; Rizwan, M.; Ashraf, M.; Khan, A.N.; Ali, Q. Seismic collapse safety of reinforced concrete moment resisting frames with/without beam-column joint detailing. *Bull. New Zealand Soc. Earthq. Eng.* **2020**, *54*, 1–20. [CrossRef]

62. Baker, J.W. Efficient analytical fragility function fitting using dynamic structural analysis. *Earthq. Spectra* **2015**, *31*, 579–599. [CrossRef]

 *buildings*

 MDPI

Article

Seismic Damage Index Spectra Considering Site Acceleration Records: The Case Study of a Historical School in Kermanshah

Mahnoosh Biglari [1],*, Marijana Hadzima-Nyarko [2] and Antonio Formisano [3]

[1] Civil Engineering Department, School of Engineering, Razi University, Kermanshah 6718773654, Iran
[2] Faculty of Civil Engineering and Architecture Osijek, The Josip Juraj Strossmayer University of Osijek, 31000 Osijek, Croatia
[3] Department of Structures for Engineering and Architecture, School of Polytechnic and Basic Sciences, University of Naples Federico II, 80131 Naples, Italy
* Correspondence: m.biglari@razi.ac.ir or mahnooshbiglari@yahoo.com

Abstract: The frequency content and time duration of earthquakes are as effective as the peak ground acceleration on structural damage. Therefore, using rapid seismic vulnerability assessment methods that consider the earthquake acceleration time history is noticeable. Kermanshah is a historical city that is generally affected by far-field earthquakes. Therefore, it is necessary to consider the effect of the low-frequency shocks in evaluating the vulnerability of buildings in this city. Herein, a historic school in Kermanshah is assumed as a case study and two well-known damage index formulas are used for determining the damage index spectra of this structure, considered as a single degree of freedom system. Then, the effective parameters of the damage index, including ductility, relative degradation of stiffness, and dissipated energy are determined from a nonlinear analysis of the structure under the effect of the most probable earthquake acceleration records. Finally, the damage index spectra can be used for rapid seismic vulnerability assessment of masonry buildings on similar sites with various fundamental periods for large-scale assessments. The result shows that the building tends to collapse at a peak ground acceleration of 0.15 g. Furthermore, results confirm the seismic resistance reduction effect of flexible floors.

Keywords: seismic vulnerability; damage index spectra; historical masonry building; nonlinear analysis

Citation: Biglari, M.; Hadzima-Nyarko, M.; Formisano, A. Seismic Damage Index Spectra Considering Site Acceleration Records: The Case Study of a Historical School in Kermanshah. *Buildings* **2022**, *12*, 1736. https://doi.org/10.3390/buildings12101736

Academic Editors: Rajesh Rupakhety and Dipendra Gautam

Received: 20 September 2022
Accepted: 17 October 2022
Published: 19 October 2022

Publisher's Note: MDPI stays neutral with regard to jurisdictional claims in published maps and institutional affiliations.

1. Introduction

The first step in reducing the seismic risk in a city is to identify the earthquake hazard and evaluate both the vulnerability of at-risk structures and the consequences of the earthquake. Although the seismic hazard of an earthquake at a site cannot be reduced, it can be determined by studying the seismic sources and determining the set of possible magnitudes of each source with deterministic and probabilistic seismic hazard analysis methods [1]. Realistic knowledge of hazards is helpful in accurately estimating seismic vulnerability and, therefore, seismic risk analysis. Additionally, if there is an accurate assessment of the seismic vulnerability of buildings, it is possible to reduce the seismic risk by retrofitting or renovating vulnerable buildings. Other steps to reduce the seismic risk are providing relief and rescue facilities, and predicting post-earthquake facilities, shelters, and settlement locations [2].

To assess the seismic vulnerability of structures, several rapid and qualitative methods, and different quantitative methods based on the structural condition and modeling of their seismic behavior are available. In the case of historical buildings, destructive identification methods and renovation are usually not available options. Therefore, in historical structures, qualitative and nondestructive identification methods are recommended to evaluate the structural seismic vulnerability. These buildings often have complex geometry and heavy mass, and are not provided with a seismic resistance system, so as a result, they show damage induced by past destructive earthquakes. Choosing the most appropriate vulnerability assessment method depends on the number of assets under investigation, the importance and nature of the structures, the level of access to information, and the available budget. Quantitative assessment methods, based on structural modeling, are time-consuming and expensive and are not recommended in the first evaluation phase. Contrarily, rapid evaluations given by qualitative methods are used in the initial phase of assessment and prioritizing rehabilitation planning. These indirect methods are based on the observation of structural elements and building configurations. An example of an indirect method that uses indicators is the vulnerability index method, introduced by [3] and then developed by other research [4] for masonry buildings. This method uses standard forms to evaluate the seismic vulnerability level of structures by considering several in situ parameters. An extension of the screening tool was presented in [5], where a vulnerability assessment form was used to consider the structural interaction among adjacent buildings. This form added to the basic parameters of the original forms [3] by including five parameters that account for interaction effects among aggregate structural units under earthquakes. Another example of a scoring method is one introduced by the Italian Directive [6], indicated as the Level of Evaluation 1, which was delivered to be used for evaluations at a territorial scale. This method requires only in situ visual inspections, providing a seismic performance in terms of acceptable peak ground acceleration depending on a global vulnerability index. Further, damage matrices and fragility curves help to predict the damage of similar structures by the damage surveys collected after destructive earthquakes. The empirical fragility curves were firstly developed after the 1971 San Fernando (Los Angeles, CA, USA) earthquake by [7], and, subsequently, some remarkable examples came from the 2009 L'Aquila, Italy, earthquake, developed by [8], the 1934 to 2015 Nepali earthquakes, introduced by [9], and the 2017 Sarpol-e-zahab, Iran, earthquake, delivered by [10,11]. Alternative methods to develop fragility curves were provided by studying the structural models of buildings through incremental dynamic analyses (e.g., [12,13]). The advantage to the basic simplified methods is their high evaluation speed. Specifically, the indirect method and empirical fragility curves do not require modeling and structural analysis. The disadvantage to these methods is that they use only the intensity or the peak ground acceleration of the earthquake, without considering the effect of other necessary parameters. Another suggestion is to use a hybrid approach, whereby empirical and analytical methods are combined to obtain quantitative and more reliable results for a group of buildings. A similar way was used by [14] to evaluate a masonry building in Osijek, Croatia.

This research presents the seismic damage index spectra for the historic center of Kermanshah city and evaluates the seismic vulnerability of a historic high school. Kermanshah is the ninth most populated city in Iran and the most significant city in the central region of western Iran. The city of Kermanshah, located in the Zagros seismic zone, has been exposed to several destructive earthquakes throughout history, of which the earthquake of 2016 November 21st is the latest example. The city of Kermanshah is one of the historical and cultural cities of Iran. Its origin dates back to the fourth century AD. It was the second capital of the Sassanid Empire until the Arab invasion of Iran. At the end of the 18th century, Kermanshah's commercial and strategic importance increased with the establishment of the customs office. Most of the attractive post-Islamic historical buildings in Kermanshah include the traditional market (Bazaar), mosques, Takaya, schools, caravanserais, and mansions from the Qajar era. Owing to the high seismicity of Kermanshah, these buildings are prone to seismic damage by the occurrence of destructive earthquakes. One of the issues of earthquake engineering is to evaluate the existing seismic resistance and ductility requirements of existing buildings, and/or prepare crisis maps using real seismic scenarios. Therefore, it is desirable to provide a relatively fast but accurate seismic damage assessment, as the method proposed in this paper. Before this study, a series of investigations were conducted on the seismic vulnerability assessment of the historical mosques of Kermanshah city, including rapid seismic assessment [15], identification of material properties [16], macroelement modeling, pushover analysis, and fragility curves [17].

In this research, the seismic vulnerability of the historic Kazazi High School is evaluated based on the most probable earthquakes at its site by using the damage indices as a measure of structural damage. For the investigated structure, first, a set of three pairs of earthquake records was determined; then, using several assumed single degree-of-freedom (SDOF) systems, the spectral functions of the damage index were determined based on the two definitions presented in the literature. Finally, by knowing the proposed parameters of an SDOF system (e.g., the natural period of investigated structure) representing an unreinforced masonry (URM) building, it is possible to determine the structure's response to a given earthquake using the damage index spectra. Since there is no history of using this methodology and presenting this type of spectra in Iran, this analysis approach can be considered a novel type for the historic area of Kermanshah city. In addition, this method accounts for all the necessary parameters of the earthquake, including the amplitude, frequency content, and duration, by considering the time-history record of the acceleration. Furthermore, equating the structure with an SDOF system facilitates nonlinear analysis. Hence, proposing the seismic vulnerability index spectra, which is not a complicated task, can be very effective. By damage index spectra, the vulnerability of buildings with similar seismic behavior, at a similar seismic site, and different fundamental frequencies (different mass, stiffness, and damping ratio), can be quickly evaluated. It is, therefore, possible to add this method to rapid assessment instructions. The rest of the paper is organized as follows: introduction to the history and architectural characteristics of the school building, description of the damage index functions and properties of the applied seismic accelerograms, presentation of the seismic damage index spectra for original records and different scaled records, and, finally, conclusions on the seismic vulnerability assessment.

2. Kazazi High School

Kazazi High School is the oldest high school in Kermanshah and one of the first Iranian modern-style high schools. The construction of this high school started in January 1921, six years before the modern-style school construction in the Pahlavi period started, by the suggestion and efforts of the late Seyyed Hossein Kazazi (1874–1923). The execution works of the structure ended in 1923. Other traditional schools were established before construction of this high school, but this was the first high school with a modern educational syllabus and equipment, such as classrooms, laboratories, libraries, drawing galleries, and a conference hall. In 1998, this school was registered by the Ministry of Cultural Heritage, Tourism, and Handicrafts (MCTH) of Iran as a national heritage construction of Iran. It operates as a museum after retrofitting in 2009.

The architecture of the building is simple, and more than beautification, a rigorous appearance of the building has been preferred (Figure 1a). Only the brickwork of the facade represents an embellishment system of the building.

In 2006, during the renovation, the building plan changed, and some parts of the transverse walls of the classrooms were removed. Currently, the plan of the building is rectangular, 36 m long in the east–west direction and 16 m wide in the north–south direction (Figure 1b). A corridor with a width of 4.65 m and a length of 36 m (the whole building length) is located in the middle of the building. Classrooms are situated on both sides of this corridor. The width of the classrooms is 4 m, and their length varies from 6.15 to 12.9 m. The main entrance of the building is located on the north side, and another door is located on the east side. Figure 1c–g show the north, south, east, and west perspective views, respectively.

There is another floor in the center of the building, where there are the manager's office, the meeting hall, and the documents archive, along with two terraces on the north and south sides. The staircase is in the middle of the terrace on the north side, which, after renovation, has been surrounded by walls. The height of the first and second floors are 4.00 and 3.80 m, respectively. The wall materials are brick and lime mortar. The ceiling is made of a wooden beam and a gable roof placed on a wooden truss.

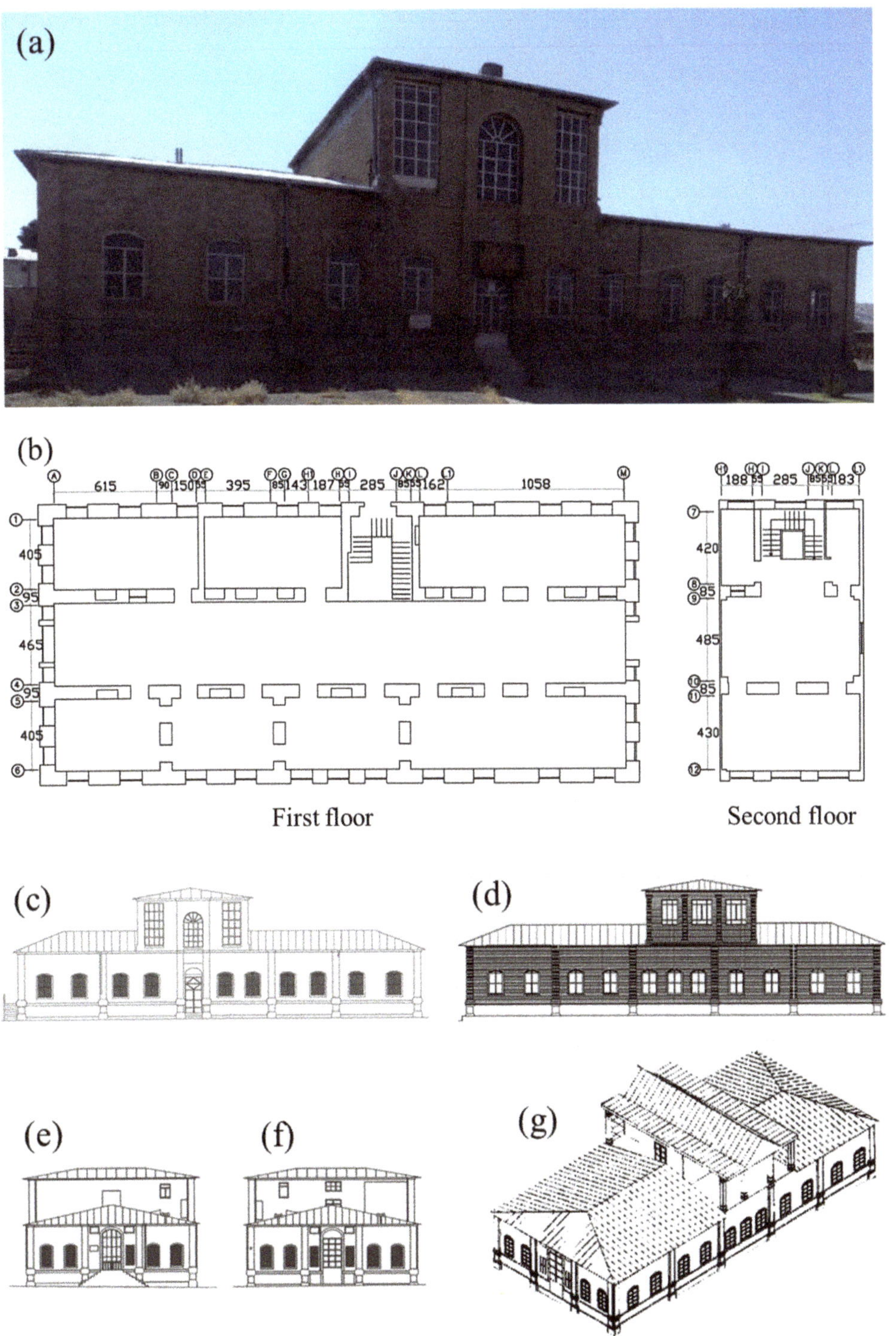

Figure 1. Kazazi High School: (**a**) north-view photo, (**b**) first-floor and second-floor plan (dimensions in centimeter), (**c**) north-view, (**d**) south-view, (**e**) east-view, (**f**) west-view, and (**g**) building perspective.

In addition to the main classroom building, the high school has two additional separate buildings that are not discussed here. These two buildings are (1) the northern building and the main entrance built in 1926 and (2) the meeting hall and theatre built in 1935.

3. Methodologies, Descriptions, and Seismic Records

To perform a seismic vulnerability assessment, the method used in this paper requires the use of appropriate damage index (DI) spectral functions. Particularly, two of these functions are herein considered. One is the function introduced by Park and Ang [18], and the other is the function introduced by Moric et al. [19]. Seismic vulnerability assessments by these methods, in addition to the high speed of the evaluation and the ability to use it for a large number of building stocks on an urban scale, are based on more than just the geometry of the structure, the quality of the materials, and the single parameter of the peak ground acceleration. In these two methods, other than considering the maximum amplitude of the accelerograms, the amplitude in all other cycles, frequency content, and time duration of the record, the force–displacement behavior, and the fundamental frequency of the structure, are also considered. The only simplification made is generalizing the structure with an SDOF system, which is not unrealistic for short buildings. As mentioned, the method of Moric et al. [19] has already been used by [14] to evaluate a historical masonry building in Osijek, named "II Gymnasium Osijek". Further, the well-known function of Park and Ang [18] has not been used to extract the damage index spectrum and is adapted in this research for proposing the damage index spectrum.

The investigated building is modeled by an SDOF system, determined by weight, elastic stiffness, post-elastic stiffness, damping ratio, and strength limit of elasticity. The nonlinear dynamic analyses are performed using the NONLIN code [20] that implements iterative time-history numerical integration. The NONLIN code [20] is a Microsoft Windows-based application for the dynamic analysis of SDOF structural systems. The structure may be modeled as elastic, elastic–plastic, or a yielding system with an arbitrary level of secondary stiffness. The secondary stiffness may be positive, to represent a strain hardening system, or negative, to model P–Delta effects. The dynamic loading may be assigned as an earthquake accelerogram acting at the building's base or as a linear combination of sine, square, or triangular waves applied at the building's roof. The program uses a step-by-step method to solve the incrementally nonlinear equations of motion based on Clough and Penzien's [21] theoretical description. The code needs elastic and post-elastic behavior, damping, and strength limit of elasticity. Then, the code provides dissipated energy, cycle numbers of yielding achieved during the earthquake, force–displacements hysteresis response, and time history of top displacements. These outputs represent inputs for the DI formulas reported in this section.

During dynamic analysis, the input seismic load is represented by a site match acceleration time-history record. DI function calculations are repeated for several SDOF systems with a wide range of fundamental periods from 0.05 to 4.5 s (i.e., T = 0.05, 0.1, 0.2, 0.3, 0.4, 0.5, 0.6, 0.7, 0.8, 0.9, 1, 1.5, 2.0, 2.5, 3.5, and 4.5 s) by nonlinear analyses. Each spectrum is determined by applying a set of original earthquake acceleration records or the scaled acceleration records to fixed values ranging from 0.075 to 0.75 g (i.e., PGA = 0.075, 0.15, 0.20, 0.30, 0.45, 0.60, and 0.75 g).

For the historical center of Kermanshah city, three pairs of records (i.e., six horizontal records) of real accelerograms recorded in the KRM1 station of ISMN [22] are selected. For any kind of dynamic analysis, the Seismic Code of Iran IRSt2800 [23] suggests applying three pairs of records (i.e., six horizontal records) or seven nonpair records compatible with the site. The KRM1 is the most seismically compatible with the historical center of Kermanshah city. The KRM1 station and the investigated structure are located in the same seismotectonic plateau in Zagros. They are located close to each other and at a similar distance from the active faults of the region. Regarding the geological conditions, both sites are located on type II soil category based on the Seismic Code of Iran IRSt2800 [23] from the average shear wave velocity of the upper 30 m of the soil, as provided by [24].

On the other hand, the Kazazi High School building is located in the vicinity of the Jame Mosque, whose ground seismic properties have been studied and reported by [16] as type II. In addition, the investigated building has experienced the KRM1 recorded events. This provides a benchmark for verification of the vulnerability assessment result with factual events. The records used in this research are presented in Table 1.

Table 1. Accelerograms used in the analysis.

Record Number	Date d/m/y	Time h:min:s	PGA (cm/s^2)	Latitude	Longitude	Epicentral Distance (km)	Focal Depth (km)	Magnitude (Mw)	Reference
2730	24/04/2002	19:48:07	12	34.65	47.47	53	14	5.3	ISMN *
7292	12/11/2017	18:18:16	55	34.81	45.91	120	18	7.3	ISMN
8250	25/11/2018	16:37:31	12	34.31	45.69	128	16	6.3	ISMN

* [22].

The damage index, which can be used for damage estimation, is defined as a mathematical function, where 0 means the undamaged state of the structure and 1 indicates failure.

Equation (1) introduces the DI function developed by Park and Ang [18], which is composed of two terms controlling the seismic behavior of a masonry structure: the first term depends on the ductility and the second term depends on the dissipated energy. This function is expressed through the following relationship:

$$DI = \frac{u_{max}}{u_{ult}} + 0.15\frac{E_H}{F_{cr}u_{ult}} \tag{1}$$

where u_{max} is the maximum displacement reached, u_{ult} is the ultimate displacement, E_H is the energy amount dissipated by hysteresis, and Fcr is the force at the cracking limit, which is considered equal to the strength limit of elasticity fy.

Moric et al. [19] proposed Equation (2) as a DI function. This function has three terms: the first term depends on the ductility, the second term depends on the relative degradation of stiffness at the end of the earthquake, and the third term depends on energy dissipation during loading and unloading cycles. It is expressed through the following relationship:

$$DI = \frac{1}{30}\left[\mu + \Delta k + \sqrt[3]{\frac{N_\gamma E_H}{W}}\right] \tag{2}$$

where μ is the required ductility displacement equal to u_{max}/u_y; $\Delta k = k_{el}/k_R$ is the relative degradation of stiffness at the end of the earthquake, $k_{el} = f_y/u_y$ the initial stiffness and $k_R = f_{max}/u_{max}$ the residual secant stiffness at the end of the earthquake; N_γ is the cycle number of yielding achieved during the earthquake; E_H is dissipated energy during the earthquake; and W is structure weight.

Moric [25] presented a modification to the DI function for unconfined masonry buildings with flexible floor structures (DI_{flex}) by using a correction coefficient, which is a function of the floor type, the masonry tensile strength (f_t), and the ratio between the wall length perpendicular to the direction of the earthquake (l) and the story height (h). For timber-floor-type ceilings and $f_t < 0.15$ MPa for Kazazi High School, DI/DI_{flex} changes according to the data shown in Table 2. Hence, depending on the maximum l/h ratio, the DI values should be modified to find the maximum DI_{flex} that corresponds to the most damaged state of the investigated structure.

Table 2. DI/DI_{flex} to l/h for URM buildings up to three stories, timber floor, and $f_t < 0.15$ MPa (after Moric [25]).

l/h	2	3	4	5	6
DI/DI_{flex}	0.70	0.30	0.18	0.12	0.10

The investigated building is modeled as an SDOF system with a constant weight $W = 1000$ kN or through the lumped weight of the structure, equal to the roof weight plus the half weight of the walls. Since the DI is expressed in terms of the fundamental period of the building (T), the combinations of mass ($m = W/g$), elastic stiffness (k_{el}), and damping ratio (ξ) are necessary. Therefore, any assumption for weight could be considered. Thus, the desired fundamental period of the SDOF system could be obtained by considering the elastic stiffness, mass, and damping ratio. A damping ratio of 5% has been considered for the investigated building.

Furthermore, since masonry structures do not have post-elastic stiffness, the elastic–perfectly plastic behavior is considered for a post-elastic branch (i.e., $k_1 = k_{el}$, and $k_2 = 0k_{el}$). Additionally, the strength limit of elasticity (f_y) is considered equal to 10% of the total weight of the structure (i.e., $f_y = 0.1\ W = 0.1\ mg$), since it represents the structure with low elastic earthquake resistance.

Finally, the DI of the investigated building is read from the DI spectra versus the fundamental period of the building. For predicting the fundamental period of the building, this study uses the last version of the Iranian seismic code IRSt2800 [23], which suggests the following empirical Equation (3) for masonry structures:

$$T = 0.05H^{0.75} \tag{3}$$

where T is the fundamental period and H is the total structure height.

To classify structural damage, Table 3 presents damage descriptions related to DI values for both DI functions.

Table 3. DI values and damage description.

DI Function	Damage Index Value (DI)	Damage Description	Damage State
Park and Ang [22]	$1.0 < \text{DI}$	Total collapse	Complete
	$0.77 < \text{DI} \leq 1.0$	The structure is near to collapse	Extensive
	$0.40 < \text{DI} \leq 0.77$	The structure has undergone severe and nonreversible damages	Moderate
	$\text{DI} \leq 0.40$	The structure has limited damage	Slight
Modified Moric et al. [23]	$1.0 < \text{DI}$	Extremely high level or collapse	Collapse
	$0.8 < \text{DI} \leq 1.0$	Heavy	Extensive
	$0.5 < \text{DI} \leq 0.8$	Severe	Moderate
	$0.3 < \text{DI} \leq 0.5$	Moderate	Light
	$\text{DI} \leq 0.3$	Insignificant	Slight

The DI classification is based on four categories from the Park and Ang [18] DI formula, while the classification of Moric et al. [19] is based on five-category criteria, the same as EMS 98 [26].

Figure 2 shows the methodology flowchart for DI spectra and the seismic vulnerability assessment procedure.

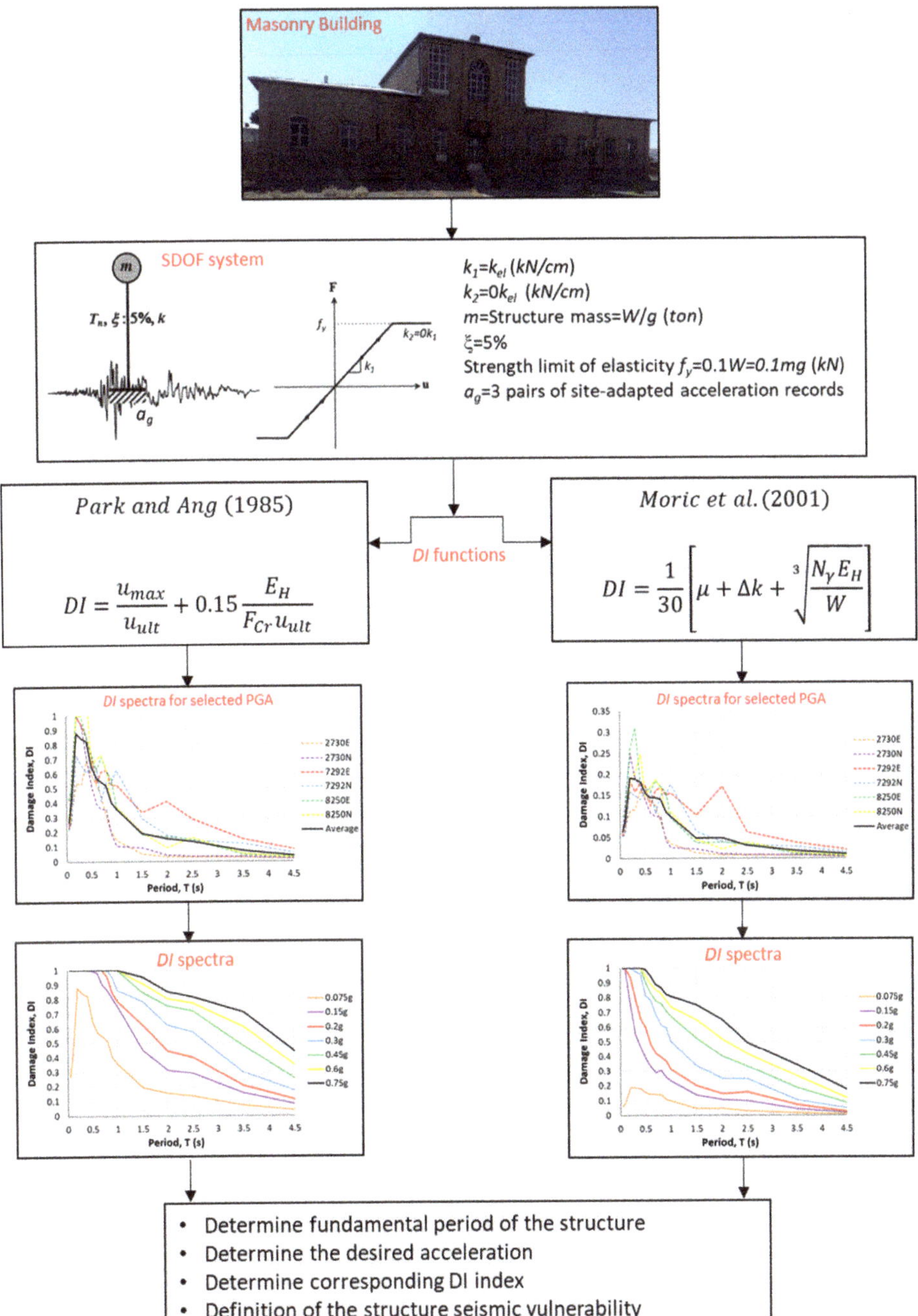

$$DI = \frac{u_{max}}{u_{ult}} + 0.15 \frac{E_H}{F_{Cr} \cdot u_{ult}}$$

$$DI = \frac{1}{30}\left[\mu + \Delta k + \sqrt[3]{\frac{N_\gamma E_H}{W}}\right]$$

Figure 2. Methodology flowchart for DI spectra and seismic vulnerability assessment procedure [18,19].

4. Seismic Damage Index Spectra

Figures 3 and 4 present spectral functions from Park and Ang [18] and Moric et al. [19] DI functions, respectively, considering the fundamental period (T) (on the x-axis) versus the obtained DI values (on the y-axis) plots for each original earthquake and scaled records to PGAs from 0.075 to 0.75 g. Dashed lines show the period-dependent damage indices from analyzing the natural and scaled earthquake records at the KRM1 station, while the black line is the average curve from all other six spectra. For original earthquakes which the Kazazi High School experienced recently, from 2002 to 2017, Park and Ang [18] and Moric et al. [19] provided limited and insignificant damage states (Figures 3a and 4a). It is confirmed by the evidence of the health state of the building after the occurrence of these earthquakes, so to show that the criteria used, despite being qualitative, are accurate in defining the seismic damage of structures.

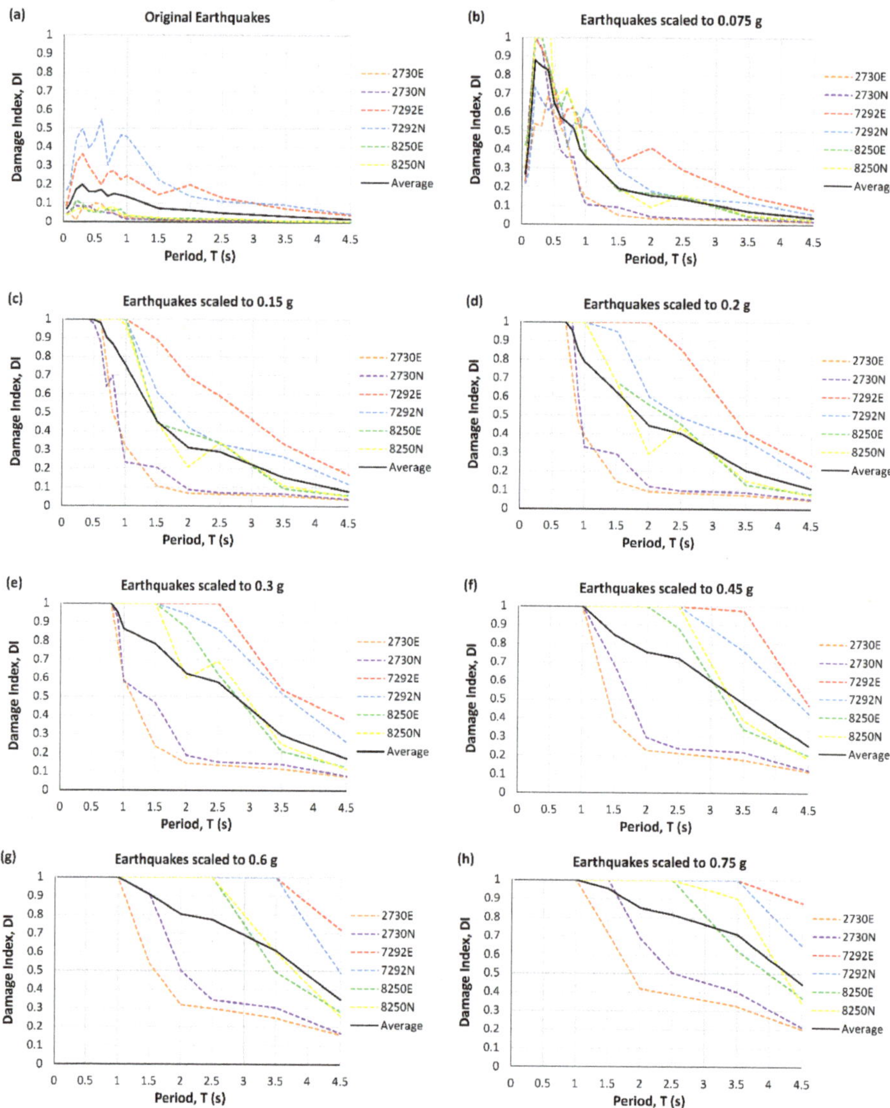

Figure 3. Park and Ang [18] DI spectral functions for seismic records: (**a**) original, (**b**) scaled to 0.075 g, (**c**) scaled to 0.15 g, (**d**) scaled to 0.2 g, (**e**) scaled to 0.3 g, (**f**) scaled to 0.45 g, (**g**) scaled to 0.6 g, and (**h**) scaled to 0.75 g.

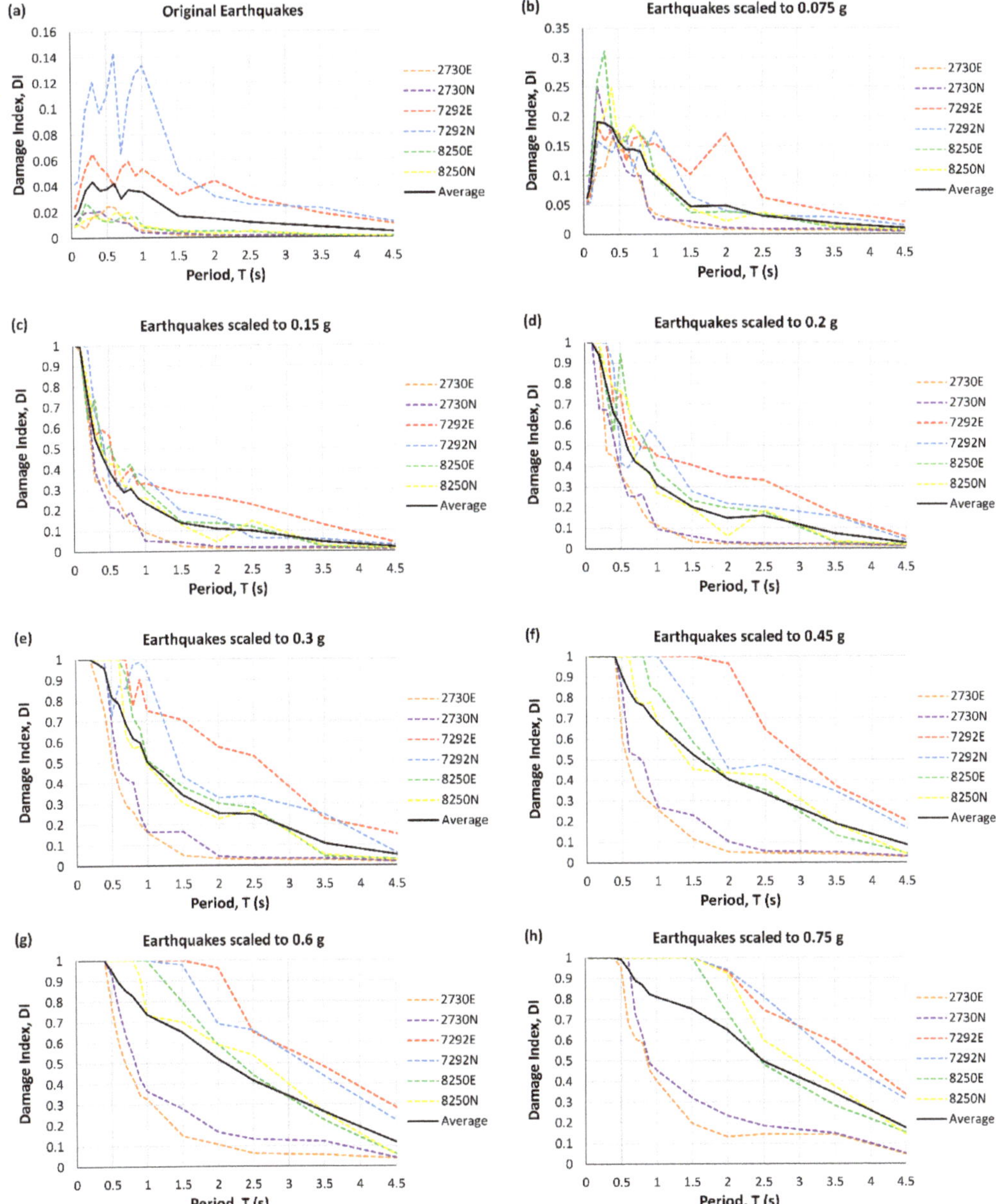

Figure 4. Moric et al. [19] DI spectral functions for seismic records: (**a**) original, (**b**) scaled to 0.075 g, (**c**) scaled to 0.15 g, (**d**) scaled to 0.2 g, (**e**) scaled to 0.3 g, (**f**) scaled to 0.45 g, (**g**) scaled to 0.6 g, and (**h**) scaled to 0.75 g.

At low values of PGA, the DI initially grows by increasing the period up to 0.2 s and then decreases (Figures 3b and 4b). For larger PGAs, the high damage index decreases from about 0.5 to 1 s (Figures 3 and 4c–h). In addition, as expected, the graphs show DI increases with the increase in PGA.

Figure 5 presents the sets of average damage spectra in other PGAs by the methods provided by Park and Ang [18] and Moric et al. [19]. These spectra were used for the rapid assessment of the investigated structure, which is discussed in the next section, and are also helpful for large-scale damage assessment of buildings located on similar ground categories in this area. The PGA considered for this area can be identified from seismic hazard analysis, or the most probable earthquake of the Iranian seismic code (IRSt2800, [23]) is considered as PGA. Alternatively, PGA can be found in the microzonation study of Kermanshah city [27].

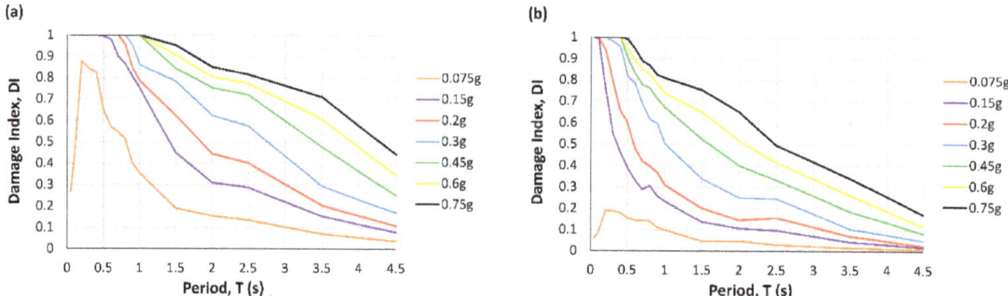

Figure 5. DI spectra for various PGA range by (**a**) Park and Ang [18] DI spectral functions and (**b**) Moric et al. [19] DI spectral functions.

5. Seismic Vulnerability Assessment

This section presents the application of DI spectra on vulnerability assessment of Kazazi High School as an unreinforced masonry historical building with flexible timber floors.

The steps of using DI spectra to assess the seismic vulnerability follow:

1. Determine the fundamental period (T) of the investigated building from empirical equations (e.g., Equation (3)).
2. Define the *PGA* (e.g., from the zoning or microzoning seismic map).
3. Read the value of *DI* from the DI spectra, which is determined according to the characteristics of the investigated building, considering site-adapted accelerograms (e.g., Figure 5).
4. Determine the ratio (l/h) and define the proposed DI_{flex} from DI/DI_{flex} (from Table 2).
5. Provide inference of possible vulnerability based on the modified DIflex (Table 3).

The accurate fundamental period of Kazazi High School can be determined by measuring the fundamental periods on the structure. For the building that has not yet been measured, it is specified from empirical Equation (3). The analyzed building has a rectangular plan with a relatively high length-to-width ratio, which leads to different fundamental periods in both directions, while this empirical equation can only provide the first natural period.

Therefore, the foundation period is calculated using Equation 3. Hence, for $H = 8.8$ m, $T = 0.25$ s. Depending on the building parameters, DI values for PGAs from 0.075 to 0.6 g are read using the generated DI spectra (Figure 4). The maximum distance of transverse walls is $l = 12.9$ m and the floor height is h = 4 m; hence, the maximum ratio l/h is $12.9/4 = 3.2$, and from Table 2 the correction coefficient for the calculated l/h ratio is $DI/DI_{flex} = 0.28$. The results of acceleration-dependent *DI* for $T = 0.25$ are presented in Table 4.

The results presented in Table 4 show that, under the influence of an earthquake with a PGA of 0.075 g, the building will suffer moderate to extensive damage, which can be displayed as severe damage. According to the formula by Moric et al. [19], the influence of rigid floors on the behavior of the URM building can be seen. For buildings with flexible floors, and depending on the l/h ratio, DI is greater, since it is divided by the coefficient of 0.28 (Table 2). Failure of the building is expected already at an acceleration of 0.15 g. If the building has rigid floors (Table 4), DI is not divided by the coefficient, and the building does not collapse up to 0.2 g, but of course, it would suffer extensive damage.

Table 4. DI values versus different PGAs at $T = 0.25$ s.

DI Function	PGAs			
	0.075 g	**0.15 g**	**0.2 g**	**$\geq$0.3 g**
Park and Ang [18]	0.86	1	1	1
	Extensive	Collapse	Collapse	Collapse
Modified Moric et al. [19]	0.19/0.28 = 0.68	0.65/0.28 considered 1	0.87/0.28 considered 1	1
	Moderate	Collapse	Collapse	Collapse

The most probable earthquake expected in Kermanshah from Iranian seismic code IRSt2800 [23] should be the one with $PGA = 0.3$ g. To perform a seismic vulnerability assessment, the method used in this paper requires the use of appropriate damage index (DI) spectral functions. Two damage index spectral functions are used in this research. One is the function introduced by Park and Ang [18], and the other is the function introduced by Moric et al. [19].

Hence, structural collapse is expected for this building. This is mainly due to the long distance of transverse walls, which will cause the out-of-plane collapse of longitudinal walls. The vulnerability would be lower if this distance is decreased.

The recent changes in the transverse walls (removing a part of the walls and opening between the transverse walls to create a connection between the spaces for the new use of the building as a museum) have made the condition even more dangerous. However, according to the results obtained from both methods, modeling of the building and its quantitative investigation are recommended to accurately identify the weak points of the structure and plan effective intervention strategies for its requalification.

6. Conclusions

This research presents the seismic vulnerability assessment of the historic Kazazi High School in Kermanshah. The damage index spectra corresponding to the school building site were determined. Then, the building damage index was evaluated according to its fundamental period. Two formulas developed and reported in the literature by Park and Ang and by Moric et al. were used to determine the damage indices in each period. The damage index parameters were determined from numerous nonlinear analyses of the single degree of freedom system under three pairs of accelerograms. Because there are no suitable natural high acceleration records, the accelerograms were scaled to various ranges of peak ground accelerations. It is worth noting that the second damage formulation considers the effect of the flexibility of the ceiling on reducing the seismic resistance of the building. The results showed that:

- Without considering the ceiling flexibility, the first damage formulation (Park and Ang) provided a higher damage index. Contrarily, the second damage formulation (Moric et al.), which considered the influence of flexible ceilings, significantly reduced the building's resistance to earthquakes.
- Although this school has not been seriously damaged in the past due to the low acceleration of the previous seismic events, in the case of an earthquake with a peak ground acceleration of 0.15 g, collapse can be expected.
- The recent structural intervention, increasing the openings in the load-bearing walls during conversion of the school building to a museum, has increased the seismic vulnerability of the building. This issue shows the prioritization for in situ study and numerical modeling of the building to find weak points in the structure and to plan appropriate retrofit interventions.

In general terms, the most important advantage of the proposed method is that it does not consider only one parameter of the peak ground acceleration/velocity for its application. The DI is obtained from the nonlinear dynamic analysis of the acceleration time history.

Hence, it can take account of the constitutive curve of the structure in a rapid assessment method, other than considering all the necessary parameters of the accelerograms, such as amplitude, frequency content, and duration.

In contrast, the generalization of the structure with an SDOF system, especially for the high-raised buildings, is a disadvantage of this method. For high-raised buildings or systems with multiple degrees of freedom (MDOF), this problem can be solved by MDOF. The problem is the inability to consider architectural diversity in structures with irregular and complex plans. This effect may be reduced by measuring the fundamental period of the building and applying it to the DI selection from the DI spectrum.

When applied to a large urban area or compartment scale, it is possible to select a suitable set of records for each region according to the seismic microzonation and propose DI spectra for each typology of buildings. The use of these spectra is fast and solely needs the fundamental period of each building. These spectra can be recommended in rapid seismic vulnerability assessment guidelines for large-scale urbanized areas.

Author Contributions: Conceptualization, M.B., M.H.-N. and A.F.; methodology, M.B. and M.H.-N.; software, M.B. and M.H.-N.; validation, M.B., M.H.-N. and A.F.; formal analysis, M.B.; investigation, M.B.; resources, M.B.; writing—original draft preparation, M.B.; writing—review and editing, M.B., M.H.-N. and A.F.; visualization, M.B. and M.H.-N.; supervision, M.B., M.H.-N. and A.F.; project administration, M.B. All authors have read and agreed to the published version of the manuscript.

Funding: This research received no external funding.

Acknowledgments: The Authors sincerely wish to thank the Education Department of Kermanshah and the Ministry of Cultural Heritage, Tourism, and Handicrafts of Iran (MCTH) for providing the required permissions and plans, and thank the Office for International Affairs of Razi Univ., Iran, and the Dept. of Structures for Engineering and Architecture of the University of Naples Federico II, Italy, for facilitating the joint research collaboration. The authors also thank Dragan Moric for his input and advice during the writing of this paper.

Conflicts of Interest: The authors declare no conflict of interest.

References

1. Green, R.A.; Hall, W.J. *An Overview of Selected Seismic Hazard Analysis Methodologies*; University of Illinois at Urbana-Champaign: Urbana, IL, USA, 1994.
2. Coburn, A.W.; Spence, R.J.S. *Earthquake Protection*; Wiley: New York, NU, USA, 2002; 424p, ISBN 978-0-470-84923-1.
3. Benedetti, D.; Petrini, V. On the seismic vulnerability of masonry buildings: An assessment method. *L'industria Delle Costr.* **1984**, *149*, 66–74. (In Italian)
4. National Group for Earthquakes Defense (GNDT). *First and Second Level form for Exposure, Vulnerability and Damage Survey (Masonry and Reinforced Concrete)*; GNDT: Rome, Italy, 1994. (In Italian)
5. Formisano, A.; Florio, G.; Landolfo, R.; Mazzolani, F.M. Numerical calibration of an easy method for seismic behavior assessment on large scale of masonry building aggregates. *Adv. Eng. Softw.* **2015**, *80*, 116–138. [CrossRef]
6. GU. n. 47 (09/02/2011). In *Directive of the Prime Minister Dated on 09/02/2011, Assessment and Mitigation of Seismic Risk of Cultural Heritage with Reference to the Technical Code for the Design of Construction*; Italian Guideline; National Guidelines System (SNLG): Rome, Italy, 2011. Issued by D.M. 14/01/2008. (In Italian)
7. Whitman, R.V.; Reed, J.W.; Hong, S.T. Earthquake damage probability matrices. In Proceedings of the 5th World Conference on Earthquake Engineering, Rome, Italy, 25–29 June 1973; pp. 2531–2540.
8. Del Gaudio, C.; De Martino, G.; Di Ludovico, M.; Manfredi, G.; Prota, A.; Ricci, P.; Verderame, G. Empirical fragility curves from damage data on RC buildings after the 2009 L'Aquila earthquake. *Bull. Earthq. Eng.* **2017**, *15*, 1425–1450. [CrossRef]
9. Gautam, D.; Fabbrocino, G.; Santucci de Magistris, F. Derive empirical fragility functions for Nepali residential buildings. *Eng. Struct.* **2018**, *171*, 617–628. [CrossRef]
10. Biglari, M.; Formisano, A. Damage probability matrices and empirical fragility curves from damage data on masonry buildings after Sarpol-e-zahab and bam earthquakes of Iran. *Front. Built. Environ.* **2020**, *6*, 2297–3362. [CrossRef]
11. Biglari, M.; Formisano, A.; Hosseini Hashemi, B. Empirical fragility curves of engineered steel and RC residential buildings after Mw 7.3 2017 Sarpol-e-zahab earthquake. *Bull. Earthq. Eng.* **2021**, *19*, 2671–2689. [CrossRef]
12. Kappos, A.J.; Panagopoulos, G.; Penelis, G.G. Development of a seismic damage and loss scenario for contemporary and historical buildings in Thessaloniki, Greece. *Soil Dynam. Earthq. Eng.* **2008**, *28*, 836–850. [CrossRef]
13. D'Ayala, D.; Ansal, A. Non linear push over assessment of heritage buildings in Istanbul to define seismic risk. *Bull. Earthq. Eng.* **2011**, *10*, 285–306. [CrossRef]

14. Hadzima-Nyarko, M.; Mišetić, V.; Morić, D. Seismic vulnerability assessment of an old historical masonry building in Osijek, Croatia, using Damage Index. *J. Cult. Heri.* **2017**, *28*, 140–150. [CrossRef]
15. Biglari, M.; D'Amato, M.; Formisano, A. Rapid seismic vulnerability and risk assessment of Kermanshah historic mosques. *Open Civ. Eng. J.* **2021**, *15*, 135–148. [CrossRef]
16. Ashayeri, I.; Biglari, M.; Formisano, A.; D'Amato, M. Ambient vibration testing and empirical relation for natural period of historical mosques; case study of eight mosques in Kermanshah, Iran. *Construct. Build. Mater.* **2021**, *289*, 123191. [CrossRef]
17. Biglari, M.; Formisano, A.; Davino, A. Seismic vulnerability assessment and fragility analysis of Iranian historical mosques in Kermanshah city. *J. Buil. Eng.* **2022**, *45*, 103673. [CrossRef]
18. Park, Y.J.; Ang, A.H. Mechanistic seismic damage model for reinforced concrete. *J. Struct. Eng.* **1985**, *111*, 722–739. [CrossRef]
19. Moric, D.; Hadzima, M.; Ivanusic, D. Seismic damage model for regular structures. *Int. J. Eng. Model.* **2001**, *14*, 29–44.
20. Charney, F.A.; Barngrover, B. NONLIN: A computer program for earthquake engineering education. In *Proceedings of the EERC-CUREe Symposium in Honor of Vitelmo V. Bertero, Berkeley, CA, USA, 31 January–1 February 1997*; Earthquake Engineering Research Institute, University of California: Berkeley, CA, USA, 1997; pp. 251–254. [CrossRef]
21. Clough, R.W.; Penzien, J. *Dynamics of Structures*; McGraw Hill: New York, NY, USA, 1993; Volume 2, ISBN 978-0070113923.
22. ISMN. Available online: https://ismn.bhrc.ac.ir/ (accessed on 15 February 2021).
23. *IRSt2800*; Iranian Code of Practice for Seismic Resistant Design of Buildings, 4th Revision. Building and Housing Research Center: Tehran, Iran, 2014.
24. Ashayeri, I.; Shahvar, M.P.; Moghofeie, A. Seismic characterization of Iranian strong motion stations in Kermanshah province (Iran) using single-station Rayleigh wave ellipticity inversion of ambient noise measurements. *Bull. Earthq. Eng.* **2022**, *20*, 3739–3773. [CrossRef]
25. Moric, D. Seismic resistance diagrams for buildings belonging to architectural heritage. *Gradjevinar* **2002**, *54*, 206–209.
26. Grünthal, G. (Ed.) European Macroseismic Scale. In *Cahiers du Centre Européen de Géodynamique et de Séismologie*; European Center for Geodynamics and Seismology: Luxembourg, 1998; p. 15.
27. Biglari, M.; Ashayeri, I.; Moftizadeh, R. Urban planning of Kermanshah city based on the seismic geotechnical hazards. *J. Seismol. Earthq. Eng.* **2015**, *17*, 203–211.

Article

Machine Learning Models for Predicting Shear Strength and Identifying Failure Modes of Rectangular RC Columns

Van-Tien Phan, Viet-Linh Tran, Van-Quang Nguyen and Duy-Duan Nguyen *

Department of Civil Engineering, Vinh University, Vinh 461010, Vietnam
* Correspondence: duyduankxd@vinhuni.edu.vn

Abstract: The determination of shear strength and the identification of potential failure modes are the crucial steps in designing and evaluating the structural performance of reinforced concrete (RC) columns. However, the current design codes and guidelines do not clearly provide a detailed procedure for governing failure types of RC columns. This study predicted the shear strength and identified the failure modes of rectangular RC columns using various Machine Learning (ML) models. Six ML models, including Multivariate Adaptive Regression Splines (MARSs), Naïve Bayes (NBs), K-nearest Neighbors (KNNs), Decision Tree (DT), Support Vector Machine (SVM), and Artificial Neural Network (ANN), were developed to calculate the shear strength and to classify the failure modes of rectangular RC columns. A total of 541 experimental data samples were collected from literature and utilized for developing the ML models. The results reveal that the ANN and KNNs models outperformed other ML models in predicting the shear strength of rectangular RC columns with the R^2 value larger than 0.99. Additionally, the KNNs model achieved the highest accuracy, mostly 100%, for identifying the failure modes of rectangular RC columns. Based on the superior performance of the ANN and KNNs models, a graphical user interface was also developed to rapidly predict the shear strength and failure modes of rectangular RC columns.

Keywords: rectangular reinforced concrete column; failure mode; shear strength; machine learning model; graphical user interface

Citation: Phan, V.-T.; Tran, V.-L.; Nguyen, V.-Q.; Nguyen, D.-D. Machine Learning Models for Predicting Shear Strength and Identifying Failure Modes of Rectangular RC Columns. *Buildings* **2022**, *12*, 1493. https://doi.org/10.3390/buildings12101493

Academic Editors: Rajesh Rupakhety, Dipendra Gautam and Humberto Varum

Received: 13 July 2022
Accepted: 14 September 2022
Published: 20 September 2022

1. Introduction

Rectangular reinforced concrete (RC) columns have been widely used in civil engineering structures. The columns are key components in ensuring the global safety of the structures. Calculating the shear strength and identifying the failure modes of such columns are crucial problems in the design process and structural analyses. Numerous experimental studies reported that the RC column could be failed in flexure, shear, or combined flexure–shear, depending on input design parameters.

Practical formulas for calculating the shear strength of RC columns were specified in design codes, such as ASCE/SEI-41-06 [1], ACI 318 [2], Eurocode-8 (EC-8) [3], CSA [4], and FEMA 273 [5]. Furthermore, many studies proposed equations for estimating the shear strength of RC columns [6–12]. However, a discrepancy was observed in comparing experimental tests and equations in the codes and published studies [13–16]. Additionally, several proposed formulas in previous works were limited in specific cases, such as short rectangular RC columns [17] and low transversal reinforcement [12]. Few models were even difficult in application practices since they contain many coefficients [18].

To overcome these drawbacks, Machine Learning (ML) techniques have been employed. Various studies applied ML models for predicting structural capacity and response of RC columns and beams [19–31]. Inel [19] estimated the ultimate flexure deformation of rectangular RC columns using Artificial Neural Networks (ANNs). A total of 237 data samples were used to construct the ANN model, which showed superior performance to other empirical equations. The flexure moment capacity of spiral RC columns was predicted using an Adaptive Fuzzy Inference System (ANFIS) model [20] and neural networks combined with particle swarm and Harris Hawks optimization techniques [21]. They proved that those ML models estimated the moment capacity with high precision. Recently, Feng et al. [22] employed the adaptive boosting model to predict the plastic hinge length of RC columns, in which 133 data sets were utilized. Lee et al. [23] collected 210 experimental tests to propose empirical equations for estimating the lumped plasticity model of circular RC columns using regression techniques. Aldabagh et al. [24] proposed simplified equations for predicting drift limits of circular RC bridge columns based on ML-based symbolic regression. Recently, Quaranta et al. [32] developed hybrid models combining mechanical concepts with machine learning-calibrated coefficients for improving shear strength equations of reinforced concrete members. They demonstrated that the accuracy of updated equations was significantly enhanced. Moreover, some studies utilized ML models to estimate the shear strength of circular [13–15,25] and rectangular RC columns [16,17,25,26]. However, they stated that a wide range of ML algorithms should be investigated and the number of data samples should be increased. Additionally, the previously developed ML models were not transferred to practical tools (e.g., mathematical formulas or graphical user interface), which can be used in design problems. Therefore, it will be very challenging to apply those ML models for practice. In addition, the influence of input variables on the predicted shear strength was not investigated systematically.

Normally, three typical failure modes of RC columns are observed under seismic loading [26,33]:

(1) Flexure failure: degradation of lateral load capacity occurred due to flexural deformation after yielding of the longitudinal reinforcing bar. This is the ductile failure, and it has a visible warning before losing capacity. In other words, it is the expected failure type in the design of the columns.

(2) Shear failure: degradation of lateral load capacity occurred due to shear distress (diagonal cracks) before yielding of the longitudinal reinforcing bar. This is the brittle failure and an unexpected type. It should be avoided in the design procedure.

(3) Flexure–shear failure: degradation of lateral load capacity occurred after yielding of the longitudinal reinforcing bar but results from shear distress.

In the last decades, the identification of the failure modes of RC columns has used conventional methods, such as the shear aspect ratio, shear strength ratio, and ductility factor. A very simplified indicator is the column aspect ratio, a/d (i.e., shear span—to-effective depth ratio) [34]. If $a/d \geq 4$, the column fails in flexure; if $2 < a/d < 4$, a flexure–shear failure is governed; otherwise, the shear failure is dominated if $a/d \leq 2$. Nevertheless, this approach does not consider the influence of material properties and reinforcement details [26]. Another parameter, the shear strength ratio (V_r), which is defined as the ratio of the shear demand to shear capacity, has also been utilized for estimating the failure modes of rectangular RC columns [34,35]. The column fails in shear if $V_r > 1$; the column suffers a flexure failure if $V_r \leq 0.6$; otherwise, a flexure–shear failure is governed. However, it was stated that this method predicted failure modes of RC columns less accurately [34,36]. Moreover, Ghee et al. [37] employed the displacement ductility factor (μ) to classify failure types of circular RC columns. If $\mu \geq 6$, the column fails in flexure; a ductile–shear failure can be experienced if $2 < \mu < 6$, and otherwise if $\mu \leq 2$, a shear failure is dominated. Nevertheless, a small data set was used; this approach is not suggested to apply for a wide range of columns. Additionally, the variation of calculated capacity and predicted failure is significant due to the complexity of mechanism.

To improve the prediction, Qi et al. [37] classified failure models of RC columns using the Fisher discriminant analysis. They gathered 111 tests of circular RC columns to derive the statistic method. However, a limited accuracy was shown for the flexure–shear failure mode. Ning and Feng [38] developed a probabilistic indicator to classify the failure mode of solid rectangular RC columns. For that, an explicit expression was derived using the simplified truss-and-arch model accounting for flexural and shear models, in which five unknown parameters were involved. However, given the complexity of failure mechanisms, prevailing uncertainties, and subjective definition of thresholds in the proposed modeling [38], the predicted failure modes were not perfectly consistent with the experimental data provided by Berry et al. [39] and Zhu et al. [33].

So far, several studies have applied ML techniques to identify the failure modes of RC columns. Mangalathu and Jeon [40] predicted failure modes of circular RC bridge piers based on six ML models, which were established based on 311 experimental results. They pointed out that ANN was the optimal approach among the investigated ML models. Feng et al. [26] applied various single and ensemble learning algorithms to classify failure modes and to predict the bearing capacity of rectangular RC columns using 254 test data samples. As a result, the adaptive boosting algorithm demonstrated better performance for classifying failures than other single learning techniques. Mangalathu et al. [41] employed the random forest model and the *SHapley Additive exPlanation* (SHAP) method [42] to predict failure modes of spiral RC columns. They concluded that the used methods provided an accuracy of 84% in identifying failure modes of the columns. Recently, Naderpour et al. [43] utilized ANN and Decision Tree (DT) models to predict failure modes of RC columns, in which 163 and 253 data sets were considered for spiral and rectangular RC columns, respectively. The aforesaid studies emphasized the high precision of ML techniques in identifying failure modes of RC columns. However, practical tools, such as mathematical equations or the graphical user interface, were not developed for design purposes.

This study employs six ML models to predict the shear strength and identify failure modes of rectangular RC columns. A total of 541 experimental data samples were selected to train the ML models. Six used ML models include Multivariate Adaptive Regression Splines (MARSs), DT, K-Nearest Neighbors (KNNs), Support Vector Machine (SVM), ANN, and Naïve Bayes (NBs). Among these, the MARSs, KNNs, DT, SVM, and ANN models were used for predicting the shear strength; meanwhile, the NBs, KNNs, DT, and SVM models are used for classifying failure modes. The optimal ML model was recognized, and then practical tools (i.e., mathematical equation and the graphical user interface) were developed for convenient design purposes of rectangular RC columns. Additionally, the effects of input parameters on the shear strength of RC columns are investigated in this study.

2. Data Collection

A significant database should be used in developing ML models to cover a wide range of input parameters of RC columns, such as geometric dimensions, material properties, and axial load effects. Additionally, the large enough sample size also enables the accuracy of prediction. A total of 541 experimental data sets of rectangular RC columns were extensively collected from the study of Ghannoum et al. [44] and other studies [45–103]. Ten input parameters, including geometric dimensions, reinforcing bar details, material properties, and axial load, need to be provided to estimate the shear strength and identify the failure mode of the RC columns. Geometric dimensions include the height of the column (L), the width of the cross-section (B), and the length of the cross-section (H). It is worth noting that the aspect ratio of columns (L/B) is varied from 1.1 to 15.3, covering short and slender RC columns. Reinforcement details contain the longitudinal reinforcement ratio (ρ_l), the transversal reinforcing bar ratio (ρ_h), and the spacing of the transversal reinforcements (s). Material properties comprise the yield strength of the longitudinal (f_{yl}) and transversal (f_{yh}) reinforcing bars and the compressive strength of the concrete (f_c'). The values of f_c' are ranged from 20 MPa to a high compressive strength of 140 MPa. Moreover, the effects of the centric axial load (P) are considered in the data sets, in which the axial compression ratio is varied from zero to 90%.

Figure 1 depicts the configurations and reinforcement properties of the rectangular RC column. The statistical properties of the experimental results are described in Table 1. In this table, ten input parameters, numbered as variables from X_1 to X_{10}, are involved in training machine learning models. The frequency histograms of input parameters and failure modes of the 541 data samples are shown in Figure 2. For this database, the number of columns that failed in flexure (F), flexure–shear combination (FS), and shear (S) is 335, 91, and 115, respectively. Figure 3 shows the correlation matrix of input and output parameters of the collected data. Based on this figure, it can be found that some parameters had a strong correlation, such as B and H. In addition, the shear strength (V) is strongly correlated with the cross-section dimensions (B and H). Meanwhile, some others were poorly correlated, such as s and ρ_h, B or H, and f'_c since their physical meanings have no connection. Moreover, the correlation among axial load (P), column height (L), and the output (V) showed to be medium.

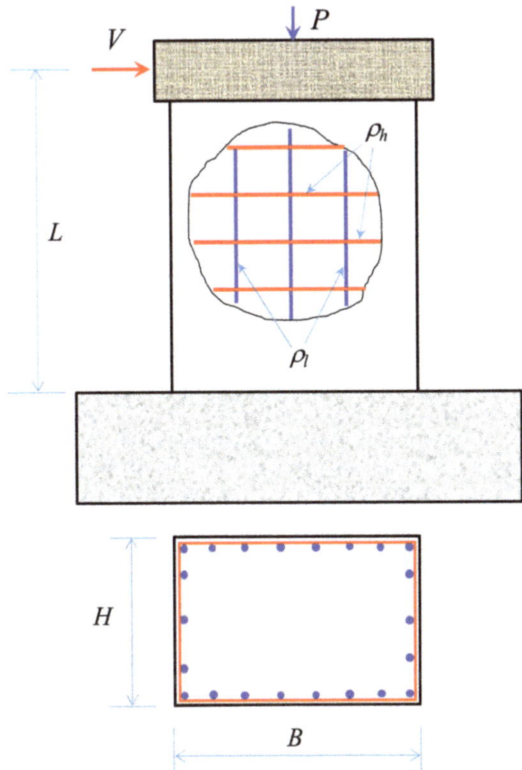

Figure 1. Configurations and properties of rectangular RC columns.

Table 1. Summary of input parameters of database.

Input Parameter	L (mm)	B (mm)	H (mm)	s (mm)	f'_c (MPa)	f_{yl} (MPa)	f_{yh} (MPa)	ρ_l (%)	ρ_h (%)	P (kN)
(Variable)	(X_1)	(X_2)	(X_3)	(X_4)	(X_5)	(X_6)	(X_7)	(X_8)	(X_9)	(X_{10})
Min	225	150	100	20	20	313	215	0.20	0.01	0.0
Mean	1286	284	301	101	49	448	496	2.15	0.94	1130
Max	3000	610	610	457	141	745	1470	4.50	4.00	5492
SD	647	109	115	77	27	77	222	0.69	0.94	1069
COV	0.53	0.38	0.38	0.76	0.55	0.17	0.45	0.32	0.99	0.95

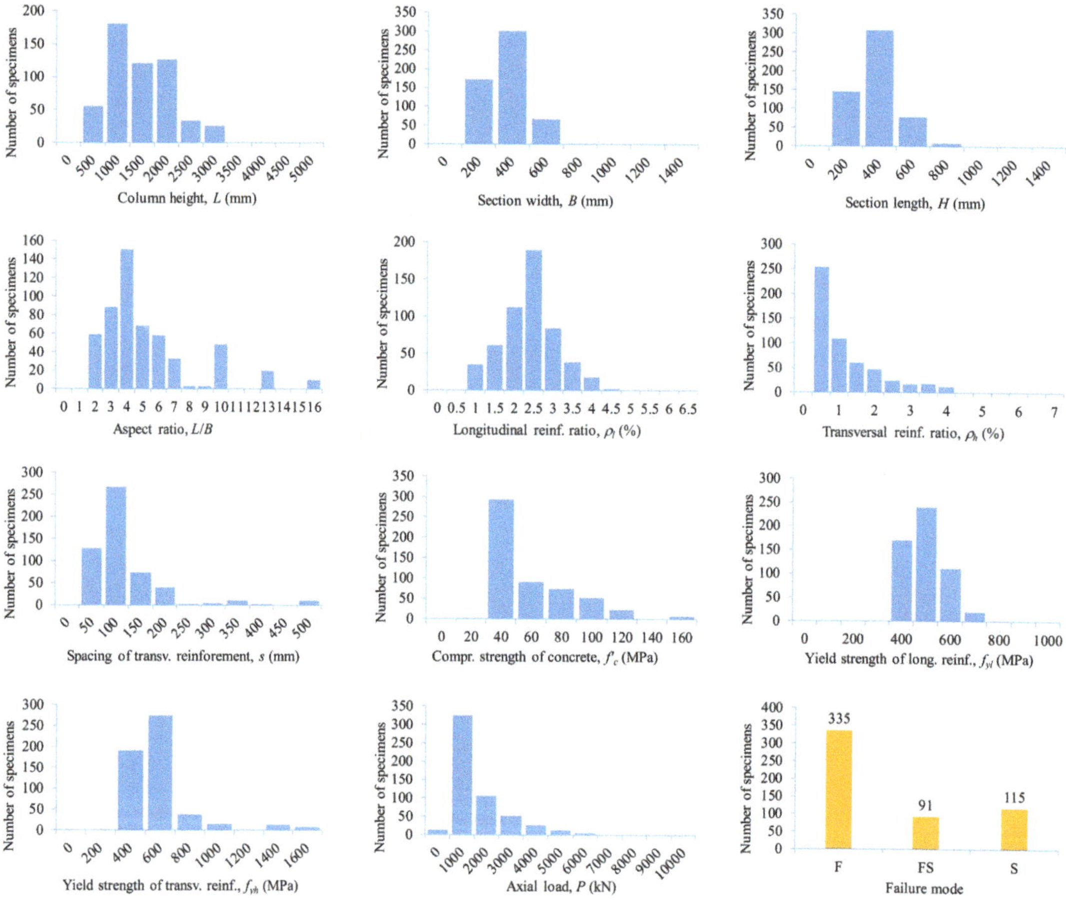

Figure 2. Frequency of input parameters and observed failure modes of database.

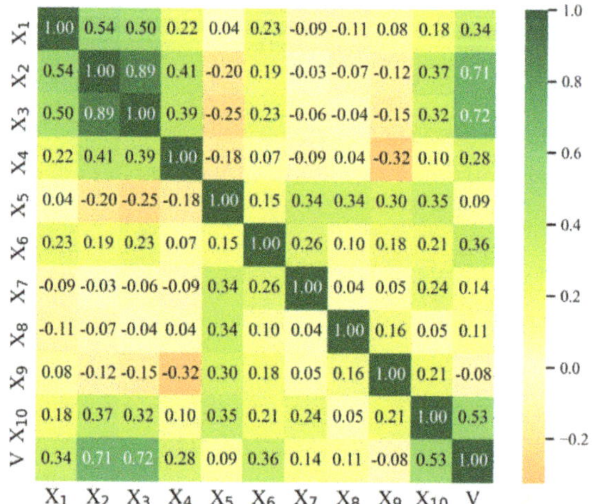

Figure 3. Correlation matrix of input and output parameters.

3. Background of Machine Learning Models

So far, there are numerous ML models, which can be applied for regression and classification problems in structural engineering. Each algorithm contains advantages and limitations. In this study, the authors selected some typical single ML models for predicting shear strength and classifying failure modes of rectangular RC columns, in which NBs, MARSs, KNNs, DT, SVM, and ANN were used. A brief description of the investigated ML models is provided as follows.

3.1. Naïve Bayes

NBs classifiers are a set of supervised ML algorithms based on the Bayes theorem. The NBs algorithm works very fast and can easily predict the class of a test data set. NBs methods assume that the value of a given class is independent of the feature vector. The Bayes theorem can be expressed as

$$P(y|x_1,\ldots,x_n) = \frac{P(y)P(x_1,\ldots,x_n|y)}{P(x_1,\ldots,x_n)} \tag{1}$$

where y is the given class variable; x_i is the dependent feature vector; $P(y)$ is the prior probability (i.e., the class probability); $P(x_i|y)$ is the conditional probability, and $P(x_i)$ is the evidence. The various Naïve Bayes classifiers differ mostly due to the assumptions of the distribution of $P(x_i|y)$. In this study, we used Gaussian naïve Bayes for classifying failure modes of rectangular RC columns. However, it should be noted that if the data sets have a categorical variable of a category that is not included in the training data set, the NBs model will assign it zero probability and will not be able to make any predictions in this regard. Thus, a smoothing technique is required to solve this problem.

3.2. K-Nearest Neighbors

KNNs, a nonparametric supervised ML method, classifies data based on K samples that are nearest to it. The K parameter is determined as the square root of the number of the training data set. This algorithm does not make any assumption about the data, and it can be applied both for classification and regression problems Additionally, KNNs is very simple to perform for multiclass problems. However, the KNNs algorithm is very sensitive to outliers as it simply chooses the neighbors based on distance criteria. Moreover, the KNNs model does not perform well on imbalanced data.

To apply KNNs for classification, it can be implemented through three main steps:

- Firstly, the KNNs determines the distance between a new data point to all other points of training data. This space can be calculated using Euclidean distance, i.e., $\sqrt{\sum_i^K (x_i - y_i)^2}$ or Manhattan distance, i.e., $\sum_i^K |x_i - y_i|$, where x_i and y_i are coordinates of data points.
- Next, the KNNs algorithm randomly selects the K nearest data points (K is an integer). The selection is based on the proximity to other data points regardless of what feature the values.
- Finally, the algorithm allocates the data point to the class where similar data points lie down.

3.3. Decision Tree

DT, a supervised ML algorithm, is known as one of the most effective techniques for prediction and classification [104]. This ML technique is a tree-like flowchart, in which each nonleaf node (i.e., internal node) represents a test on a feature. The training data from the root node are recursively partitioned into subsets or branches using the Gini or entropy index criterion [105]. Each branch denotes an output of the test, and each leaf node (i.e., terminal node) indicates a classification. The advantages of DT are easy to understand and interpret, and it has no assumptions about data sets. Additionally, this algorithm can operate with numerical and categorical features. However, this model can easily create the overfitting problem during training process.

3.4. Support Vector Machine

SVM is a supervised ML algorithm, which is popularly used in performing classification and regression [106]. SVM is also a fast and dependable classification technique. It is effective in high dimensional spaces and works well with a clear margin of separation. SVM can perform very well with a limited amount of used data. However, SVM does not work well when the database is large and has more noise, i.e., target classes are overlapping. Normally, the procedure for classification using SVM can be implemented by the following steps:

- Creating a hyperplane or decision boundary that separates the features (i.e., classes). It can be a linear or nonlinear hyperplane. A good decision boundary is achieved when it contains the largest space to the adjacent training data point of classes.
- Using a kernel function, $K(x,y)$, to facilitate the computation of dot products of pairs of input data vectors, which are designed by mapping from the original finite-dimensional space to a higher-dimensional space.

Kernel functions play a crucial role in SVM to connect from linearity to nonlinearity. Three typical kernel functions, which are Linear, Gaussian, and Polynomial, can be employed to evaluate the classification performance. Those functions defined on Euclidean space, $\mathbb{R}^d$, are expressed as

$$\text{Linear kernel function}: \; K(x,y) = x^T y, \; \left(x,\, y \in \mathbb{R}^d\right) \tag{2}$$

$$\text{Gaussian kernel function}: \; K(x,y) = e^{-\frac{\|x-y^2\|}{2\sigma^2}}, \; (x,\, y \in \mathbb{R}^d,\, \sigma > 0) \tag{3}$$

$$\text{Polynomial kernel function}: \; K(x,y) = \left(x^T y + r\right)^n, \; \left(x,\, y \in \mathbb{R}^d,\, r \geq 0,\, n \geq 1\right) \tag{4}$$

3.5. Artificial Neural Networks

ANNs are flexible and can be used for both regression and classification problems. It is good to work with nonlinear data containing a large number of inputs. Neural networks can be trained with any number of inputs and layers, and the predictions are obtained very fast. However, ANNs depend on training data significantly, and sometimes this can create an overfitting problem.

An ANN model comprises three components:

- Input layer, where input parameters are entered;
- Hidden layer(s);
- Output layer, where the predicted result is obtained.

The neurons in the network are bridged in some forms, in which the signal is transferred from neurons to other neurons. These connections hold a weight, and each neuron has a bias and an activation. The input vector (i.e., signal) of the neuron is represented by $x = [x_1, x_2, \ldots, x_m]$, while the weighted sum of the input vector is determined by $z \in \mathbb{R}$ as follows:

$$z = \sum_{i=1}^{d} w_i x_i = w^T x + b \tag{5}$$

where $w = [w_1, w_2, \ldots, w_d] \in \mathbb{R}^d$ denotes the weight vector in the d-dimension; $b \in \mathbb{R}$ is the bias. To consider the nonlinear relationship between the input and output vectors, a nonlinear processing with respect to z is performed in the form of

$$y = f(z) \tag{6}$$

where f represents the activation function; y denotes the activation value of the neuron. In this study, the *tansig* and *purelin* functions were employed for making a smooth transition during training the network [107], expressed by Equations (7) and (8). This approach is also consistent with studies elsewhere [108–117].

$$y = tansig(x) = \frac{2}{1 + e^{-2x}} - 1 \tag{7}$$

$$y = purelin(x) = x \tag{8}$$

To perform the ANN algorithm, the following processes are required:

- Firstly, the input signals (i.e., data) are entered to the input layer, and the signals are transferred from one node (neuron) to another through the connections in the network. This is called the forward pass.
- Secondly, after obtaining the output from the forward pass, it is required to evaluate this output by comparing it with the target using the Mean Squared Error (MSE), as expressed in Equation (9). This is called the backward pass.
- Moreover, it is needed to minimize the error by iteratively updating those processes until the MSE is converged.

$$MSE = \frac{1}{N} \sum_{i=1}^{N} (p_i - t_i)^2 \tag{9}$$

where N is the number of samples; t_i and p_i are the target and predicted values of the ith sample, respectively.

3.6. Multivariate Adaptive Regression Splines

MARSs is a flexible nonparametric regression method, which was proposed by Friedman [118]. This model is more flexible than linear regression ones. In addition, it is simple to understand and interpret. Moreover, MARSs can deal with both numerical and categorical databases. However, the disadvantage of MARSs is that the fitting function is not smooth. The general expression of nonparametric regression is represented by the following form:

$$y_i = f(x_{i1}, x_{i2}, \ldots, x_{ij}) + \varepsilon_i = f(X) + \varepsilon_i \tag{10}$$

where $X = (x_{i1}, x_{i2}, \ldots, x_{ij})$ is an $i \times j$ matrix of j input variables and i samples; ε_i is the error of the ith sample. A MARSs model is constructed using basis functions to approximate the $f(X)$, expressed by

$$f(X) = c_0 + \sum_{n=1}^{N} c_n B_n(x) \tag{11}$$

where c_0 is a constant; c_n is the coefficients of basis functions $B_n(x)$; N is the number of basis functions. Basic functions are splines, which normally have piece-wise linear functions. A a basis function can be one of the three forms: (1) a constant, (2) a hinge function, which yields the kink, and (3) a product of two or more hinge functions. A hinge function is expressed in the forms of $\max(0, x - c)$ or $\max(0, c - x)$, in which c is a constant, the so-called *knot* or cutpoint value. Generally, to perform MARSs, the following steps are conducted [119]:

- Construct a forward stepwise algorithm to select spline basis functions.
- Develop a backward stepwise algorithm to delete unnecessary basis functions until the optimal set is obtained.

4. Prediction of Shear Strength of Rectangular RC Columns

4.1. Existing Formulas for Calculating Shear Strength of RC Columns

So far, numerous studies have proposed equations for calculating the shear strength of RC columns [1–10]. In this study, we employed six typical equations for calculating the shear strength of RC columns, in which equations in current design codes and well-known previous studies are considered. Six formulas included ACI 318 [2], Eurocode-8 (EC-8) [3], CSA [4], FEMA 273 [5], Ascheim and Moehle [6], and Sezen and Moehle [9] (ASCE/SEI 41-06 [1]), as described in Table 2.

Table 2. Formulas for calculating shear strength of rectangular RC columns.

ID	Model	Expression	Equation
1	ACI 318 [2]	$V_1 = 0.166\left(1 + \frac{P}{13.8A_g}\right)b_w d\sqrt{f'_c} + \frac{A_{sh}f_{yh}d}{s}$	(12)
2	CSA [4]	$V_2 = \min\left(\beta b_w d_v \sqrt{f'_c} + \frac{A_{sh}f_{yh}d}{s}\cot\theta;\ 0.25f'_c bd\right)$ $d_v = 0.9d$	(13)
3	FEMA 273 [5]	$V_3 = 0.29\lambda\left(k + \frac{P}{13.8A_g}\right)bd\sqrt{f'_c} + \frac{A_{sh}f_{yh}d}{s}$ $k = 1.0$ for low ductility demand. $k = 0$ for moderate and high ductility demand.	(14)
4	EC8 [3]	$V_4 = k(V_c + V_w) + V_P$ $V_c = 0.16\max(0.5; 100\rho_l\left(1 - 0.16\min(5;\frac{a}{d})\right)A_c\sqrt{f'_c}$ $V_w = \frac{A_{sw}}{s}(d - d\prime)f_{yw}$ $V_P = \frac{(D-x)}{2a}\min(N; 0.55A_c f'_c)$	(15)
5	Ascheim and Moehle [6]	$V_5 = 0.3\left(k + \frac{P}{13.8A_g}\right)0.8A_g\sqrt{f'_c} + \frac{A_{sh}f_{yh}d}{s\tan(30^0)}$ $k = \frac{4-\mu}{3}$, μ is the displacement ductility $d = 0.8H$	(16)
6	Sezen and Moehle [9] (ASCE/SEI 41-06 [1])	$V_6 = k\left(\frac{0.5\sqrt{f'_c}}{a/d}\sqrt{1 + \frac{P}{0.5A_g\sqrt{f'_c}}}\right)0.8A_g + k\frac{A_{sh}f_{yh}d}{s};$ $d = D - \text{cover}$ $k = 1$ for $\mu < 2.0$; $k = 0.7$ for $\mu > 6.0$; $0.7 \leq k = 1.15 - 0.075\mu \leq 1.0$ for $2.0 \leq \mu \leq 6.0$ a is the shear span, (i.e., the distance from loading point to the boundary).	(17)

4.2. Performance of ML Models

Five ML models, which are MARSs, DT, KNNs, SVM, and ANN, were employed to predict the shear strength of RC columns. For each ML model, large and wide-ranging training–testing ratios were tested to identify the optimal model. The ratios include 0.6–0.4, 0.65–0.35, 0.7–0.3, 0.75–0.25, 0.8–0.2, 0.85–0.15, and 0.9–0.1. In the current study, we used three statistical parameters, which are the coefficient of determination (R^2), root-mean-square error ($RMSE$), and $a20 - index$, to evaluate the performance of the ML models. The definitions of these indicators are expressed by following equations:

$$R^2 = 1 - \left(\frac{\sum_{i=1}^{N}(t_i - o_i)^2}{\sum_{i=1}^{N}(t_i - \bar{o})^2}\right) \tag{18}$$

$$RMSE = \sqrt{\left(\frac{1}{N}\right)\sum_{i=1}^{N}(t_i - o_i)^2} \tag{19}$$

$$a20 - index = \frac{n20}{N} \tag{20}$$

where t_i and o_i represent the target and output of ith data point, respectively; $\bar{o}$ is the mean of output data samples; N is the total number of data set; $n20$ is the number of data statisfied $0.8 \leq \left|\frac{V_{exp}}{V_{predict}}\right| \leq 1.2$, in which V_{exp} and $V_{predict}$ are the shear strengths obtained from experiments and predictions, respectively.

Figures 4–8 show the performance of ML models, in which R^2, $RMSE$, and $a20 - index$ are measured for various scenarios in training ML models. It should be noted that 210 cases, which combine 7 training ratios (i.e., 0.6, 0.65, 0.7, 0.75, 0.8, 0.85, and 0.9) and 30 values of the model parameter, were trained for each ML model. The ranking of 210 training–testing ratios in each ML model is shown in Figure 9. For MARSs, 30 basis functions were tested, and the best model was obtained with 29 basis functions and a training data ratio of 0.8. For DT, the minimum leaf size ranged from 1 to 30, and the best model was achieved at a leaf size of 1, and the training data ratio was 0.7. For KNNs, the number of nearest neighbors were tested up to 30, and the best model was found with 1-nearest neighbor and a training data ratio of 0.85. For SVM, the ε-insensitive zone, which determines the deviation margin of the predicted point, ranged from 0.01 to 0.3 with an interval of 0.01. As a result, the optimum SVM model achieved at ε was 0.01, and the training data ratio was 0.8. Meanwhile, the best ANN model was obtained when the number of neurons in the hidden layer were 27 and the training data ratio was 0.9. Figures 4–8 also demonstrate that the accuracy of the ANN and MARSs models is improved as the number of key parameters of ML models increased. By contrast, the predicting performance of the other ML models was lessened as the number of key parameters increased.

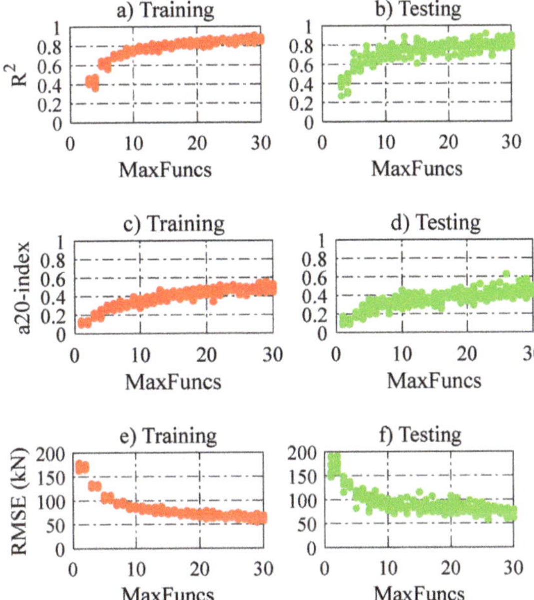

Figure 4. The performance of 210 MARSs models.

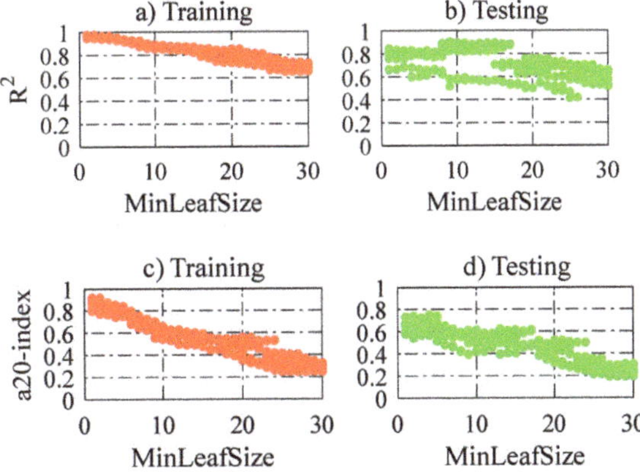

Figure 5. *Cont.*

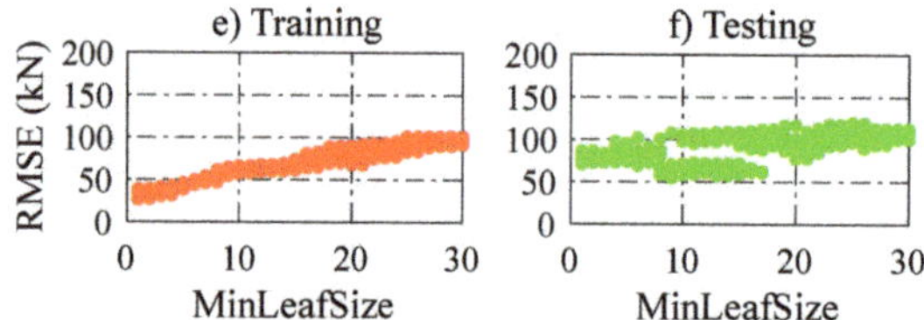

Figure 5. The performance of 210 DT models.

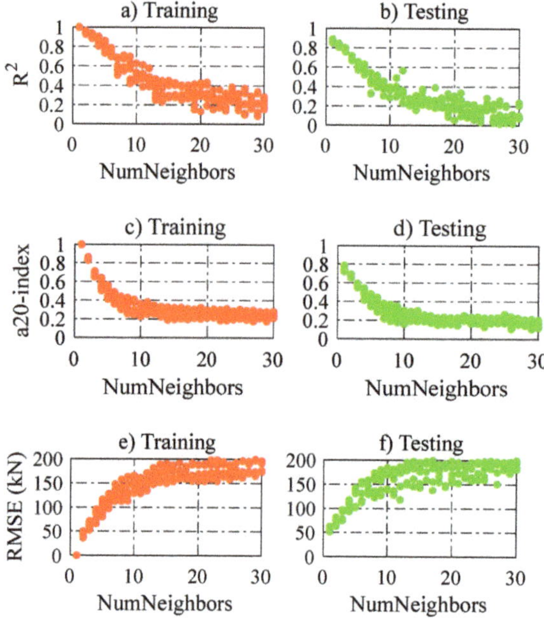

Figure 6. The performance of 210 KNNs models.

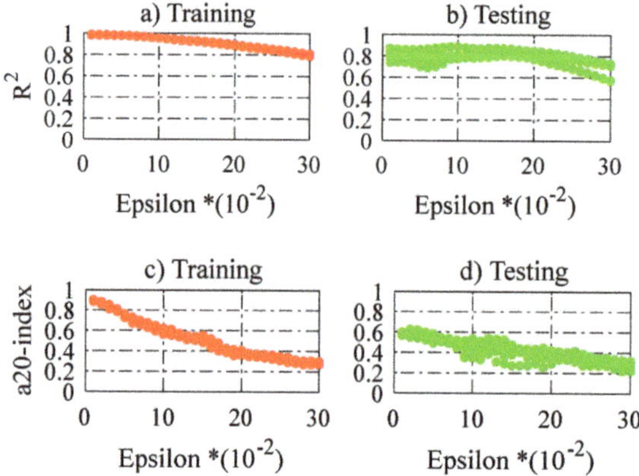

Figure 7. *Cont.*

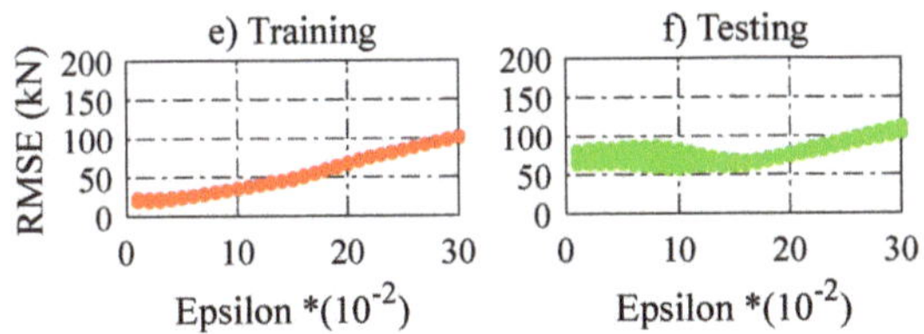

Figure 7. The performance of 210 SVM models.

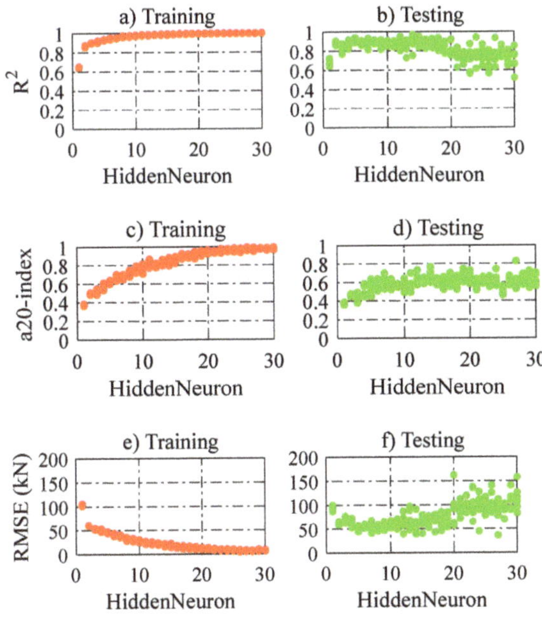

Figure 8. The performance of 210 ANN models.

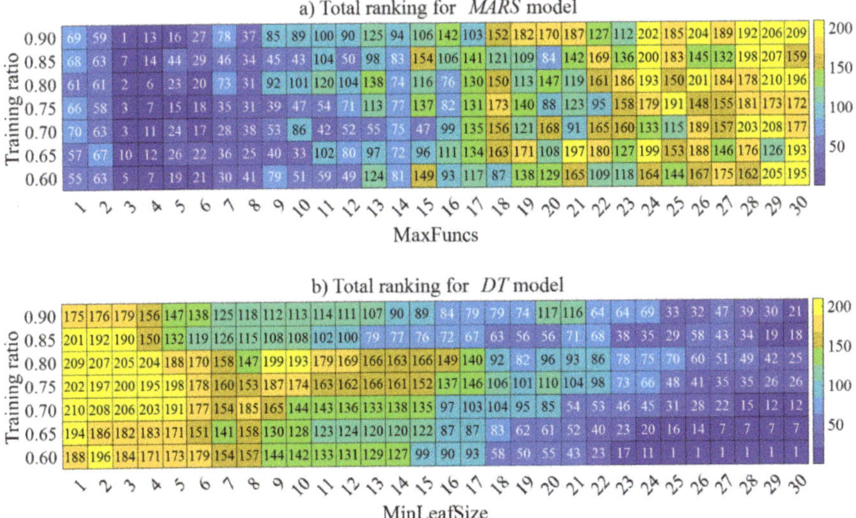

Figure 9. *Cont.*

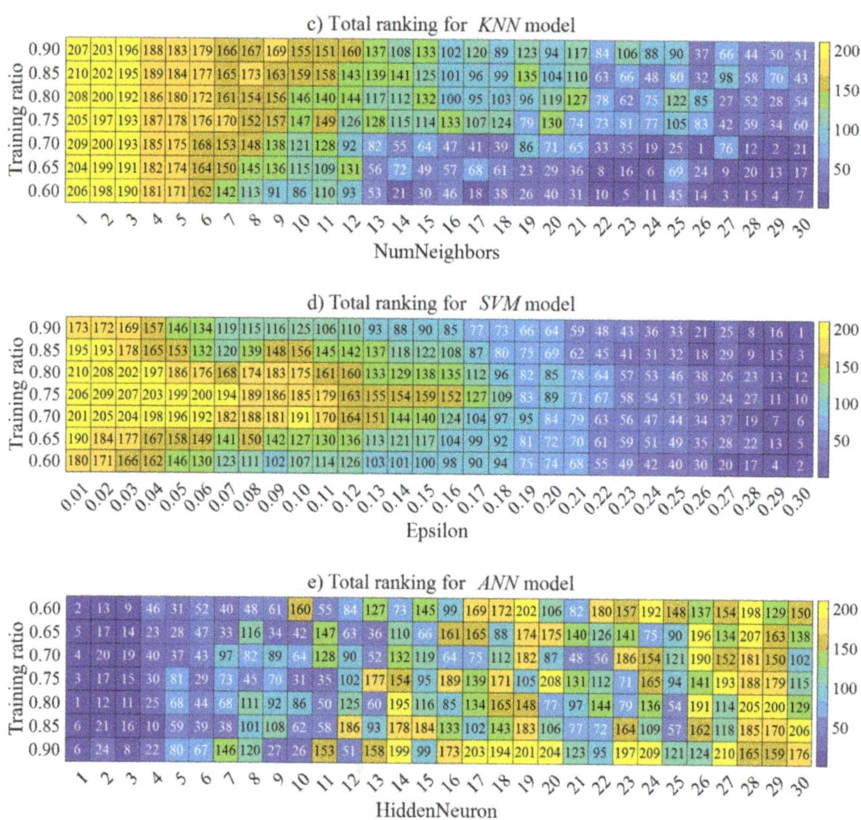

Figure 9. The ranking of ML models with 210 training-testing ratios.

4.3. Comparison of Performance between ML Models and Existing Equations

The comparisons of predictive models (i.e., existing equations and ML models) and experimental results are shown in Figure 10. In this figure, the horizontal axis represents the shear strength provided by the experiments, while the calculated and predicted values are shown in the vertical axis. It was obviously demonstrated that the ML models outperformed the empirical equations, which are proposed by design codes and previous studies. The data scattering of ML models is significantly smaller than that obtained by existing formulas. Among these, the ANN and KNNs models showed to be the best options in predicting the shear strength of rectangular RC columns, followed by the SVM and DT models.

Table 3 shows the calculated statistical indicators (R^2, $a20 - index$, $RMSE$) from eleven predicted models in this study. It should be noted that the values in parentheses are for the testing data. The values of R^2 and $a20 - index$ obtained from the ML models showed to be noticeably higher than those from existing equations. Both the ANN and KNNs models had a superior performance among the used ML models with an R^2 of 0.998 and 1.0 for training data, respectively, and an R^2 of 0.938 and 0.876 for testing data, respectively. Additionally, the $RMSE$ values calculated from the ML techniques were significantly smaller compared to those from empirical equations, specifically ANN and KNNs had the smallest errors. This again implies that ANN and KNNs are the efficient models in predicting the shear strength of RC columns.

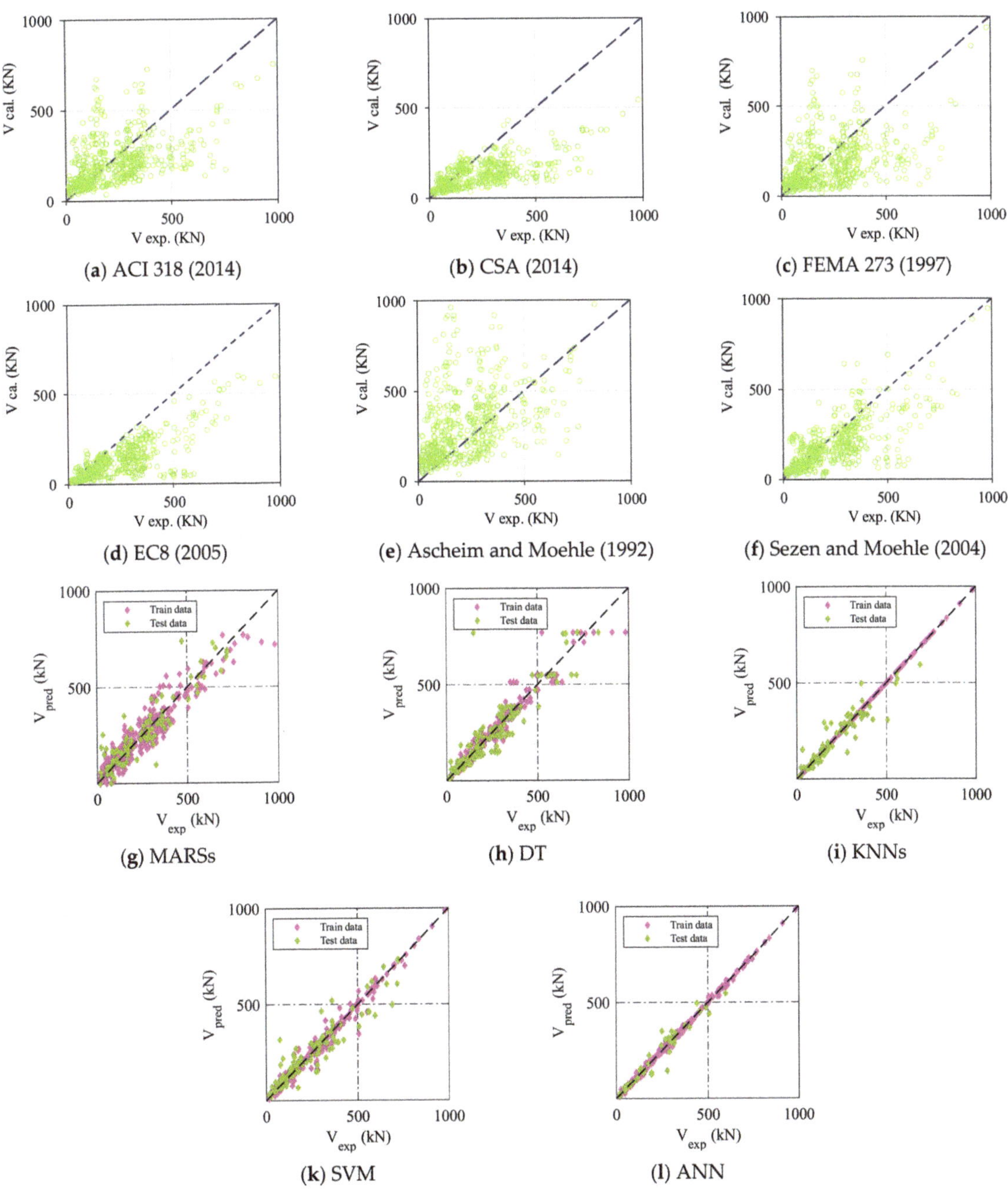

Figure 10. Comparison of shear strengths between experiments and various predicted models.

Table 3. Statistical indicators of different predictive models.

Model	R^2	$a20-Index$	$RMSE$ (kN)
ACI 318 [2]	0.254	0.197	133.8
CSA [4]	0.463	0.211	136.9
FEMA 273 [5]	0.180	0.187	149.0
EC8 [3]	0.627	0.174	128.3
Ascheim and Moehle [6]	0.186	0.194	208.5
Sezen and Moehle [9]	0.529	0.338	108.5
MARSs	0.897 (0.845)	0.503 (0.537)	54.65 (71.99)
DT	0.962 (0.847)	0.855 (0.660)	33.59 (70.18)
KNNs	1.000 (0.876)	1.000 (0.765)	0.000 (51.54)
SVM	0.982 (0.874)	0.894 (0.602)	22.78 (63.86)
ANN	0.998 (0.938)	0.976 (0.830)	8.08 (36.00)

5. Failure Mode Classification Using Machine Learning Models

Four ML techniques including NBs, DT, KNNs, and SVM were applied for classifying the failure modes of rectangular RC columns. Ten input parameters in Table 1 were used for implementing those ML algorithms. All ML algorithms in this study were developed using MATLAB. For that, the trial-and-error method was also employed for training models. Again, it is noted that the number of experimental columns that failed in flexure (F), flexure–shear combination (FS), and shear (S) was 335, 115, and 91, respectively. Figure 11a shows the distribution of failure modes in the collected data set. It can be found that the data points of three failure modes were not balanced, in which the number of flexural failure modes predominated in the data samples. The imbalanced data set affects the decision boundary of ML models during the learning process. To overcome this problem, the synthetic minority oversampling technique (SMOTE) [120] was employed to balance the distribution of classes by replicating samples of the minority classes. As a result, the instances of failure modes were equal after balancing, as shown in Figure 11b. The duplicating process was conducted using Scikit-learn module with following steps:

(1) Randomly select a data point (i.e., sample) from the minority class.
(2) Determine the nearest neighbors (e.g., 5 data points) of the selected sample.
(3) Create synthetic samples between two data points in the feature space.

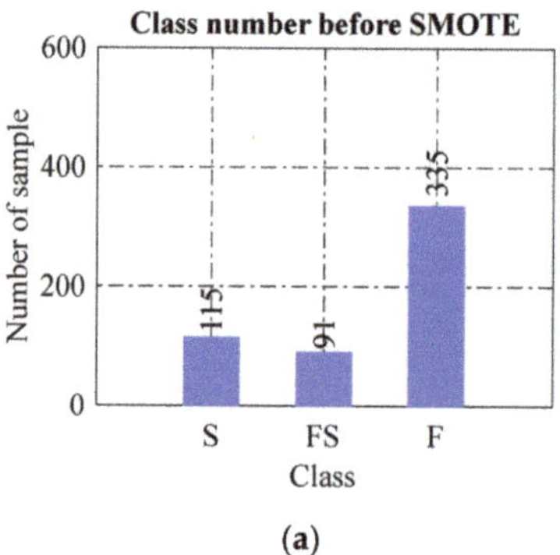

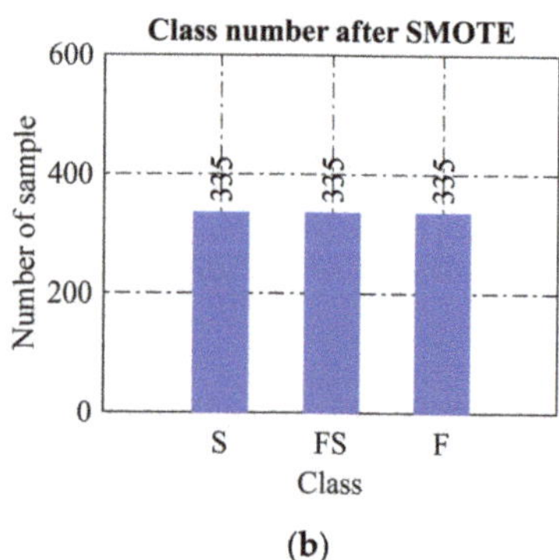

Figure 11. Distribution of failure modes in the database (**a**) before and (**b**) after performing SMOTE.

To evaluate the performance of the ML techniques, typical measures consisting of accuracy, sensitivity, specificity, and area under the curve (AUC) were used.

- **Accuracy** is expressed as the ratio of the sum of correct classifications to the total number of classes.
- **Specificity** represents the proportion of the negative class correctly classified.

- **Sensitivity** (i.e., recall) is measured as the ratio of the number of accurately predicted failure modes to the total number of failure modes considered in the data.
- **AUC** is the indicator for measuring the ability of an ML model to distinguish between classes. AUC is the area under the Receiver Operating Characteristic (ROC) curve, which is plotted by *Sensitivity* in the y-axis against $1 - Specificity$ in the x-axis. The higher the AUC, the better the performance of an ML model.

Figure 12 depicts the concept of the confusion matrix and how to calculate those measuring indicators. It should be noted that the higher values of accuracy, sensitivity, specificity, F1-score, and AUC, the better the performance of the machine learning techniques. However, the sensitivity and specificity should be simultaneously going up to indicate the good performance of the models.

$$Accuracy = \frac{TP + TN}{TP + TN + FP + FN}$$

$$Sensitivity = \frac{TP}{TP + FN}$$

$$Specificity = \frac{TN}{TN + FP}$$

Figure 12. Concept of confusion matrix and definition of measuring indicators.

Figures 13–16 show the performance of four classifying ML models, in which 210 cases were investigated for each model. It should be noted that the key parameter for optimizing NBs is the equal bin width value. Meanwhile, the minimum leaf size was used for DT; the number of nearest neighbors was for KNNs; the kernel scale was for SVM, and the number of neurons in the hidden layer was for ANN. It can be observed that the highest accuracy of the NBs, DT, and KNNs models was obtained at the smallest value of the model parameters, whereas SVM was optimized at the kernel scale of 0.12 and the training data ratio of 0.8.

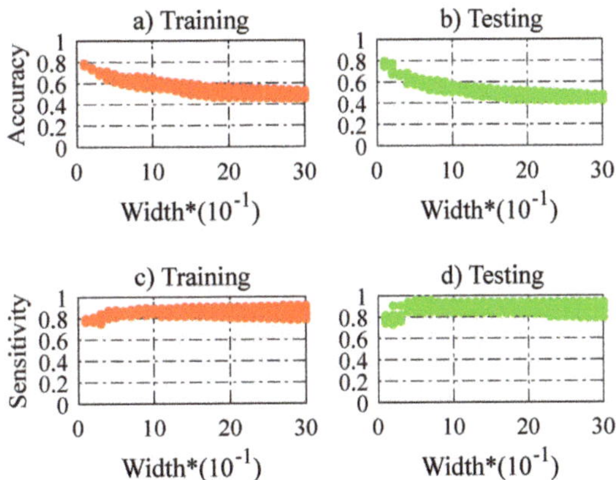

Figure 13. *Cont.*

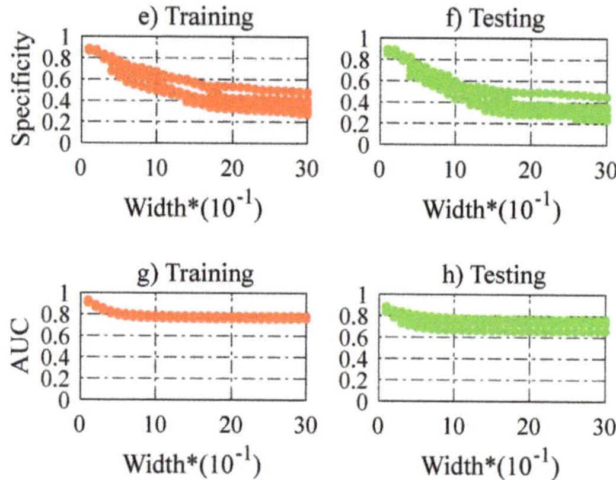

Figure 13. Performance of 210 NBs models for classifying failure modes.

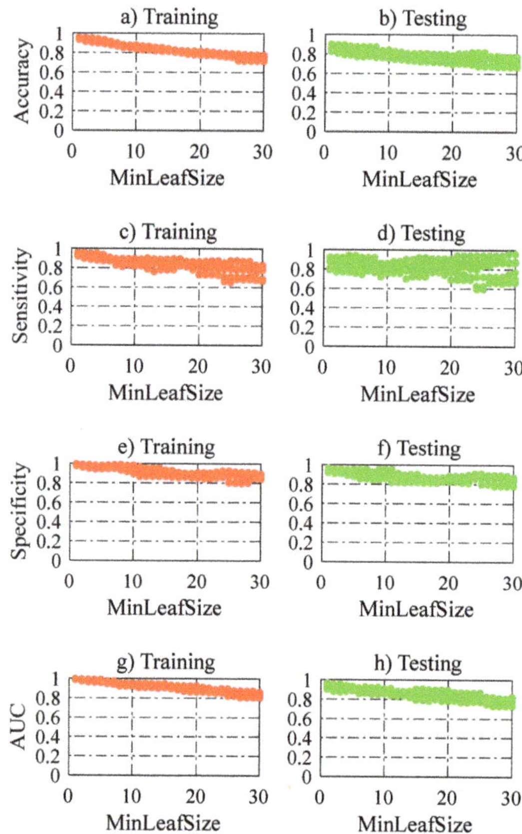

Figure 14. Performance of 210 DT models for classifying failure modes.

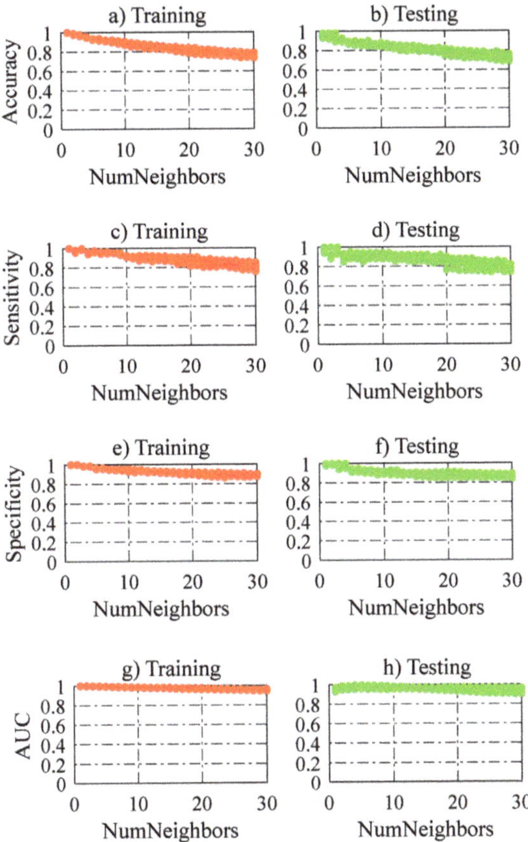

Figure 15. Performance of 210 KNNs models for classifying failure modes.

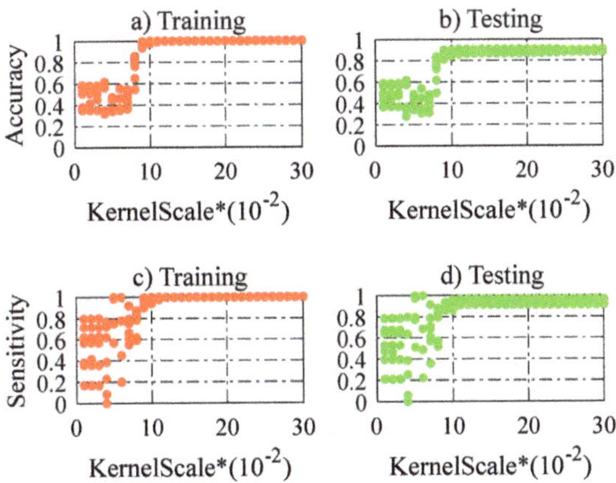

Figure 16. *Cont.*

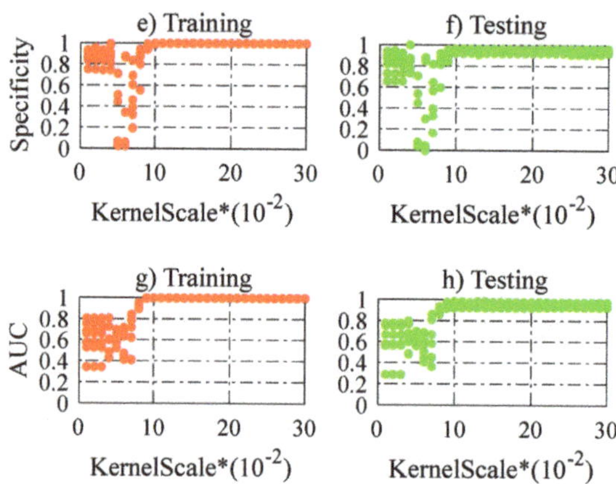

Figure 16. Performance of 210 SVM models for classifying failure modes.

The ranking of 210 cases in each classifying model is shown in Figure 17. For NBs, 30 bin width values were tested, and the best model was obtained with 0.4 and at a training data ratio of 0.85. For DT, the minimum leaf size ranged from 1 to 30, and the best model was achieved at the leaf size of 1, and the training data ratio was 0.9. For KNNs, the number of nearest neighbors was tested up to 30, and the best model was found with 1-nearest neighbor and a training data ratio of 0.9. For SVM, the kernel scale ranged from 0.01 to 0.3 with an interval of 0.01. As a result, the optimum SVM model was achieved at a scale of 0.12 and a training ratio of 0.8.

Figure 18 shows the performance of the best ML models in terms of confusion matrices for identifying failure modes of rectangular RC columns after performing SMOTE. The confusion matrix represents the number of correctly and incorrectly predicted classes for each class and based on observed experiment results (i.e., true class) are plotted versus the predicted failure modes (i.e., predicted class) in the confusion matrix. Each element in the matrix, A_{ij}, expresses the number of true classes i, but predicted to class j. As a result, the diagonal elements of the matrix (from the upper left to the lower right) are failure modes, which were accurately predicted by the ML techniques. In other words, the off-diagonal elements denote failure modes, which were misclassified. Among the investigated ML techniques, KNNs showed to be the best model for classifying failure modes of rectangular RC columns with a very high accuracy, mostly reaching 100%. Furthermore, the SVM and DT models also demonstrated a good prediction of failure modes with an accuracy greater than 87%. Meanwhile, the NBs model was not a good option for identifying failure modes of rectangular RC columns.

a) Total ranking for *NB* model

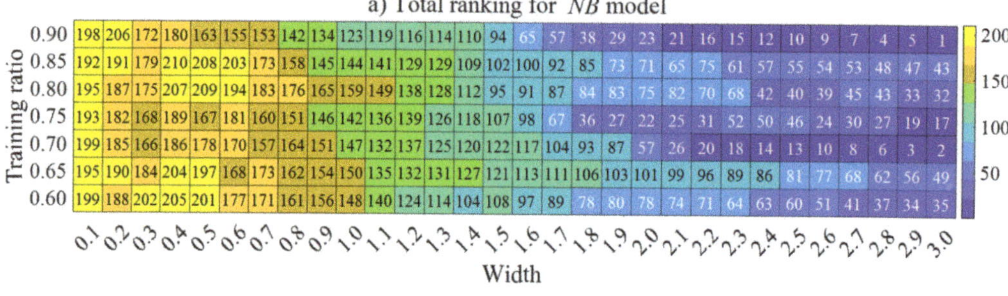

Training ratio	0.1	0.2	0.3	0.4	0.5	0.6	0.7	0.8	0.9	1.0	1.1	1.2	1.3	1.4	1.5	1.6	1.7	1.8	1.9	2.0	2.1	2.2	2.3	2.4	2.5	2.6	2.7	2.8	2.9	3.0
0.90	198	206	172	180	163	155	153	142	134	123	119	116	114	110	94	65	57	38	29	23	21	16	15	12	10	9	7	4	5	1
0.85	192	191	179	210	208	203	173	158	145	144	141	129	129	109	102	100	92	85	73	71	65	75	61	57	55	54	53	48	47	43
0.80	195	187	175	207	209	194	183	176	165	159	149	138	128	112	95	91	87	84	83	75	82	70	68	42	40	39	45	43	33	32
0.75	193	182	168	189	167	181	160	151	146	142	136	139	126	118	107	98	67	36	27	22	25	31	52	50	46	24	30	27	19	17
0.70	199	185	166	186	178	170	157	164	151	147	132	137	125	120	122	117	104	93	87	57	26	20	18	14	13	10	8	6	3	2
0.65	195	190	184	204	197	168	173	162	154	150	135	132	131	127	121	113	111	106	103	101	99	96	89	86	81	77	68	62	56	49
0.60	199	188	202	205	201	177	171	161	156	148	140	124	114	104	108	97	89	78	80	78	74	71	64	63	60	51	41	37	34	35

Figure 17. *Cont.*

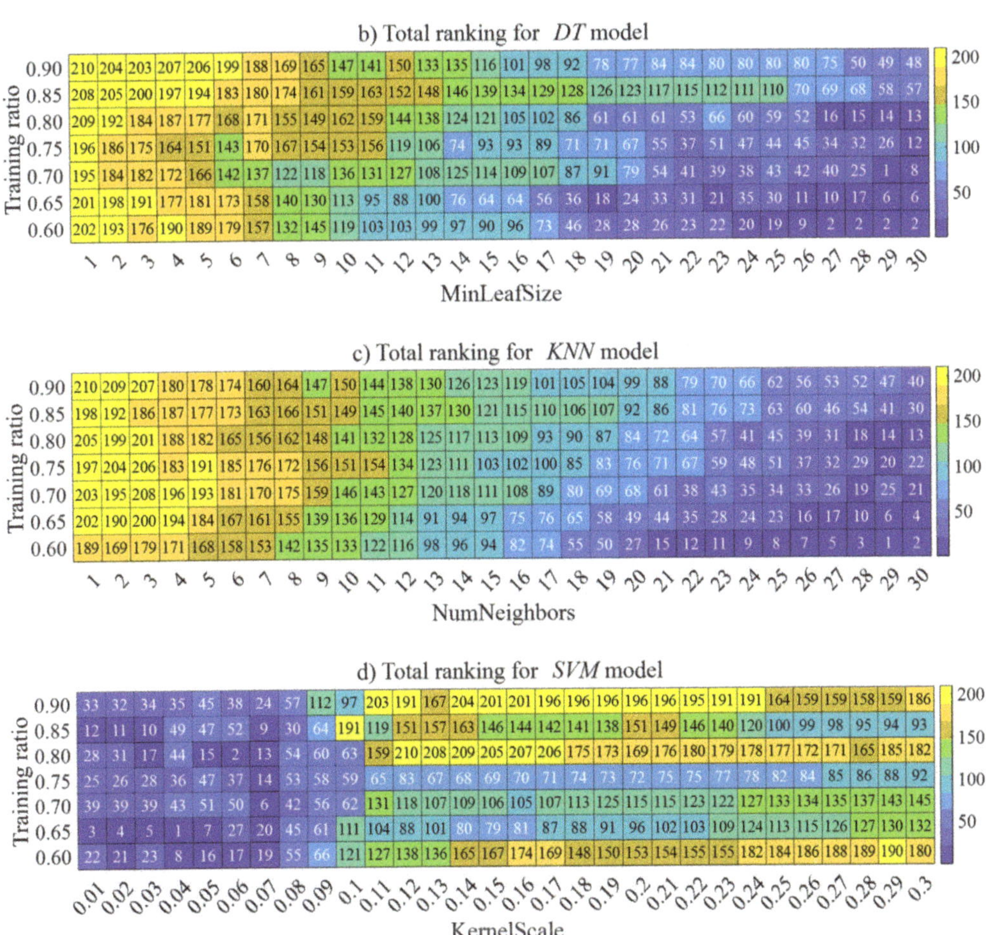

Figure 17. The ranking of 210 cases in each ML model for classifying failure modes.

Table 4 summarizes the calculated indicators for measuring the performance of different ML models in identifying the failure modes of RC columns. Noting that the values outside parentheses were for the training phase, while the values in parentheses were for the testing phase. Once again, it can be observed that the KNNs model showed to be the optimal technique with the accuracy, sensitivity, specificity, and AUC mostly close to 1.0. Additionally, SVM was also a good model for identifying failure modes of the columns.

Table 4. Measuring indicators of classifying models.

Model	Accuracy	Sensitivity	Specificity	AUC
NBs	0.667 (0.653)	0.838 (0.940)	0.754 (0.710)	0.666 (0.807)
DT	0.954 (0.890)	0.957 (0.909)	0.986 (0.955)	0.954 (0.966)
KNNs	1.000 (0.980)	1.000 (1.000)	1.000 (1.000)	1.000 (0.970)
SVM	1.000 (0.905)	1.000 (0.970)	1.000 (0.940)	1.000 (0.951)

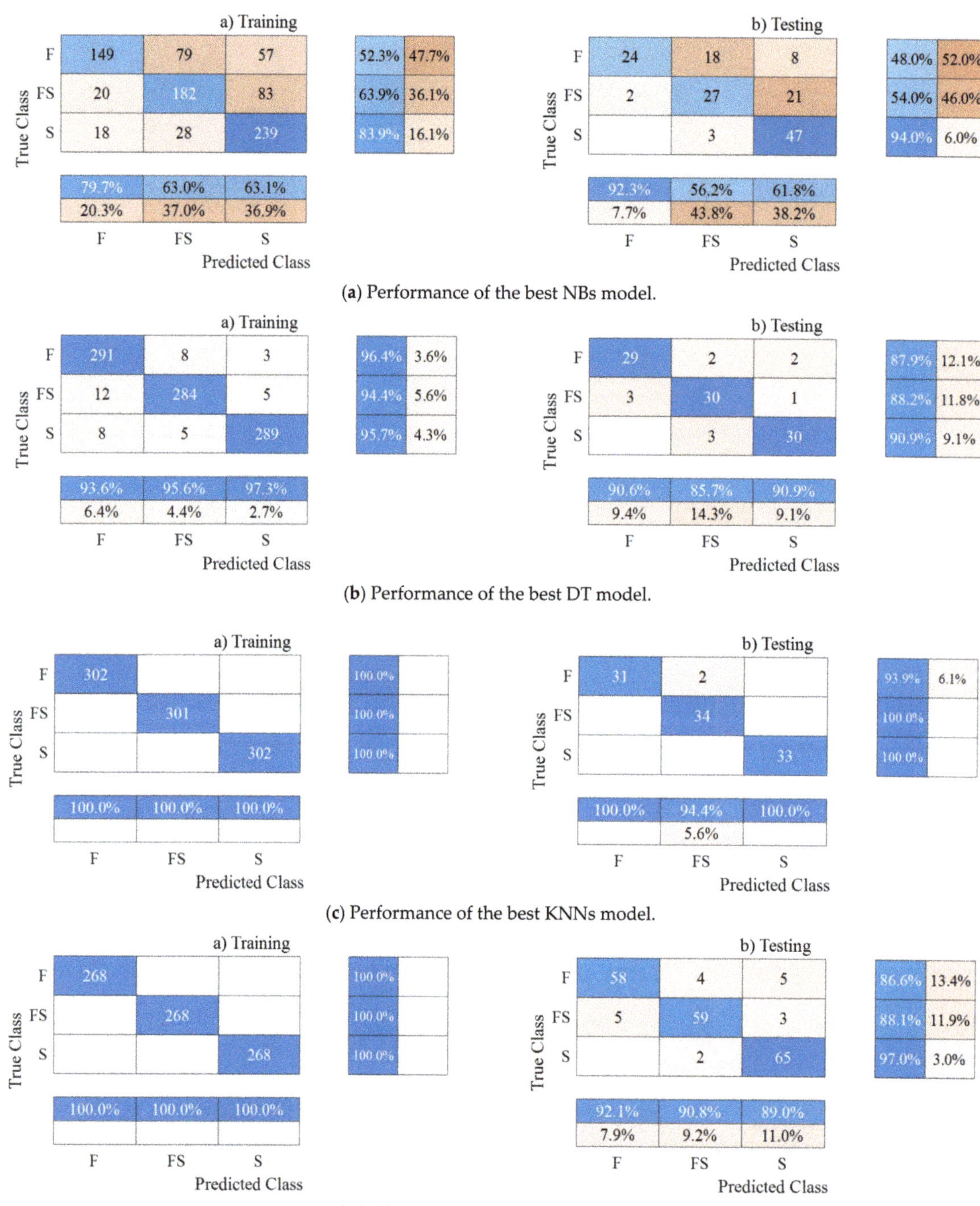

Figure 18. Confusion matrix of classified failure modes using different models.

6. GUI Tool for Predicting Shear Strength and Failure Modes of Rectangular RC Columns

To apply the proposed ML models in specific problems, a convenient tool needs to be established. We developed a Graphical User Interface (GUI) in MATLAB for facilitating failure mode identification as well as for the prediction of the shear strength of rectangular RC columns, as shown in Figure 19. Ten input parameters need to be provided. The shear strength and failure modes of the column are readily obtained by clicking on the 'Start Predict' button after entering the input parameters. It takes a few seconds to achieve the predictive results. It should be noted that the GUI tool is provided freely at the link: https://github.com/duyduan1304/GUI_RC_Columns (accessed on 12 July 2022).

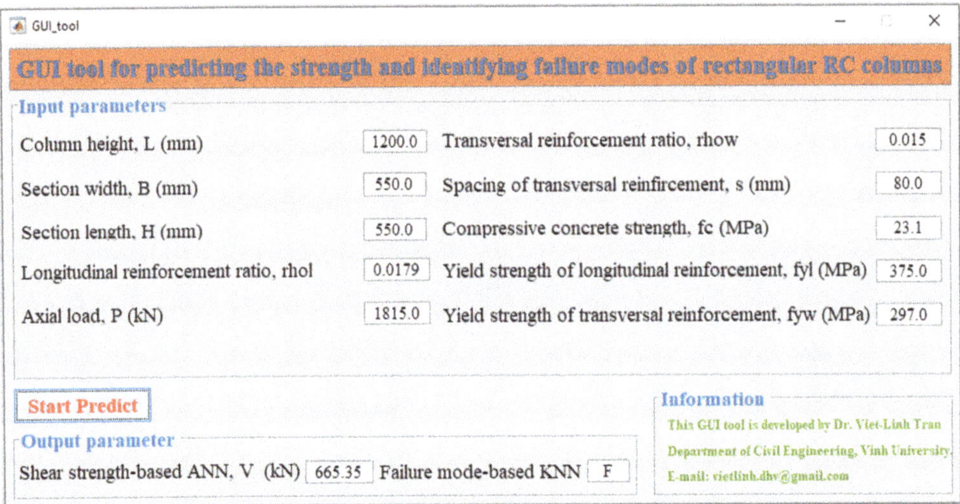

Figure 19. GUI for predicting the shear strength and identifying failure modes of rectangular RC columns.

7. Conclusions

The shear strength and failure modes of rectangular reinforced concrete (RC) columns were predicted using six novel Machine Learning (ML) techniques, which were developed based on a set of 541 experimental results. The six used ML models included Multivariate Adaptive Regression Splines (MARSs), Naïve Bayes (NBs), K-nearest Neighbors (KNNs), Decision Tree (DT), and Support Vector Machine (SVM), and Artificial Neural Network (ANN). Among these, the MARSs, KNN, DT, SVM, and ANN models were employed to predict the shear strength, while the NBs, KNNs, DT, and SVM models were used for classifying failure modes. The following conclusions are drawn:

- The ANN and KNNs models predicted the shear strength of rectangular RC columns more accurately than that of existing formulas with an R^2 value larger than 0.99.
- Used ML models in this study identified the failure modes of rectangular RC columns precisely. Among them, the KNNs algorithm showed to be the optimal method in classifying the failure modes of rectangular RC column with a high accuracy of almost 100%.
- A practical GUI tool was developed and readily applied for predicting the shear strength and identifying failure modes of rectangular RC columns in the design process and structural performance evaluation.

It should be noted that the findings of this study focus on the rectangular columns. A consideration of circular columns should be conducted in a future work.

Author Contributions: Conceptualization, V.-L.T. and D.-D.N.; methodology, V.-L.T. and D.-D.N.; software, V.-L.T.; validation, D.-D.N., V.-T.P. and V.-Q.N.; formal analysis, V.-L.T. and V.-T.P.; investigation, V.-Q.N.; resources, D.-D.N.; data curation, V.-L.T. and V.-Q.N.; writing—original draft preparation, V.-L.T. and D.-D.N.; writing—review and editing, V.-L.T. and D.-D.N.; visualization, V.-T.P. and V.-Q.N.; supervision, D.-D.N.; project administration, D.-D.N. All authors have read and agreed to the published version of the manuscript.

Funding: This research was supported by the Ministry of Education and Training of Vietnam, grant number B2022-TDV-09.

Data Availability Statement: All the data supporting the key findings of this paper are presented in the figures and tables of the article. The list of databases can be downloaded at: https://github.com/duyduan1304/recRC-Columns_541data, accessed on 12 July 2022. Requests for other data will be considered by the corresponding author.

Acknowledgments: The authors acknowledge the support of the Ministry of Education and Training of Vietnam, grant number B2022-TDV-09.

Conflicts of Interest: The authors declare no conflict of interest.

References

1. *ASCE/SEI-41-06*; Seismic Rehabilitation of Existing Buildings. Seismic Rehabilitation Standards Committee, American Society of Civil Engineers: Reston, VA, USA, 2007.
2. *ACI-318-14*; Building Code Requirements for Structural Concrete and Commentary. American Concrete Institute: Farmington Hills, MI, USA, 2014.
3. *EC8*; Eurocode 8: Design of Structures for Earthquake Resistance—Part 1: General Rules. Seismic Actions and Rules for Buildings. European Committee for Standardization: Brussels, Belgium, 2004.
4. CSA. *Design of Concrete Structures A23 3-14*; Canadian Standards Association: Rexdale, ON, Canada, 2014.
5. FEMA. *NEHRP 273: Guidelines for the Seismic Rehabilitation of Buildings*; Federal Emergency Management Agency: Washington, DC, USA, 1997.
6. Ascheim, M.; Moehle, J. *Shear Strength and Deformability of RC Bridge Columns subjected to Inelastic Cyclic Displacements*; Technical report no. UCB/EERC-92/04; University of California at Berkeley: Berkeley, CA, USA, 1992.
7. Priestley, M.N.; Verma, R.; Xiao, Y. Seismic shear strength of reinforced concrete columns. *J. Struct. Eng.* **1994**, *120*, 2310–2329. [CrossRef]
8. Kowalsky, M.J.; Priestley, M.N. Improved analytical model for shear strength of circular reinforced concrete columns in seismic regions. *ACI Struct. J.* **2000**, *97*, 388–396.
9. Sezen, H.; Moehle, J.P. Shear strength model for lightly reinforced concrete columns. *J. Struct. Eng.* **2004**, *130*, 1692–1703. [CrossRef]
10. Biskinis, D.E.; Roupakias, G.K.; Fardis, M.N. Degradation of shear strength of reinforced concrete members with inelastic cyclic displacements. *ACI Struct. J.* **2004**, *101*, 773–783.
11. Cassese, P.; De Risi, M.T.; Verderame, G.M. A modelling approach for existing shear-critical RC bridge piers with hollow rectangular cross section under lateral loads. *Bul. Earthq. Eng.* **2019**, *17*, 237–270. [CrossRef]
12. Tran, C.T.N.; Li, B. Shear strength model for reinforced concrete columns with low transverse reinforcement ratios. *Adv. Struct. Eng.* **2014**, *17*, 1373–1385. [CrossRef]
13. Caglar, N. Neural network based approach for determining the shear strength of circular reinforced concrete columns. *Constr. Build. Mater.* **2009**, *23*, 3225–3232. [CrossRef]
14. Ketabdari, H.; Karimi, F.; Rasouli, M. Shear strength prediction of short circular reinforced-concrete columns using soft computing methods. *Adv. Struct. Eng.* **2020**, *23*, 3048–3061. [CrossRef]
15. Fiore, A.; Marano, G.C.; Laucelli, D.; Monaco, P. Evolutionary modeling to evaluate the shear behavior of circular reinforced concrete columns. *Adv. Civil. Eng.* **2014**, *2014*, 1–14. [CrossRef]
16. Said, A.; Gordon, N. Predicting Shear Strength of RC Columns Using Artificial Neural Networks. *J. Build. Mater. Struct.* **2019**, *6*, 64–76. [CrossRef]
17. Aval, S.B.; Ketabdari, H.; Gharebaghi, S.A. Estimating shear strength of short rectangular reinforced concrete columns using nonlinear regression and gene expression programming. *Structures* **2017**, *12*, 13–23. [CrossRef]
18. Yu, B.; Liu, S.; Li, B. Probabilistic calibration for shear strength models of reinforced concrete columns. *J. Struct. Eng.* **2019**, *145*, 04019026. [CrossRef]
19. Inel, M. Modeling ultimate deformation capacity of RC columns using artificial neural networks. *Eng. Struct.* **2007**, *29*, 329–335. [CrossRef]
20. Naderpour, H.; Mirrashid, M. Moment capacity estimation of spirally reinforced concrete columns using ANFIS. *Complex Intell. Syst.* **2020**, *6*, 97–107. [CrossRef]

21. Naderpour, H.; Parsa, P.; Mirrashid, M. Innovative Approach for Moment Capacity Estimation of Spirally Reinforced Concrete Columns Using Swarm Intelligence–Based Algorithms and Neural Network. *Pract. Period. Struct. Des. Constr.* **2021**, *26*, 04021043. [CrossRef]

22. Feng, D.-C.; Cetiner, B.; Azadi Kakavand, M.R.; Taciroglu, E. Data-driven approach to predict the plastic hinge length of reinforced concrete columns and its application. *J. Struct. Eng.* **2021**, *147*, 04020332. [CrossRef]

23. Lee, C.S.; Park, Y.; Jeon, J.-S. Model parameter prediction of lumped plasticity model for nonlinear simulation of circular reinforced concrete columns. *Eng. Struct.* **2021**, *245*, 112820. [CrossRef]

24. Aldabagh, S.; Hossain, F.; Alam, M.S. Simplified Predictive Expressions of Drift Limit States for Reinforced Concrete Circular Bridge Columns. *J. Struct. Eng.* **2022**, *148*, 04021285. [CrossRef]

25. Azadi Kakavand, M.R.; Sezen, H.; Taciroglu, E. Data-driven models for predicting the shear strength of rectangular and circular reinforced concrete columns. *J. Struct. Eng.* **2021**, *147*, 04020301. [CrossRef]

26. Feng, D.-C.; Liu, Z.-T.; Wang, X.-D.; Jiang, Z.-M.; Liang, S.-X. Failure mode classification and bearing capacity prediction for reinforced concrete columns based on ensemble machine learning algorithm. *Adv. Eng. Inf.* **2020**, *45*, 101126. [CrossRef]

27. Asteris, P.G.; Armaghani, D.J.; Hatzigeorgiou, G.D.; Karayannis, C.G.; Pilakoutas, K. Predicting the shear strength of reinforced concrete beams using Artificial Neural Networks. *Comput. Concr.* **2019**, *24*, 469–488.

28. Cakiroglu, C.; Islam, K.; Bekdaş, G.; Kim, S.; Geem, Z.W. Interpretable Machine Learning Algorithms to Predict the Axial Capacity of FRP-Reinforced Concrete Columns. *Materials* **2022**, *15*, 2742. [CrossRef] [PubMed]

29. Marani, A.; Nehdi, M.L. Predicting shear strength of FRP-reinforced concrete beams using novel synthetic data driven deep learning. *Eng. Struct.* **2022**, *257*, 114083. [CrossRef]

30. Wakjira, T.G.; Ebead, U.; Alam, M.S. Machine learning-based shear capacity prediction and reliability analysis of shear-critical RC beams strengthened with inorganic composites. *Case Stud. Construct. Mater.* **2022**, *16*, e01008. [CrossRef]

31. De Domenico, D.; Ricciardi, G. Shear strength of RC beams with stirrups using an improved Eurocode 2 truss model with two variable-inclination compression struts. *Eng. Struct.* **2019**, *198*, 109359. [CrossRef]

32. Quaranta, G.; De Domenico, D.; Monti, G. Machine-learning-aided improvement of mechanics-based code-conforming shear capacity equation for RC elements with stirrups. *Eng. Struct.* **2022**, *267*, 114665. [CrossRef]

33. Zhu, L.; Elwood, K.; Haukaas, T. Classification and seismic safety evaluation of existing reinforced concrete columns. *J. Struct. Eng.* **2007**, *133*, 1316–1330. [CrossRef]

34. Qi, Y.-l.; Han, X.-l.; Ji, J. Failure mode classification of reinforced concrete column using Fisher method. *J. Cent. South Univ.* **2013**, *20*, 2863–2869. [CrossRef]

35. Pekelnicky, R.; Engineers, S.; Chris Poland, S.; Engineers, N. ASCE 41-13: Seismic evaluation and retrofit rehabilitation of existing buildings. In Proceedings of the SEAOC 2012, Santa Fe, NM, USA, 12–15 September 2012.

36. Ma, Y.; Gong, J.-X. Probability identification of seismic failure modes of reinforced concrete columns based on experimental observations. *J. Earthq. Eng.* **2018**, *22*, 1881–1899. [CrossRef]

37. Ghee, A.B.; Priestley, M.N.; Paulay, T. Seismic shear strength of circular reinforced concrete columns. *ACI Struct. J.* **1989**, *86*, 45–59.

38. Ning, C.-L.; Feng, D.-C. Probabilistic indicator to classify the failure mode of reinforced-concrete columns. *Mag. Concr. Res.* **2019**, *71*, 734–748. [CrossRef]

39. Berry, M.; Parrish, M.; Eberhard, M. *PEER Structural Performance Database User's Manual (Version 1.0)*; University of California: Berkeley, CA, USA, 2004.

40. Mangalathu, S.; Jeon, J.-S. Machine learning–based failure mode recognition of circular reinforced concrete bridge columns: Comparative study. *J. Struct. Eng.* **2019**, *145*, 04019104. [CrossRef]

41. Mangalathu, S.; Hwang, S.-H.; Jeon, J.-S. Failure mode and effects analysis of RC members based on machine-learning-based SHapley Additive exPlanations (SHAP) approach. *Eng. Struct.* **2020**, *219*, 110927. [CrossRef]

42. Lundberg, S.M.; Lee, S.-I. A unified approach to interpreting model predictions. *Adv. Neural Inf. Process. Syst.* **2017**, *30*, 4765–4774.

43. Naderpour, H.; Mirrashid, M.; Parsa, P. Failure mode prediction of reinforced concrete columns using machine learning methods. *Eng. Struct.* **2021**, *248*, 113263. [CrossRef]

44. Ghannoum, W.; Sivaramakrishnan, B.; Pujol, S.; Catlin, A.; Fernando, S.; Yoosuf, N. ACI 369 rectangular column database. Network for Earthquake Engineering Simulation (Database), Dataset. 2012. Available online: https://datacenterhub.org/dataviewer/view/neesdatabases:db/aci_369_rectangular_column_database/ (accessed on 13 February 2021).

45. Belkacem, M.A.; Bechtoula, H.; Bourahla, N.; Belkacem, A.A. Effect of axial load and transverse reinforcements on the seismic performance of reinforced concrete columns. *Front. Struct. Civ. Eng.* **2019**, *13*, 831–851. [CrossRef]

46. Wang, D.; Li, H.-N.; Li, G. Experimental study on dynamic mechanical properties of reinforced concrete column. *J. Reinf. Plast. Compos.* **2013**, *32*, 1793–1806. [CrossRef]

47. Xiao, J.; Zhang, C. Seismic behavior of RC columns with circular, square and diamond sections. *Constr. Build. Mater.* **2008**, *22*, 801–810. [CrossRef]

48. Rodrigues, H.; Furtado, A.; Arêde, A. Behavior of rectangular reinforced-concrete columns under biaxial cyclic loading and variable axial loads. *J. Struct. Eng.* **2016**, *142*, 04015085. [CrossRef]

49. Melo, J.; Varum, H.; Rossetto, T. Experimental cyclic behaviour of RC columns with plain bars and proposal for Eurocode 8 formula improvement. *Eng. Struct.* **2015**, *88*, 22–36. [CrossRef]

50. Yun, H.W. *Full-Scale Experimental and Analytical Studies on High-Strength Concrete Columns*; University of Southern California: Los Angeles, CA, USA, 2003.
51. Ongsupankul, S.; Kanchanalai, T.; Kawashima, K. Behavior of reinforced concrete bridge pier columns subjected to moderate seismic load. *Sci. Asia* **2007**, *33*, 175–185. [CrossRef]
52. Ho, J.C.M. Experimental tests on high-strength concrete columns subjected to combined medium axial load and flexure. *Adv. Struct. Eng.* **2012**, *15*, 1359–1374. [CrossRef]
53. Mo, Y.-L.; Wang, S. Seismic behavior of RC columns with various tie configurations. *J. Struct. Eng.* **2000**, *126*, 1122–1130. [CrossRef]
54. Wu, D.; Ding, Y.; Su, J.; Li, Z.-X.; Zong, L.; Feng, K. Effects of tie detailing configurations on reinforcement buckling and seismic performance of high-strength RC columns. *Soil Dyn. Earthq. Eng.* **2021**, *147*, 106791. [CrossRef]
55. Woodward, K.A.; Jirsa, J.O. Influence of reinforcement on RC short column lateral resistance. *J. Struct. Eng.* **1984**, *110*, 90–104. [CrossRef]
56. Lam, S.S.E.; Wu, B.; Wong, Y.; Wang, Z.; Liu, Z.; Li, C. Drift capacity of rectangular reinforced concrete columns with low lateral confinement and high-axial load. *J. Struct. Eng.* **2003**, *129*, 733–742. [CrossRef]
57. Hwang, S.-K.; Yun, H.-D. Effects of transverse reinforcement on flexural behaviour of high-strength concrete columns. *Eng. Struct.* **2004**, *26*, 1–12. [CrossRef]
58. Ahn, J.-M.; Shin, S.-W. An evaluation of ductility of high-strength reinforced concrete columns subjected to reversed cyclic loads under axial compression. *Mag. Conc. Res.* **2007**, *59*, 29–44. [CrossRef]
59. Woods, J.M.; Kiousis, P.D.; Ehsani, M.R.; Saadatmanesh, H.; Fritz, W. Bending ductility of rectangular high strength concrete columns. *Eng. Struct.* **2007**, *29*, 1783–1790. [CrossRef]
60. Marefat, M.; Khanmohammadi, M.; Bahrani, M.; Goli, A. Experimental assessment of reinforced concrete columns with deficient seismic details under cyclic load. *Adv. Struct. Eng.* **2006**, *9*, 337–347. [CrossRef]
61. Xiao, X.; Guan, F.; Yan, S. Use of ultra-high-strength bars for seismic performance of rectangular high-strength concrete frame columns. *Mag. Conc. Res.* **2008**, *60*, 253–259. [CrossRef]
62. Bae, S.; Bayrak, O. Plastic hinge length of reinforced concrete columns. *ACI Struct. J.* **2008**, *105*, 290.
63. Tran, C.T.N. *Experimental and Analytical Studies on the Seismic Behavior of Reinforced Concrete Columns with Light Transverse Reinforcement*; Nanyang Technological University Singapore: Singapore, 2010.
64. Ou, Y.-C.; Kurniawan, D.P.; Handika, N. Shear behavior of reinforced concrete columns with high-strength steel and concrete under low axial load. *ACI Spec. Publ.* **2013**, *293*, 1–12.
65. Martirossyan, A.; Xiao, Y. Flexural-shear behavior of high-strength concrete short columns. *Earthq. Spectra* **2001**, *17*, 679–695. [CrossRef]
66. Li, Y.-A.; Huang, Y.-T.; Hwang, S.-J. Seismic response of reinforced concrete short columns failed in shear. *ACI Struct. J.* **2014**, *111*, 945. [CrossRef]
67. Nakamura, T.; Yoshimura, M. Gravity load collapse of reinforced concrete columns with brittle failure modes. *J. Asian Archit. Build. Eng.* **2002**, *1*, 21–27. [CrossRef]
68. Nakamura, T.; Yoshimura, M. Gravity load collapse of reinforced concrete columns with decreased axial load. In Proceedings of the 2nd European Conference on Earthquake Engineering and Seismology, Istanbul, Turkey, 25–29 August 2014.
69. Popa, V.; Cotofana, D.; Vacareanu, R. Effective stiffness and displacement capacity of short reinforced concrete columns with low concrete quality. *Bull. Earthq. Eng.* **2014**, *12*, 2705–2721. [CrossRef]
70. Jin, C.; Pan, Z.; Meng, S.; Qiao, Z. Seismic behavior of shear-critical reinforced high-strength concrete columns. *J. Struct. Eng.* **2015**, *141*, 04014198. [CrossRef]
71. EL-Attar, M.M.; El-Karmoty, H.Z.; EL-Moneim, A.A. The behavior of ultra-high-strength reinforced concrete columns under axial and cyclic lateral loads. *HBRC. J.* **2016**, *12*, 284–295. [CrossRef]
72. Eom, T.-S.; Kang, S.-M.; Park, H.-G.; Choi, T.-W.; Jin, J.-M. Cyclic loading test for reinforced concrete columns with continuous rectangular and polygonal hoops. *Eng. Struct.* **2014**, *67*, 39–49. [CrossRef]
73. Elwood, K.J.; Moehle, J.P. Drift capacity of reinforced concrete columns with light transverse reinforcement. *Earthq. Spectra.* **2005**, *21*, 71–89. [CrossRef]
74. Sezen, H. Seismic Response and Modeling of Reinforced Concrete Building Columns. Ph.D. Thesis, Department of Civil and Environmental Engineering, University of California, Berkeley, CA, USA, 2002.
75. Esaki, F. Reinforcing effect of steel plate hoops on ductility of R/C square columns. In Proceedings of the 11th World Conference on Earthquake Engineering, Acapulco, Mexico, 23–28 June 1996; pp. 23–29.
76. Li, X.; Park, R.; Tanaka, H. Effects of variations in axial load level on the strength and ductility of reinforced concrete columns. In Proceedings of the Pacific Conference on Earthquake Engineering, Auckland, New Zealand, 20–23 November 1991.
77. Yalcin, C. Seismic evaluation and retrofit of existing reinforced concrete bridge columns. Ph.D. Thesis, Department of Civil Engineering, University of Ottawa, Ottawa, ON, USA, 1997.
78. Opabola, E.A.; Elwood, K.J.; Oliver, S. Deformation capacity of reinforced concrete columns with smooth reinforcement. *Bull. Earthq. Eng.* **2019**, *17*, 2509–2532. [CrossRef]
79. Goksu, C.; Yilmaz, H.; Chowdhury, S.; Orakcal, K.; Ilki, A. The effect of lap splice length on the cyclic lateral load behavior of RC members with low-strength concrete and plain bars. *Adv. Struct. Eng.* **2014**, *17*, 639–658. [CrossRef]

80. Zhang, Y.; Zheng, S.; Rong, X.; Dong, L.; Zheng, H. Seismic performance of reinforced concrete short columns subjected to freeze–thaw cycles. *Appl. Sci.* **2019**, *9*, 2708. [CrossRef]

81. Verderame, G.M.; Fabbrocino, G.; Manfredi, G. Seismic response of rc columns with smooth reinforcement. Part I: Monotonic tests. *Eng. Struct.* **2008**, *30*, 2277–2288. [CrossRef]

82. Verderame, G.M.; Fabbrocino, G.; Manfredi, G. Seismic response of rc columns with smooth reinforcement. Part II: Cyclic tests. *Eng. Struct.* **2008**, *30*, 2289–2300. [CrossRef]

83. Bousias, S.; Spathis, A.-L.; Fardis, M.N. Seismic retrofitting of columns with lap spliced smooth bars through FRP or concrete jackets. *J. Earthq. Eng.* **2007**, *11*, 653–674. [CrossRef]

84. Arani, K.K.; Di Ludovico, M.; Marefat, M.S.; Prota, A.; Manfredi, G. Lateral response evaluation of old type reinforced concrete columns with smooth bars. *ACI Struct. J.* **2014**, *111*, 827–838. [CrossRef]

85. Di Ludovico, M.; Verderame, G.; Prota, A.; Manfredi, G.; Cosenza, E. Cyclic behavior of nonconforming full-scale RC columns. *J. Struct. Eng.* **2014**, *140*, 04013107. [CrossRef]

86. Ilki, A.; Demir, C.; Bedirhanoglu, I.; Kumbasar, N. Seismic retrofit of brittle and low strength RC columns using fiber reinforced polymer and cementitious composites. *Adv. Struct. Eng.* **2009**, *12*, 325–347. [CrossRef]

87. Pham, T.P.; Li, B. Seismic performance of reinforced concrete columns with plain longitudinal reinforcing bars. *ACI Struc. J.* **2014**, *111*, 561. [CrossRef]

88. Arani, K.K.; Marefat, M.S.; Amrollahi-Biucky, A.; Khanmohammadi, M. Experimental seismic evaluation of old concrete columns reinforced by plain bars. *The Struct. Des. Tall Spec. Build.* **2013**, *22*, 267–290. [CrossRef]

89. Shi, Q.; Ma, L.; Wang, Q.; Wang, B.; Yang, K. Seismic performance of square concrete columns reinforced with grade 600 MPa longitudinal and transverse reinforcement steel under high axial load. *Structures* **2021**, *32*, 1955–1970. [CrossRef]

90. Zhang, J.; Cai, R.; Li, C.; Liu, X. Seismic behavior of high-strength concrete columns reinforced with high-strength steel bars. *Eng. Struct.* **2020**, *218*, 110861. [CrossRef]

91. Dinh, N.H.; Park, S.-H.; Choi, K.-K. Seismic performance of reinforced concrete columns retrofitted by textile-reinforced mortar jackets. *Struct. Infrastruct. Eng.* **2020**, *16*, 1364–1381. [CrossRef]

92. Kim, C.-G.; Park, H.-G.; Eom, T.-S. Effects of Type of Bar Lap Splice on Reinforced Concrete Columns Subjected to Cyclic Loading. *ACI Struct. J.* **2019**, *116*, 183–194. [CrossRef]

93. Hwang, H.-J.; Noh, J.-O.; Park, H.-G. Structural capacity of reinforced concrete columns with U-shaped transverse bars. *Eng. Struct.* **2020**, *216*, 110686. [CrossRef]

94. Choi, K.-K.; Truong, G.T.; Kim, J.-C. Seismic performance of lightly shear reinforced RC columns. *Eng. Struct.* **2016**, *126*, 490–504. [CrossRef]

95. Kim, C.-G.; Eom, T.-S.; Park, H.-G. Cyclic Load Test of Reinforced Concrete Columns with V-Shaped Ties. *ACI Struct. J.* **2020**, *117*, 91–102.

96. Sezen, H.; Moehle, J.P. Seismic tests of concrete columns with light transverse reinforcement. *ACI Struct. J.* **2006**, *103*, 842.

97. Li, Y.; Cao, S.; Jing, D. Concrete Columns Reinforced with High-Strength Steel Subjected to Reversed Cycle Loading. *ACI Struct. J.* **2018**, *115*, 10378–11048. [CrossRef]

98. Barrera, A.; Bonet, J.; Romero, M.L.; Miguel, P. Experimental tests of slender reinforced concrete columns under combined axial load and lateral force. *Eng. Struct.* **2011**, *33*, 3676–3689. [CrossRef]

99. Melek, M.; Wallace, J.W. Cyclic behavior of columns with short lap splices. *ACI Struct. J.* **2004**, *101*, 802–811.

100. Kim, C.-G.; Park, H.-G.; Eom, T.-S. Seismic performance of reinforced concrete columns with lap splices in plastic hinge region. *ACI Struct. J.* **2018**, *115*, 235–245. [CrossRef]

101. Yang, W.X.; Shi, Q.X.; Sun, H.X. Experimental Studies on Seismic Performance of High Strength Reinforced Concrete Columns. *Appl. Mech. Mater. Trans. Tech. Publ.* **2012**, *166*, 919–926. [CrossRef]

102. Zhang, X.; Li, T.; ZHang, L. Experimental study on the seismic behavior of reinforced concrete short columns with high-strength longitudinal reinforcements. In Proceedings of the 2011 Second International Conference on Mechanic Automation and Control Engineering, Inner Mongolia, China, 15–17 July 2011; IEEE: Piscataway, NJ, USA, 2011; pp. 5807–5810.

103. Nojavan, A. *Performance of Full-Scale Reinforced Concrete Columns Subjected to Extreme Earthquake Loading*; University of Minnesota: Minneapolis, MN, USA, 2015.

104. Kotu, V.; Deshpande, B. *Data Science: Concepts and Practice*; Morgan Kaufmann: Campbridge, MA, USA, 2018.

105. Breiman, L.; Friedman, J.; Stone, C.J.; Olshen, R.A. *Classification and Regression Trees*; CRC Press: Boca Raton, FL, USA, 1984.

106. Vapnik, V. The support vector method of function estimation. In *Nonlinear Modeling*; Springer: Berlin/Heidelberg, Germany, 1998; pp. 55–85.

107. Nikbin, I.M.; Rahimi, S.; Allahyari, H. A new empirical formula for prediction of fracture energy of concrete based on the artificial neural network. *Eng. Fract. Mech.* **2017**, *186*, 466–482. [CrossRef]

108. Tran, V.-L.; Thai, D.-K.; Kim, S.-E. Application of ANN in predicting ACC of SCFST column. *Compos. Struct.* **2019**, *228*, 111332. [CrossRef]

109. Tran, V.-L.; Thai, D.-K.; Kim, S.-E. A new empirical formula for prediction of the axial compression capacity of CCFT columns. *Steel Compos. Struct.* **2019**, *33*, 181–194.

110. Tran, V.-L.; Kim, S.-E. A practical ANN model for predicting the PSS of two-way reinforced concrete slabs. *Eng. Comput* **2021**, *37*, 2303–2327. [CrossRef]

111. Tran, V.-L.; Thai, D.-K.; Nguyen, D.-D. Practical artificial neural network tool for predicting the axial compression capacity of circular concrete-filled steel tube columns with ultra-high-strength concrete. *Thin-Walled Struct.* **2020**, *151*, 106720. [CrossRef]
112. Nguyen, M.-S.T.; Thai, D.-K.; Kim, S.-E. Predicting the axial compressive capacity of circular concrete filled steel tube columns using an artificial neural network. *Steel Compos. Struct.* **2020**, *35*, 415–437.
113. Nguyen, D.-D.; Tran, V.-L.; Ha, D.-H.; Nguyen, V.-Q.; Lee, T.-H. A machine learning-based formulation for predicting shear capacity of squat flanged RC walls. *Structures* **2021**, *29*, 1734–1747. [CrossRef]
114. Silva, F.A.; Delgado, J.M.; Cavalcanti, R.S.; Azevedo, A.C.; Guimarães, A.S.; Lima, A.G. Use of nondestructive testing of ultrasound and artificial neural networks to estimate compressive strength of concrete. *Buildings* **2021**, *11*, 44. [CrossRef]
115. Sirimontree, S.; Keawsawasvong, S.; Ngamkhanong, C.; Seehavong, S.; Sangjinda, K.; Jearsiripongkul, T.; Thongchom, C.; Nuaklong, P. Neural Network-Based Prediction Model for the Stability of Unlined Elliptical Tunnels in Cohesive-Frictional Soils. *Buildings* **2022**, *12*, 444. [CrossRef]
116. Almasabha, G.; Alshboul, O.; Shehadeh, A.; Almuflih, A.S. Machine Learning Algorithm for Shear Strength Prediction of Short Links for Steel Buildings. *Buildings* **2022**, *12*, 775. [CrossRef]
117. Luat, N.V.; Lee, K.; Thai, D.K. Application of artificial neural networks in settlement prediction of shallow foundations on sandy soils. *Geomech. Eng.* **2020**, *20*, 385–397.
118. Friedman, J.H. Multivariate adaptive regression splines. *Ann. Stat.* **1991**, *19*, 1–67. [CrossRef]
119. Luat, N.V.; Nguyen, V.Q.; Lee, S.; Woo, S.; Lee, K. An evolutionary hybrid optimization of MARS model in predicting settlement of shallow foundations on sandy soils. *Geomech. Eng.* **2020**, *21*, 583–598.
120. Chawla, N.V.; Bowyer, K.W.; Hall, L.O.; Kegelmeyer, W.P. SMOTE: Synthetic minority over-sampling technique. *J. Artif. Intell. Res.* **2002**, *16*, 321–357. [CrossRef]

Article

An Ontology-Based Holistic and Probabilistic Framework for Seismic Risk Assessment of Buildings

Minze Xu, Peng Zhang, Chunyi Cui * and Jingtong Zhao

Department of Civil Engineering, Dalian Maritime University, Dalian 116026, China
* Correspondence: cuichunyi@dlmu.edu.cn

Abstract: To avoid over-reliance on the identification of building damage states post-earthquake in the seismic risk assessment process, an ontology-based holistic and probabilistic framework is proposed here for seismic risk prediction of buildings with various purposes and different damage states. Based on vulnerability analysis, the seismic risk probabilities of buildings are first obtained by considering the on-site seismic hazard. Taking economic losses and casualties as assessment indicators, a system for seismic risk assessment of buildings, OntoBSRA (Ontology for Building Seismic Risk Assessment), is then developed by combining ontology and semantic web rule language. A case study is carried out to demonstrate the application of the proposed framework and further validate the semantic web rules. The results show that the proposed framework can provide a holistic knowledge base that allows risk assessors or asset managers to predict the consequences of earthquakes effectively, thereby improving efficiency in decision-making.

Keywords: seismic vulnerability; seismic hazard; seismic risk probability; ontology; semantic web rule

Citation: Xu, M.; Zhang, P.; Cui, C.; Zhao, J. An Ontology-Based Holistic and Probabilistic Framework for Seismic Risk Assessment of Buildings. *Buildings* **2022**, *12*, 1391. https://doi.org/10.3390/buildings 12091391

Academic Editors: Rajesh Rupakhety and Dipendra Gautam

Received: 23 July 2022
Accepted: 31 August 2022
Published: 5 September 2022

Publisher's Note: MDPI stays neutral with regard to jurisdictional claims in published maps and institutional affiliations.

1. Introduction

Earthquake, as one of the most destructive natural disasters, causes huge economic losses and casualties annually. Taking the USA as an example, the annual economic loss caused by earthquakes is estimated to be USD 4.5 billion, and this estimate does not include casualties [1,2] Therefore, to analyze the seismic performance of structures for decision-making for more effective disaster prevention and mitigation measures, researchers have carried out extensive research using shaking table tests and time-history analyses on the deformation mechanism, weak links, and characteristics of mechanical responses of structures subjected to earthquakes [3–5].

It should be noted that the shaking table tests and the time-history analysis method can only obtain the mechanical performance of a structure under specific ground motions; they have limitations in analyzing the randomness of the seismic response caused by the uncertainty in both the ground motion and the structure (e.g., design, construction, size and material strength, etc.) [6]. Therefore, in order to fully consider the uncertainties in multiple factors, performance-based earthquake engineering (PBEE) research has been conducted based on probability theory, in which probability is used as the basic measure to estimate the seismic performance of a structure [7–10]. Koutsourelakis [11] used fragility curves to evaluate the seismic performance of a structure constructed on a saturated sand deposit and provided confidence intervals of the vulnerability by combining Bayesian theory with the Markov Chain Monte Carlo technique. Cimellaro and Reinhorn [12] used the combination of acceleration and inter-story drift as a response variable and defined a generalized multidimensional limit state function. Then, multidimensional fragility curves were established for a California hospital considering the correlations among the thresholds. Michel et al. [13] obtained fragility curves for a reinforced concrete (RC) structure based on two complete methodologies. One is to use a multiple degrees of freedom system considering higher modes, while the other is to use a single degree of freedom model considering the fundamental mode. Furthermore, Ruggieri and Gentile et al. [14–16] discussed the trade-off between complexity (modelling effort and computational time) and accuracy in seismic fragility analysis of RC structures so as to provide proper methods of seismic fragility for practical PBEE assessment. Bakhshi and Asadi [17] quantified the impact of overall structural ductility on failure probability using vulnerability curves. Karapetrou et al. [18] studied the influence of corrosion on the seismic performance of RC structures using time-dependent fragility curves based on the incremental dynamic analysis (IDA) method. Based on vulnerability analysis, Yu et al. [19] took four groups of buildings with low-to-medium heights corresponding to 3, 5, 8, and 10 stories as examples in order to investigate the influence of seismic design levels on the seismic performance of RC moment-resisting frame buildings designed according to the provisions of the current Chinese code for seismic design of buildings. Generally speaking, according to Dowrick [20], seismic risks can be divided into three parts: seismic vulnerability, seismic hazard, and seismic loss. Therefore, assuming that the seismic hazard function can be approximated by the extreme value type II distribution, Cornell et al. [21] conducted vulnerability analysis and derived an analytical solution to the seismic risk probability for structures with different damage states using convolving seismic vulnerability function and seismic hazard function based on the full probability theory. Along with later collaborators, they further established the second-generation PBEE theory to establish a probabilistic framework for seismic performance evaluation of structures [22–24]. Furthermore, Lu et al. [25] clarified the distinctions and connections of five seismic fragility models in the second-generation PBEE theory, namely, the seismic demand fragility model, the seismic capacity fragility model, the seismic damage fragility model, the seismic loss fragility model, and the seismic decision fragility model, and put forward the concepts of forward and backward PBEE theory to integrate the traditional seismic risk theory and the second-generation PBEE theory into a unified framework.

According to ISO 31000-2018 [26], risk is defined as the effect of uncertainty on the objectives. However, the previous theoretical studies on seismic risk theory for structures only predict the probability of a structure with different damage states subjected to random seismic loads. They cannot quantitatively evaluate the consequences caused by earthquakes from a macroscopic point of view, such as economic losses and casualties, while traditional seismic risk assessment heavily relies on the investigation of actual earthquake damage. Wang et al. [27] conducted statistical analysis of the measured damage loss from the Tianjin and Lancang–Gengma earthquakes, determining the ratio of the indirect losses to the initial construction cost of buildings with different damage states. Spence et al. [28] proposed a global earthquake vulnerability estimation system to determine the mean damage ratio for buildings with different purposes under earthquake loads. Sahar et al. [29] developed an algorithm for automatic extraction and identification of two-dimensional building shape information using aerial images and geographic information systems, and further evaluated the seismic risks of cities. Lu et al. [30] proposed a near real-time method for estimation of building seismic losses based on combined satellite or aerial images and dynamic nonlinear time history analysis. Xiong et al. [31] put forward a seismic damage evaluation method for regional buildings on the basis of drones and a convolutional neural network, which was capable of accurately evaluating the seismic damage in regional buildings through collected detailed damage information on buildings. Using photographs of buildings, Ruggieri et al. [32] proposed a VULMA (vulnerability analysis using machine-learning) framework based on machine learning to evaluate seismic vulnerability of existing buildings. More accurate seismic risk assessment results can be obtained by applying the methods mentioned above. However, these methods are heavily dependent on the identification of structural damage states after a specific earthquake, and the statistical process is tedious and complicated. In addition, these methods cannot take into account earthquake uncertainty, and there is a lack of a unified seismic risk assessment framework to predict earthquake losses for existing buildings with various purposes and different damage states.

Ontology, as a new semantic technology, can be used for knowledge sharing in different areas. Its semantic structures and ability for logical inference provide an effective method for integrated decision-making based on multi-objective knowledge. Ever since its emergence, ontology technology has been widely applied in many aspects, such as agriculture, biology, economy, medicine, construction, etc. [33]. Tserng et al. [34] proposed an ontology-based risk management method for managing the risks in construction stages. Taking into account the relationship between risk sources and cost overruns, Fidan et al. [35] proposed an ontology-based model to predict cost overruns. Scheuer et al. [36] established an ontology-based knowledge base for flood risk management using the accessibility and repeatability of multi-criteria risk assessment of floods. Du et al. [37] combined the hierarchical clustering method with ontology and developed an integrated system for risk information of surface subsidence of underground tunnels. Ding et al. [38] proposed an information management framework for construction risks in the Building Information Modeling (BIM) environment by making full use of the advantages of the BIM, ontology, and semantic web technologies. Meng et al. [39] developed an ontology of a pile integrity evaluation system for quantitative identification and qualitative evaluation of piles with defects combined with an analytical methodology for pile vibrations. Ontology is capable of integrating multi-objective knowledge into a unified system. However, no studies have been conducted to apply ontology in order to develop a knowledge base for seismic risk assessment of buildings.

Based on an extensive literature review, it is paramount to develop an efficient building seismic risk prediction framework based on probability and ontology to predict economic losses and casualties for better disaster prevention and mitigation. This paper aims to develop an ontology-based probabilistic framework for seismic risk assessment of buildings. In the developed framework, seismic risk probabilities of buildings with different damage states are first derived based on seismic vulnerability analysis and seismic hazard analysis. On this basis, an Ontology for Building Seismic Risk Assessment (OntoBSRA) system is developed to integrate knowledge on seismic risk assessments of buildings into a unified knowledge base by combining ontology and semantic web rule language (SWRL). Thus, automated seismic risk prediction including direct losses, indirect losses, and casualties related to buildings with various purposes and different damage states can be realized. The flow chart of the seismic risk prediction framework is shown in Figure 1. In addition, a case study is conducted in order to illustrate the application of the framework for seismic risk prediction.

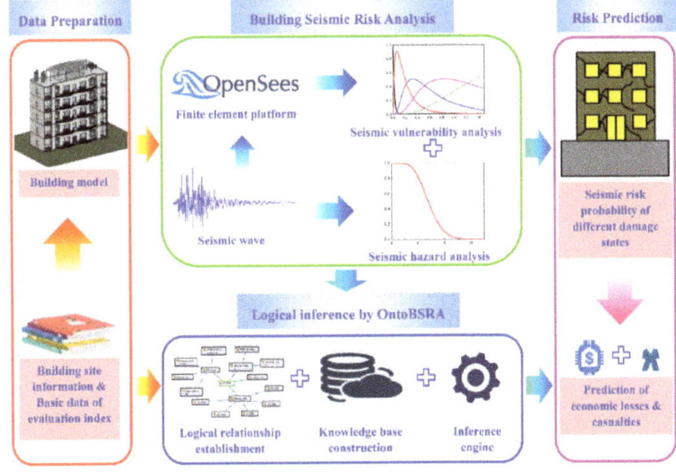

Figure 1. Seismic risk prediction framework flow chart.

The remainder of this paper is organized as follows: Section 2 presents the method for seismic risk probability analysis; Section 3 discusses the details of the OntoBSRA; a case study is conducted to illustrate the details of the application of the proposed framework in Section 4; finally, in Section 5 the major findings and limitations of this study are summarized in the course of concluding this paper.

2. Method for Seismic Risk Probability Analysis

Because of the need to take into account the uncertainty in both the earthquake and the structure while meeting the explicit requirements of various stakeholders in terms of performance targets seismic risk-oriented performance-based earthquake engineering has attracted the attention of many researchers, and full-probability seismic performance assessment methods have been proposed for engineering structures. Seismic risk probability analysis mainly includes seismic vulnerability analysis and seismic hazard analysis.

2.1. Seismic Vulnerability Analysis

In vulnerability analysis, the first step is to conduct a probabilistic seismic demand analysis. Considering the randomness of earthquake loads, a large number of selected seismic waves are input into the structural model for nonlinear dynamic time-history analysis in order to obtain the engineering demand parameter (*EDP*). Then, the relationship between the *EDP* and the ground motion intensity measure (*IM*) is obtained through data fitting. In this paper, the maximum inter-story drift of a structure is selected as the *EDP* and the peak ground acceleration (*PGA*) is used as the *IM*. According to Cornell and Krawinkler [19], the relationship between the *EDP* and *IM* is as follows:

$$EDP = a(IM)^b \quad \text{or} \quad \ln(EDP) = \ln a + b \ln IM \tag{1}$$

where a and b are the fitting parameters.

Seismic fragility quantitatively describes the ability of a structure to resist a certain level of seismic damage based on probability theory, and is defined as the conditional probability of the *EDP* of the structure subjected to an earthquake exceeding a certain limit state. It is generally assumed that seismic fragility follows a lognormal distribution, which can be expressed as follows [25]:

$$P_f = P(EDP \geq DI | IM) = 1 - \Phi \left(\frac{\ln(DI) - \ln\left(a(IM)^b\right)}{\sqrt{\beta_d^2 + \beta_c^2}} \right) \tag{2}$$

where P_f is the probability of the limit being exceeded, *DI* is the damage state thresholds, and β_d and β_c are the logarithmic standard deviations of the engineering demand parameter and the seismic capacity, respectively. According to literature [40], when the ground peak acceleration *PGA* is selected as the *IM*, $\sqrt{\beta_d^2 + \beta_c^2}$ is equal to 0.5.

According to GB50011-2010 [41] and Mwafy and Almorad [42], the structural performance level in this paper is categorized into five groups, namely, Normal Occupancy (OP), Immediate Occupancy (IO), Life Safety (LS), Collapse Prevention (CP), and Destruction (DS). The corresponding damage states are termed as no damage, slight damage, moderate damage, extensive damage, and complete collapse. The thresholds of these damage states are shown in Table 1 [41,43].

Table 1. The thresholds of the different damage states.

No Damage	Slight Damage	Moderate Damage	Extensive Damage	Complete Collapse
$EDP \leq 1/550$	$1/550 < EDP \leq 1\%$	$1\% < EDP \leq 2\%$	$2\% < EDP \leq 4\%$	$EDP > 4\%$

Based on the above analysis, the occurrence probability curves for various damage states of a structure can be obtained by Equation (3):

$$F_{ds,j} = P(EDP = ds_j | IM) = \begin{cases} 1 - P_{f,j} & j = 1 \\ P_{f,j-1} - P_{f,j} & j = 2,3,4 \\ P_{f,j-1} & j = 5 \end{cases} \tag{3}$$

where ds denotes the damage state and j = 1, 2, 3, 4, 5 represent no damage, slight damage, moderate damage, extensive damage, and complete collapse, respectively.

Moreover, seismic intensity can intuitively reflect the severity of the seismic damage and accelerate the process of assessing seismic risks. To this end, in this paper the occurrence probability curves are converted into probability matrixes related to the seismic intensity according to the relationship between the seismic intensity and the PGA, as shown in Equation (4) [44]:

$$I = 3.70 \log(PGA) - 1.60 \tag{4}$$

where I is the seismic intensity level.

2.2. Seismic Hazard Analysis

Seismic hazard refers to the probability distribution of the ground motion in the studied area within a certain period, among which the exceeding or occurrence probability of the seismic intensity is an important index. It is assumed that the seismic intensity follows a Weibull distribution [45]. Hence, the exceeding probability of the seismic intensity within 50 years is expressed as follows:

$$P(I \geq i) = 1 - F_{III}(i) = 1 - \exp\left(-\left(\frac{\omega - i}{\omega - I_0}\right)^k / 10^{0.9773}\right) \tag{5}$$

where ω is the upper limit value of the seismic intensity (generally, ω = 12), I_0 is the basic intensity, i is the specific value of seismic intensity, and k is the shape parameter, which can be determined by the least-square method. For regions with a seismic precautionary intensity of grades VI, VII, VIII, and IX, the values of k are 9.7932, 8.3339, 6.8713, and 5.4028, respectively.

2.3. Seismic Risk Probability Analysis

Seismic risk probability refers to the probability of certain disaster consequences in the area of interest caused by earthquakes. In this paper, it is defined as the occurrence probability of a structure in different damage states, which is based on the seismic vulnerability analysis and the seismic hazard analysis. The seismic risk probability can be expressed as follows:

$$P_{ds,j} = \sum_i F(ds_j | I_i) P(I_i) \tag{6}$$

where $P_{ds,j}$ is the seismic risk probability of a structure in the jth damage state, $F(ds_j | I_i)$ is the conditional probability of the jth damage state when the seismic intensity is equal to I_i (which is determined by Equations (3) and (4)), and $P(I_i)$ is the occurrence probability of the seismic intensity I_i, also termed the seismic hazard.

3. Design and Development of the OntoBSRA

3.1. Framework of the OntoBSRA

The developed OntoBSRA includes a knowledge base, ontology management system, rule editor, and query function. The ontology knowledge base stores all seismic risk-related knowledge in the form of an Ontology Web Language (OWL) file, which plays an important role in the OntoBSRA. The ontology management system provides the rule-editing function, which can achieve the deductive reasoning ability of the developed ontology, and has the function of creating as well as updating the ontology. Moreover, users can obtain the final reasoning results by editing the query language rules, (such as the simple protocol and RDF query language (SPAQRL) and the semantic query-enhanced web rule language (SQWRL), using the query function according to their demands [46]. In addition, as Protégé software [36] can realize the establishment of classes, logical relationships, attributes, and instances as well as provide the SWRLTab and SQWRLQueryTab interfaces for SWRL and SQWRL editing, respectively, the OntoBSRA proposed in this paper was developed based on Protégé. The schematic diagram of the OntoBSRA system is shown in Figure 2.

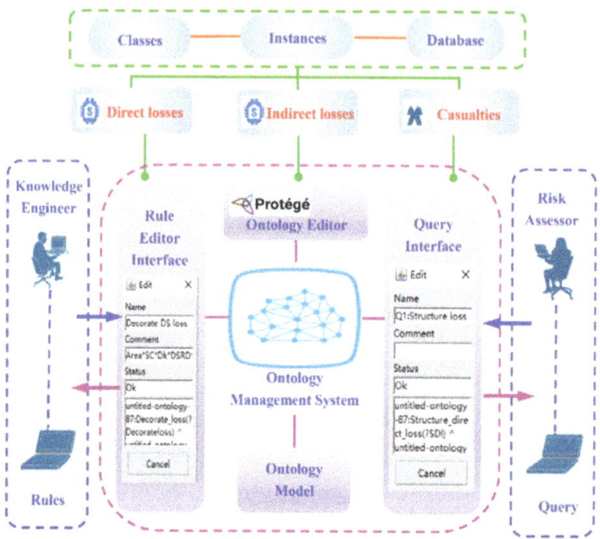

Figure 2. Schematic diagram of the OntoBSRA system.

3.2. Determination of Primary Indicators for the OntoBSRA

Seismic risk prediction of a building mainly includes direct losses, indirect losses, and casualties, among which direct losses include structural losses, indoor and outdoor property losses, and decoration property losses.

3.2.1. Direct Losses

(1) Structural losses

Structural losses can be obtained using the following equation:

$$SL_{ds,j} = SC \times A \times P_{ds,j} \times RB_{ds,j} \tag{7}$$

$$SL = \sum_{j=1}^{5} SL_{ds,j} \tag{8}$$

where SL denotes the total structural damage losses, $SL_{ds,j}$ indicates the loss of the total value of a structure ($SC \times A$) in the jth damage state, SC and A represent the unit construction cost and total area of the RC structure, respectively, $P_{ds,j}$ is the seismic risk probability, and $RB_{ds,j}$ is the direct loss ratio of the structure in the jth damage state.

(2) Indoor and outdoor property losses

Indoor and outdoor property losses caused by an earthquake can be determined using the following equation:

$$CL_{ds,j} = SC \times A \times Ck \times P_{ds,j} \times RC_{ds,j} \tag{9}$$

$$CL = \sum_{j=1}^{5} CL_{ds,j} \tag{10}$$

where CL denotes the total indoor and outdoor property losses, $CL_{ds,j}$ represents the losses of total indoor and outdoor property ($SC \times A \times Ck$) in the jth damage state, Ck is the ratio of the indoor and outdoor property replacement cost to the construction cost of the RC structure, and $RC_{ds,j}$ is the direct loss ratio of the indoor and outdoor property in the jth damage state.

(3) Decoration property losses

Decoration property losses can be obtained using the following equation:

$$DL_{ds,j} = SC \times A \times Dk \times P_{ds,j} \times RD_{ds,j} \tag{11}$$

$$DL = \sum_{j=1}^{5} DL_{ds,j} \tag{12}$$

where DL denotes the total decoration property losses, $DL_{ds,j}$ represents the total losses of decoration property ($SC \times A \times Dk$) in the jth damage state, Dk is the ratio of the decoration property replacement cost to the construction cost of the RC structure, and $RD_{ds,j}$ is the direct loss ratio of the decoration property in the jth damage state.

(4) Total direct losses

Total direct losses can be calculated using the following equation:

$$DirL_{ds,j} = SL_{ds,j} + CL_{ds,j} + DL_{ds,j} \tag{13}$$

$$DirL = \sum_{j=1}^{5} DirL_{ds,j} \tag{14}$$

where $DirL$ denotes the total direct losses and $DirL_{ds,j}$ represents the direct losses in the jth damage state.

The direct loss ratio and the ratio of the indoor and outdoor property and decoration property replacement cost to the construction cost of the RC structure according to the relevant literature [43,47] are shown in Tables 2 and 3, respectively.

Table 2. Direct loss ratio.

Damage States	No Damage	Slight Damage	Moderate Damage	Extensive Damage	Complete Collapse
Structural losses	0	0.02	0.10	0.50	1.00
Indoor and Outdoor property losses	0	0.01	0.05	0.20	0.60
Decoration property losses	0.1	0.25	0.6	0.85	1

Table 3. Ratio of the replacement cost to the construction cost.

Building Types	Residential Building	Commercial Building	Medical Building	Office Building	Educational Building
Indoor and Outdoor property	0.2	0.1	1.5	1.0	1.0
Decoration property	0.3	0.43	0.25	0.35	0.25

3.2.2. Indirect Losses

Indirect losses can be calculated using the following equation:

$$IndL_{ds,j} = DirL_{ds,j} \times R_{ds,j} \tag{15}$$

$$IndL = \sum_{j=1}^{5} IndL_{ds,j} \tag{16}$$

where *IndL* denotes the total indirect losses, while $IndL_{ds,j}$ and $R_{ds,j}$ are the indirect losses and the ratio of the indirect losses to the direct losses in the *j*th damage state, respectively. The values of $R_{ds,j}$ are shown in Table 4 [27].

Table 4. Ratio of indirect losses to direct losses.

Damage States	No Damage	Slight Damage	Moderate Damage	Extensive Damage	Complete Collapse
Ratio	0	0	0.50–1.00	3.00–6.00	8.00–20.00

3.2.3. Casualties

(1) Number of Deaths

The number of deaths in a building as a result of an earthquake can be obtained using the following equation:

$$DN_{ds,j} = PD \times A \times DR_{ds,j} \times P_{ds,j} \tag{17}$$

$$DN = \sum_{j=1}^{5} DN_{ds,j} \tag{18}$$

where *DN* denotes the total number of deaths, $DN_{ds,j}$ represents the number of deaths in the *j*th damage state, *PD* is the personnel density, and $DR_{ds,j}$ is the death rate in the *j*th damage state.

(2) Number of injuries

The number of injuries in a building as a result of an earthquake can be determined using the following equation:

$$IN_{ds,j} = PD \times A \times IR_{ds,j} \times P_{ds,j} \tag{19}$$

$$IN = \sum_{j=1}^{5} IN_{ds,j} \tag{20}$$

where *IN* denotes the total number of injuries, $IN_{ds,j}$ is the number of injuries in the *j*th damage state, and $IR_{ds,j}$ is the injury rate in the *j*th damage state.

The death rate and injury rate in different damage states of buildings and the personnel density of buildings with various purposes according to Comerio [48] and Wang [49] are shown in Tables 5 and 6, respectively.

Table 5. Deaths and injury rates.

Damage States	No Damage	Slight Damage	Moderate Damage	Extensive Damage	Complete Collapse
Death rate	0	0	0–0.001	0.001–0.01	0.02–0.3
Injury rate	0	0–0.0005	0.0002–0.03	0.001–0.05	0.05–0.7

Table 6. Personnel density.

Building Types	Residential Building	Commercial Building	Medical Building	Office Building	Educational Building
Personnel density (person/m^2)	0.33	0.72	0.91	0.4	1.12

3.3. Development of the OntoBSRA

The methods used for developing the ontology include the Uschold and King method, the Gruninger and Fox method, the Methontology method, the KACTUS method, and the Ontology Development 101 method [50]. In this paper, the Ontology Development 101 method is employed, as shown in Figure 3. According to this method, new ontologies can be developed using the Protégé software by following specified steps or reusing the existing semantic resources and ontologies. The steps for developing the OntoBSRA are explained in details below.

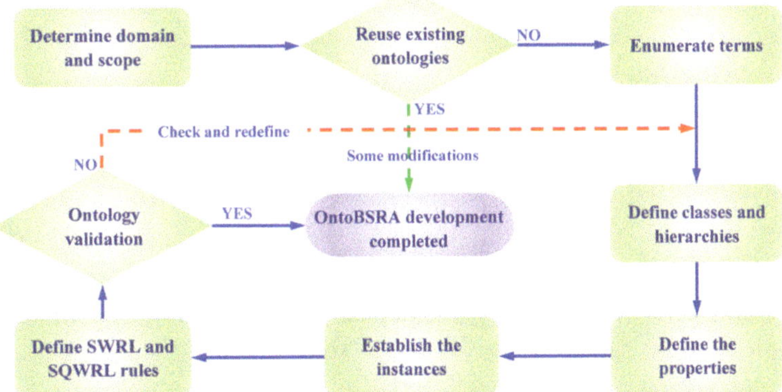

Figure 3. Procedure of the OntoBSRA development.

Step 1. Determine the domain and scope of the OntoBSRA.

In the early design stage of the ontology, the following questions in Table 7 are raised to check whether the ontology involves enough information to correct any missing and wrong information.

Step 2. Consider reusing the existing ontologies.

Table 7. Question table for determining the domain and scope of the OntoBSRA.

Questions	Answers
What is the purpose of developing this ontology?	To establish a unified knowledge base to enable rapid seismic risk assessments of buildings.
Who are the users of the developed ontology?	The engineers with responsibility for seismic risk evaluation.
What is the premise behind OntoBSRA?	Nonlinear time-domain analysis using finite element software, seismic vulnerability analysis, and seismic hazard analysis.
What types of structures is OntoBSRA developed for?	Reinforced concrete structures.
How is seismic risk quantified by OntoBSRA?	By economic losses (direct losses and indirect losses) and casualties.

Newly developed ontologies can share knowledge information with existing ontology models owing to the interactivity of the OWL language. Therefore, ontologies can be extended across multiple disciplines for wider applications. In this study, the content structure of the OntoBSRA is designed based on the common characteristics of existing ontology frameworks and the semantic rule language [33–35,39,50] in order to avoid unnecessary mistakes in developing a new ontology. The content structure of the OntoBSRA is shown in Figure 4.

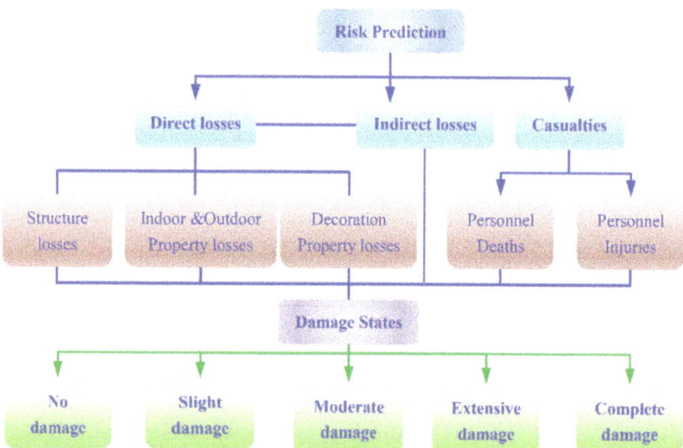

Figure 4. Content structure of OntoBSRA.

Step 3. Enumerate important terms for the OntoBSRA.

In this step, a glossary of knowledge fields such as seismic risk probabilities, economic losses, and casualties is obtained by review and analysis of the basic terms in the relevant literature. Moreover, through extensive research, basic data in the relevant knowledge field, such as loss ratios, ratios of indirect losses to direct losses, casualty rates, personnel densities, etc., are summarized in this paper in tables, as shown in Section 3.2.

Step 4. Define classes and class hierarchies.

Defining the classes and class hierarchies is the primary stage in the process of developing an ontology. In this paper, a top-down method is adopted to define the classes. The superclasses for seismic intensities, damage states, direct losses, indirect losses, casualties, etc., are first created. Each superclass is then refined to establish subclasses. The specific details are shown in Figure 5a.

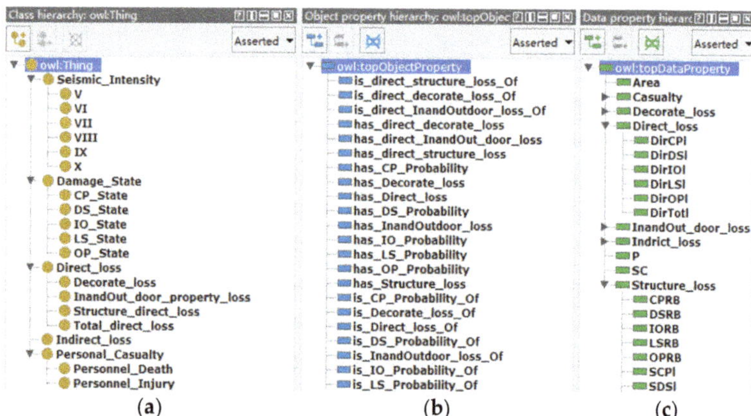

Figure 5. Development of OntoBSRA in Protégé-OWL 5.2. (**a**) Classes; (**b**) Object properties; (**c**) Data properties.

Step 5. Define the properties of classes.

The OntoBSRA includes two types of properties, namely, the object property and the data property. The object property represents the relationship among different classes, such as 'has OP Probability' and 'is OP Probability Of'. The data property represents the characteristics of instances quantitatively and qualitatively; its data type includes Number,

String, Boolean and Enumerated. In the OntoBSRA, the data property is adopted to describe the created instances quantitatively, and the data format adopts the "float" type, e.g., the risk probability of the LS performance level is 0.260f. Figure 5b,c shows the detailed object and data properties.

Step 6. Establish instances.

The instances in the classes have their own locations and hierarchies, and the object property and data property of instances must be defined in OntoBSRA. In OntoBSRA, different building types such as residential buildings, medical buildings, commercial buildings, office buildings, and educational buildings are established as instances in the classes. The basic data, such as loss ratios and casualty rates, are manually input in the established instances, while the evaluation indices, such as direct losses and casualties, need to be inferred by the inference machine using the SWRL rules. Figure 6 shows the instances in the class of the structural losses.

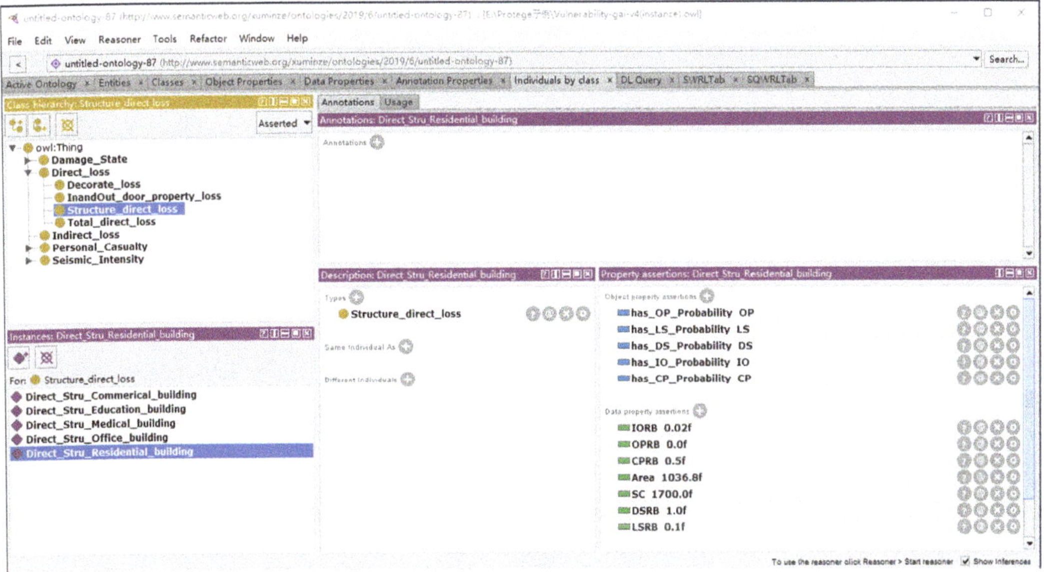

Figure 6. Create instances.

Step 7. Define SWRL rules.

SWRL rules can represent the relationship among the classes and meet the reasoning requirements of the ontology. There are Class atoms, Individual Property atoms, Data Valued Property atoms, and Built-in atoms in the OntoBSRA, all of which are connected by the symbol "^". The symbol "?" represents variables, and the antecedent and the consequence are connected by the symbol "→". Furthermore, computing ability can be realized through the SWRL rules in SWRLTab interface of the Protégé software. For example, the SWRL rule of the structural losses in the OP performance level is shown below.

Equation	$SL_{ds,1} = A \times SC \times P_{ds,1} \times RB_{ds,1}$
SWRL	Structure_direct_loss(?SDl)^SC(?SDl,?sc)^Area(?SDl,?A)^OPRB(?SDl,?S_OPR-B) ^ has_OP_Probability(?SDl, ?OPstate) ^ OP_State(?OPstate) ^P(?OPstate,?O-P_Probability) ^ swrlb: multiply (?S_OPl, ?sc, ?A, ?S_OPRB, ?OP_Probability) ->SOPl(?SDl, ?S_OPl)

Step 8. Define SQWRL rules

The query function is implemented by the SQWRL rules. In the Protégé software, the SQWRLTab interface is adopted to compile the SQWRL rules, compare the results of the ontology inference, and query and filter out the information of interest. For example, the SQWRL rule of the structural losses in different damage states is shown below.

SQWRL	Structure_direct_loss(?SDl) ˆ SOPl (?SDl, ?S_OPl) ˆ SIOl (?SDl, ?S_IOl) ˆ SLSl (?SDl, ?S_LSl) ˆ SCPl (? SDl, ?SCPl) ˆ SDSl (?SDl, ?S_DSl) ˆ STotl (?SDl, ?S_Totl) -> sqwrl: select(?SDl, ?S_OPl, ?S_IOl, ?S_LSl, ?S_CPl, ?S_DSl, ?S_Totl)

Step 9. Ontology validation.

Syntactical validation.

Syntactical validation is conducted to ensure a correct hierarchical structure and logical relationship which can infer and calculate the explicit and implicit relationships and data accurately in the developed ontology [51]. In this study, the pellet reasoner in the Protégé software is used to complete the syntactical validation. The schematic diagram of successful syntactical validation is shown in Figure 7.

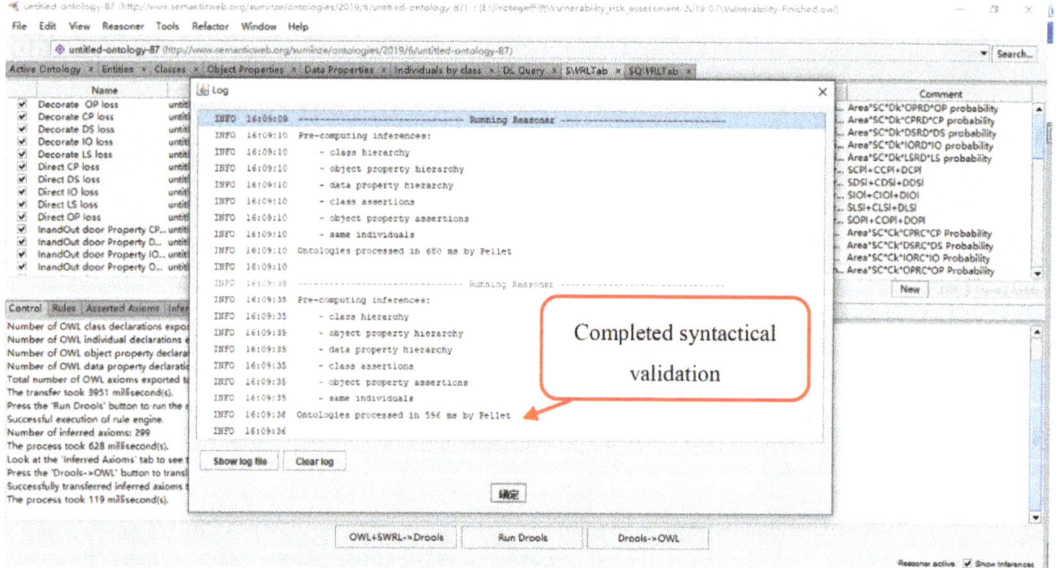

Figure 7. Syntactical validation of the OntoBSRA.

Rule validation.

Rule validation is conducted to make sure that the developed rules are compatible with the OntoBSRA and can carry out logical inference and data calculation correctly. In this study, the SWRLTab plug-in in the Protégé software is adopted for the rule-checking. The schematic diagram of successful rule validation is shown in Figure 8.

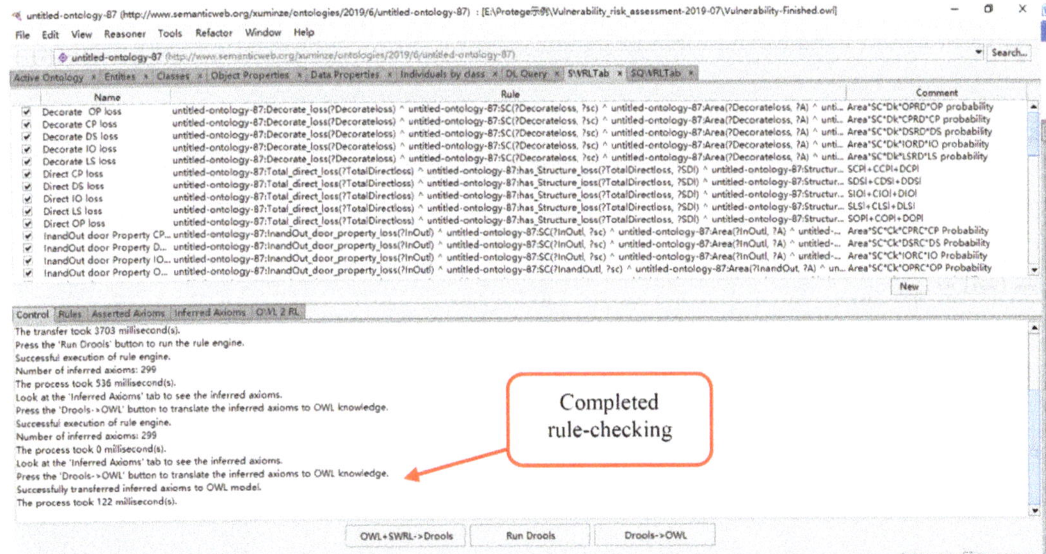

Figure 8. Rule-checking.

4. Case Study

In this section, an example using a single residential building is provided to illustrate the application of the presented framework for predicting the seismic risk of a building and to demonstrate the validity of the semantic web rules. The residential building is a five-story RC frame structure with an area of 1036.8 m^2 and a unit cost of 1700 USD/m^2. The seismic precautionary intensity of the region is grade VIII. The building model is shown in Figure 9.

Figure 9. Model of the residential building.

4.1. Seismic Risk Probability of the Building

Current methods used for probabilistic seismic demand analysis include the cloud method, strip method, IDA method, etc. Of those methods, the IDA method can simulate the whole collapse process of a structure subjected to seismic action [52], and is consequently adopted here for seismic demand analysis.

On the basis of the IDA of the structure determined by the OpenSees finite element platform, the exceeding probability and occurrence probability curves in various damage states were obtained using Equations (1)–(3) and the thresholds of the damage states specified in Table 1, as shown in Figures 10 and 11, respectively.

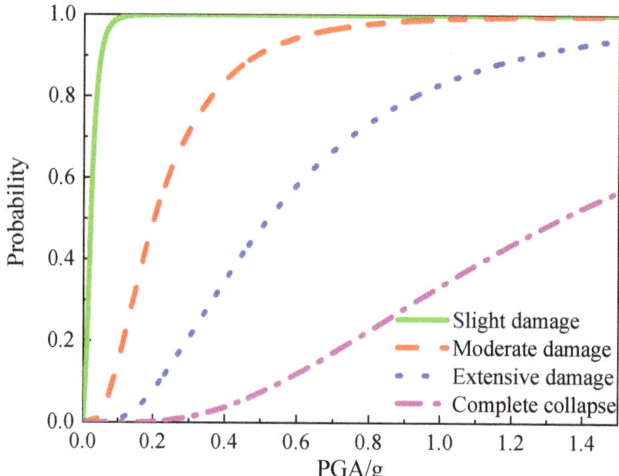

Figure 10. Exceeding probability curves.

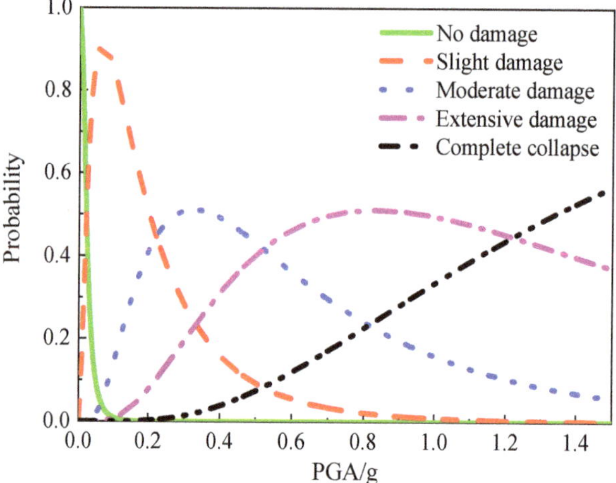

Figure 11. Occurrence probability curves.

According to Equation (4), the occurrence curves were converted into a probability matrix related to the seismic intensity, as shown in Table 8.

Table 8. Vulnerability matrix.

Intensity	V	VI	VII	VIII	IX	X
None	0.0482	0.0049	0.0002	0	0	0
Slight	0.9115	0.7903	0.4661	0.1580	0.0274	0.0023
Moderate	0.0395	0.1913	0.4372	0.4914	0.2691	0.0712
Extensive	0.0008	0.0134	0.0929	0.3122	0.5063	0.4010
Complete	0	0.0001	0.0036	0.0384	0.1972	0.5255

Furthermore, through seismic hazard analysis using Equation (5), the occurrence probabilities of the seismic intensity over 50 years were determined as summarized in Table 9.

Table 9. Occurrence probability of the seismic intensity.

Intensity	V	VI	VII	VIII	IX	X
Probability	0.1738	0.4327	0.2863	0.0855	0.0136	0.0009

Considering the vulnerability and the seismic hazard, the seismic risk probabilities of the building in different damage states were obtained according to Equation (6) as shown in Table 10.

Table 10. Seismic risk probability.

Damage States	No Damage	Slight Damage	Moderate Damage	Extensive Damage	Complete Collapse
Risk probability	0.0106	0.6477	0.2605	0.0665	0.0075

4.2. Application of the OntoBSRA

The seismic risk probabilities obtained from the analysis in Section 4.1 were manually input into the OntoBSRA, and new facts were generated by running the pre-set SWRL rules. The main SWRL rules based on the evaluation index equation in Section 3.2 are shown in Table 11. Taking the decoration property losses as an example, the new facts shown in Figure 12 were deduced by running the pre-set SWRL rules, thus validating the correction of syntax and SWRL rules for decoration property losses.

Table 11. Main SWRL rules for the seismic risk assessment.

Rule Number	SWRL Rules
Rule 1	Calculation of the total structural damage losses: Structure_direct_loss(?SDl) ^ SOPl(?SDl, ?S_OPl) ^ SIOl(?SDl, ?S_IOl) ^ SLSl(?SDl, ?S_LSl) ^ SCPl(?SDl,?S_CPl) ^ DSl(?SDl,?S_DSl) ^ swrlb: add(?S_Totl, ?S_OPl, ?S_IOl, ?S_LSl, ?S_CPl,?S_DSl) ->STotl(?SDl,?S_Totl)
Rule 2	Calculation of the total indoor and outdoor property losses: InandOut_door_property_loss(?InOutl) ^ COPl(?InOutl, ?InOut_OPl) ^ CIOl(?InOutl, ?InOut_IOl) ^ CLSl(?InOutl, ?InOut_LSl) ^ CCPl(?InOutl, ?InOut_CPl) ^ CDSl(?InOutl, ?InOut_DSl) ^ swrlb:add(?InOut_Totl, ?InOut_OPl,?InOut_IOl,?InOut_LSl,?InOut_CPl,?InOut_DSl)->CTotl(?InOutl, ?InOut_Totl)
Rule 3	Calculation of the total decoration property losses: Decoration_loss(?Decorationloss) ^ DOPl(?Decorationloss, ?D_OPl) ^ DIOl(?Decorationloss, ?D_IOl) ^ DLSl(?Decorationloss, ?D_LSl) ^ DCPl(?Decorationloss, ?D_CPl) ^ DDSl(?Decorationloss, ?D_DSl) ^ swrlb:add(?D_Totl, ?D_OPl, ?D_IOl, ?D_LSl, ?D_CPl, ?D_DSl)->DTotl(?Decorationloss, ?D_Totl)
Rule 4	Calculation of the total direct losses: Total_direct_loss(?TotalDirectloss) ^ DirOPl(?TotalDirectloss, ?DirectOPloss) ^ DirIOl(?TotalDirectloss, ?DirectIOloss) ^ DirLSl(?TotalDirectloss, ?DirectLSloss) ^ DirCPl(?TotalDirectloss, ?DirectCPloss) ^ DirDSl(?TotalDirectloss, ?DirectDSloss) ^ swrlb:add(?DirectTotalloss, ?DirectOPloss, ?DirectIOloss, ?DirectLSloss, ?DirectCPloss, ?DirectDSloss) ->DirTotl(?TotalDirectloss, ?DirectTotalloss)
Rule 5	Calculation of the total indirect losses: Indirect_loss(?Indirectloss) ^ IndOPl(?Indirectloss, ?IndirectOPloss) ^ IndIOl(?Indirectloss, ?IndirectIOloss) ^ IndLSl(?Indirectloss, ?IndirectLSloss) ^ IndCPl(?Indirectloss, ?IndirectCPloss) ^ IndDSl(?Indirectloss, ?IndirectDSloss) ^ swrlb:add(?IndirectTotalloss, ?IndirectOPloss, ?IndirectIOloss, ?IndirectLSloss, ?IndirectCPloss, ?IndirectDSloss)->IndTotl(?Indirectloss, ?IndirectTotalloss)

Table 11. *Cont.*

Rule Number	SWRL Rules
Rule 6	Calculation of the total number of deaths: Personnel_Death(?PersonnelDeath) ^ OPDN(?PersonnelDeath, ?DeathOPNumber) ^ IODN(?PersonnelDeath, ?DeathIONumber) ^ LSDN(?PersonnelDeath, ?DeathLSNumber) ^ CPDN(?PersonnelDeath, DeathCPNumber)^DSDN(?PersonnelDeath,?DeathDSNumber)^swrlb:add(?DeathTotalNumber, ?DeathOPNumber,?DeathIONumber,?DeathLSNumber,?DeathCPNumber,?DeathDSNumber)-> TotDN(? PersonnelDeath, ?DeathTotalNumber)
Rule 7	Calculation of the total number of injuries: Personnel_Injury(?PersonnelInjury) ^ OPIN(?PersonnelInjury, ?InjuryOPNumber) ^ IOIN(?PersonnelInjury, ?InjuryIONumber) ^ LSIN(?PersonnelInjury, ?InjuryLSNumber) ^ CPIN(?PersonnelInjury, ?InjuryCPNumber) ^ DSIN(?PersonnelInjury, ?InjuryDSNumber) ^ swrlb:add(?InjuryTotalNumber, ?InjuryOPNumber, ?InjuryIONumber, ?InjuryLSNumber, ?InjuryCPNumber, ?InjuryDSNumber) -> TotIN(?PersonnelInjury, ?InjuryTotalNumber)

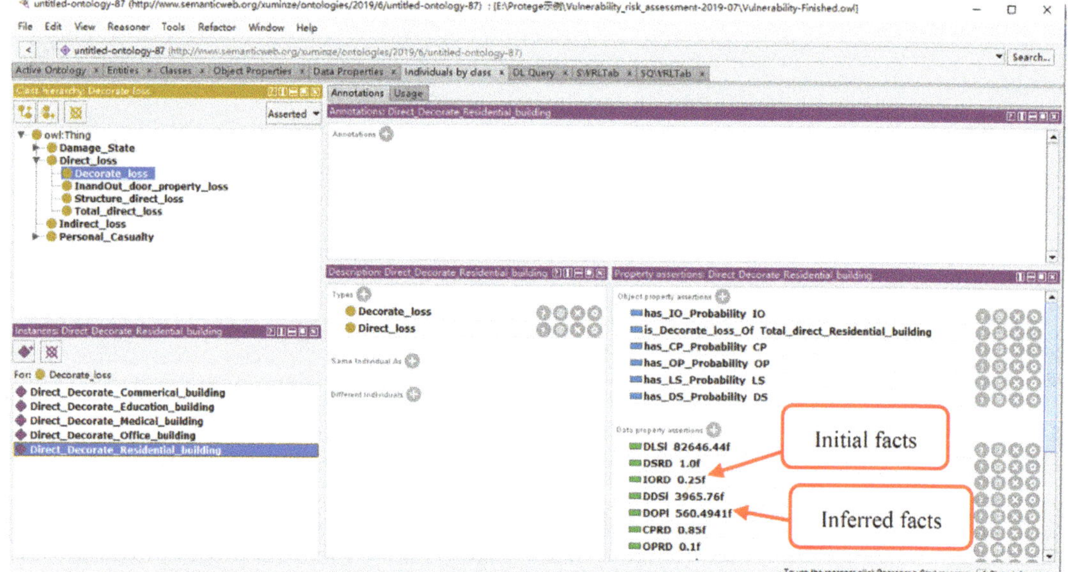

Figure 12. Inferred facts.

Moreover, the specialized assessment results can be queried using the SQWRL rules according to the user's demand. Table 12 shows the SQWRL querying command for the assessment indicators of the OntoBSRA. In the application of the OntoBSRA, if the users/stakeholders attach importance to the economic losses (such as direct losses), the assessment results can be obtained by running the pre-set SQWRL rule Q4 in Table 12. The querying results are illustrated in Figure 13. Furthermore, if users/stakeholders pay more attention to the casualties, they can run the SQWRL rules Q6 and Q7 to obtain the assessment results. The querying results of the number of injuries is shown in Figure 14. It can be seen from Figures 13 and 14 that the OntoBSRA system can provide the total economic losses or casualties as well as the related assessment indicators of buildings in different damage states. This function can assist engineers, provide guidance on earthquake hazard mitigation, and improve efficiency in decision-making.

Table 12. SQWRL rules.

Rule Number	SQWRL Rules
Q1: Structural losses	Structure_direct_loss(?SDl)^SOPl(?SDl,?S_OPl) ^IOl(?SDl,?S_IOl)^SCPl(?SDl,?S_CPl)^SDSl(?SDl,?S_DSl)^STotl(?SDl,?S_Totl)->sqwrl:select(?SDl,?S_OPl,?S_IOl,?S_LSl,?S_CPl,?S_DSl,?S_Totl)
Q2: Indoor and Outdoor property losses	InandOut_door_property_loss(?InOutl)^COPl(?InOutl,?InOut_OPl)^CIOl(?InOutl,?InOut_IOl)^CLSl(?InOutl,?InOut_LSl)^CCPl(?InOutl,?InOut_CPl)^CDSl(?InOutl,?InOut_DSl)^CTotl(?InOutl,?InOut_Totl)->sqwrl:select(?InOutl,?InOut_OPl,?InOut_IOl,?InOut_LSl,?InOut_CPl,?InOut_DSl,?InOut_Totl)
Q3: Decoration property losses	Decoration_loss(?Decorationloss)^DOPl(?Decorationloss,?D_OPl)^DIOl(?Decorationloss,?D_IOl)^DLSl(?Decorationloss,?D_LSl) ^ DCPl(?Decorationloss, ?D_CPl) ^ DDSl(?Decorationloss,?D_DSl)^DTotl(?Decorationloss,?D_Totl)->sqwrl:select(?Decorationloss,?D_OPl,?D_IOl,?D_LSl,?D_CPl,?D_DSl,?D_Totl)
Q4: Total direct losses	Total_direct_loss(?TotalDirectloss)^DirOPl(?TotalDirectloss, ?DirectOPloss) ^ DirIOl(?TotalDirectloss,?DirectIOloss)^DirLSl(?TotalDirectloss,?DirectLSloss) ^ DirCPl ?TotalDirectloss,?DirectCPloss)^DirDSl(?TotalDirectloss,?DirectDSloss) ^ DirTotl(?TotalDirectloss,?DirectTotalloss)->sqwrl:select(?TotalDirectloss,?DirectOPloss,?DirectIOloss,?DirectLSloss,?DirectCPloss,?DirectDSloss,?DirectTotalloss)
Q5: Indirect losses	Indirect_loss(?Indirectloss)^IndOPl(?Indirectloss,?IndirectOPloss) ^ IndIOl(?Indirectloss,?IndirectIOloss)^IndLSl(?Indirectloss,?IndirectLSloss) ^ IndCPl(?Indirectloss, ?IndirectCPloss) ^ IndDSl(?Indirectloss, ?IndirectDSloss) ^ IndTotl(?Indirectloss, ?IndirectTotalloss) -> sqwrl: select(?Indirectloss, ?IndirectOPloss, ?IndirectIOloss,?IndirectLSloss,?IndirectCPloss,?IndirectDSloss,?IndirectTotalloss)
Q6: Personnel deaths	Personnel_Death(?PersonnelDeath)^OPDN(?PersonnelDeath,?DeathOPNumber) ^ IODN(?PersonnelDeath,?DeathIONumber)^LSDN(?PersonnelDeath,?DeathLSNumber)^CPDN(?PersonnelDeath,?DeathCPNumber)^DSDN(?PersonnelDeath,?DeathDSNumber)^TotDN(?PersonnelDeath,?DeathTotalNumber)->sqwrl:select(?PersonnelDeath,?DeathOPNumber,?DeathIONumber,?DeathLSNumber,?DeathCPNumber,?DeathDSNumber,?DeathTotalNumber)
Q7: Personnel injuries	Personnel_Injury(?PersonnelInjury)^OPIN(?PersonnelInjury,?InjuryOPNumber) ^ IOIN(?PersonnelInjury,?InjuryIONumber)^LSIN(?PersonnelInjury, ?InjuryLSNumber)^CPIN(?PersonnelInjury,?InjuryCPNumber)^DSIN(?PersonnelInjury, ?InjuryDSNumber)^TotIN(?PersonnelInjury,?InjuryTotalNumber)->sqwrl: select(?PersonnelInjury?InjuryOPNumber,?InjuryIONumber,?InjuryLSNumber, ?InjuryCPNumber,?InjuryDSNumber,?InjuryTotalNumber)

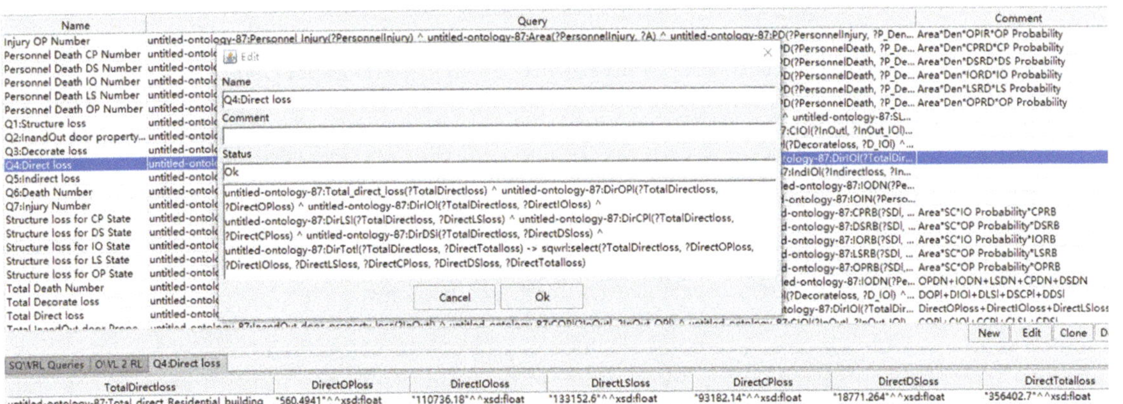

Figure 13. Inference results of querying according to the SQWRL rule Q4 in Table 12.

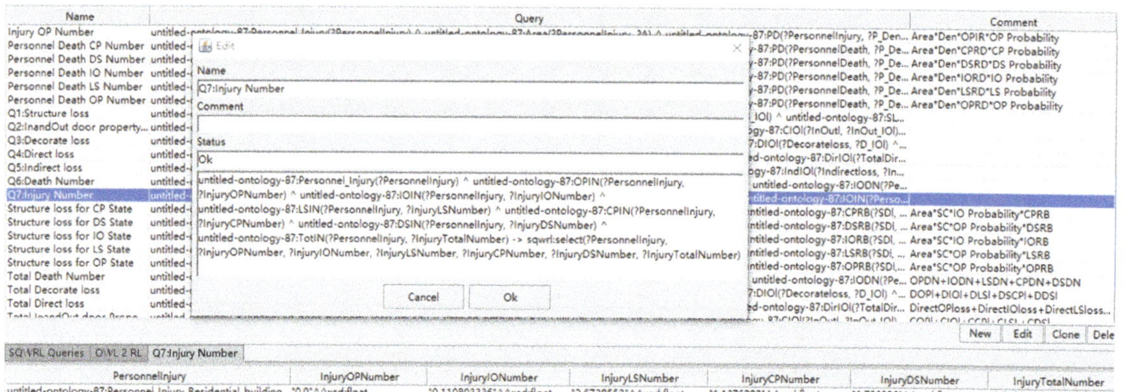

Figure 14. Inference results of querying according to the SQWRL rule Q7 in Table 12.

To assess the seismic risk of the building comprehensively, users/stakeholders can query direct losses, indirect losses, and casualties by running all the SQWRL rules in Table 12 according to building type. The querying results for the building in this case study are shown in Table 13.

Table 13. Seismic risk assessment results.

		No Damage	Slight Damage	Moderate Damage	Extensive Damage	Complete Collapse	Total Losses
Direct losses ($)	Structure losses	0	22,832.2	45,914.6	58,605.1	13,219.2	140,571.1
	Indoor and outdoor property losses	0	2283.2	4591.4	4688.4	1586.3	13,149.3
	Decoration property losses	560.4	85,620.76	82,646.4	29,888.6	3965.7	202,681.86
	Total direct losses	560.4	110,736.1	133,152.6	93,182.1	18,771.2	356,402.4
Indirect losses ($)		0	0	133,152.6	559,092.8	375,425.2	1,067,670.6
Casualties	Personnel deaths	0	0	0.1	0.2	0.8	1.1
	Personnel injuries	0	0.1	2.7	1.1	1.8	5.7

In addition, it should be noted that although numerical simulation-based seismic vulnerability analysis can evaluate the seismic performance of buildings, it cannot assess seismic risk from the macro perspective. This has been verified by comparison between numerical and ontology-based results. On the other hand, the developed OntoBSRA system can achieve seismic risk assessment of buildings with various functions in different damage states (such as direct losses, indirect losses and casualties) from the macro perspective, and is thus beneficial in decision-making based on the principle of targeted-risk.

5. Conclusions

This paper has developed an ontology-based probabilistic framework for seismic risk assessment of buildings. In the developed framework, seismic risk probabilities are first obtained based on vulnerability analysis and seismic hazard analysis. Then, the developed OntoBSRA system is used to integrate the knowledge on seismic risk assessments of buildings into a unified knowledge base by combining the ontology with SWRL, which achieved automated seismic risk prediction (such as direct losses, indirect losses, and casualties) of buildings with various purposes in different damage states. In this way, over-reliance on identification of the damage states of buildings after earthquake can be avoided. Based on the developed framework, risk assessors and asset managers can estimate the consequences of earthquakes effectively within a certain period considering the actual construction cost, thus providing guidance on earthquake prevention and mitigation and improving efficiency in decision-making.

A case study of a residential RC frame structure has been conducted to illustrate the detailed application of the proposed framework to predict seismic risk as quantified by indexes including direct losses, indirect, losses and casualties, demonstrating the validity of the semantic web rules.

It should be noted that the OntoBSRA proposed in this paper was developed based only on the seismic damage information of RC structures. Our future work will focus on the extension of OntoBSRA using knowledge information related to other structures, such as wood structures, brick–concrete structures, etc., and the development of data interfaces between the BIM, Protégé software, and finite element software in order to avoid tedious manual data input in OntoBSRA.

Author Contributions: Conceptualization and methodology, M.X.; software, C.C. and P.Z.; validation, M.X., P.Z. and J.Z.; data curation, J.Z.; writing—original draft preparation, M.X.; writing—review and editing, C.C.; supervision, C.C. and P.Z.; funding acquisition, C.C. All authors have read and agreed to the published version of the manuscript.

Funding: This research is funded by the National Key Research and Development Program of China (Grant No. 2021YFB2601102), the National Natural Science Foundation of China (Grant No. 51878109 & 51578100), and the Special Foundation for 'Double First-Class' Construction Project (Grant No. BSCXXM022).

Institutional Review Board Statement: Not applicable.

Informed Consent Statement: Not applicable.

Data Availability Statement: The data reported in this article are available from the corresponding author upon request.

Conflicts of Interest: The authors declare no conflict of interest.

References

1. Jaiswal, K.S.; Bausch, D.; Chen, R.; Bouabid, J.; Seligson, H. Estimating Annualized Earthquake Losses for the Conterminous United States. *Earthq. Spectra* **2015**, *31*, S221–S243. [CrossRef]
2. Cui, C.; Meng, K.; Xu, C.; Wang, B.; Xin, Y. Vertical vibration of a floating pile considering the incomplete bonding effect of the pile-soil interface. *Comput. Geotech.* **2022**, *150*, 104894. [CrossRef]
3. Dai, K.-Y.; Liu, C.; Lu, D.-G.; Yu, X.-H. Experimental investigation on seismic behavior of corroded RC columns under artificial climate environment and electrochemical chloride extraction: A comparative study. *Constr. Build. Mater.* **2020**, *242*, 118014. [CrossRef]
4. Tian, K.; Liu, W.; Feng, D.; Yang, Q. Dynamic characteristic analysis and shaking table test for a curved surface isolated structure. *Eng. Struct.* **2019**, *203*, 109847. [CrossRef]
5. Meng, K.; Cui, C.; Liang, Z. A new approach for longitudinal vibration of a large-diameter floating pipe pile in visco-elastic soil considering the three-dimensional wave effects. *Comput. Geotech.* **2020**, *128*, 103840. [CrossRef]
6. Chen, J.; Wan, Z. A compatible probabilistic framework for quantification of simultaneous aleatory and epistemic uncertainty of basic parameters of structures by synthesizing the change of measure and change of random variables. *Struct. Saf.* **2019**, *78*, 76–87. [CrossRef]
7. Celik, O.C.; Ellingwood, B.R. Seismic fragilities for non-ductile reinforced concrete frames—Role of aleatoric and epistemic uncertainties. *Struct. Saf.* **2010**, *32*, 1–12. [CrossRef]
8. Gautam, D.; Rupakhety, R.; Adhikari, R. Empirical fragility functions for Nepali highway bridges affected by the 2015 Gorkha Earthquake. *Soil. Dyn. Earthq. Eng.* **2019**, *126*, 105778. [CrossRef]
9. Lu, D.; Yu, X.; Jia, M.; Wang, G. Seismic risk assessment for a reinforced concrete frame designed according to Chinese codes. *Struct. Infrastruct. Eng.* **2013**, *10*, 1295–1310. [CrossRef]
10. Gautam, D.; Rupakhety, R. Empirical seismic vulnerability analysis of infrastructure systems in Nepal. *Bull. Earthq. Eng.* **2021**, *19*, 6113–6127. [CrossRef]
11. Koutsourelakis, P. Assessing structural vulnerability against earthquakes using multi-dimensional fragility surfaces: A Bayesian framework. *Probabilistic Eng. Mech.* **2009**, *25*, 49–60. [CrossRef]
12. Cimellaro, G.P.; Reinhorn, A.M. Multidimensional Performance Limit State for Hazard Fragility Functions. *J. Eng. Mech.* **2011**, *137*, 47–60. [CrossRef]

13. Michel, C.; Guéguen, P.; Causse, M. Seismic vulnerability assessment to slight damage based on experimental modal parameters. *Earthq. Eng. Struct. Dyn.* **2011**, *41*, 81–98. [CrossRef]

14. Ruggieri, S.; Porco, F.; Uva, G.; Vamvatsikos, D. Two frugal options to assess class fragility and seismic safety for low-rise reinforced concrete school buildings in Southern Italy. *Bull. Earthq. Eng.* **2021**, *19*, 1415–1439. [CrossRef]

15. Gentile, R.; Galasso, C. Simplicity versus accuracy trade-off in estimating seismic fragility of existing reinforced concrete buildings. *Soil Dyn. Earthq. Eng.* **2021**, *144*, 106678. [CrossRef]

16. Silva, V.; Akkar, S.; Baker, J.; Bazzurro, P.; Castro, J.M.; Crowley, H.; Dolsek, M.; Galasso, C.; Lagomarsino, S.; Monteiro, R.; et al. Current Challenges and Future Trends in Analytical Fragility and Vulnerability Modeling. *Earthq. Spectra* **2019**, *35*, 1927–1952. [CrossRef]

17. Bakhshi, A.; Asadi, P. Probabilistic evaluation of seismic design parameters of RC frames based on fragility curve. *Sci. Iran.* **2013**, *20*, 231–241. [CrossRef]

18. Karapetrou, S.; Fotopoulou, S.; Pitilakis, K. Seismic Vulnerability of RC Buildings under the Effect of Aging. *Procedia Environ. Sci.* **2017**, *38*, 461–468. [CrossRef]

19. Yu, X.-H.; Lu, D.-G.; Li, B. Relating Seismic Design Level and Seismic Performance: Fragility-Based Investigation of RC Moment-Resisting Frame Buildings in China. *J. Perform. Constr. Facil.* **2017**, *31*, 4017075. [CrossRef]

20. Dowrick, D.J. *Earthquake Resistant Design and Risk Reduction*; Wiley: Hoboken, NJ, USA, 2009.

21. Cornell, C.A.; Jalayer, F.; Hamburger, R.O.; Foutch, D.A. Probabilistic Basis for 2000 SAC Federal Emergency Management Agency Steel Moment Frame Guidelines. *J. Struct. Eng.* **2002**, *128*, 526–533. [CrossRef]

22. Cornell, C.A.; Krawinkler, H. Progress and challenges in seismic performance assessment. *PEER Cent. News* **2000**, *3*, 1–4. [CrossRef]

23. Moehle, J.; Deierlein, G.G. A framework for performance-based earthquake resistive design. In Proceedings of the 13th World Conference on Earthquake Engineering, Vancouver, BC, Canada, 1–6 August 2004.

24. Federal Emergency Management Agency (FEMA). *Seismic Performance Assessment of Buildings*; Applied Technology Council: Redwood City, CA, USA, 2012.

25. Lu, D.-G.; Liu, Y.; Yu, X.-H. Seismic fragility models and forward-backward probabilistic risk analysis in second-generation performance-based earthquake engineering. *Eng. Mech.* **2019**, *36*, 1–11. (In Chinese) [CrossRef]

26. *ISO 31000*; Risk management-Guidelines. ISO: Geneva, Switzerland, 2018.

27. Wang, G.-Y.; Cheng, G.-D.; Shao, Z.-M. *Optimal Fortification Intensity and Reliability of Anti-Seismic Struct*; Science Press: Beijing, China, 1999.

28. Spence, R.; So, E.; Jenny, S.; Castella, H.; Ewald, M.; Booth, E. The Global Earthquake Vulnerability Estimation System (GEVES): An approach for earthquake risk assessment for insurance applications. *Bull. Earthq. Eng.* **2008**, *6*, 463–483. [CrossRef]

29. Sahar, L.; Muthukumar, S.; French, S.P. Using Aerial Imagery and GIS in Automated Building Footprint Extraction and Shape Recognition for Earthquake Risk Assessment of Urban Inventories. *IEEE Trans. Geosci. Remote Sens.* **2010**, *48*, 3511–3520. [CrossRef]

30. Lu, X.; Zeng, X.; Xu, Z.; Guan, H. Improving the Accuracy of near Real-Time Seismic Loss Estimation using Post-Earthquake Remote Sensing Images. *Earthq. Spectra* **2018**, *34*, 1219–1245. [CrossRef]

31. Xiong, C.; Li, Q.; Lu, X. Automated regional seismic damage assessment of buildings using an unmanned aerial vehicle and a convolutional neural network. *Autom. Constr.* **2020**, *109*, 102994. [CrossRef]

32. Ruggieri, S.; Cardellicchio, A.; Leggieri, V.; Uva, G. Machine-learning based vulnerability analysis of existing buildings. *Autom. Constr.* **2021**, *132*, 103936. [CrossRef]

33. Hou, S.; Li, H.; Rezgui, Y. Ontology-based approach for structural design considering low embodied energy and carbon. *Energy Build.* **2015**, *102*, 75–90. [CrossRef]

34. Tserng, H.P.; Yin, S.Y.L.; Dzeng, R.-J.; Wou, B.; Tsai, M.D.; Chen, W.Y. A study of ontology-based risk management framework of construction projects through project life cycle. *Autom. Constr.* **2009**, *18*, 994–1008. [CrossRef]

35. Fidan, G.; Dikmen, I.; Tanyer, A.M.; Birgonul, M.T. Ontology for Relating Risk and Vulnerability to Cost Overrun in International Projects. *J. Comput. Civ. Eng.* **2011**, *25*, 302–315. [CrossRef]

36. Scheuer, S.; Haase, D.; Meyer, V. Towards a flood risk assessment ontology—Knowledge integration into a multi-criteria risk assessment approach. *Comput. Environ. Urban Syst.* **2013**, *37*, 82–94. [CrossRef]

37. Du, J.; He, R.; Sugumaran, V. Clustering and ontology-based information integration framework for surface subsidence risk mitigation in underground tunnels. *Clust. Comput.* **2016**, *19*, 2001–2014. [CrossRef]

38. Ding, L.; Zhong, B.; Wu, S.; Luo, H. Construction risk knowledge management in BIM using ontology and semantic web technology. *Saf. Sci.* **2016**, *87*, 202–213. [CrossRef]

39. Meng, K.; Cui, C.; Li, H. An Ontology Framework for Pile Integrity Evaluation Based on Analytical Methodology. *IEEE Access* **2020**, *8*, 72158–72168. [CrossRef]

40. Li, H.; Cheng, H.; Wang, D. A review of advances in seismic fragility research on bridge structures. *Eng. Mech.* **2018**, *35*, 1–16. (In Chinese) [CrossRef]

41. Ministry of Transport of the People's Republic of China. *Code for Seismic Design of Buildings*; China Architecture & Building Press: Beijing, China, 2010.

42. Mwafy, A.; Almorad, B. Verification of performance criteria using shake table testing for the vulnerability assessment of reinforced concrete buildings. *Struct. Des. Tall Spéc. Build.* **2019**, *28*, e1601. [CrossRef]

43. Federal Emergency Management Agency. *NEHRP Guidelines for the Seismic Rehabilitation of Buildings*; Federal Emergency Management Agency: Redwood City, CA, USA, 2009.

44. Worden, C.B.; Gerstenberger, M.C.; Rhoades, D.; Wald, D. Probabilistic Relationships between Ground-Motion Parameters and Modified Mercalli Intensity in California. *Bull. Seism. Soc. Am.* **2012**, *102*, 204–221. [CrossRef]

45. Gao, X.-W.; Bao, A.-B. Probabilistic model and its statistical parameters for seismic load. *Earthq. Eng. Eng. Vib.* **1985**, *5*, 13–22. (In Chinese) [CrossRef]

46. Ma, Z.; Liu, Z. Ontology- and freeware-based platform for rapid development of BIM applications with reasoning support. *Autom. Constr.* **2018**, *90*, 1–8. [CrossRef]

47. Kircher, C.A.; Whitman, R.V.; Holmes, W.T. HAZUS Earthquake Loss Estimation Methods. *Nat. Hazards Rev.* **2006**, *7*, 45–59. [CrossRef]

48. Comerio, M. Earthquake Protection, 2nd Edition. *Earthq. Spectra* **2003**, *19*, 731–732. [CrossRef]

49. Wang, D. Seismic Fragility Analysis and Probabilistic Risk Analysis of Steel Frame Structures. Ph.D. Thesis, Harbin Institute of Technology, Harbin, China, 2006.

50. Zhang, J.; Li, H.; Zhao, Y.; Ren, G. An ontology-based approach supporting holistic structural design with the consideration of safety, environmental impact and cost. *Adv. Eng. Softw.* **2018**, *115*, 26–39. [CrossRef]

51. Sirin, E.; Parsia, B.; Grau, B.C.; Kalyanpur, A.; Katz, Y. Pellet: A practical OWL-DL reasoner. *SSRN Electron. J. SSRN Electron. J.* **2007**, *5*, 51–53. [CrossRef]

52. Pang, Y.; Wang, X. Cloud-IDA-MSA Conversion of Fragility Curves for Efficient and High-Fidelity Resilience Assessment. *J. Struct. Eng.* **2021**, *147*, 4021049. [CrossRef]

Parameter Optimization and Application for the Inerter-Based Tuned Type Dynamic Vibration Absorbers

Xiaoxiang Wu [1], Xinnan Liu [2], Jian Chen [1], Kan Liu [3],* and Chongan Pang [1]

[1] Construction Engineering Construction Technology, Architectural Engineering Institute, Zhejiang Tongji Vocational College of Science and Technology, Hangzhou 311231, China; z20120170902@zjtongji.edu.cn (X.W.); z20120160201@zjtongji.edu.cn (J.C.); z20120060704@zjtongji.edu.cn (C.P.)
[2] Department of Equipment and Engineering Management, North China University of Technology, Beijing 100144, China; xinnanliu@ncut.edu.cn
[3] College of architecture and enviroment, Sichuan University, Chengdu 610065, China
* Correspondence: supkan007@hotmail.com

Abstract: As an acceleration-type mechanical element, inerter element has been widely used in the dynamic suppressing field. In this paper, a tuned mass damper with inerter (TMDI) is presented for vibration control and energy dissipation. To evaluate the effectiveness of the TMDI, the simplified model of TMDI coupled with a single-degree-of-freedom (SDOF) structure has been established. Numerical optimization has been conducted with the goal of minimizing the maximum transfer function amplitude of displacement for the damped primary structure. The control performance and robustness for TMDI has been evaluated with the SDOF system in the frequency and time domain, compared with the classical TMD device. Lately, multiple active TMDI (MATMDI) has been proposed as a vibration suppression strategy for a multi-story steel structure. The performances of passive and active control methods have been evaluated in the time domain via real earthquake excitations, and it has proven that the MATMDI is more effective at reducing the response of the structure and the stroke of devices. The results show that the proposed optimal TMDI system can sufficiently harvest vibrational energy and enhance the robustness of structure.

Keywords: tuned-mass damper; inerter; passive control; numerical optimization; active control; earthquake excitation

Citation: Wu, X.; Liu, X.; Chen, J.; Liu, K.; Pang, C. Parameter Optimization and Application for the Inerter-Based Tuned Type Dynamic Vibration Absorbers. *Buildings* **2022**, *12*, 703. https://doi.org/10.3390/buildings12060703

Academic Editor: Gianfranco De Matteis

Received: 20 April 2022
Accepted: 18 May 2022
Published: 24 May 2022

Publisher's Note: MDPI stays neutral with regard to jurisdictional claims in published maps and institutional affiliations.

1. Introduction

Lately, with the development of enormous activities, an ever-increasing number of structures are being constructed in the seismically dynamic regions. One of the most efficient methods to reduce earthquake-induced structural vibrations is to employ external vibration-control system on the structures. For instance, the base isolation [1], additional buckling-restraint bracing [2], and auxiliary dampers [3] can be added to the structure. Among them, the tuned mass damper (TMD) is a device with easy manufacture and disassembly, that is widely utilized in the structural vibration mitigation field [4]. It can be found in the existing large-scale and ultra-high structures, such as Taipei 101 [5], Citicorp Center in New York City [6], the Sydney TV Tower in Australia [7], and the Shanghai Center Tower in China [8]. It works on a simple principle. By altering the frequency of TMD, so that it could oscillate with the same period as the host structure, it absorbs and dissipates energy [9,10] from the main structure.

The TMD system is composed of the mass block, spring, and damping element . The component of TMD is pretty convenient. However, it is a challenges for the parameter design concerning obtaining the most reliable control efficacy. The fixed-point theory was first proposed by Den Hartog [11] to obtain the optimal parameters of TMD when it is attached to the undamped single-degree-of-freedom (SDOF) system. Further, Warburton et al. [12] summarized the analytical expressions for the optimal design parameters under various excitation conditions. Asami et al. [13] used the H_∞ and H_2 optimization approaches to acquire the analytical TMD solutions, considering the structural response under random excitation. While, for damped structures, it is hard to achieve the ideal design parameters by theoretical derivation, numerical approaches are commonly used to obtain values of the needed parameters [14,15]. In addition, due to the limitation of frequency band controlled by a single TMD, multiple TMDs have been employed to broaden the control frequency range [16,17]. Considering the complexity of the structure, environmental conditions and nonlinear characteristics of spring and damper element, the classical design approach makes the device too conservative and challenging to adapt to complex working conditions [18,19]. Besides, multiple objectives need to be regarded to optimization of parameters, in order to enhance the economy and adaptability of the device [20,21].

Inerter components, which are employed in a variety of applications in automobile suspension [22] and structural vibration suppression systems [23], have gotten a lot of attention in recent years. The conception of inerter was first proposed by the Professor Smith [24]. Similar to the damping and spring elements, the inerter is a mechanical element that is relevant to the acceleration of the terminals. The inerter can be simply equated to a system with large inertia, while its own mass value is quite small. A number of ways have been put forward to realize this mechanism, including ball-screw mechanical [25], hydraulic mechanism [26] and electromagnetic approach [27]. In the TMD system, the inerter element can be used to improve the vibration control efficiency of the device. For example, it could be used to minimize the required additional mass [28] or enhance energy dissipation capability [29].

As for the TMDI, it is much more difficult to obtain the optimum parametric resolution for each element compared with the classical TMD, especially for the damped structure under complex excitation situations. At the same time, it is also tough to obtain the appropriate parameters when more than four elements are included. Additionally, in most cases, the TMDI is used in the passive control strategies. However, there are negative aspects in passive control, such as the start-up lag, huge mass and inaccurate tuning. Apart from this, the variation of external environmental excitation could cause some impacts on the effectiveness of the device, such as the uncertainty of seismic excitation [30,31], the influence of the interaction between the foundation and the structure [32], and the degression of the structure [33]. Meanwhile, the location of the attached TMDI is needed for optimization, which is a significant factor for the vibration mitigation performance of the device [34,35]. In this case, passive control methods can hardly serve their function well. As a method of real-time regulation, active control can effectively provide feedback forces corresponding to the external excitation and structural response, which can significantly improve the robustness and reliability of the system. However, there are fewer reports [36] of active control being used to improve the control performance of TMDI systems.

The content of this paper is organized as follows. In Section 2, the numerical optimization approach is utilized to get the optimal TMD and TMDI parameters for an SDOF system under base acceleration excitation, with the optimization target of minimizing the amplitude of displacement transfer function for the main structure. In Section 3, an actual SDOF system is utilized to assess the vibration suppression efficacy under the ground motions. The response amplitude of the primary structure, energy dissipation ability and the stroke of the device are analyzed and compared. In Section 4, a multi-story steel structure is adopted as the multiple-degree-of-freedom system(MDOF) benchmark model for verification. Furthermore, both passive and active control strategies of multiple TMDIs are proposed to alleviate the response of the primary system. In Section 5, different vibration control strategies are compared and examined in the time and frequency domains in terms of various seismic records. Finally, in Section 6 some conclusions are given.

2. Model Establishment and Parameter Optimization

2.1. Numerical Model of SDOF System

The reaction forces of the spring and damping element are related to the relative displacement and velocity, respectively. While for the inerter element, the force magnitude is proportional to the relative acceleration between the endpoints. A typical ball-screw mechanical model of the ineter is presented in Figure 1. The force magnitude of the inerter can be noted as $F = m_{in}(\ddot{x}_i - \ddot{x}_j)$, where the constant of proportionality m_{in} is named as the inertance and has kilograms as units. Furthermore, the amount of energy it stores is equal to $1/2 m_{in}(\dot{x}_i^2 - \dot{x}_j^2)$.

The mass, spring, and damper element make up the basic TMD, as shown in Figure 2a. However, when the inerter element is added, it can perform superior in terms of energy dissipation. In the real applications, the components of TMDI systems may also have more complex behavior. Thus, various mechanical topology layouts are put forward [23,37]. While in this paper, a typical linear model of TMDI is adopted to demonstrate the ability of inerter element, as shown in Figure 2b, which is mentioned in reference [38].

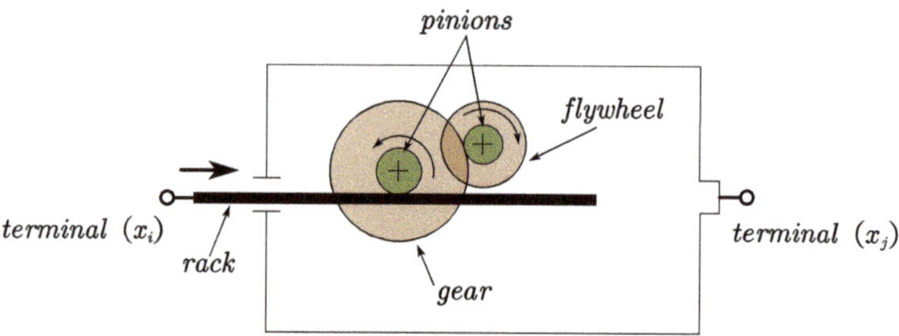

Figure 1. Schematic of a ball-screw mechanical model of an inerter [39].

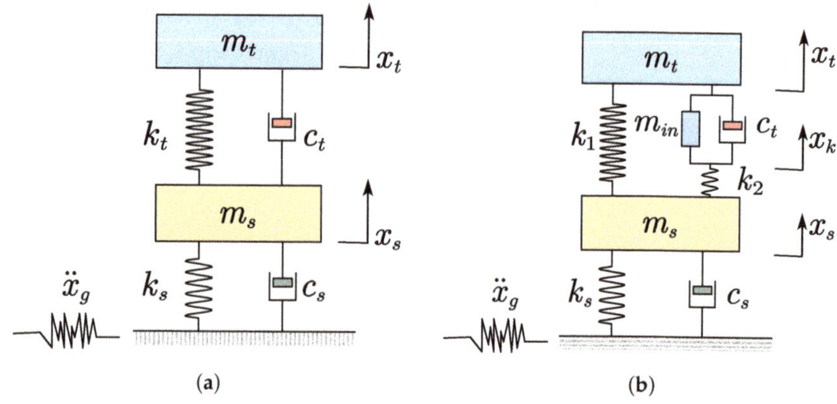

Figure 2. The schematic diagram of TMD and TMDI: (**a**) TMD and (**b**) TMDI.

It is assumed that the whole system is excited by the base acceleration $\ddot{x}_g$, the equation of motion for the dynamic system indicated above could be obtained with the equilibrium of forces. Accordingly, the kinetic equations for both the TMD and TMDI are expressed in Equations (1) and (2), respectively.

$$
\begin{bmatrix} m_s & 0 \\ 0 & m_t \end{bmatrix}
\begin{bmatrix} \ddot{x}_s \\ \ddot{x}_t \end{bmatrix} +
\begin{bmatrix} c_s + c_t & -c_t \\ -c_t & c_t \end{bmatrix}
\begin{bmatrix} \dot{x}_s \\ \dot{x}_t \end{bmatrix} +
\begin{bmatrix} k_s + k_t & -k_t \\ -k_t & k_t \end{bmatrix}
\begin{bmatrix} x_s \\ x_t \end{bmatrix} =
-\begin{bmatrix} m_s \\ m_t \end{bmatrix} \ddot{x}_g
\tag{1}
$$

$$\begin{bmatrix} m_s & 0 & 0 \\ 0 & m_t + m_{in} & -m_{in} \\ 0 & -m_{in} & m_{in} \end{bmatrix} \begin{bmatrix} \ddot{x}_s \\ \ddot{x}_t \\ \ddot{x}_k \end{bmatrix} + \begin{bmatrix} c_s & 0 & 0 \\ 0 & c_t & -c_t \\ 0 & -c_t & c_t \end{bmatrix} \begin{bmatrix} \dot{x}_s \\ \dot{x}_t \\ \dot{x}_k \end{bmatrix} + \begin{bmatrix} k_s + k_1 + k_2 & -k_1 & -k_2 \\ -k_1 & k_1 & 0 \\ k_2 & 0 & k_2 \end{bmatrix} \begin{bmatrix} x_s \\ x_t \\ x_k \end{bmatrix} = - \begin{bmatrix} m_s \\ m_t \\ 0 \end{bmatrix} \ddot{x}_g \tag{2}$$

where m_s, k_s, and c_s are the mass, spring, and damping coefficient of the host structure; m_t, c_t are the mass and damping of the addition attached damper; m_{in} is the inertance of inerter element; k_t is the spring stiffness of TMD; k_1, k_2 are the two spring stiffnesses of TMDI.

The transfer function (TF) reflects the output process of structural response under excitation, which presents the dynamic characteristics of the structure. As long as the transfer function of the structure is suppressed, the ideal control effect can be achieved regardless of the external excitation. For this reason, the optimization objective is to reduce the displacement transfer function of the main structure as much as possible. Meanwhile, in an effort to make the control effect more clearly reflected and representative, the parameters should be dimensionless. For this regard, the nondimensional expressions for the parameters listed above are expressed as follows:

$$\begin{aligned} &\zeta_s = c_s/(2m_s\omega_s), \omega_s = \sqrt{k_s/m_s}, \omega_1 = \sqrt{k_1/m_t}, \omega_2 = \sqrt{k_2/m_t}, \omega_t = \sqrt{k_t/m_t}, \\ &\zeta_t = c_t/(2m_t\omega_t) = c_t/(2m_t\omega_1), u_t = m_t/m_s, u_{in} = m_{in}/m_t, \\ &f_d = \omega_t/\omega_s, f_1 = \omega_1/\omega_s, f_2 = \omega_2/\omega_s \end{aligned} \tag{3}$$

where ω_s, ζ_s are the circular frequency and damping ratio of the primary structure; ω_1, ω_2 and ω_t are the circular frequency of spring k_1, k_2, and k_t, respectively; u_t and u_{in} are the mass ratio and interance-mass ratio; ω is the frequency of the external excitation; f_1, f_2, and f_d represent the tuning frequency ratio of the spring k_1, k_2, and k_t corresponding to the primary system.

Assuming that the base acceleration is simplified as a harmonic excitation, i.e., $\ddot{x}_g = e^{i\omega t}$, the displacement of the main structure can be denoted as $x_s = H_m(\omega)e^{i\omega t}$. Then, the dimensionless transfer function for the structural displacement (H_m) can be found in Equation (4). While the parameters of $A/B/C/D$ for TMD and TMDI can be obtained, as shown in Equations (5) and (6).

$$H_m(\beta) = \frac{x_s\omega_s^2}{\ddot{x}_g} = \left| \frac{A_i + B_i}{C_i + D_i} \right| \tag{4}$$

For TMD:

$$\begin{cases} A_1 = u_t\left[\beta^2 - f^2(1 + u_t)\right] \\ B_1 = 2i\zeta_t f\beta u_t(u_t + 1) \\ C_1 = -u_t\left[\beta^2 - \beta^4 + f^2(-1 + \beta^2(1 + u_t)) + 2i\beta(-f^2 + p^2)\zeta_s\right] \\ D_1 = -2f\beta u_t\zeta_t\left[i(-1 + \beta^2(1 + u_t)) + 2\beta\zeta_s\right] \end{cases} \tag{5}$$

For TMDI:

$$\begin{cases} A_2 = -2f_2\zeta_t\beta u_t^2\left[-\beta^2 + f_1^2(1 + u_t) + f_2^2(1 + u_t)\right] \\ B_2 = -iu_t^2\left[-\beta^4 u_{in} - f_1^2\left(f_2^2 - p^2 u_{in}\right)(1 + u_t) + f_2^2\beta^2(1 + u_{in} + u_{in}u_t)\right] \\ C_2 = -2f_2\beta u_t^2\zeta_t\left[\begin{matrix} -2i\beta\zeta_s\left(f_1^2 + f_2^2 - \beta^2\right) + f_1^2(\beta^2(u_t + 1) - 1) \\ +f_2^2\beta^2 u_t + f_2^2\beta^2 - f_2^2 - \beta^4 + \beta^2 \end{matrix}\right] \\ D_2 = -u_t^2\left[i\left(\begin{matrix} -f_1^2(\beta^2(u_t + 1) - 1)\left(f_2^2 - \beta^2 u_{in}\right) \\ +f_2^2\beta^2(\beta^2(u_{in}u_t + u_{in} + 1) - u_{in} - 1) \\ -\beta^4(\beta^2 - 1)u_{in} \end{matrix}\right) + 2\beta\zeta_s\left(\begin{matrix} f_1^2\left(\beta^2 u_{in} - f_2^2\right) \\ +f_2^2\beta^2(u_{in} + 1) \\ -\beta^4 u_{in} \end{matrix}\right)\right] \end{cases} \tag{6}$$

where $i = \sqrt{-1}$ is the imaginary number; $\beta = \omega/\omega_s$, which is the ratio between the frequency of base acceleration and main structure.

2.2. Parameters Optimization

The optimal design is needed to get the best performance of the dynamical vibration absorbers (DVAs). In the previous optimization of TMD, the fixed-point theory was commonly used [40]. For the undamped host structure, it is assumed that the transfer function curve always passes through a fixed point, when the damping ratio (ζ_t) changes only. By keeping the amplitude equal at the fixed point, the optimal tuning frequency of the device can be found. The optimal damping ratio is obtained by holding the curve at this fixed point until it reaches its peak. The analytical expression of the optimal parameters can be obtained through this method.

It is, however, incapable of handling a sophisticated model with more than three mechanical elements in the device. Furthermore, the damping ratio of the primary structure is ignored in this fixed-point theory, which does not correlate to reality. As a result, a general approach to solving these problems is required. The numerical searching method is performed, regarding peak relative displacement responses of the primary structure as the optimization objectives, which is indicated in the following expression.

$$\begin{cases} \text{Minimize}: & \max(H_m(\beta)) \\ \text{Objective}: & \zeta_{t,opt}, f_{t,opt}, f_{1,opt}, f_{2,opt}, u_{in,opt} \\ \text{Subject to}: & 0 < \zeta_t, f_d, f_1, f_2, u_{in}, u_t \leq 1 \end{cases} \tag{7}$$

These values are optimized in MATLAB software using the ***fmincon*** command [41] based on the above boundary conditions. Besides this, the damping ratios (ζ_s) for the primary structure are set as 0.01, 0.03, and 0.05 to give an example. The optimal parameters of TMD and TMDI with varied mass ratios (u_t) are obtained using this numerical method for various damping ratios (ζ_s) of the main structure, as illustrated in Figures 3 and 4.

From the above figures, it can be seen that the damping ratio of the structure (ζ_s) affects the optimal design parameters of the DVA device. For both different devices, the value of ζ_d increased a little with the increase of ζ_s, while f_1 and f_d decreased with ζ_s. In addition, as the mass ratio u_t increases, the tuning frequency ratio of f_d and f_1 decreases, while f_2 keeps rising when u_t is less than 0.4 and then starts to fall. Meanwhile, the inertance – mass ratio (u_{in}) of TMDI has the same tendency as f_2. For convenience in practical application, all the results are fitted with the quartic polynomial, as written in Equation (8). In addition, the coefficient of the polynomial is reported in Table 1.

$$P_{opt}(u_t) = p_1 u_t{}^4 + p_2 u_t{}^3 + p_3 u_t{}^2 + p_4 u_t + p_5 \tag{8}$$

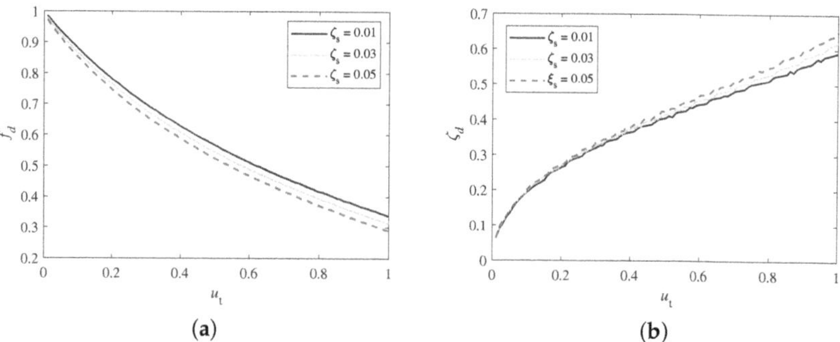

Figure 3. The optimal design parameters of TMD with the various mass ratios u_t: (**a**) Tuning frequency (f_d) and (**b**) Damping ratio (ζ_d).

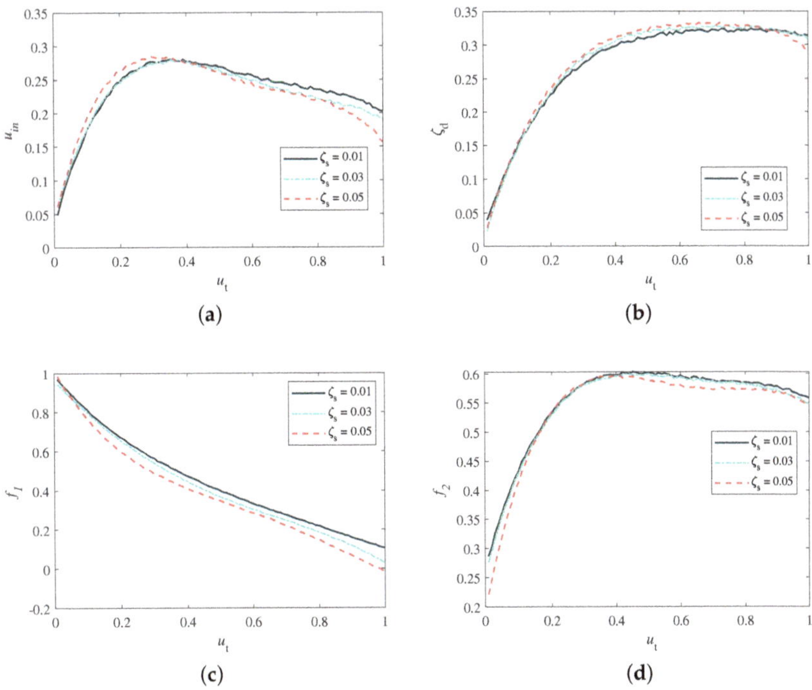

Figure 4. The optimal design parameters of TMDI with the various mass ratios u_t: (**a**) Inertance - mass ratio u_{in}; (**b**) Damping ratio ζ_d; (**c**) Tuning frequency f_1; (**d**) Tuning frequency f_2.

Table 1. The coefficient of the polynomial for the optimal parameters.

		TMDI				TMD	
		u_{in}	ζ_t	f_1	f_2	f_d	ζ_t
$\zeta_s = 0.01$	p_1	−1.969	−0.907	0.578	−1.800	0.220	−1.043
	p_2	5.114	2.606	−1.923	4.915	−0.746	2.708
	p_3	−4.832	−2.896	2.458	−4.981	1.138	−2.535
	p_4	1.855	1.485	−1.997	2.160	−1.269	1.389
	p_5	0.030	0.022	0.986	0.262	0.997	0.070
$\zeta_s = 0.03$	p_1	−1.547	−1.139	−0.403	−2.032	0.309	−1.022
	p_2	4.261	3.166	−0.133	5.379	−0.968	2.703
	p_3	−4.271	−3.394	1.407	−5.302	1.339	−2.543
	p_4	1.703	1.669	−1.806	2.249	−1.353	1.404
	p_5	0.043	0.006	0.962	0.251	0.990	0.073
$\zeta_s = 0.05$	p_1	−2.770	−1.404	1.982	−3.056	0.398	−0.981
	p_2	6.724	3.583	−5.619	7.908	−1.200	2.646
	p_3	−5.880	−3.596	5.567	−7.391	1.549	−2.507
	p_4	2.042	1.696	−2.955	2.891	−1.438	1.413
	p_5	0.039	0.012	1.009	0.191	0.983	0.074

2.3. Performance Comparison

To visually show the difference between the two oscillators, the displacement transfer functions of the main structure are plotted. The mass ratio u_t for both TMD and TMDI is set as 0.10. The difference between them can be seen in Figure 5. Compared with the TMD, it could be found that the maximum amplitude with TMDI is significantly decreased. However, the TMDI has three peaks in its curve, which is one more than the TMD. This is because TMDI has three DOFs, as can be seen from the configuration, which means that there are more alternatives for adjusting the overall performance of device. In the process of functioning, the two springs of TMDI could play different roles. The spring k_1 is used for the primary tuning, while the spring k_2 is responsible for increasing the stroke and deformation between the damping element in order to absorbing more energy. As illustrated in Figure 6, the maximum amplitudes of the transfer function of the structural displacement at various mass ratios (u_t) are also compared. It can be observed that with the increasing of additional mass, the control effectiveness of the two devices improves, while the TMDI outperform the TMD.

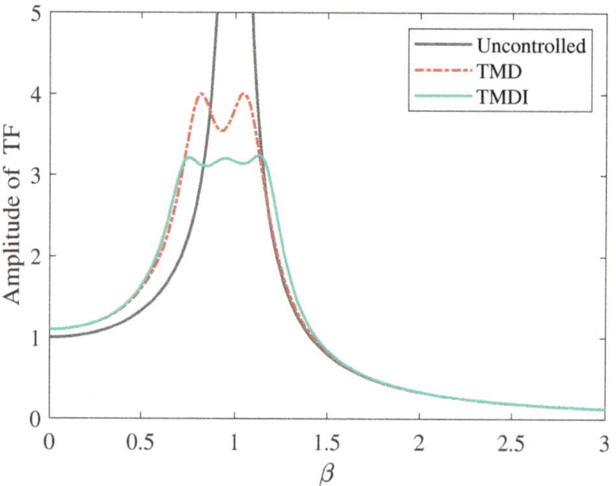

Figure 5. The TF amplitude of the structural displacement under different devices ($u_t = 0.10$).

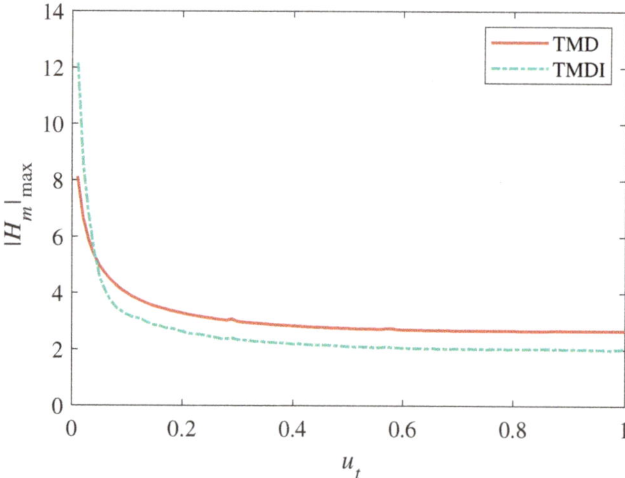

Figure 6. The maximum value of displacement TF with different mass ratios.

What is more, the additional influence of the DVAs on the structure can be attributed to a higher damping ratio in the primary system. By comparing the altered damping ratio of the structure, the influence of the DVA on the structure can be explicitly represented. As for this respect, various methods are presented to evaluate the overall damping of a structure with the addition of dampers. As derived by reference [42], when the mass ratio is small enough, the equivalent damping ratio (ζ_e) of the whole system can be estimated by comparing the amplitude of the transfer function, as described in Equation (9). In accordance with this method, the variation trend of the equivalent damping ratio of the whole system with different mass ratios is plotted in Figure 7, with an assumed damping ratio (ζ_s) of 0.03. It can be found that the equivalent damping ratio (ζ_e) of the system extended as the mass ratio u_t increases. Compared to the TMD, the TMDI system could consume more energy with the assistance of inerter element.

$$\zeta_e \approx \frac{1}{2H_{m,\max}} \tag{9}$$

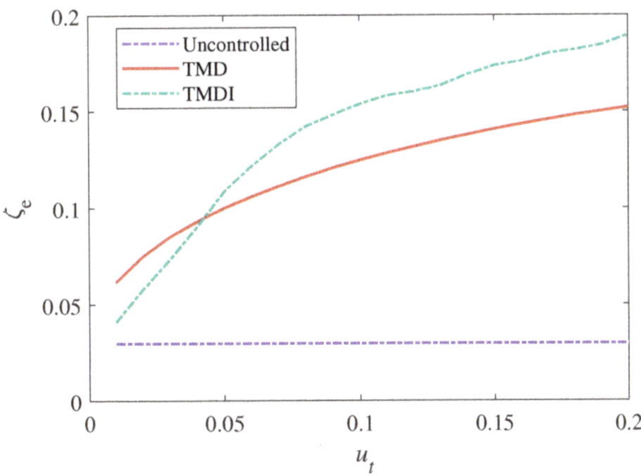

Figure 7. The equivalent damping ratio of the whole system varying with u_t ($\zeta_s = 0.03$).

3. Case Study 1—A SDOF Oscillator

3.1. Parameters of SDOF Structure

In this section, an actual SDOF structure is adopted for time-domain analysis to evaluate the differences between two DVAs more intuitively. The mass of the structure is assumed to be 2000 kg, with a stiffness of 20,000 N/m and 379.47 N · m/s for the damping coefficient ($\zeta_s = 0.03$). In this section, the mass ratio u_t for both TMD and TMDI are all set as 0.10. The specific values of each parameter can be obtained using Equation (8)), with the coefficients in Table 1. The parameters of the two equipments are listed in the table below.

Table 2. Parameters of TMDI and TMD for the specific SDOF system.

		TMDI			TMD	
m_t (kg)	u_{in} (kg)	c_t (N·m/s)	k_1 (N/m)	k_2 (N/m)	c_t (Nm/s)	k_t (N/m)
200.00	34.98	77.78	1272.47	372.24	214.15	1500.50

3.2. Reaction of the Primary Structure

To compare the performance of the two devices, the typical ground motion records of EL Centro and Kobe waves are taken for verification. The reactions of the structure under seismic loading are obtained using the Newmark-β algorithm [43], which is one of the most adopted time-integration methods [44,45]. The relative displacement and absolute acceleration of the main structure are depict in Figures 8 and 9. It can be noticed that, for the same mass ratio, both devices perform well in the whole process. However, in comparison with the classical TMD, the TMDI can further decrease the response of the results over the entire time range.

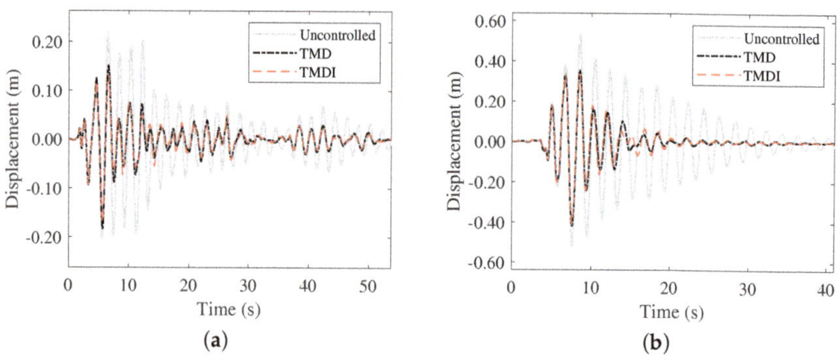

Figure 8. The relative displacement of the primary structure under earthquake: (**a**) EL Centro and (**b**) Kobe.

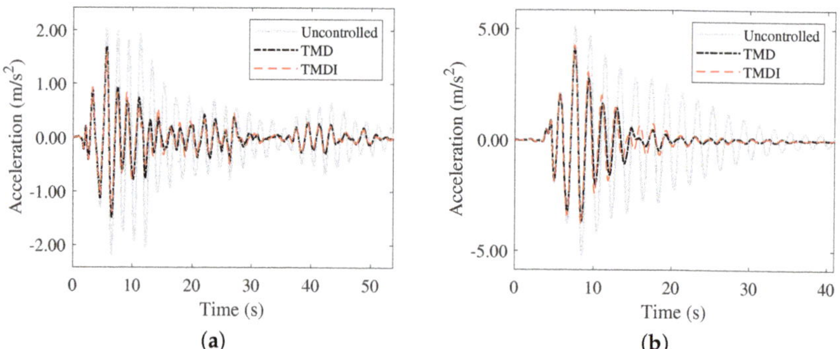

Figure 9. The absolute acceleration of the primary structure under earthquake: (**a**) EL Centro and (**b**) Kobe.

3.3. Stroke of Damping elements

Apart from the responsiveness of the main structure, the stroke of devices is also the essential factor to consider. Especially in the buildings with limited space, the stroke of the additional devices is a critical restraint. In this SDOF system, the relative displacement is compared under different ground motions, as illustrated in Figure 10. In this regard, the TMD has the advantage of a shorter stroke, whereas the maximal relative displacement of the TMDI is around 1.3 times as much as the TMD. Additionally, the deformation of the two spring for the TMDI are also varied. The relative motion of k_1 is larger than k_2 because of smaller stiffness compared with the classical TMD. Whereas, the extra spring k_2 is utilized to increase the deformation of the damping element.

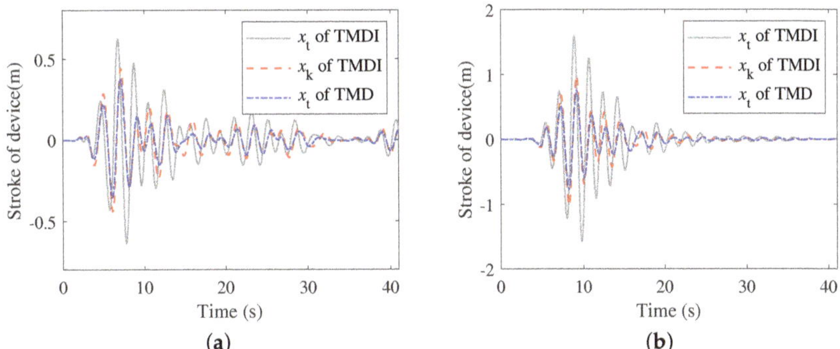

Figure 10. The stroke of different devices under earthquake: (**a**) EL Centro and (**b**) Kobe.

3.4. Energy Dissipation

As indicated in the Table 2, the damping coefficient of TMDI is much smaller than that of TMD, which is around 1/3 of it. Therefore, a greater deformation is required to output larger damping forces to achieve the same desired energy dissipation. The damping forces and relative displacement for the damping element for each device are depicted in Figure 11 when activated by the earthquake. It can be seen that the hysteresis loops of the damping force and relative displacement for the TMD are broadly circular, whereas the curve of TMDI is almost elliptical. In contrast, the damping force of TMDI is much smaller, with a more extensive deformation compared with TMD.

The energy consumption curves of TMD and TMDI are given in Figure 12, with the example of EL Centro wave. It consumed about 538.74 and 568.64 kJ of energy for TMD and TMDI, accounting for 68.02% and 72.24% of the toltal energy harvest by the entire system, respectively. As can be observed, the TMDI could absorb more energy than the traditional TMD. At the same time, it is possible to select a damping element with a lower damping coefficient, hence reducing the requirements of the auxiliary device and cost. Similarly, a lower mass ratio of TMDI can be considered to achieve the same vibration suppression effectiveness.

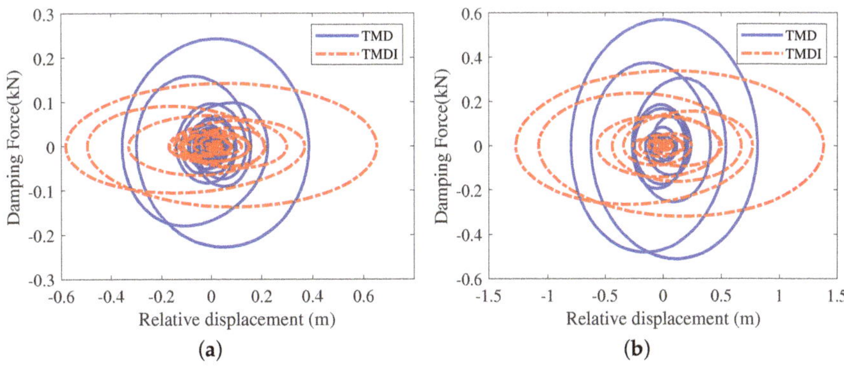

Figure 11. The hysteretic curve of damping element under different ground motion: (**a**) EL Centro and (**b**) Kobe.

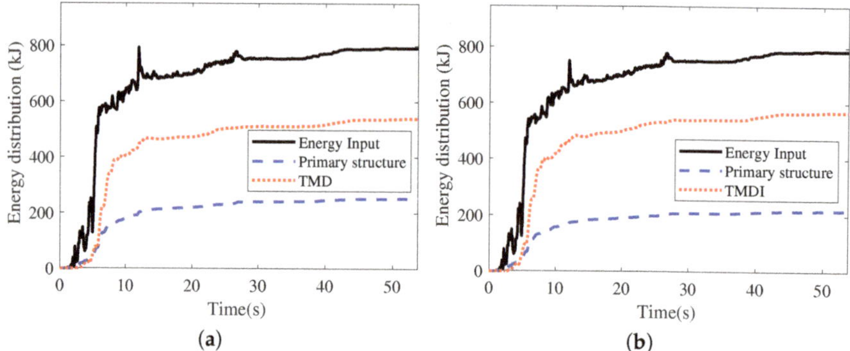

Figure 12. The energy consumption for TMD and TMDI under EL Centro wave: (**a**) TMD and (**b**) TMDI.

4. Case Study 2– A MDOF Benchmark Model

4.1. Introduction of Model

In this section, a typical multi-story steel building located in Los Angeles is taken as a benchmark model [46] in this paper. It has commonly been used as a benchmark model to assess the effectiveness of various control strategies [47,48]. The planar dimensions of this structure are 30.48 m by 36.58 m, with a height of 80.77 m. There are 20 stories above the ground surface, and the roof denotes the 21st level, as depicted in Figure 13. It includes two extra basement floors, and each floor averages 3.96 m in elevation. The two basement floors have a height of 3.65 m, while the first floor has a height of 5.49 m. The horizontal displacement of the first floor is restricted by the concrete foundation walls and surrounding soil. The mass of the first and second floors is 2.66×10^5 kg and 2.83×10^5 kg, the mass of the third to 20th floors is 2.76×10^5 kg, and the mass of the roof is 2.92×10^5 kg. In this section, we only concentrate on the plane of this structure, as well as the short direction.

The structure is modeled as a plane-frame elements, and the mass and stiffness matrices for the structure are determined. There are 180 nodes interconnected by 284 elements in this model. The beam and column members are modeled as the plane element, with two nodes and six degrees of freedom. The detailed properties of these elements can be found in reference [49], including the length, area, the moment of inertia, modulus of elasticity, and mass density.

The structure is modelled by the finite element method (FEM), with 540 DOFs for the mass and stiffness matrix. After that, 14 DOFs on the first floor are eliminated, considering the boundary constraints on the bottom. In addition, the system can be further simplified because the floor slab can be assumed as rigid in the horizontal plane, and the nodes associated with each floor have the same horizontal displacement. Then the Guyan reduction method [50] is used to decrease the rotational and most of the vertical DOFs. In the end, the mass matrix $\mathbf{M}_s$ and stiffness matrix $\mathbf{K}_s$ have the dimensions 106×106. The modal parameters, including the frequencies and mode shapes, are obtained by solving the following eigenvalue problem. The modal shapes of the first three modes are plotted in Figure 14.

$$\left(\mathbf{K}_s - \omega^2 \mathbf{M}_s\right)\mathbf{\Phi} = 0 \tag{10}$$

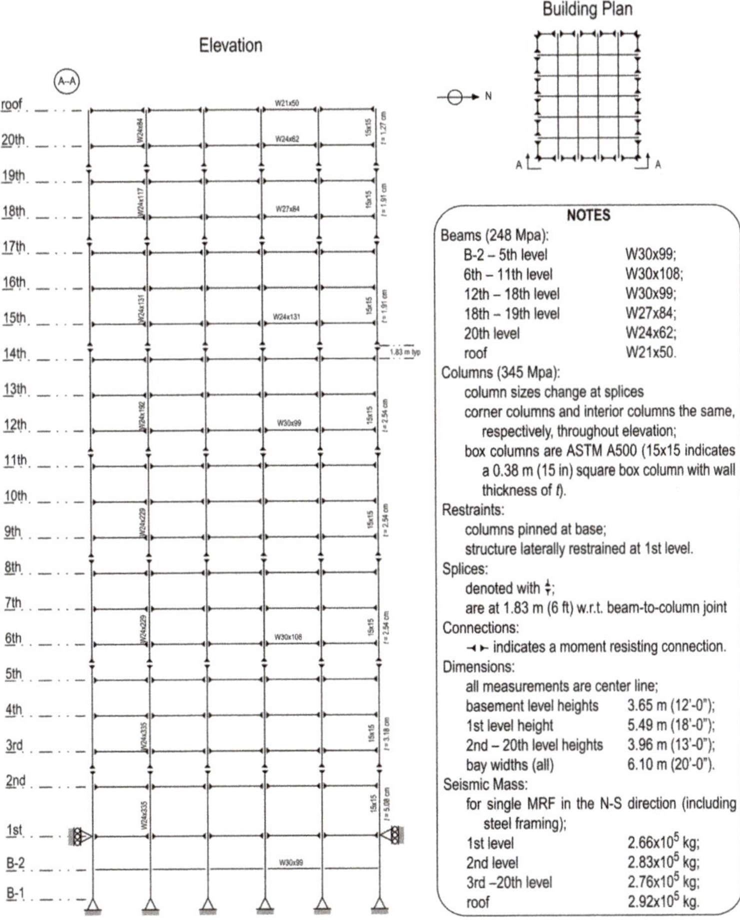

Figure 13. Cross-section frame in the north–south direction [46].

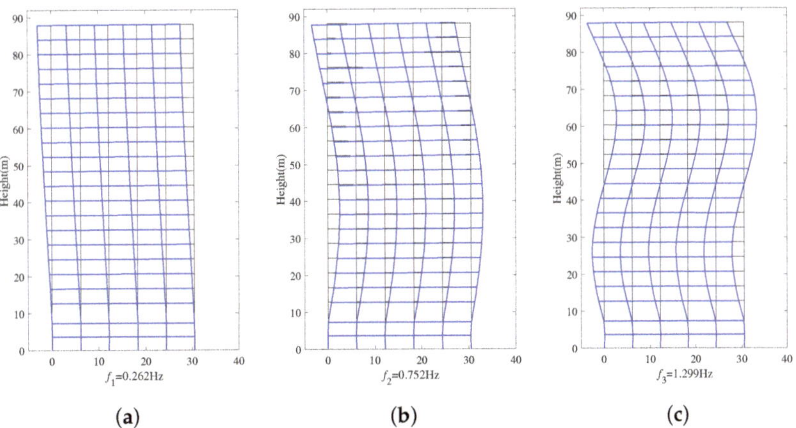

Figure 14. The first three mode shapes of the structure: (**a**) Fundamental mode; (**b**) Second mode; (**c**) Third mode.

The damping matrix $\mathbf{C}_s$ is obtained on the base of the reduced system with the assumption of modal damping. The damping of each mode ζ_i is assumed to be proportional to the associated frequency, with a maximum critical damping of 10%. Then, the values of ζ_i are calculated according to Equation (11), and the global damping matrix can be acquired as in Equation (12).

$$\zeta_i = \min\left\{\frac{\omega_i}{50\omega_1}, 0.1\right\}. \tag{11}$$

$$\mathbf{C}_s = \mathbf{M}_s\mathbf{\Phi}(2\zeta\omega)\mathbf{\Phi}^{-1} \tag{12}$$

where ω_i and ζ_i are the i_{th} modal circular frequency and damping ratio of the structure, respectively; $\zeta = diag\{\zeta_1, \zeta_2, \cdots, \zeta_n\}$; $\omega = diag\{\omega_1, \omega_2, \cdots, \omega_n\}$.

4.2. External Loads Acting on the Structure

In order to further verify the dynamic performance of the high–rise buildings with the designed proposed control strategy. It is supposed that this structure is excited by the earthquake. In addition, it is assumed that the base accleration excitation is non-stationary in nature and has a more complicated spectral content than the white–noise random process. In this respect, the colored power spectral density (PSD) of the Clough–Penzien model [51] is employed as the base excitation, which is represented as shown in Equation (13). According to the theory of earthquake engeering, the amplitude of ground motion is filtered by the soil when it originates at the bed rocket. The soil effect could be converted to a two DOF linear filter that filters out the low-frequency component.

$$S_{\ddot{x}_g}(\omega) = S_0\frac{\omega_g^4 + 4\zeta_g^2\omega_g^2\omega^2}{\left(\omega_g^2 - \omega^2\right)^2 + 4\zeta_g^2\omega_g^2\omega^2}\frac{\omega^4}{\left(\omega_f^2 - \omega^2\right)^2 + 4\zeta_f^2\omega_f^2\omega^2} \tag{13}$$

where S_0 is a scale factor; ζ_g, ω_g denote the equivalent damping ratio and fundamental frequency of the single-layer soil of the local site; ζ_f, ω_f denote the equivalent damping ratio and fundamental frequency of the second-layer soil of the local site.

In the power spectral density (PSD), the values of the parameters considering different types of soils are indicated in Table 3, as illustrated in reference [51]. The short period of ground motion is related to the firm soil site, while the long period of ground motion refers to the soft soil site. The structure is considered to be at the firm soil site in this region, for the reason that the response would be more severe with the short period of base excitation. The scale factor S_0 demonstrated in Equation (13) is considered as 1 $g^2 \cdot s$, while other parameters are selected from Table 3 according to the firm soil type. After that, the base acceleration excitation is a non-stationary random process specified by the Clough–Penzien spectra. Three authentic acceleration signals are then generated in accordance with this PSD model and the use of inverse Fourier transformation, which is named W1 $\sim$ W3.

Table 3. The parameters of the Clough–Penzien model for different soil types.

Soil Type	ω_g (rad/s)	ζ_g	ω_f (rad/s)	ζ_f
Firm	15.0	0.6	1.5	0.6
Medium	10.0	0.4	1.0	0.6
Soft	5.0	0.2	0.5	0.6

Moreover, the efficacy of the two DVAs is tested with the actual ground motions. The construction is assumed to be placed on a class I_1 site with firm soils in the second group, according to the Chinese standard code for seismic design of buildings [52]. The shear velocity v_{s20} is 500$\sim$800 m/s for the firm soil site, which is transformed to 510$\sim$760 m/s for v_{s30} according to the standard of ASCE [53]. Following that, the target acceleration response spectra are obtained, as shown in Figure 15. Seven records are obtained from the database of PEER [54], and the selected ground motion data are listed in Table 4. After that, the response spectrum for all the ground motions are also plotted in Figure 15.

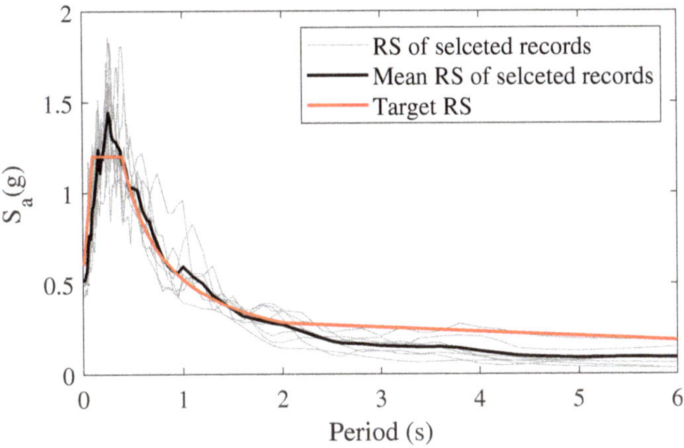

Figure 15. The acceleration response spectrum of the selected records

Table 4. The information of the selected earthquake records

Series	Earthquake Name	Year	Magnitude	Station Name	V_{s30} (m/s)	R_{jb} (km)
R1	Chi-Chi_Taiwan	1999	7.62	CHY019	573.04	46.59
R2	Chi-Chi_Taiwan	1999	7.62	HWA035	677.49	44.02
R3	Chi-Chi_Taiwan	1999	7.62	ILA050	621.06	63.82
R4	Chi-Chi_Taiwan	1999	7.62	TCU071	624.85	0.00
R5	Chi-Chi_Taiwan	1999	7.62	TCU089	535.13	83.38
R6	Chi-Chi_Taiwan	1999	7.62	TTN051	665.20	30.77
R7	Chi-Chi_Taiwan	1999	7.62	TCU129	511.18	1.83

4.3. Dynamic Modal Establishment for Passive Control

In the previous study, it is normally organized at the top of the structure if only a single tuned-type oscillator is considered [32]. The first-order frequency of the structure is mainly considered for the single TMDI to achieve the best control effect. However, this usually causes some problems. The mass and volume of the device will be so large that the structure needs to be specially designed to consider the arrangement of the attached device, which may result in additional construction and costs. For this reason, in this section, multiple TMD and TMDIs are considered to be arranged in the structure to reduce the huge additional mass of a single device. Besides, the designed frequency of each DVA is adjusted to control to enhance the resilience and dependability of the equipment.

When considering the arrangement of DVAs, it can be considered according to the distribution of vibration shapes for different modes. From Figure 14, it can be found that the largest amplitude of the fundamental mode is at the top of the building, whereas the peak position of the second mode is on the ninth floor. According to this, multiple DVAs are employed to suppress the first two modes of the structure. Then, eight devices are organized in this section, and they are located between the 7th and 21st floors, with intervals of two stories. The schematic of the entire system is shown in Figure 16. In addition, the frequencies are uniformly dispersed from the first to the second model, with the frequency for each device computed using Equation (14). At the same time, the other parameters can be acquired by Equation (8) when the frequency is definite.

$$\omega_{t,i} = \omega_{s,1} + \frac{\omega_{s,2} - \omega_{s,1}}{n_t - 1}(i-1)(i=1,2,\cdots,n_t) \tag{14}$$

where $\omega_{s,1}$ and $\omega_{s,2}$ are the fundamental and second model frequency of the main structure, respectively; $\omega_{t,i}$ is the frequency that the i_{th} TMD and TMDI need to be controlled; n_t is the total number of DVAs.

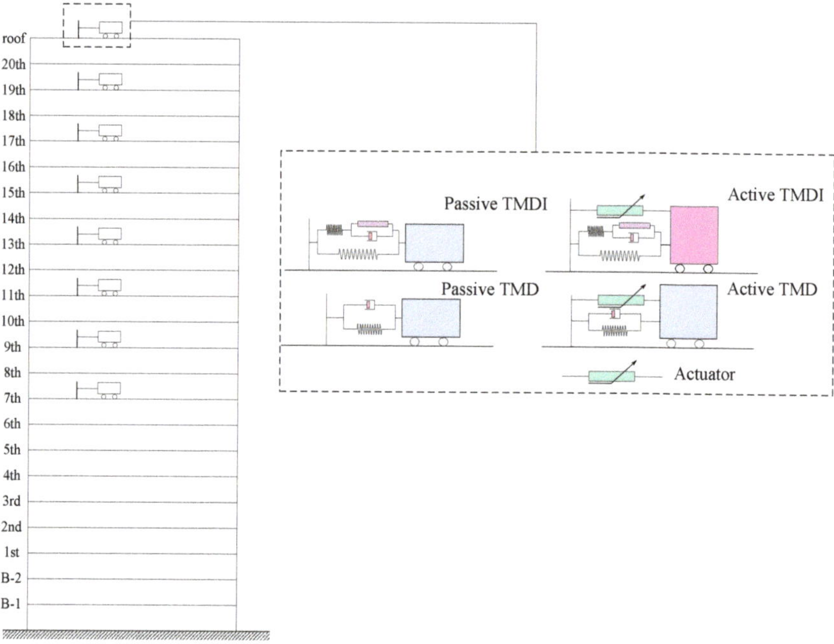

Figure 16. The entire structure assembled with multiple passive and active DVAs.

When the parameters of each device are determined, the equation of motion of the whole system under the base acceleration excitation is expressed as in the following formulation.

$$M\ddot{X} + C\dot{X} + KX = -ML\ddot{x}_g \tag{15}$$

For the multiple TMDs:

$$M = \begin{bmatrix} M_s & 0 \\ 0 & M_t \end{bmatrix}, C = \begin{bmatrix} C_s + P_1 C_t P_1^T & -P_1 C_t \\ -C_t P_1^T & C_t \end{bmatrix}, K = \begin{bmatrix} K_s + P_1 K_t P_1^T & -P_1 K_t \\ -K_t P_1^T & K_t \end{bmatrix} \tag{16}$$

For the multiple TMDIs:

$$M = \begin{bmatrix} M_s & 0 & 0 \\ 0 & M_t + M_{in} & -M_{in} \\ 0 & -M_{in} & M_{in} \end{bmatrix}, C = \begin{bmatrix} C_s & 0 & 0 \\ 0 & C_t & -C_t \\ 0 & -C_t & C_t \end{bmatrix}$$

$$K = \begin{bmatrix} K_s + P_2(K_1 + K_2)P_2^T & -P_2 K_1 & -P_2 K_2 \\ -K_1 P_2^T & K_1 & 0 \\ -K_2 P_2^T & 0 & K_2 \end{bmatrix} \tag{17}$$

where $\ddot{X}, \dot{X}$ and X are the relative acceleration, velocity, and displacement of the global system; M, C and K are the mass, stiffness, and damping matrix of the whole system, respectively; M_s, C_s and K_s are the mass, stiffness, and damping matrix of the original main system, respectively; $M_t = diag\{m_{t1}, m_{t2}, \cdots, m_{tn}\}$ is the diagonal matrix of additional mass of TMD and TMDI; $M_{in} = diag\{m_{in1}, m_{in2}, \cdots, m_{inn}\}$ is the diagonal matrix of inerterance; $C_t = diag\{c_{t1}, c_{t2}, \cdots, c_{tn}\}$ is the diagonal matrix of damping of

TMD and TMDI; $\mathbf{K_1} = diag\{k_{11}, k_{12}, \cdots, k_{1n}\}$ is the diagonal matrix of spring k_1 of TMDI; $\mathbf{K_2} = diag\{k_{21}, k_{22}, \cdots, k_{2n}\}$ is the diagonal matrix of spring k_2 of TMDI; $\mathbf{K_t} = diag\{k_{t1}, k_{t2}, \cdots, k_{2tn}\}$ is the diagonal matrix of spring k_t of TMD; $\mathbf{P_1}$ and $\mathbf{P_2}$ are the Boolean matrix for the location of TMD and TMDI, respectively; $\mathbf{L}$ is the distribution matrix for the inertia forces under earthquake; n_s is the total number of DOFs of the original structure; $\ddot{x}_g$ is the acceleration of ground motion.

4.4. Active Control with Multiple DVAs

Passive control does not require external energy and can maintain process stability. However, some shortages of the passive control strategy exist in the operation period, such as start-up lag and large stroke. While the active control method is a useful technique in addressing such issues, which have been frequently employed in practice. To implement the effective control, multiple active TMDs and TMDIs are adopted in this section. The placement is identical to that of passive control, with the exception of the addition of an active actuator between each additional mass of devices and the main structure, as shown in Figure 16. Consequently, the equation of motion for the entire system has altered as a result of the inclusion of active forces, as illustrated in the equation below.

$$\mathbf{M\ddot{X}} + \mathbf{C\dot{X}} + \mathbf{KX} = -\mathbf{ML}\ddot{x}_g + \mathbf{B_s U}_e \tag{18}$$

where $\mathbf{U}_e$ is the external active forces; $\mathbf{B}_s$ is the Boolean matrix of the active force distribution in global coordinates.

When the actuator is added, the challenge of determining how to design the actuator output forces should be examined. In this part, the design method is based on the linear quadratic regulator (LQR) [55]. Firstly, the equation of motion can be transformed into the form of state space, as shown in Equation (19). Furthermore, for the sake of computation, it is assumed that the displacements and accelerations are measurable for all the DOFs.

$$\begin{cases} \mathbf{\dot{Z}}(t) = \mathbf{AZ}(t) + \mathbf{BU}(t) + \mathbf{D}\ddot{x}_g \\ \mathbf{Y}(t) = \mathbf{CZ}(t) + \mathbf{H}\ddot{x}_g \end{cases} \tag{19}$$

$$\mathbf{A} = \begin{bmatrix} \mathbf{0} & \mathbf{I} \\ -\mathbf{M}^{-1}\mathbf{K} & -\mathbf{M}^{-1}\mathbf{C} \end{bmatrix}, \mathbf{B} = \begin{bmatrix} \mathbf{0} \\ \mathbf{M}^{-1}\mathbf{B_s} \end{bmatrix}, \mathbf{D} = \begin{bmatrix} \mathbf{0} \\ \mathbf{L} \end{bmatrix},$$
$$\mathbf{C} = \begin{bmatrix} \mathbf{I} & \mathbf{0} \\ -\mathbf{M}^{-1}\mathbf{K} & -\mathbf{M}^{-1}\mathbf{C} \end{bmatrix}, \mathbf{H} = \begin{bmatrix} \mathbf{0} \\ \mathbf{L} \end{bmatrix} \tag{20}$$

where $\mathbf{Z} = \{\mathbf{X}, \mathbf{\dot{X}}\}^{\mathrm{T}}$ is the state response of the whole system.

After that, the control objective is to satisfy a quadratic performance generalization, as indicated in Equation (21). In addition, the optimal control force $\mathbf{U}$ can be expressed as a function of the state of system $\mathbf{Z}$, as presented in Equation (21).

$$J = \frac{1}{2} \int_{t_0}^{\infty} \left[\mathbf{Z}^T(t)\mathbf{QZ}(t) + \mathbf{U}^T(t)\mathbf{RU}(t) \right] dt \tag{21}$$

$$\mathbf{U}(t) = \mathbf{GZ}(t) \tag{22}$$

where $\mathbf{Q}$ and $\mathbf{R}$ are the weight matrices of the structural state response and the output forces, and they are semi-positive and positive definite matrices, respectively; $\mathbf{G}$ is the optimal state feedback gain matrix.

The weights matrix of $\mathbf{Q}$ is presumed to be a function of the mass and stiffness matrices of the whole system. While the weights matrix $\mathbf{R}$ for the active force is assumed to be the identity matrix, as shown in Equation (23). The weight coefficient of α and β are assumed to be 10^3 and 10^{-3} through constant trial, respectively, and it can also be modified according to actual needs. According to stochastic control theory, the optimal feedback force gain $\mathbf{G}$ can be obtained by solving the Riccati matrix algebraic equation [55], as demonstrated in Equations (24) and (25).

$$Q = \alpha \begin{bmatrix} \mathbf{M} & 0 \\ 0 & \mathbf{K} \end{bmatrix}, \mathbf{R} = \beta \mathbf{I} \tag{23}$$

$$- \mathbf{PA} - \mathbf{A}^{\mathrm{T}}\mathbf{P} + \mathbf{PBR}^{-1}\mathbf{B}^{\mathrm{T}}\mathbf{P} - \mathbf{Q} = 0 \tag{24}$$

$$\mathbf{G} = \mathbf{R}^{-1}\mathbf{B}^{\mathrm{T}}\mathbf{P} \tag{25}$$

5. Verification for Vibration Suppression

The time histories under the specified ground motion are estimated using the Newmark-β method, based on the benchmark model stated in the preceding section and the control strategy. In this case, eight TMDs and TMDIs are organized in accordance with Figure 16, with a mass of 21.783 t for each device, assuming a total mass of 3% of the main structure. The remaining spring stiffness and damping coefficient are determined using the approach described in Section 2.2. The results of both the passive and active control systems have been computed and are presented in this section.

5.1. The Top Response of This Building

Under external excitation, the reaction of the top of this high-rise building is the prominent factor that needs to be taken into account. By adopting parameters designed according to the proposed passive and active method, a series of time history analyses were conducted for verification. Taking the ground motion data of R2 as an example, as can be seen in Table 4, the response of the building is plotted, as shown in Figure 17. When the seismic acted, it can be seen that the structural response under passive control is reduced, but there is still great potential for the improvement of mitigation effectiveness. Meanwhile, it can be found that TMDI can further improve the suppression effect during the earthquake with the same mass ratio. In addition, the application of active control can drastically diminish the responsiveness of the structure, compared with the passive control methods.

In addition, the PSD of the top structural response is plotted to assess differences in the frequency domain, as shown in Figure 18. It can be observed that, except for the fundamental mode, higher–order modes also have the impacts on the displacement and acceleration response of the structure under the uncontrolled cases. The passive control mainly affects the first- and second-order structure modes but has limited influence on the remainder of the higher-order modes. However, the active control performs better, with reduced frequency sensitivity and significant suppression of all modes. This shows that active control features the strength of being less sensitive to excitation and that it can still provide effective control forces in situations where the external input is varied.

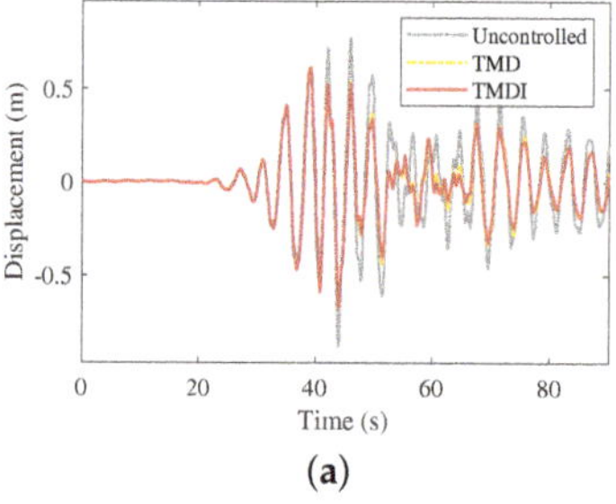

(a)

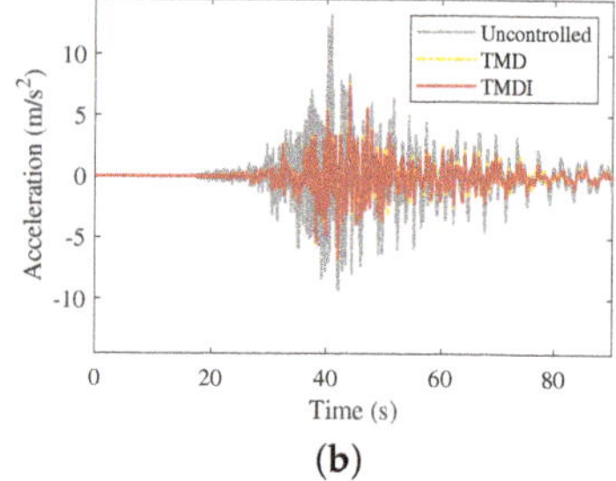

(b)

Figure 17. *Cont.*

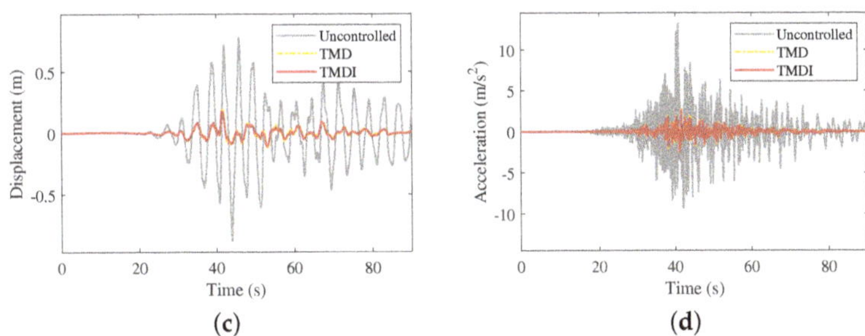

(c)

(d)

Figure 17. The top response of the main structure under ground motion of R2: (**a**) Displacement under passive control; (**b**) Acceleration under passive control; (**c**) Displacement under active control; (**d**) Acceleration under active control.

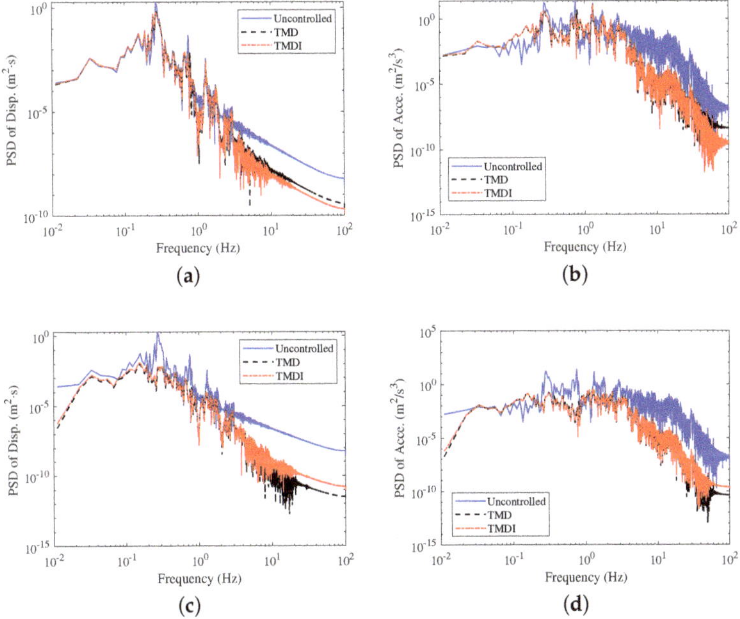

(a)

(b)

(c)

(d)

Figure 18. The PSD value of the top response of the main structure under earthquake of R2: (**a**) Top displacement under passive control; (**b**) Top acceleration under passive control; (**c**) Top displacement under active control; (**d**) Top acceleration under active control.

5.2. Statistical Characteristics of the Response

The time-dependent reaction shown above is the consequence of assuming the structure is in the undamaged condition. After a system is subjected to a strong earthquake, some non–structural components may be damaged and no longer be able to offer auxiliary stiffness to the structure, resulting in stiffness decline. Considering this situation, the stiffness of the structure is discounted in this situation for the further analysis, and it is supposed to be 0.9 times that of the undamaged condition. Figures 19 and 20 compare the maximum amplitude of the structural response before and after being damage under various ground motions. It can be seen that the amplitude of the main structure decreased after the structure got damaged. This could be resulted by the extension of structural period. With approximately constant mass, the reduction in stiffness causes a reduction in the frequency of the structure, resulting in a longer period. Then the structure will capture less energy, resulting in smaller responses, as demonstrated in Figure 14.

When the stiffness of the structure declined, the control effects of TMDI and TMD in the passive state deteriorated considerably, because the parameter of DVAs are designed based on the structural frequencies before being damaged. As for the passive control strategy, the counterproductive result could happen in some cases, as shown in Figure 20a. The amplitude is amplified compared with the uncontrolled situation. Although some reduction in vibration can be achieved with passive control, there is still much to be done to improve the performance. The efficacy of the multiple active TMDI (MATMDI) is more stable than the given multiple active TMD (MATMD), and the reaction of the structure is always kept as low as possible for the given ground motion data.

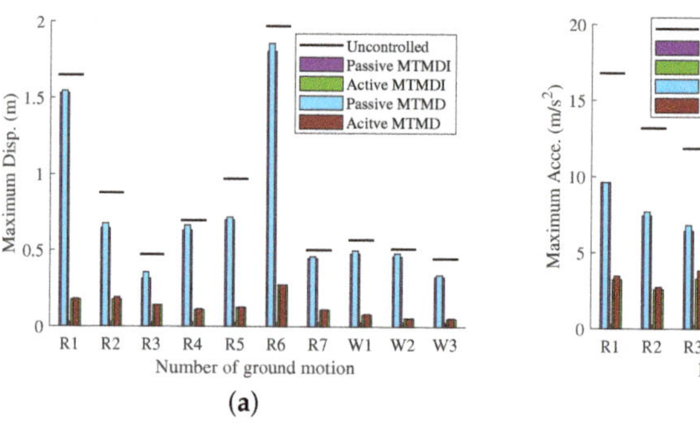

Figure 19. The maximum response of the top of the main structure before being damaged: (**a**) The maximum relative displacement and (**b**) The maximum absolute acceleration.

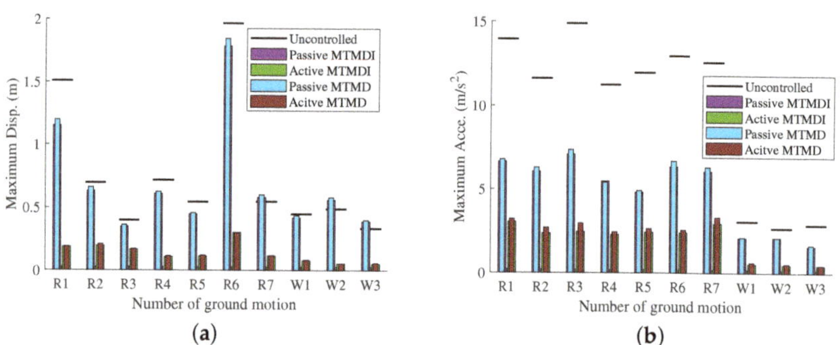

Figure 20. The maximum response of the top of the main structure after being damaged: (**a**) The maximum relative displacement and (**b**) The maximum absolute acceleration.

5.3. Stroke of the Devices

The amplitude of the stroke of every DVA is additionally an essential aspect to consider. The outcomes of identifying the maximum relative displacement per device in terms of each floor movement are obtained. And the records under all the ground motions are count out throught the box–plot, as depicted in Figure 21. Before the structure got damaged, the stroke of the device could be significantly reduced under active control compared to passive control. In most cases, the relative motion of MATDMI increased compared with MATMD in order to achieve the superior vibration mitigation effects. However, after the structure being damaged, the stiffness of the structure decreased. The frequencies of some devices altered in the passive control mode, making it difficult to resonate with the host structure. Then the stroke of devices have been reduced.

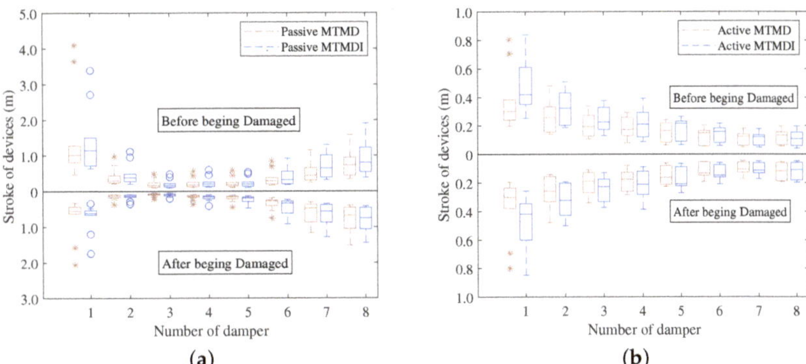

Figure 21. The statistics value for the stroke of different DVAs: (**a**) Passive control and (**b**) Active control.

Additionally, because of the first mode that the devices controlled, the stroke of the first device is the greatest in passive control strategy, and the value of the second to the fifth device steadily diminishes. While the stroke of the sixth to eighth devices increased at the same time, due to the location close to the peak position of the second mode shape. In the active control, the relative movements significantly decreased, particularly for the foremost device, and the lower position the device located, the smaller the strokes it makes. This once again highlights the value of active control and is not sensitive to the structural variation.

6. Conclusions

In this study, a dynamic vibration absorber is integrated with a traditional TMD by adding an inerter element. According to the quantity and features of the various mechanical elements, a simple and effective topology of the system is adopted. To address the shortcomings of the classical fixed-point theory, the optimal design parameters of the TMDI are calculated for the damped SDOF system under the base acceleration by using the numerical optimization method. Moreover, the variation between the conventional TMD and TMDI are compared in the frequency and time domains. On this basis, an actual multi-story steel structure model is used to test the mitigation effect of TMDI under severe earthquakes. In addition, multiple passive and active control strategies of TMDI are proposed, considering the distribution characteristics of the structure. The effectiveness of vibration suppression is verified through time history analysis. From the results, some conclusions can be drawn as follows.

1. The deficiency of classical fixed-point theory can be compensated by employing numerical searching, and the optimal design parameters of TMD and TMDI under external excitation can be effectively found.
2. The TMDI performs better than the classical TMD, with the lower amplitude of the transfer function for the primary system. Meanwhile, the TMDI can provide more extra additional damping to the structure and increase the overall system energy dissipation, but the stroke has to be increased.
3. With the contribution of extra actuators, the multiple active TMDI could greatly alleviate the response of the structure and stroke of attached DVAs. Besides, it also has the strength of being insensitive to structural and environmental changes, with stronger robustness and stability.

Author Contributions: Conceptualization, methodology, validation, investigation, X.W. and X.L.; software, X.W. and J.C.; formal analysis, K.L. and X.L.; resources, X.W., K,L. and K.L.; data curation, X.W. and J.C.; writing—original draft preparation, X.W. and K.L.; editing, C.P., X.L., J.C. and K.L.;visualization, K.L., X.W., C.P. and X.L.; supervision, X.L. and K.L. All authors have read and agreed to the published version of the manuscript.

Funding: This research was funded by Reliability optimization of high–rise Structure considering comfort level OF Youth Foundation of Zhejiang Tongji Vocational College of Science and Technology, Grant No. FRF20QN002 and Structural wind resistance reliability optimization design considering comfort level OF Science and Technology Planning project of Zhejiang Provincial Water Resources Department with Technology, Grant No. RC2004.

Institutional Review Board Statement: Not applicable.

Informed Consent Statement: Not applicable.

Data Availability Statement: The data presented in this study are available on request from the corresponding author.

Conflicts of Interest: The authors declare no conflict of interest.

References

1. Ismail, M. Seismic isolation of structures. Part I: Concept, review and a recent development. *Hormigón Acero* **2018**, *69*, 147–161. [CrossRef]
2. Zhou, Y.; Shao, H.; Cao, Y.; Lui, E.M. Application of buckling-restrained braces to earthquake-resistant design of buildings: A review. *Eng. Struct.* **2021**, *246*, 112991. [CrossRef]
3. De Domenico, D.; Ricciardi, G.; Takewaki, I. Design strategies of viscous dampers for seismic protection of building structures: A review. *Soil Dyn. Earthq. Eng.* **2019**, *118*, 144–165. [CrossRef]
4. Elias, S.; Matsagar, V. Research developments in vibration control of structures using passive tuned mass dampers. *Annu. Rev. Control* **2017**, *44*, 129–156. [CrossRef]
5. Haskett, T.; Breukelman, B.; Robinson, J.; Kottelenberg, J. *Tuned Mass Dampers under Excessive Structural Excitation*; Report; Motioneering Inc.: Guelph, ON, Canada, 2004.
6. McNamara, R.J. Tuned mass dampers for buildings. *J. Struct. Div.* **1977**, *103*, 1785–1798. [CrossRef]
7. Kwok, K.C. Damping increase in building with tuned mass damper. *J. Eng. Mech.* **1984**, *110*, 1645–1649. [CrossRef]
8. Lu, X.; Chen, J. Parameter optimization and structural design of tuned mass damper for Shanghai centre tower. *Struct. Des. Tall Spec. Build.* **2011**, *20*, 453–471. [CrossRef]
9. Yang, F.; Sedaghati, R.; Esmailzadeh, E. Vibration suppression of structures using tuned mass damper technology: A state-of-the-art review. *J. Vib. Control* **2021**, *28*, 812–836. [CrossRef]
10. Rahimi, F.; Aghayari, R.; Samali, B. Application of tuned mass dampers for structural vibration control: A state-of-the-art review. *Civ. Eng. J.* **2020**, *6*, 1622–1651. [CrossRef]
11. Den Hartog, J.P. *Mechanical Vibrations*; Courier Corporation: North Chelmsford, USA, 1985.
12. Warburton, G.B. Optimum absorber parameters for various combinations of response and excitation parameters. *Earthq. Eng. Struct. Dyn.* **1982**, *10*, 381–401. [CrossRef]
13. Asami, T.; Nishihara, O.; Baz, A.M. Analytical solutions to H_∞ and H_2 optimization of dynamic vibration absorbers attached to damped linear systems. *J. Vib. Acoust.* **2002**, *124*, 284–295. [CrossRef]
14. Hoang, N.; Warnitchai, P. Design of multiple tuned mass dampers by using a numerical optimizer. *Earthq. Eng. Struct. Dyn.* **2005**, *34*, 125–144. [CrossRef]
15. Nishihara, O.; Asami, T. Closed-form solutions to the exact optimizations of dynamic vibration absorbers (minimizations of the maximum amplitude magnification factors). *J. Vib. Acoust.* **2002**, *124*, 576–582. [CrossRef]
16. Li, C.; Liu, Y. Optimum multiple tuned mass dampers for structures under the ground acceleration based on the uniform distribution of system parameters. *Earthq. Eng. Struct. Dyn.* **2003**, *32*, 671–690. [CrossRef]
17. Elias, S.; Matsagar, V.; Datta, T. Effectiveness of distributed tuned mass dampers for multi-mode control of chimney under earthquakes. *Eng. Struct.* **2016**, *124*, 1–16. [CrossRef]
18. Vaiana, N.; Sessa, S.; Marmo, F.; Rosati, L. A class of uniaxial phenomenological models for simulating hysteretic phenomena in rate-independent mechanical systems and materials. *Nonlinear Dyn.* **2018**, *93*, 1647–1669. [CrossRef]
19. Vaiana, N.; Sessa, S.; Rosati, L. A generalized class of uniaxial rate-independent models for simulating asymmetric mechanical hysteresis phenomena. *Mech. Syst. Signal Process.* **2021**, *146*, 106984. [CrossRef]
20. Zhang, M.; Xu, F. Tuned mass damper for self-excited vibration control: Optimization involving nonlinear aeroelastic effect. *J. Wind Eng. Ind. Aerodyn.* **2022**, *220*, 104836. [CrossRef]
21. Chen, X.; Kareem, A.; Xu, G.; Wang, H.; Sun, Y.; Hu, L. Optimal tuned mass dampers for wind turbines using a Sigmoid satisfaction function-based multiobjective optimization during earthquakes. *Wind Energy* **2021**, *24*, 1140–1155. [CrossRef]

22. Tseng, H.E.; Hrovat, D. State of the art survey: Active and semi-active suspension control. *Veh. Syst. Dyn.* **2015**, *53*, 1034–1062. [CrossRef]
23. Ma, R.; Bi, K.; Hao, H. Inerter-based structural vibration control: A state-of-the-art review. *Eng. Struct.* **2021**, *243*, 112655. [CrossRef]
24. Smith, M.C. The inerter: A retrospective. *Annu. Rev. Control. Robot. Auton. Syst.* **2020**, *3*, 361–391. [CrossRef]
25. Makris, N.; Kampas, G. Seismic protection of structures with supplemental rotational inertia. *J. Eng. Mech.* **2016**, *142*, 4016089. [CrossRef]
26. Smith, M.C.; Houghton, N.E.; Long, P.J.; Glover, A.R. Force-Controlling Hydraulic Device. US Patent 8,881,876, 11 November 2014.
27. Luo, Y.; Sun, H.; Wang, X.; Chen, A.; Zuo, L. Parametric optimization of electromagnetic tuned inerter damper for structural vibration suppression. *Struct. Control Health Monit.* **2021**, *28*, e2711. [CrossRef]
28. Barredo, E.; Blanco, A.; Colín, J.; Penagos, V.M.; Abúndez, A.; Vela, L.G.; Meza, V.; Cruz, R.H.; Mayén, J. Closed-form solutions for the optimal design of inerter-based dynamic vibration absorbers. *Int. J. Mech. Sci.* **2018**, *144*, 41–53. [CrossRef]
29. Zhang, L.; Xue, S.; Zhang, R.; Xie, L.; Hao, L. Simplified multimode control of seismic response of high-rise chimneys using distributed tuned mass inerter systems (TMIS). *Eng. Struct.* **2021**, *228*, 111550. [CrossRef]
30. Hoang, N.; Fujino, Y.; Warnitchai, P. Optimal tuned mass damper for seismic applications and practical design formulas. *Eng. Struct.* **2008**, *30*, 707–715. [CrossRef]
31. Rahman, M.M.; Nahar, T.T.; Kim, D. Effect of Frequency Characteristics of Ground Motion on Response of Tuned Mass Damper Controlled Inelastic Concrete Frame. *Buildings* **2021**, *11*, 74. [CrossRef]
32. Jia, F.; Jianwen, L. Performance degradation of tuned-mass-dampers arising from ignoring soil-structure interaction effects. *Soil Dyn. Earthq. Eng.* **2019**, *125*, 105701. [CrossRef]
33. Pinkaew, T.; Lukkunaprasit, P.; Chatupote, P. Seismic effectiveness of tuned mass dampers for damage reduction of structures. *Eng. Struct.* **2003**, *25*, 39–46. [CrossRef]
34. Su, N.; Xia, Y.; Peng, S. Filter-based inerter location dependence analysis approach of Tuned mass damper inerter (TMDI) and optimal design. *Eng. Struct.* **2022**, *250*, 113459. [CrossRef]
35. Shi, X.; Zhu, S. Dynamic characteristics of stay cables with inerter dampers. *J. Sound Vib.* **2018**, *423*, 287–305. [CrossRef]
36. Li, C.; Cao, L. High performance active tuned mass damper inerter for structures under the ground acceleration. *Earthquakes Struct.* **2019**, *16*, 149–163.
37. Li, Y.Y.; Park, S.; Jiang, J.Z.; Lackner, M.; Neild, S.; Ward, I. Vibration suppression for monopile and spar-buoy offshore wind turbines using the structure-immittance approach. *Wind Energy* **2020**, *23*, 1966–1985. [CrossRef]
38. Hu, Y.; Chen, M.Z. Performance evaluation for inerter-based dynamic vibration absorbers. *Int. J. Mech. Sci.* **2015**, *99*, 297–307. [CrossRef]
39. Smith, M.C. Synthesis of mechanical networks: The inerter. *IEEE Trans. Autom. Control* **2002**, *47*, 1648–1662. [CrossRef]
40. Shen, W.; Niyitangamahoro, A.; Feng, Z.; Zhu, H. Tuned inerter dampers for civil structures subjected to earthquake ground motions: Optimum design and seismic performance. *Eng. Struct.* **2019**, *198*, 109470. [CrossRef]
41. Messac, A. *Optimization in Practice with MATLAB®: For Engineering Students and Professionals*; Cambridge University Press: Cambridge, UK, 2015.
42. Jerome, J.C. Introduction. In *Structural Motion Control*; Prentice Hall Pearson Education, Inc.: Upper Saddle River, NJ, USA, 2002; pp. 217–285.
43. Mario, P.; Young, H.K. *Structural Dynamics: Theory and Computation*; Springer: Berlin, Germany, 2019.
44. Vaiana, N.; Sessa, S.; Marmo, F.; Rosati, L. Nonlinear dynamic analysis of hysteretic mechanical systems by combining a novel rate-independent model and an explicit time integration method. *Nonlinear Dyn.* **2019**, *98*, 2879–2901. [CrossRef]
45. Chang, S.Y. Family of structure-dependent explicit methods for structural dynamics. *J. Eng. Mech.* **2014**, *140*, 6014005. [CrossRef]
46. Spencer Jr, B.F.; Christenson, R.E.; Dyke, S.J. Next generation benchmark control problem for seismically excited buildings. In Proceedings of the Second World Conference on Structural Control, Kyoto Japan, **1998**, *2*, 1135–1360.
47. Shen, W.; Zhu, S.; Zhu, H. Unify energy harvesting and vibration control functions in randomly excited structures with electromagnetic devices. *J. Eng. Mech.* **2019**, *145*, 4018115. [CrossRef]
48. Rahmani, H.R.; Wiering, M.M. Artificial Intelligence Approach for Seismic Control of Structures. Ph.D. Thesis, Bauhaus-Universität Weimar, Weimar, Germany, 2020.
49. Cook, R.D.; et al. *Concepts and Applications of Finite Element Analysis*; John Wiley & Sons: Hoboken, NJ, USA, 2007.
50. Roy, R.; Craig, J. *Structural Dynamics: An Introduction to Computer Methods*; John Wiley & Sons: Hoboken, NJ, USA, 1981.
51. Kiureghian, A.D.; Neuenhofer, A. Response spectrum method for multi-support seismic excitations. *Earthq. Eng. Struct. Dyn.* **1992**, *21*, 713–740. [CrossRef]
52. Ministry of Housing and Urban-Rural Development of the People's Republic of China. *GB 50011–2010*; Code for Seismic Design of Buildings. China Building Industry Press: Beijing, China, 2010. (In Chinese)
53. ASCE. *Minimum Design Loads for Buildings and Other Structures*; American Society of Civil Engineers: Reston, USA, 2013.
54. Ancheta, T.D.; Darragh, R.B.; Stewart, J.P.; Seyhan, E.; Silva, W.J.; Chiou, B.S.J.; Wooddell, K.E.; Graves, R.W.; Kottke, A.R.; Boore, D.M.; et al. NGA-West2 database. *Earthq. Spectra* **2014**, *30*, 989–1005. [CrossRef]
55. Tewari, A. *Modern Control Design with MATLAB and SIMULINK*; Wiley: Chichester, UK, 2002.

Review

Sliding Isolation Systems: Historical Review, Modeling Techniques, and the Contemporary Trends

A. R. Avinash, A. Krishnamoorthy, Kiran Kamath and M. Chaithra *

Department of Civil Engineering, Manipal Institute of Technology, Manipal Academy of Higher Education, Manipal 576104, India
* Correspondence: chaithra.mitthur@manipal.edu

Abstract: Base isolation techniques have emerged as the most effective seismic damage mitigation strategies. Several types of aseismic devices for base isolation have been invented, studied, and used. Out of several isolation systems, sliding isolation systems are popular due to their operational simplicity and ease of manufacturing. This article discusses the historical development of passive sliding isolation systems, such as pure friction systems, friction pendulum systems, and isolators with other sliding surface geometries. Moreover, multiple surface isolation systems and their behavior as well as the effectiveness of using complementary devices with standalone passive isolation devices are examined. Furthermore, the article explored the various modeling techniques adopted for base-isolated single and multi-degree freedom building structures. Special attention has been given to the techniques available for modeling the complex phenomena of sliding and non-sliding phases of sliding bearings. The discussion is further extended to the development in the contemporary areas of seismic isolation, such as active and hybrid isolation systems. Although a significant amount of research is carried out in the area of active and hybrid isolation systems, the passive sliding isolation system still has not lost its appeal due to its ease of adaptability to the structures.

Keywords: base isolation; friction pendulum system; multi-surface isolation system; passive isolation system; active isolation system; semi-active isolation system

Citation: Avinash, A.R.; Krishnamoorthy, A.; Kamath, K.; Chaithra, M. Sliding Isolation Systems: Historical Review, Modeling Techniques, and the Contemporary Trends. *Buildings* **2022**, *12*, 1997. https://doi.org/10.3390/buildings12111997

Academic Editors: Rajesh Rupakhety and Dipendra Gautam

Received: 7 October 2022
Accepted: 3 November 2022
Published: 16 November 2022

Publisher's Note: MDPI stays neutral with regard to jurisdictional claims in published maps and institutional affiliations.

1. Introduction

Earthquakes are one of the most catastrophic natural events, often resulting in the loss of lives and structures. Humans have been trying to reduce the harmful effects of earthquakes on structures, such as buildings, bridges, and tanks. The main challenge faced in reducing the harmful effect of an earthquake is to arrive at a suitable method to dissipate or offset the vast amount of energy imparted to the structure during a seismic event. In this regard, researchers have been developing some mechanical appurtenances over the years. As a result, several mechanical devices, such as fluid viscous dampers [1], visco-elastic dampers [2], and yielding-type dampers [3], have been invented. Depending on their location and type of arrangement within the structure, these devices absorb seismic energy locally. Over the years, the research focus shifted to developing mechanical devices that can significantly reduce the transfer of seismic energy to the structure, which led to the invention of base isolation systems. In principle, a base isolation system must have very high stiffness in the vertical direction and low stiffness in the lateral direction. High vertical stiffness enables the transfer of gravity loads, whereas low lateral stiffness ensures that the structure behaves as a rigid unit for the lateral load. The main issue to be addressed in the earthquake-resistant design of structures is the resonance problem. During a seismic event, structures generally are prone to frequency amplification as their fundamental frequency typically lies in the range of earthquake frequencies. Base isolation systems minimize the resonance issue by significantly altering the fundamental frequency of the structure. The concept of base isolation is not new; historically, the earliest known patent on the base isolation technique can be traced back to 1870 [4]. In the patent application, Jules Touaillon of California proposed a seismic isolation system which uses a combination of spheres and spherical surfaces to achieve seismic isolation (Figure 1).

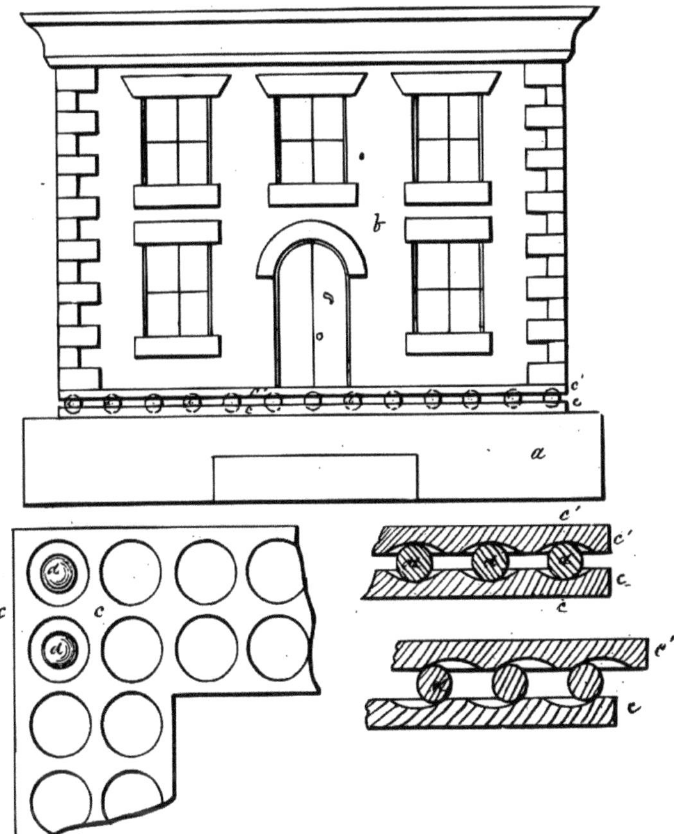

Figure 1. Isolation system patented by Touaillon [4].

Furthermore, in the year 1909, a doctor named Calantarients of Great Britain proposed an isolation technique [5], wherein a talc or mica layer is used between the foundation and superstructure to safeguard the building against earthquakes. One of the earlier known examples of seismic isolation can be found in a multi-story building in Ljubljana, Slovenia. The building was built in 1933 by adhering to the seismic isolation philosophy. The building's foundation is separated from the deep file foundation using 2.5 mm thick layers of metals and asphalt [6,7]. To date, several review articles on various base isolation systems are available [8–15]. In 1969, an elementary school in Skopje, Yugoslavia, was base-isolated with unreinforced rubber blocks [16]. Therefore, the base isolation project of the Pestalozzi school is one of the earliest known implementations of the base isolation technique in an actual structure. Since the rubber blocks were unreinforced, the isolators lacked sufficient vertical stiffness, resulting in some amount of horizontal acceleration converting into vertical acceleration. This acceleration produces an additional bouncing effect which is highly undesired. As a result, these blocks are not used for base isolation at the present time. However, the experiences with these rubber blocks have led to considerable research, and since then, researchers have developed several types of rubber bearings [11].

One of the simplest ways to achieve base isolation is to implement some sliding mechanism between the superstructure and substructure. In its simplest form, this type of isolation can use a sand layer between the superstructure and substructures [17]. Although this approach is simple and reasonably effective for small structures, it is not suitable for multi-story buildings. More recently, Azinović et al. [18] proposed the use of thermal insulation boards installed beneath the foundations of a contemporary energy-efficient building to allow for controlled lateral sliding between the individual layers of the board. Out of the various scenarios considered by these researchers, the sliding prevention scenario was found to be the most cost-effective. To date, several varieties of sliding isolation systems have been developed. These systems are discussed in detail in the subsequent sections. Although the idea of seismic isolation is more than a century old, research and wide practical implementation started about 40 years ago. Since its inception, the concept of base isolation has significantly matured, and various base isolators have been developed, patented, and implemented.

Modern isolation systems can be broadly classified into elastomer base isolators, rolling-type isolators, and sliding isolators. Elastomer isolators use rubber as the base material, which may have metal insertions to increase the vertical stiffness. The details regarding elastomer isolators and their applications to the structures can be found in [19]. The rolling-type isolators use balls or rolling rods which can roll on concave surfaces. During an earthquake event, the energy is dissipated through rolling friction. Several research articles on various rolling-type isolations are available [20–23]. A detailed review of these isolators is also available [24]. The sliding isolators have two or more contact surfaces and work by mutual sliding of one or multiple surfaces. Although many review articles on various base isolation systems are available, the authors found only a few review articles on sliding isolators. Moreover, review articles discussing the various modeling techniques are still rare. Furthermore, review articles addressing the use of complementary devices along with sliding isolators are limited. In this context, the present review paper mainly focuses on sliding isolation systems, with special attention given to various modeling techniques for isolated structures and the complementary devices used with the sliding isolators.

2. Sliding Isolation Systems—History of Various Types

In this section, the historical development of sliding isolators based on the number of sliding surfaces is discussed. Several important studies and key findings of these studies are also briefly examined.

2.1. Sliding Isolation Systems with a Single Sliding Surface

In Bihar (India), during the 1934 earthquakes, many buildings survived as they developed cracks running longitudinally below the superstructure. These cracks permitted the sliding of the superstructure over the foundation, preventing damage to the super-structure [25]. This observation became the topic of interest, and researchers proposed sliding-type joints [26–29]. Few authors proposed a pure friction type (P-F) joint, where materials, such as graphite powder, sand or engine oil, are used between the superstructure and substructure to achieve the required coefficient of friction [25,26]. These structures significantly reduced the spectral acceleration when subjected to dynamic loading, indicating the effectiveness of P-F base isolation. Since these studies were restricted to simple one-story buildings, Nikolić-Brzev studied [30] a multiple-level P-F isolation scheme for multi-story buildings. The system was found to be effective in reducing the dynamic response of the buildings when compared with fixed-base buildings. Several other detailed studies and modeling techniques are also available on P-F isolation systems [31–35]. The details of these modeling techniques are discussed in Section 7.

Due to the lack of restoring mechanism in a P-F system, the superstructure may permanently shift from its original position at the end of a seismic event. This residual shift may affect the functional use of the building. Therefore, researchers shifted their focus toward sliding systems with restoring capabilities. Zayas et al. [36] proposed a simple system known as the friction pendulum system (FPS). Details of a typical friction pendulum isolator are shown in Figure 2.

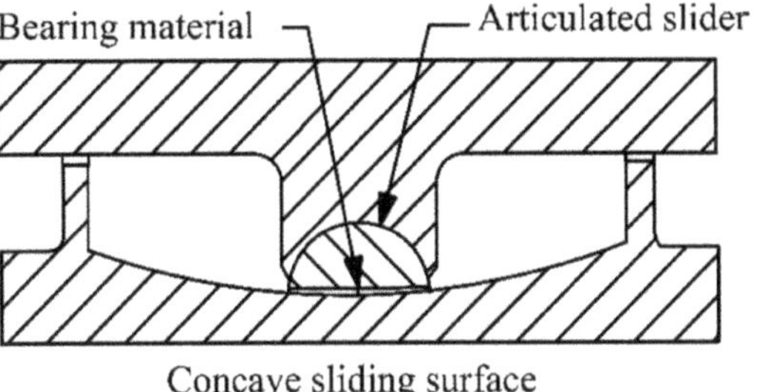

Figure 2. Schematics of a typical friction pendulum bearing.

The isolator consists of an articulated slider, which moves on a stainless-steel spherical surface. The slider is coated with stainless steel and encased in the cavity of a spherical plate. The slider portion in contact with the spherical surface is coated with a composite material with low friction, usually Teflon. Teflon is the trade name of polytetrafluoroethylene (PTFE). During a seismic excitation, the slider moving on the spherical surface may lift the mass, and the gravitational force provides the necessary restoring force. The friction between the slider and surface provides the necessary damping. This type of isolator has been extensively studied, and the results of detailed experimental studies are available [37,38]. Jangid [39] developed a mathematical model for a multi-story building and a bridge isolated with FPS. The researcher found the existence of optimal friction values, which can significantly reduce sliding displacements and floor accelerations. More recently, researchers proposed a regression expression to obtain the optimal value for the coefficient of friction [40]. It was found that the optimal value is dependent on the intensities of earthquakes. Vibhute et al. [41] proposed a graphical approach to obtain the optimal friction value. This method was found to be intuitive and simpler than the optimization algorithms. The effect of local bending effects on the response of FPS isolated buildings under seismic loading was studied by Tsai [42]. Through the study, the researcher found that ignoring the local bending effect may be unrealistic and should be included in the formulation phase. A study on FPS-isolated structures indicates the importance of considering bi-directional earthquake excitations [43]. It was found that the sliding displacement increases between 20% and 38% in the bi-directional excitation case when compared with unidirectional excitation. Several researchers studied various FPS-isolated structures, such as buildings [44], bridges [45], liquid storage tanks [46], and nuclear power plants [47]. Typically for analysis purposes, FPS is modeled in 1D or 2D; although this is agreeable under normal loading, it is not realistic in extreme loads, which may include uplifting [48]. In this context, several researchers have studied the behavior of FPS under extreme tri-axial loading by considering 3D models [49,50]. It is a well-established fact that the response of a structure is different for a near-fault than for a far-field earthquake. Therefore, base-isolated structures may show large displacements for near-fault pulses [51]. The frequency of an FPS is a function of its sliding geometry alone. Since the curvature of FPS remains the same over the sliding surface, the isolation frequency is constant. As a result, a low-frequency earthquake might induce resonance in the structure resting on FPS. Therefore, Pranesh and Sinha [52] proposed a variable frequency pendulum isolator (VFPI) whose frequency varies along the geometrical surface. The surface considered for the isolator is derived based on the modified expression of an ellipse. Since the curvature changes along the surface, the period of the isolator varies throughout the sliding. As a result, the matching of isolator and earthquake frequency is avoided, effectively curbing the resonance issue. Furthermore, the isolator is found to be effective for high-intensity earthquakes. A similar isolator known as a variable curvature friction pendulum system (VCFPS) was proposed by Tsai [53]. Here, the concave surface of the isolator is derived by subtracting a function from the equation for the sliding surface of FPS. The author studied the effectiveness of VCFPS by considering a numerical model of a building subjected to various earthquakes. The study indicated that the isolator could significantly reduce the base shear even in a near-fault earthquake characterized by low frequency. Lu et al. [54] introduced an isolator with a spherical curvature at the central portion with linearly varying geometry beyond this central region. The isolator is called a conical friction pendulum isolator (CFPI). Due to the geometry, the isolator acts similar to FPS within some threshold, and thus isolation frequency remains constant. Beyond this threshold, the frequency of isolation varies linearly along the geometry owing to the linearly varying surface. A comparison of the FPS, CFPI, and VFPI is shown in Figure 3.

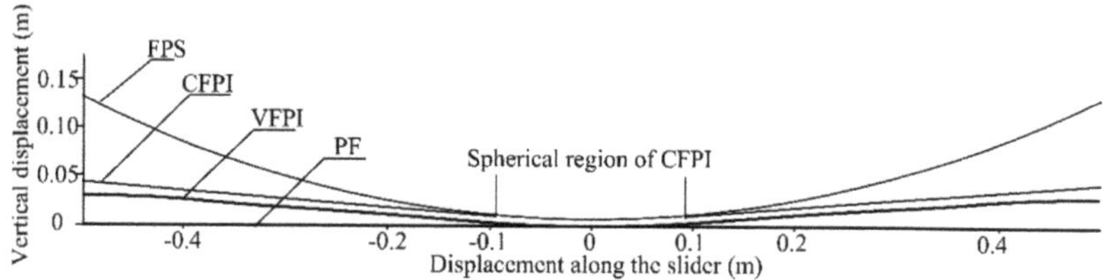

Figure 3. Comparison of various sliding isolation system geometries [10].

Herein, it was found that all the isolators performed well in a far-field earthquake. However, for near-fault earthquakes, VFPI and CFPI are found to be more effective than FPS. Except for FPS, all other isolation systems mentioned earlier have flatter sliding surfaces, resulting in large residual displacements. In an attempt to reduce this residual displacement, Lu et al. [55] developed a sliding isolator known as a polynomial friction pendulum isolator (PFPI). The surface geometry of this isolator is governed by a fifth-order polynomial. The response of PFPI in terms of displacements was significantly lower than FPS for near-fault earthquakes. Krishnamoorthy [56] proposed an isolator whose geometry, as well as the coefficient of friction, varies with isolator displacement, named variable radius friction pendulum system (VRFPS). The geometry of VRFPS varies exponentially along the sliding surface, whereas the coefficient of friction varies linearly along the surface. The isolator was effective for a wide range of earthquake frequencies and significantly reduced residual displacement. Malu and Murnal [57] considered VFPI varying the coefficient of friction along the geometry to reduce the harmful effects of near-fault earthquakes. Rather than linearly varying the coefficient of friction as in VRFPS, the authors varied the coefficient of friction only in two specific regions. Authors claimed that restricting the variation of friction coefficient only in two regions significantly reduces the difficulties involved in manufacturing these isolation bearings. Moreover, the isolator was effective in reducing both acceleration and displacement. Furthermore, Calvi et al. [58] proposed two more varieties of these isolation systems. One system uses a flat surface known as BowTie (BT), and the other with a curved surface similar to FPS is known as BowC (BC). In both cases, the coefficient of friction varies along the surface. Authors argued that this variable friction coefficient could be practically obtained by creating concentric bands of different materials. Due to the flat geometry, BT lacks the recentering ability, but this could be used as a cost-effective solution for temporary buildings. However, BC has a re-centering capability and thus can be used for more permanent structures. Although analytical studies on BC isolators showed promising results, the authors suggested additional experimental studies to check the practical feasibility of these isolators. The sliding isolation systems generally use costly materials to achieve a low coefficient of friction for the surface. In this regard, a low-cost alternative for typical isolation bearing was suggested by Brito et al. [59]. The isolation system uses typical construction materials, such as concrete and steel, and does not use any replaceable mechanical parts. The authors explored the possibility of combining both convex and concave surfaces for isolation purposes.

2.2. Sliding Isolation Systems with Multiple Sliding Surfaces

Although the earliest literature on isolators with multiple sliding surfaces dates back to the 19th century [4], a systematic study in this area started only in this century. The effectiveness of sliding isolator systems for a wide range of earthquake frequencies encouraged the researchers to focus on isolators with multiple sliding surfaces. Tsai et al. [60] proposed a multiple friction pendulum system (MFPS) with two concave sliding surfaces (Figure 4). Due to its unique design, this isolator can accommodate large sliding displacements. Furthermore, multiple frictional surfaces provide additional damping, thus offsetting the harmful accelerations.

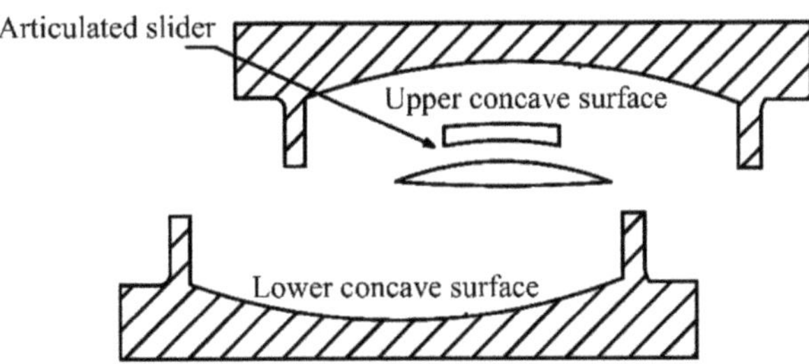

Figure 4. Schematics of a double pendulum friction bearing.

Fenz and Constantinou [61] studied various parameters of double concave friction bearings by varying the coefficient of friction, radii of sliding surfaces, and the height of articulates and sliders. The authors proposed that with several of these variables, designers can arrive at upper and lower isolator surfaces, whose radii can be varied depending on the requirement. Further research in base isolation has led to the development of another multi-stage friction pendulum bearing known as a triple pendulum (TP) bearing [62]. The system makes use of three independent pendulum mechanisms and four concave surfaces, as shown in Figure 5.

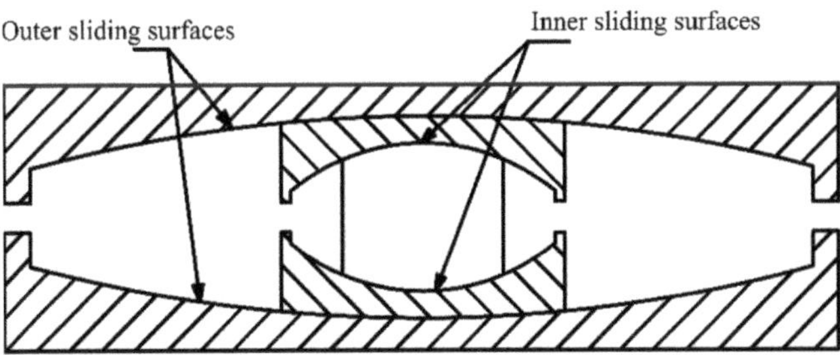

Figure 5. Schematics of a triple pendulum friction bearing.

TP bearings are capable of achieving different hysteresis properties when being displaced. This property of variable hysteresis enables the isolator to adapt to different earthquake frequencies. Detailed theoretical and experimental studies of these multi-stage frictional pendulum isolators were carried out by Fenz and Constantinou [63,64]. Since sliding displacements are distributed on all the sliding surfaces, the heat generated during the high-velocity movement is also minimized. Various modeling techniques of these isolators are also available in detail [65,66]. A detailed study has been conducted by Morgan and Mahin [67] to assess the reliability of TP bearings. This probability-based study indicated that TP bearings are effective in various seismic hazard levels. A detailed study of these isolators subjected to extreme forces conducted by Becker et al. [68] showed that the damages are generally limited to inner sliding surfaces. Most recently, Lee and Constantinou [69] developed a quintuple friction pendulum isolator (Figure 6).

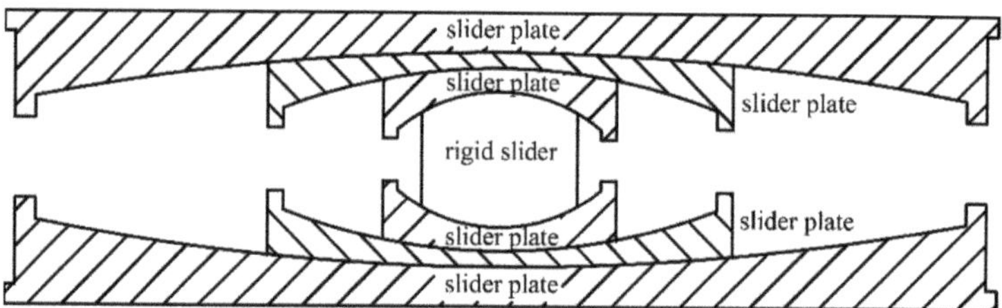

Figure 6. Schematics of a quintuple pendulum friction bearing.

The isolator has six spherical sliding surfaces, which provides designers with multiple options to achieve complex seismic isolation requirements. A detailed mathematical model and finite element technique to model this type of isolator are proposed by Keikha and Ghodrati [70] and Sodha et al. [71], respectively.

A comparison of the typical force (f)-displacement (d) behavior and stiffness at various regimes (k) for various sliding isolators discussed to date is shown in Figure 7.

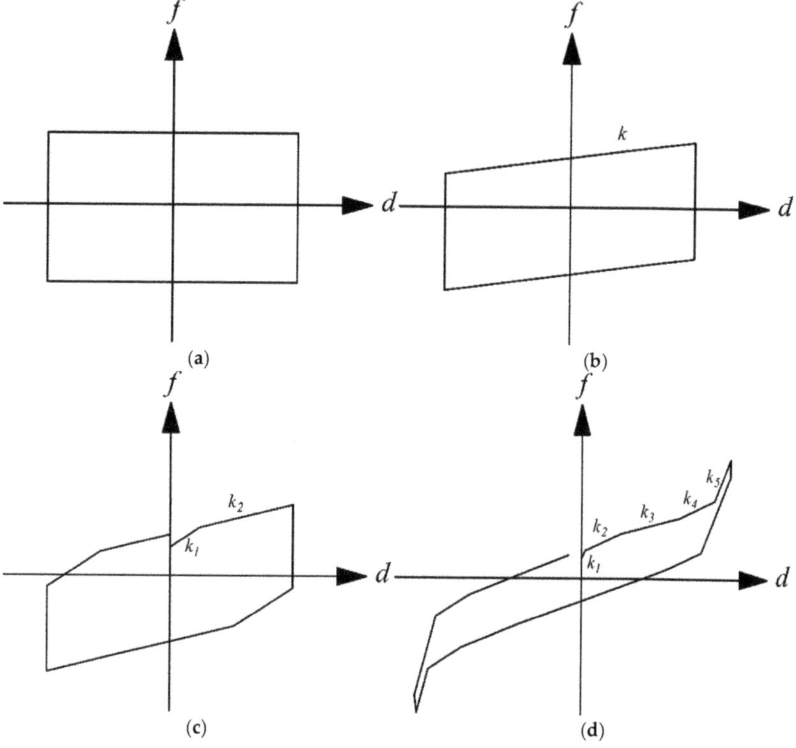

Figure 7. *Cont.*

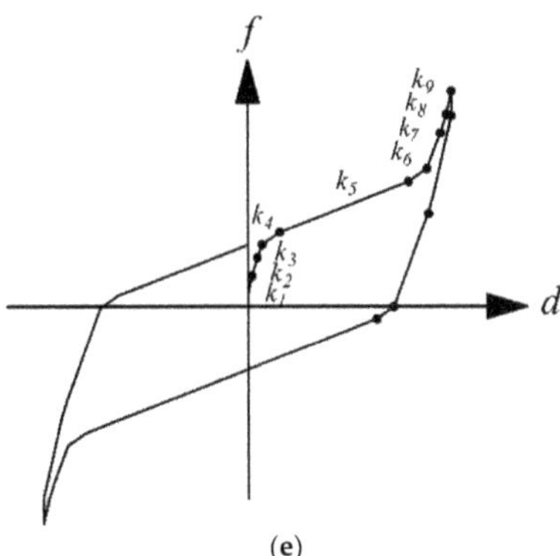

(e)

Figure 7. Idealized force-displacement relationship for various isolators. (**a**) P-F isolator, (**b**) FP isolator, (**c**) double concave friction bearing, (**d**) triple pendulum bearing, (**e**) quintuple friction pendulum isolator.

As seen in these figures, with the evolution of sliding isolators from single surface pure friction form to quintuple form, plenty of control points are now available to the designers. Each form shows better force-displacement behavior and energy dissipating capabilities than the previous one. At present, with many variables to choose from, the designers can aim at achieving very complex requirements of modern high-rise buildings. Recently, a sliding isolator known as XY-FP is gaining popularity among researchers [72]. The construction details of the isolator can be seen in Figure 8.

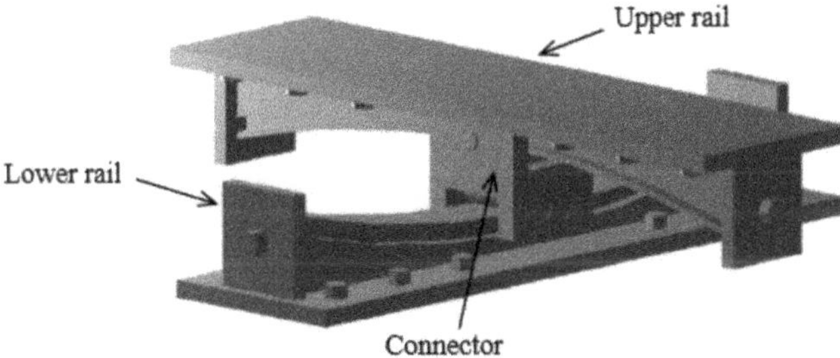

Figure 8. Schematics of a XY-FP [72].

The unique feature of this isolator is that it has two different sliding surfaces perpendicular to each other, which permits independent sliding. Due to the arrangement, the isolator can have uncoupled behavior in two directions. Furthermore, the isolator can be designed to have two separate isolation periods by providing different curvatures in each of the directions. This feature benefits designers who wish to address displacement demands in the principal directions. Moreover, due to the vertical connector, this isolator can prevent the uplift, which can be caused by unforeseen vertical accelerations due to earthquakes. Most of the isolators mentioned earlier in this section have curved surfaces which may lead to manufacturing difficulties. In this context, a sloped sliding-type bearing (SSB) has been proposed recently [73]. This isolator has two orthogonal railings similar to XY-FP bearing; however, in contrast to the curved surface, this isolator has two discontinuous slopes. Since the isolator does not have a fixed curvature, it can efficiently avoid the resonance issue. However, this isolator lacks a fillet between two sloping surfaces. As a result, the isolator is prone to impact effects during the transition from one slope to the other.

3. Parameters Affecting the Frictional Coefficient in Sliding Isolators

The contact pressure is one of the parameters which can significantly affect the coefficient of friction. Studies have shown that friction is dependent on the apparent area of bearing (A_a), shear strength (S_0), and the normal load (W) [74]. Accordingly, some researchers proposed a modified equation [75]. The equation for the coefficient of friction considering the contact pressure effect is as follows:

$$\mu = A_a\, S_0 / W \tag{1}$$

Experimental studies have shown that friction reduces with the increase in contact pressure [76–78]. A detailed mathematical model which incorporates the dependency of sliding velocity on bearing pressure was developed by some authors [78].

Sliding isolators generally use Teflon-steel interfaces, which support the weight of the superstructure. Studies have shown that the friction developed at the interfaces of Teflon-steel does not remain constant during sliding as in Coulomb friction but varies with the sliding velocity and contact pressure [79,80]. Researchers have found that friction increases with the sliding velocity up to a certain value (0.1 to 0.2 m/s) but remains constant beyond that [77]. Furthermore, these researchers found that this reduction is prominent in the case of high-velocity motions. It has been reported that ignoring the velocity-dependent friction model for sliding isolations would result in an unrealistic estimation of the forces [81]. The coefficient of sliding friction can be approximated by the equation given by Mokha et al. [82]:

$$\mu_s = \mu_{max} - (\mu_{max} - \mu_{min})\exp(-aV) \tag{2}$$

where

μ_{max} = friction coefficient corresponding to large velocity.
μ_{min} = friction coefficient corresponding to low velocity.
a = a constant varies between 20 and 100 s/m depending on the contact pressure and interface conditions.
V = velocity of sliding motion.

Equation (2) represents a simplified form of a complex behavior but has been found to produce agreeable results, and thus even incorporated in the structural analysis software SAP2000 [83]. The dependence of friction on the velocity of motion implies that the coefficient of sliding friction is directly related to the characteristics of earthquakes. Therefore, the velocity aspect needs to be carefully addressed, especially when analyzing structures subjected to high-frequency earthquakes.

Another parameter which significantly influences the friction coefficient is the heat generated during the sliding movement of the isolation components. Several authors have reported that at high velocities, the friction coefficient slightly reduces [75,84,85]. This reduction has been attributed mainly to the melting of the bearing material at elevated temperatures generated during high-velocity motion [86]. Through experimentation, researchers found that the decay in friction could be significant; for non-lubricated surfaces, this could be as high as 25% and 30% [87]. Although the reduction in friction coefficient may reduce the base shear, this may warrant undesired large displacements. Some researchers have proposed 3D models of FPS bearings capable of simulating heat flux generated depending on the sliding velocity and contact pressure [49,88]. Design tools for the estimation of temperature-characteristics are also available [89]. Recently, De Dominico et al. [90] have conducted detailed experimentation to assess the effect of temperature on the friction coefficient of a double-curved sliding isolator. This experimental regime included the use of several thermocouples to sense the temperature variation within the sliding surface. It was found that lower sliding velocities associated with a temperature rise up to 45 °C had an insignificant effect on friction.

To date, the discussion on friction variation has been dealt with by considering velocity and temperature separately in this section. However, velocity, contact pressure, and temperature are all interdependent in the actual scenario. In this context, the authors Kumar et al. [91] proposed a unified model that can account for the various parameters mentioned earlier. The study indicated that temperature has the highest effect on the isolator displacement out of various parameters. Additional parameters which affect the isolation performance are surface roughness, corrosion, contamination, lubrication, and wear, etc. The effect of these parameters is discussed in detail in the technical report [86].

4. Parameters for the Design of Sliding Isolators

Some essential design parameters related to the sliding isolators are discussed in this section. Although the discussion is mainly related to curved surface isolators, except for the sliding curvature details, other parameters are also applicable to flat surface isolators. The usage of sliding isolators in real-life structures is governed by several factors, such as the type of material and thickness required for the sliding surface, the thickness of the steel plate, the dynamic coefficient of friction, and the radius of the sliding surface in case of curved surfaces, etc. Although sliding isolators mostly use a Teflon-steel interface, some researchers have also considered polyamide (PA) and polyethylene (PE) for the sliding surface [92]. The nominal coefficient of friction can be chosen based on Equation (2). The typical specification for a FPS is given in Table 1.

Table 1. Typical specification for FPS bearings.

Description	Specifications
Thickness of Teflon layer	1.6 to 2.2 mm [93]
Thickness of stainless-steel plate	1.5 to 2.3 mm [93]
Arithmetic average surface roughness	0.05 to 0.8 μm [93]
Stainless steel type	316 type (ASTM A 240) or type 5 CrNiMo (DIN 17440) [94]
Radius of curvature	1555 to 6045 mm [94]
Diameter of concave surface	356 to 3652 mm [94]
Lubricant	Silicone grease, effective at low temperature [94]

Interested readers can refer to detailed design examples of sliding isolators for the bridges in [93,95]. The design parameters are to be chosen carefully based on the standard code of practices relevant to the country. To date, several countries have released standard codes of practice related to base isolation, such as American codes [96,97], European codes [98,99], and Japanese codes [100]. Recently, a draft version of the New Zealand code has also been released [101]. The behavior of the sliding isolators is also affected by several other aspects, such as ageing, contamination, cumulative movement, and temperature. To account for these aspects, upper and lower bound values of modification factors must also be incorporated into the design. Detailed discussions of these factors can be found in [75].

5. Behavior of Sliding Isolators under Extreme Loading

Sliding isolators have been proven to be very effective against earthquake excitations over the years. However, due to their non-adaptive characteristics, they may demonstrate undesirable behavior when subjected to certain excitations. If the sliding isolator is subjected to large unwarranted displacement, the slider can move beyond the edge and may even eject out of the housing unit. This issue can be addressed using a retainer ring around the sliding surfaces. However, this may lead to reduced displacement. Moreover, the slider may collide with the rim, damaging the bearing. This aspect should be taken into account during the design process. A sliding isolator may experience uplifting when the lifting forces overcome the compression on the bearing due to the structure. Depending on the amount of decoupling of the isolator components, the effect of uplifting can range from mild to catastrophic. Sarlis and Constantinou [102] have provided detailed modeling techniques for TP bearings, including the uplift phenomenon. This model is further used in a study conducted on double pendulum friction bearing subjected to pulse-type motion [103]. Results indicated that under large masses, the retainer failure is more significant than the uplift failure. Moreover, the study showed that providing a large radius of curvature was detrimental to the isolator. Furthermore, it was found that a short-period pulse results in failure due to impact. Recently, some researchers studied the effect of the hardening stage in a TP bearing subjected to extreme loading [68]. The hardening stage significantly reduced the force of impact within the isolator but had a negligible effect on the force transfer to the superstructure. The possible uplift issues can be avoided using uplift restrainers [104–106]. The restrainers are usually not integral to isolators and may need additional space and separate maintenance. In this regard, a more practical isolation system with uplift restraint known as XP-FP isolation bearing has been invented [72]. This bearing works similar to FPS under horizontal loading but restrains uplift when subjected to vertical excitations [107].

6. Sliding Isolators with Complementary Devices

The sliding isolators have some limitations when used as standalone devices. In the case of pure friction devices, these limitations include large sliding displacements and significant residual displacements. Furthermore, although friction pendulum isolators effectively control residual displacements, they may be subjected to large sliding displacements in a low-frequency earthquake. Some researchers tried to address these issues by combining sliding and rubber isolators [108–110] Constantinou et al. [111] proposed a sliding isolation system which uses steel springs at the base. The energy dissipation in this type of system is provided by the friction generated between the sliding interfaces, and the restoring capability is provided by the springs. A similar concept has been explored by several researchers [112,113]. The sliding isolators with pendulum mechanisms are quite effective in controlling residual displacements. Although these isolators produce less displacement when compared with P-F isolators, the magnitude of this displacement is deemed to be quite large. Therefore, Tsopelas et al. [114] considered a sliding isolator with additional energy dissipation devices known as fluid dampers for bridges. This system reduced the sliding displacement and was effective for a wide frequency range of seismic excitation. Since energy dissipation by the isolator plays a major role in reducing the harmful seismic effects, Makris and Chang [115] discussed various energy dissipation models in their study. The study indicates that, with a suitable viscous energy dissipation mechanism, it is possible to reduce large displacements without a further increase in base shear and accelerations. Chang et al. [116] carried out studies by considering viscous and friction damping in base-isolated structures. A friction pendulum bearing has been considered for the study with additional fluid dampers. The fluid dampers provide viscous damping, and the sliders provide friction damping. Experimental results indicated that viscous damping is most effective in moderately low frequencies, whereas friction damping is effective in low-frequency excitations. Soneji and Jangid [117] studied the effectiveness of a combination of base isolation and viscous fluid dampers (VFD) for flexible bridges. Due to its construction, a VFD produces the reaction out of phase with the forces on the structure. Therefore, this hybrid system effectively reduced the sliding displacement without appreciably increasing the base shear. Furthermore, the authors found an optimal viscous damping value for which the base isolation is most effective. The researcher Providakis [118] extended this study to seek the effectiveness of this hybrid base isolation system on multi-story buildings. The author studied the effect of the VFD damping coefficient on the response of the building subjected to earthquakes. The study indicated that an optimal value for the damping coefficient exists and any higher value than this has a detrimental effect on inter-story drifts and floor accelerations. As indicated by the researchers, the base isolation system scheme, which uses FPS and VFDs, is quite effective in controlling the structural response. However, the system cannot adapt to various earthquake frequencies due to a well-known issue of FPS, which may show resonant behavior in near-fault excitations. Krishnamoorthy [119,120] proposed a passive combination of VFD with VRFPS for the seismic isolation. Due to the exponential sliding geometry of the VRFPS, the system avoids the resonance issue. Moreover, in general, the sliding displacement of a VRFPS is slightly more than FPS, but the usage of VFD reduces this significantly. With this combination, the researcher was able to avoid the resonant response of the structure as well as control the sliding displacement. The previous researchers mainly used linear viscous dampers in their study; however, studies on the effect of non-linear dampers are limited. Wolff et al. [121] studied the effect of a non-linear viscous damper on the friction pendulum isolated structure. The authors concluded that linear VFDs are more beneficial than non-linear VFDs for reducing sliding displacements. Some studies conducted by authors Zhou and Chen [122] indicated that the dampers used in conjunction with isolators are more effective in strong seismic events than small magnitude earthquakes. Additional detailed modeling techniques for these base isolation systems are discussed by some researchers [123].

7. Modeling Techniques of Base-Isolated Building Structures

Modeling a base-isolated structure is rather difficult than conventional structures as it involves modeling a sliding joint. The process of obtaining a solution for governing equations is fairly complex due to the discontinuity between the foundation and the structure. One of the earliest known models capable of modeling base isolation is a coulomb friction model proposed by Hartog [124]. The author provided a solution for a single-degree of freedom system (SDOF) (Figure 9) subjected to excitation from an external source.

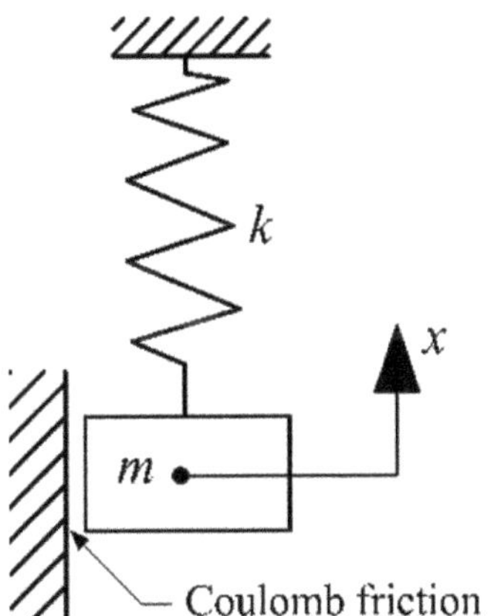

Figure 9. Coulomb friction model of a SDOF.

An improved two-degree of freedom model with the coulomb friction and viscous damping subjected to harmonic excitation was also proposed [125]. A similar modeling technique was later used to study a masonry building resting on a sliding joint [25,26]. The researchers proposed a mathematical model to represent the sliding friction effect and further verified it by conducting tests on experimental models of masonry buildings. The mathematical model is a two-degree freedom (two-DOF) model consisting of a top (m_1) and base mass (m_2), as shown in Figure 10.

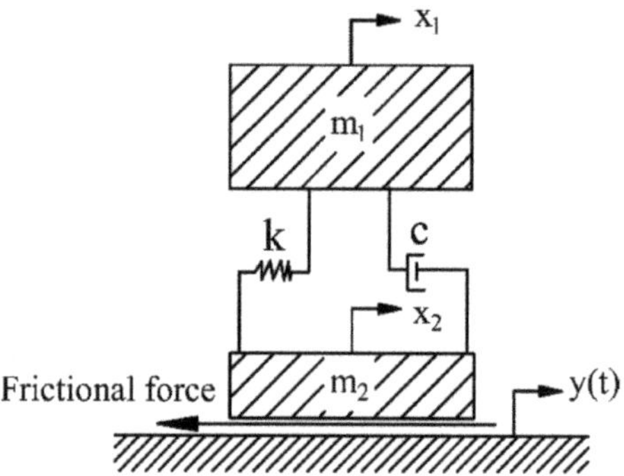

Figure 10. Mathematical model of a two-DOF sliding system.

An isolated structure on a sliding joint passes through sliding and non-sliding phases in a short time, and mathematically modeling these phases is quite a challenge. Some authors modeled a basic two-DOF oscillator system subjected to harmonic excitation and provided solutions by considering sliding and non-sliding phases [28,29]. Modeling of sliding and static phases is more complex for seismic excitations than harmonic excitations as these phases alter within a short time. To deal with this complex problem, Mostaghel and Tanbakuchi [31] proposed a mathematical model where they developed the conditions for the beginning and end of the sliding phases. The authors used the end condition of a sliding phase as the initial condition for the next phase. Although the entire system is non-linear, it acts linearly during the sliding and non-sliding phases, enabling the authors to provide solutions to the non-linear system using linear equations. The approaches mentioned to date apply to small structures (only two-DOF) and would be computationally intensive if modified to suit larger structures. One of the earlier attempts to address these limitations is made by Yang et al. [32]. Earlier researchers made use of two-DOF with base mass at the plinth level; however, the authors here considered masses of stories at floor levels. The authors proposed a multi-degree freedom system (MDOF) with a fictitious spring (K) at the base, as shown in Figure 11a. In the non-sliding phase, the fictitious spring's stiffness is very high, whereas, in the sliding phase, the stiffness considered is zero (Figure 11b).

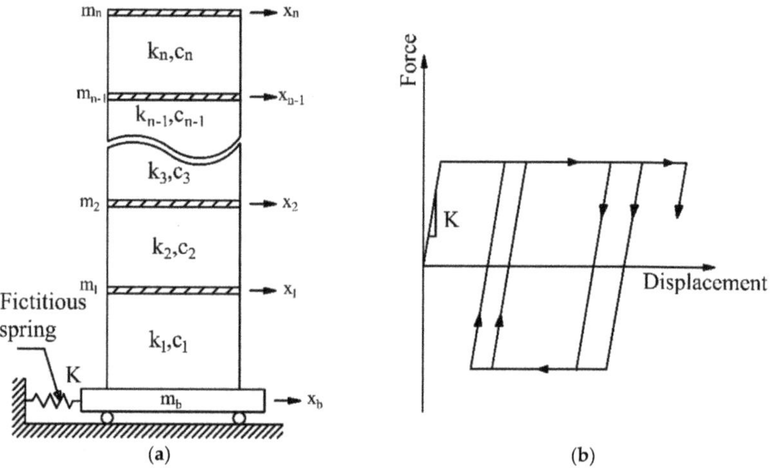

Figure 11. Structure considered by Yang et al. [32]. (**a**) MDOF system with fictitious spring, (**b**) force-displacement bbehavior of the spring.

Due to this unique technique, the solution for the sliding structure can utilize numerical methods, which are well-suited for computers. Although this method is simple, a very small velocity is expected in the non-sliding phase, which can produce errors in the solution. Vafai et al. [33] proposed a modified method using a rigid link, as shown in Figure 12a, to overcome this issue. This link assumes a value of zero stiffness during the sliding phase and infinity during the non-sliding phase (Figure 12b). Therefore, this method solves the issue of unpredicted oscillations during the transition phase from sliding to non-sliding. The method is computationally efficient and requires less time for the solutions.

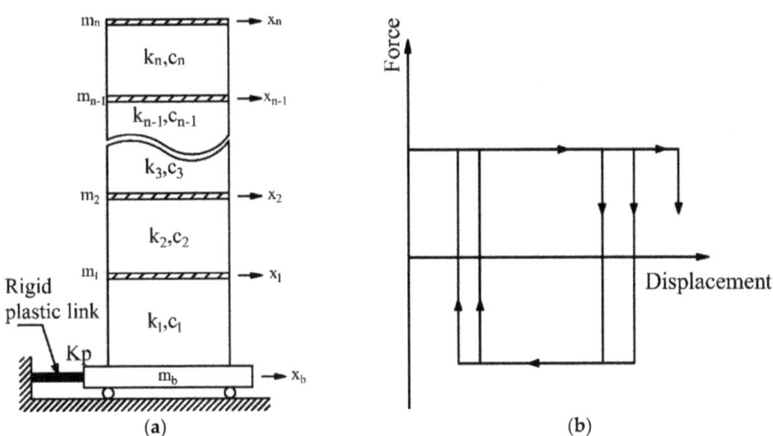

Figure 12. Structure considered by Vafai et al. [33]. (**a**) MDOF system with fictitious spring, (**b**) force-displacemen behavior of rigid plastic link.

In all these models, researchers considered only a horizontal degree of freedom at each story level, and this idealization may affect the reliability of the models proposed. Therefore, Krishnamoorthy and Saumil [34] proposed an MDOF structure resting on a P-F sliding surface considering all six degrees of freedom. The authors used the fictitious spring model proposed by Yang et al. [32] to simulate the sliding and non-sliding phases. Most of the approaches are based on relative displacements of the masses. Moreover, Krishnamoorthy [35] proposed a model based on the absolute displacement approach. This approach reduces the computational effort significantly as the matrices related to the stiffness and damping do not change when the structure undergoes sliding and non-sliding phases. Furthermore, the author studied the effects of providing a restoring mechanism on the structure's response.

8. Other Trends in Sliding Isolation Systems

The standalone sliding devices or devices connected with supplemental viscous dampers are passive, i.e., they cannot adapt and change their behavior depending on the input frequency. The general issues with passive systems are: (i) Lack of adaptability—PF and FPS isolators work only for a limited type of earthquake frequency, (ii) large isolator displacements—common in PF and pendulum isolators with flatter surfaces, (iii) residual displacement—common with most of the sliding isolators, except for FPS. Furthermore, the occurrence of high magnitude earthquakes is rare, due to which isolators might not activate for a long time. Therefore, the coefficient of friction is likely to vary and may not achieve the desired level of isolation. Moreover, passive isolators need a certain threshold frequency to activate, i.e., for minor vibrations, the isolator might not come into effect at all, which could be detrimental to sensitive instruments within the structure. Therefore, to achieve the required degree of structural control, researchers started focusing on methods to instantly control the behavior of structures depending on the response of a structure subjected to an earthquake [126–129]. This active control system consists of controllers, sensors, and actuators to achieve the required level of seismic control. The development of these active control systems gained momentum with the advancement in computer technology, and many buildings in Japan incorporated this system [130]. To date, several review literatures are available on active control systems [131,132]. Although an active control system can adapt and control the response in real-time, researchers have observed several issues in the real-world implementation of the system, such as substantial initial and maintenance costs and demand for vast amounts of electricity. Moreover, the controller design is very complex, and the system requires large actuators to cope with seismic forces [133]. Furthermore, active systems need large electric power, which may not be available during an earthquake, rendering the entire system useless.

Given these major drawbacks, researchers have developed semi-active systems. In this system, actuators provide one part of the control, and passive systems provide the remaining part. As a result, the electricity demands reduce considerably and in case of a total power failure, the passive system can still provide a reasonable amount of control. A semi-active system typically consists of passive devices, such as P-F isolator, FPS, and rubber bearings used in conjunction with controllable devices, such as variable friction, variable stiffness, and variable damping devices. Several researchers developed systems capable of varying the friction coefficient of isolators depending on earthquake intensity [134,135]. The friction coefficient varied by adjusting fluid pressure in a chamber within the isolator. In an earthquake event, the computer senses the structure's response and sends appropriate signals to the control valve of the fluid intake chamber. Nagarajiah et al. [136] proposed a semi-active system comprising a passive sliding isolator and an active hydraulic actuator with a parallel re-centering spring. The researcher developed a control algorithm to give instant feedback to the actuator to control the high-frequency region. Therefore, the system can control the acceleration and sliding displacement better than a passive sliding isolator. For a lower magnitude earthquake, the controller reduces the friction coefficient, which helps the isolator system to activate even for minor vibrations. During a large earthquake, the coefficient of friction of the isolator increases, effectively mitigating the effect of major vibrations. The effectiveness of a semi-active system of sliding isolators with actuators for a bridge was studied by Yang et al. [137]. The study showed that the system could control the large sliding displacement. This study was further extended to multi-story building models [138] to indicate the robustness of the control technique. Dyke et al. [139] studied the effectiveness of using magneto-rheological (MR) dampers (Figure 13) in structural control.

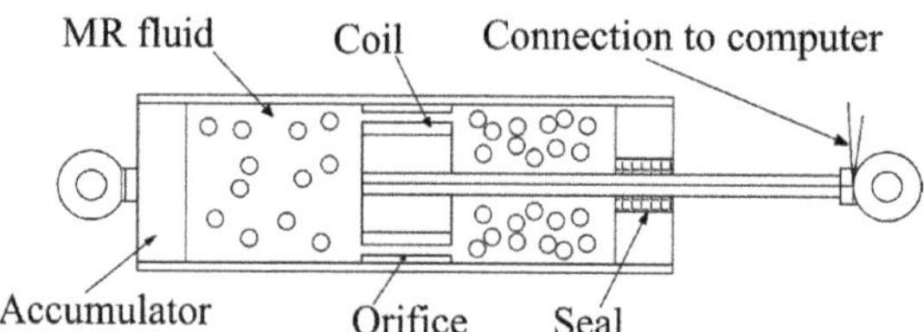

Figure 13. Schematics of an MR damper.

The damping coefficient of these dampers can be controlled instantly by passing an electric current through the MR liquid, which is contained inside the damper. Due to the availability of this control over the damping coefficient, these dampers can be effective for a wide range of excitations. Details of these dampers and the effectiveness of these dampers in controlling the behavior of the structure are available in [140,141]. Some researchers have further studied the effectiveness of using a combination of MR dampers with sliding isolation systems in reducing the earthquake response of bridges [142,143] and buildings [144].

Wongprasert and Symans [145] proposed an isolation system consisting of passive FPS and rubber bearing in conjunction with adaptable fluid dampers. The authors developed a control algorithm based on fuzzy logic for varying the fluid pressure within these dampers, which can alter the damping properties of the system depending on the excitation. A similar system, which is a combination of a curved surface isolator and a controllable fluid damper, was studied by Krishnamoorthy et al. [146]. The authors used a neural network algorithm based on a radial basis function for the dynamic control of the system. Moreover, researchers have developed devices whose stiffness can vary continuously during an earthquake to reduce the large displacements induced in bridges due to near-fault earthquakes. This system, when used along with passive sliding isolators, was found to be effective in reducing the displacement without imparting additional forces to the structure [147]. More recently, some researchers developed an FPS which can be tuned suitably depending on the type of excitation [148–150]. These isolators can change their friction coefficient depending on the seismic inputs. A fluid chamber within the isolator controls this variation of friction. Tunable FPS are found to be very effective in a variety of earthquake excitations due to their adaptability.

9. Challenges in the Worldwide Implementation of Isolation Schemes

Although base isolation techniques have become mature enough to be used in real-life structures, they are still not used in many countries. Especially in developing countries, such as India, the application of these technologies is difficult to achieve. To date, one hospital building in Bhuj, Gujrat, India, has been retrofitted with rubber bearings [151]. This reluctance could be mainly attributed to the perceived higher cost of these technologies [101]. Furthermore, the lack of understanding of the long-term benefits of isolation and complicated design code documents are other aspects which affect the implementation of isolation techniques [152]. Several researchers have shown that choosing an appropriate isolation system can reduce the life-cycle cost by 20% when compared with a rigid base structure [153,154]. When considering life-cycle costs, the major focus is on reducing the damage and downtime costs. It has been found that even for minor earthquakes providing seismic isolation could be economically viable [155]. Some researchers have shown that although the direct construction cost of the base-isolated structure is high, the post-earthquake loss cost would be about 82% to 84% lower compared with a fixed base structure [156]. Mayes et al. [157] suggested that the initial cost of the base isolation scheme mainly depends on the design forces prescribed by the codes and the location of the isolators (floor level and top of basement level). More conservative codes lead to higher force estimation and increase the cost. Furthermore, if the isolators are planned at the floor level, it may lead to higher costs due to the requirement of additional structural slabs. However, even with the high initial costs, cost savings can be achieved due to savings in non-structural component bracing. Moreover, these researchers pointed out that for the actual cost comparison, it is necessary to assess the conventional and isolated structures under the same performance criteria. In one example, where both the conventional and isolated structures are evaluated under the same performance level, the isolation cost was reduced by 6% [158]. A recent study by Egbelakin et al. [158] identified the hurdles in implementing base isolation strategies. Although the researchers discussed the context of the implementation scenario in New Zealand, most of the addressed issues are common in many developing countries. The researchers pointed out that the lack of awareness and market availability are the key reasons behind the reluctance to adopt base isolation schemes. Furthermore, the researchers suggest that the incorporation of base isolation schemes in the code of practice is the key to global acceptance of these technologies. To date, the discussion in this section has been on isolation systems in general rather than on specific isolators. A study conducted by Clemente and Buffarini [159] on buildings isolated with elastomer and sliding isolators shows a negligible cost difference between conventional and base-isolated buildings. Moreover, these researchers found that the base isolation scheme is less expensive than conventional buildings when the design is for higher seismic forces. Furthermore, a recent study on the life-cycle cost of FPS isolated structure indicates that the life-cycle cost of isolation is inversely proportional to the degree of isolation required [160].

These studies have indicated that the real-life cost of base isolation is not higher than the conventional buildings; therefore, competent code authorities and governments should address any reluctance to use these technologies.

10. Conclusions

This paper discusses the development in the field of base isolation in general and from a historical perspective. The focus has been on the development of sliding isolation systems. Various types of friction isolator systems based on their geometry have been discussed. Furthermore, various modeling techniques of sliding base-isolated structures have been reviewed. The various other developments in the area of active and semi-active isolation systems have been reviewed, as well. The following conclusions may be summarized based on the previous discussions:

(1)   Base isolation is one of the oldest techniques and has been proven effective over the centuries. However, this technique is still not widely practiced in many parts of the world [151].

(2)   Sliding isolation systems are being researched exhaustively due to their simplicity and cost-effectiveness. In fact, the cost of a curved surface isolator is lower than high-damping rubber bearings [161]. Although the sliding isolators are very effective in seismic isolation, some inherent issues, such as dependency of friction on velocity, heat, and contact pressure, need special attention during the design.

(3)   FP isolator is one of the simplest forms of isolator with a restoring capability. Although the isolator effectively controls the harmful effects of earthquakes, it is prone to pulse-like seismic excitation.

(4)   The resonance issue of the FP isolator lead the researchers to look for alternative isolator surfaces, which can provide the restoring capacity similar to the FP isolator and still avoid the resonance issue.

(5)   Various isolators, such as VFPI, VRFPS, CFPI, and VFPI have been developed, and detailed studies have proven that these isolators are capable of providing sufficient seismic isolation without inducing resonance.

(6)   Compared with FP isolators, these isolators have flatter geometry, which may lead to larger sliding displacement and higher residual displacements.

(7)   Several researchers have proposed multiple sliding surface isolators, such as double pendulum, triple pendulum, and quintuple bearings to accommodate large sliding displacement. Due to the multiple sliding surfaces, it is possible to achieve higher energy dissipation. Moreover, the hysteresis behavior of multiple surface isolators provides many control points for designers to achieve the desired isolation level.

(8)   Some researchers used supplemental systems, such as VFDs to control the large displacement of sliding isolation systems. These dampers are capable of reducing the sliding displacement without inducing higher forces on the structures. This unique feature of VFDs is attributed to its out-of-phase behavior during seismic excitations.

(9)   Modeling a sliding system is challenging as it involves obtaining the solutions for a structure which is disconnected at the base.

(10)   The solution technique for the isolated structure is further complicated due to sliding and non-sliding phases. Several researchers proposed different techniques to tackle this problem.

(11)   The standalone sliding isolators are incapable of adapting to various seismic excitations, which prompted researchers to propose active structural control systems.

(12)   Active control systems can respond to seismic excitations in real-time and thus are suitable for a wide range of frequencies. However, this system is very costly, and electricity demand is very high. Moreover, in the event of a major power failure, the entire system may be ineffective.

(13)   Several semi-active systems were proposed and tested to alleviate the drawbacks of an active system, such as a controllable friction system in conjunction with a sliding

isolator and MR dampers with PF isolators. These devices considerably reduce the cost without compromising the adaptability to wide frequency ranges.

(14) Although active and semi-active base isolation systems are quite effective in controlling the seismic behavior of the structure, due to their simplicity, the passive sliding isolation systems still appeal to the research community. Therefore, new varieties of sliding systems are still being developed and studied by researchers worldwide.

Author Contributions: Conceptualization, A.R.A., A.K. and K.K.; resources, M.C.; writing—original draft preparation, A.R.A.; writing—review and editing, A.R.A. and M.C.; visualization, A.R.A.; supervision, A.K. and K.K.; project administration, A.K. All authors have read and agreed to the published version of the manuscript.

Funding: The APC was funded by Manipal Academy of Higher Education, Manipal.

Data Availability Statement: No new data were created or analyzed in this study. Data sharing is not applicable to this article.

Conflicts of Interest: The authors declare no conflict of interest.

References

1. Symans, M.D.; Constantinou, M.C. Passive Fluid Viscous Damping Systems for Seismic Energy Dissipation. *J. Earthq. Technol.* **1998**, *35*, 185–206.
2. Chang, K.C.; Soong, T.T.; Oh, S.-T.; Lai, M.L. Seismic Behavior of Steel Frame with Added Viscoelastic Dampers. *J. Struct. Eng.* **1995**, *121*, 1418–1426. [CrossRef]
3. Skinner, R.I.; Kelly, J.M.; Heine, A.J. Hysteretic Dampers for Earthquake-Resistant Structures. *Earthq. Eng. Struct. Dyn.* **1974**, *3*, 287–296. [CrossRef]
4. Touaillon, J. Improvement in Buildings. US Patent No. 99. 973, 1870.
5. Calantarients, J.A. *Improvements in and Connected with Building and Other Works and Appurtenances to Resist the Action of Earthquakes and the Like*; Paper No. 325371; Stanford University: Stanford, CA, USA, 1909.
6. Fajfar, P. Ljubljana skyscraper. Concern for earthquake safety in the thirties. *Gradb. Vestn.* **1995**, *34*, 119–122. (In Slovene)
7. Bek, M.; Oseli, A.; Saprunov, I.; Zhumagulov, B.T.; Mian, S.M.; Gusev, B.V.; Žarnić, R.; von Bernstorff, B.; Holeček, N.; Emri, I. High Pressure Dissipative Granular Materials for Earthquake Protection of Houses. *Anal. Pazu* **2013**, *3*, 75–86. [CrossRef]
8. Kelly, J.M. Aseismic Base Isolation: Review and Bibliography. *Soil Dyn. Earthq. Eng.* **1986**, *5*, 202–216. [CrossRef]
9. Buckle, I.G.; Mayes, R.L. Seismic Isolation: History, Application, and Performance—A World View. *Earthq. Spectra* **1990**, *6*, 161–201. [CrossRef]
10. Malu, G.; Pranesh, M. Sliding Isolation Systems: State-of-the-Art Review. *IOSR J. Civ. Eng.* **2013**, *6*, 30–35.
11. Jangid, R.S.; Datta, T.K. Seismic Behaviour of Base-Isolated Buildings: A State-of-the-Art Review. *Proc. Inst. Civ. Eng.-Struct. Build.* **1995**, *110*, 186–203. [CrossRef]
12. Warn, G.P.; Ryan, K.L. A Review of Seismic Isolation for Buildings: Historical Development and Research Needs. *Buildings* **2012**, *2*, 300–325. [CrossRef]
13. Calvi, P.M.; Calvi, G.M. Historical Development of Friction-Based Seismic Isolation Systems. *Soil Dyn. Earthq. Eng.* **2018**, *106*, 14–30. [CrossRef]
14. Ismail, M. Seismic Isolation of Structures. Part I: Concept, Review and a Recent Development. *Hormig. Acero* **2018**, *69*, 147–161. [CrossRef]
15. De Luca, A.; Guidi, L.G. State of Art in the Worldwide Evolution of Base Isolation Design. *Soil Dyn. Earthq. Eng.* **2019**, *125*, 105722. [CrossRef]
16. Kelly, J.M. Vertical Flexibility in Isolation Systems. *Civ. Eng. Res. J.* **2018**, *4*, 555629. [CrossRef]
17. Li, L. Base Isolation Measure for Aseismic Buildings in China. In Proceedings of the 8th World Conference on Earthquake Engineering, San Francisco, CA, USA, 21–28 July 1984; Volume 6, pp. 791–798.
18. Azinović, B.; Kilar, V.; Koren, D. Energy-Efficient Solution for the Foundation of Passive Houses in Earthquake-Prone Regions. *Eng. Struct.* **2016**, *112*, 133–145. [CrossRef]
19. Kelly, J.M. *Earthquake-Resistant Design with Rubber*, 2nd ed.; Springer: Berlin/Heidelberg, Germany, 1997; ISBN 978-1-4471-0971-6.
20. Jangid, R.S. Stochastic Seismic Response of Structures Isolated by Rolling Rods. *Eng. Struct.* **2000**, *22*, 937–946. [CrossRef]
21. Zhou, Q.; Lu, X.; Wang, Q.; Feng, D.; Yao, Q. Dynamic Analysis on Structures Base-Isolated by a Ball System with Restoring Property. *Earthq. Eng. Struct. Dyn.* **1998**, *27*, 773–791. [CrossRef]
22. Yang, C.-Y.; Hsieh, C.-H.; Chung, L.-L.; Chen, H.-M.; Wu, L.-Y. Effectiveness of an Eccentric Rolling Isolation System with Friction Damping. *J. Vib. Control* **2012**, *18*, 2149–2163. [CrossRef]
23. Rawat, A.; Matsagar, V. Seismic Analysis of Liquid Storage Tank Using Oblate Spheroid Base Isolation System Based on Rolling Friction. *Int. J. Non-Linear Mech.* **2022**, *147*, 104186. [CrossRef]

24. Harvey, P.S.; Kelly, K.C. A Review of Rolling-Type Seismic Isolation: Historical Development and Future Directions. *Eng. Struct.* **2016**, *125*, 521–531. [CrossRef]
25. Qamaruddin, M. Development of Brick Building Systems for Improved Earthquake Performance. Ph.D. Thesis, University of Roorkee, Roorkee, India, 1978.
26. Arya, A.S.; Chandra, B.; Qamaruddin, M. A New Building System for Improved Earthquake Performance. In Proceedings of the Sixth Symposium on Earthquake Engineering, IIT Roorkee (Formerly, University of Roorkee), Roorkee, India, 5–7 October 1978; Volume 1, pp. 499–504.
27. Arya, A.S.; Qamaruddin, M.; Chandra, B. New System of Brick Buildings for Improved Behaviour during Earthquakes. In Proceedings of the Seventh World Conference on Earthquake Engineering, Istanbul, Turkey, 8–13 September 1980; pp. 225–232.
28. Mostaghel, N.; Hejazi, M.; Tanbakuchi, J. Response of Sliding Structures to Harmonic Support Motion. *Earthq. Eng. Struct. Dyn.* **1983**, *11*, 355–366. [CrossRef]
29. Westermo, B.; Udwadia, F. Periodic Response of a Sliding Oscillator System to Harmonic Excitation. *Earthq. Eng. Struct. Dyn.* **1983**, *11*, 135–146. [CrossRef]
30. Nikolić-Brzev, S. An Innovative Seismic Protection Scheme for Masonry Buildings. In Proceedings of the 10th International Conference on Brick/Block Masonry, Calgary, AB, Canada, 5–7 July 1994; pp. 273–282.
31. Mostaghel, N.; Tanbakuchi, J. Response of Sliding Structures to Earthquake Support Motion. *Earthq. Eng. Struct. Dyn.* **1983**, *11*, 729–748. [CrossRef]
32. Yang, Y.-B.; Lee, T.-Y.; Tsai, I.-C. Response of Multi-Degree-of-Freedom Structures with Sliding Supports. *Earthq. Eng. Struct. Dyn.* **1990**, *19*, 739–752. [CrossRef]
33. Vafai, A.; Hamidi, M.; Ahmadi, G. Numerical Modeling of MDOF Structures with Sliding Supports Using Rigid-Plastic Link. *Earthq. Eng. Struct. Dyn.* **2001**, *30*, 27–42. [CrossRef]
34. Krishnamoorthy, A.; Saumil, P. In-Plane Response of a Symmetric Space Frame with Sliding Upports. *Int. J. Appl. Sci. Eng.* **2005**, *3*, 1–11.
35. Krishnamoorthy, A. An Absolute Displacement Approach for Modeling of Sliding Structures. *Struct. Eng. Mech.* **2008**, *29*, 659–671. [CrossRef]
36. Zayas, V.A.; Low, S.S.; Mahin, S.A. A Simple Pendulum Technique for Achieving Seismic Isolation. *Earthq. Spectra* **1990**, *6*, 317–333. [CrossRef]
37. Mokha, A.; Constantinou, M.C.; Reinhorn, A.M.; Zayas, V.A. Experimental Study of Friction-pendulum Isolation System. *J. Struct. Eng.* **1991**, *117*, 1201–1217. [CrossRef]
38. Gilberto, M.; Whittaker, A.S.; Fenves, G.L.; Mahin, S.A. *Experimental and Analytical Studies of the Friction Pendulum System for the Seismic Protection of Simple Bridges*; Earthquake Engineering Research Center: Berkeley, CA, USA, 2004; p. 110.
39. Jangid, R.S. Optimum Friction Pendulum System for Near-Fault Motions. *Eng. Struct.* **2005**, *27*, 349–359. [CrossRef]
40. Castaldo, P.; Amendola, G. Optimal Sliding Friction Coefficients for Isolated Viaducts and Bridges: A Comparison Study. *Struct. Contr. Health Monit.* **2021**, *28*, e2838. [CrossRef]
41. Vibhute, A.S.; Bharati, S.D.; Shrimali, M.K.; Datta, T.K. Optimum Coefficient of Friction in FPS for Base Isolation of Building Frames. *Pract. Period. Struct. Des. Constr.* **2022**, *27*, 04022042. [CrossRef]
42. Tsai, C.S. Finite Element Formulations for Friction Pendulum Seismic Isolation Bearings. *Int. J. Numer. Methods Eng.* **1997**, *40*, 29–49. [CrossRef]
43. Ryan, K.L.; Chopra, A.K. Estimating the Seismic Displacement of Friction Pendulum Isolators Based on Non-Linear Response History Analysis. *Earthq. Eng. Struct. Dyn.* **2004**, *33*, 359–373. [CrossRef]
44. Mazza, F.; Sisinno, S. Nonlinear Dynamic Behavior of Base-Isolated Buildings with the Friction Pendulum System Subjected to near-Fault Earthquakes. *Mech. Based Des. Struct. Mach.* **2017**, *45*, 331–344. [CrossRef]
45. Wang, Y.-P.; Chung, L.-L.; Liao, W.-H. Seismic Response Analysis of Bridges Isolated with Friction Pendulum Bearings. *Earthq. Eng. Struct. Dyn.* **1998**, *27*, 1069–1093. [CrossRef]
46. Seleemah, A.A.; El-Sharkawy, M. Seismic Response of Base Isolated Liquid Storage Ground Tanks. *Ain Shams Eng. J.* **2011**, *2*, 33–42. [CrossRef]
47. Whittaker, A.S.; Kumar, M.; Kumar, M. Seismic Isolation of Nuclear Power Plants. *Nucl. Eng. Technol.* **2014**, *46*, 569–580. [CrossRef]
48. Almazán, J.L.; De La Llera, J.C.; Inaudi, J.A. Modelling Aspects of Structures Isolated with the Frictional Pendulum System. *Earthq. Eng. Struct. Dyn.* **1998**, *27*, 845–867. [CrossRef]
49. Monti, G.; Petrone, F. Analytical Thermo-Mechanics 3D Model of Friction Pendulum Bearings. *Earthq. Eng. Struct. Dyn.* **2016**, *45*, 957–977. [CrossRef]
50. Oliveto, N.D. Geometrically Nonlinear Analysis of Friction Pendulum Systems under Tri-Directional Excitation. *Eng. Struct.* **2022**, *269*, 114770. [CrossRef]
51. Hall, J.F.; Heaton, T.H.; Halling, M.W.; Wald, D.J. Near-Source Ground Motion and Its Effects on Flexible Buildings. *Earthq. Spectra* **1995**, *11*, 569–605. [CrossRef]
52. Pranesh, M.; Sinha, R. VFPI: An Isolation Device for Aseismic Design. *Earthq. Eng. Struct. Dyn.* **2000**, *29*, 603–627. [CrossRef]
53. Tsai, C.S.; Chiang, T.-C.; Chen, B.-J. Finite Element Formulations and Theoretical Study for Variable Curvature Friction Pendulum System. *Eng. Struct.* **2003**, *25*, 1719–1730. [CrossRef]

54. Lu, L.-Y.; Shih, M.-H.; Wu, C.-Y. Near-Fault Seismic Isolation Using Sliding Bearings with Variable Curvatures. In Proceedings of the 13th World Conference on Earthquake Engineering, Vancouver, BC, Canada, 1–6 August 2004.

55. Lu, L.-Y.; Wang, J.; Hsu, C.-C. Sliding Isolation Variable Frequency Bearings for near—Fault Ground Motions. In Proceedings of the 4th International Conference on Earthquake Engineering, Taipei, Taiwan, 12–13 October 2006.

56. Krishnamoorthy, A. Seismic Isolation of Bridges Using Variable Frequency and Variable Friction Pendulum Isolator System. *Struct. Eng. Int.* **2010**, *20*, 178–184. [CrossRef]

57. Malu, G.; Murnal, P. Variable Coefficient of Friction: An Effective VFPI Parameter to Control near-Fault Ground Motions. *ISET J. Earthq. Technol.* **2012**, *49*, 73–87.

58. Calvi, P.M.; Moratti, M.; Calvi, G.M. Seismic Isolation Devices Based on Sliding between Surfaces with Variable Friction Coefficient. *Earthq. Spectra* **2016**, *32*, 2291–2315. [CrossRef]

59. Brito, M.B.; Ishibashi, H.; Akiyama, M. Shaking Table Tests of a Reinforced Concrete Bridge Pier with a Low-cost Sliding Pendulum System. *Earthq. Eng. Struct. Dyn.* **2019**, *48*, 366–386. [CrossRef]

60. Tsai, C.S.; Chiang, T.-C.; Chen, B.-J. Experimental Study for Multiple Friction Pendulum System. In Proceedings of the 13th World Conference on Earthquake Engineering, Vancouver, BC, Canada, 1–6 August 2004.

61. Fenz, D.M.; Constantinou, M.C. Behaviour of the Double Concave Friction Pendulum Bearing. *Earthq. Eng. Struct. Dyn.* **2006**, *35*, 1403–1424. [CrossRef]

62. Fenz, D.M.; Constantinou, M.C. Modeling Triple Friction Pendulum Bearings for Response-History Analysis. *Earthq. Spectra* **2008**, *24*, 1011–1028. [CrossRef]

63. Fenz, D.M.; Constantinou, M.C. Spherical Sliding Isolation Bearings with Adaptive Behavior: Theory. *Earthq. Eng. Struct. Dyn.* **2008**, *37*, 163–183. [CrossRef]

64. Fenz, D.M.; Constantinou, M.C. Spherical Sliding Isolation Bearings with Adaptive Behavior: Experimental Verification. *Earthq. Eng. Struct. Dyn.* **2008**, *37*, 185–205. [CrossRef]

65. Becker, T.C.; Mahin, S.A. Experimental and Analytical Study of the Bi-Directional Behavior of the Triple Friction Pendulum Isolator. *Earthq. Eng. Struct. Dyn.* **2012**, *41*, 355–373. [CrossRef]

66. Sarlis, A.A.; Constantinou, M.C. A Model of Triple Friction Pendulum Bearing for General Geometric and Frictional Parameters: Revised Triple Friction Pendulum Bearing Model. *Earthq. Eng. Struct. Dyn.* **2016**, *45*, 1837–1853. [CrossRef]

67. Morgan, T.A.; Mahin, S.A. Achieving Reliable Seismic Performance Enhancement Using Multi-Stage Friction Pendulum Isolators. *Earthq. Eng. Struct. Dyn.* **2010**, *39*, 1443–1461. [CrossRef]

68. Becker, T.C.; Bao, Y.; Mahin, S.A. Extreme Behavior in a Triple Friction Pendulum Isolated Frame. *Earthq. Eng. Struct. Dyn.* **2017**, *46*, 2683–2698. [CrossRef]

69. Lee, D.; Constantinou, M.C. Quintuple Friction Pendulum Isolator: Behavior, Modeling, and Validation. *Earthq. Spectra* **2016**, *32*, 1607–1626. [CrossRef]

70. Keikha, H.; Amiri, G.G. Seismic Performance Assessment of Quintuple Friction Pendulum Isolator with a Focus on Frictional Behavior Impressionability from Velocity and Temperature. *J. Earthq. Eng.* **2019**, *25*, 1256–1286. [CrossRef]

71. Sodha, A.; Vasanwala, S.A.; Soni, D. Probabilistic Evaluation of Seismically Isolated Building Using Quintuple Friction Pendulum Isolator. In *Advances in Intelligent Systems and Computing*; Springer: Singapore, 2019; pp. 149–159. ISBN 9789811319655.

72. Marin-Artieda Claudia, C.; Whittaker Andrew, S. Theoretical Studies of the XY-FP Seismic Isolation Bearing for Bridges. *J. Bridge Eng.* **2010**, *15*, 631–638. [CrossRef]

73. Yang, C.-Y.; Wang, S.-J.; Lin, C.-K.; Chung, L.-L.; Liou, M.-C. Analytical and Experimental Study on Sloped Sliding-Type Bearings. *Struct. Control Health Monit.* **2021**, *28*, e2828. [CrossRef]

74. Tabor, D. Friction—The Present State of Our Understanding. *J. Lubr. Technol.* **1981**, *103*, 169–179. [CrossRef]

75. Constantinou, M.; Tsopelas, P.; Kasalanati, A.; Wolff, E. *Property Modification Factors for Seismic Isolation Bearings*; MCEER-99-0012; State University of New York at Buffalo (NY): Buffalo, NY, USA, 1999.

76. Constantinou, M.C.; Caccese, J.; Harris, H.G. Frictional Characteristics of Teflon–Steel Interfaces under Dynamic Conditions. *Earthq. Eng. Struct. Dyn.* **1987**, *15*, 751–759. [CrossRef]

77. Mokha, A.; Constantinou, M.; Reinhorn, A. Teflon Bearings in Base Isolation I: Testing. *J. Struct. Eng.* **1990**, *116*, 438–454. [CrossRef]

78. Mokha, A.; Constantinou, M.; Reinhorn, A. Teflon Bearings in Base Isolation II: Modeling. *J. Struct. Eng.* **1990**, *116*, 455–474. [CrossRef]

79. Flom, D.G.; Porile, N.T. Friction of Teflon Sliding on Teflon. *J. Appl. Phys.* **1955**, *26*, 1088–1092. [CrossRef]

80. Makinson, K.R.; Tabor, D. The Friction and Transfer of Polytetrafluoroethylene. *Proc. R. Soc. London. Ser. A. Math. Phys. Sci.* **1964**, *281*, 49–61. [CrossRef]

81. Castaldo, P.; Tubaldi, E. Influence of FPS Bearing Properties on the Seismic Performance of Base-Isolated Structures. *Earthq. Eng. Struct. Dyn.* **2015**, *44*, 2817–2836. [CrossRef]

82. Mokha, A.; Constantinou, M.C.; Reinhorn, A.M. *Teflon Bearings in Aseismic Base Isolation: Experimental Studies and Mathematical Modeling*; NCEER-88-0038; State University of New York at Buffalo (NY): Buffalo, NY, USA, 1988.

83. Computer and Structures Inc. *CSI Analysis Reference Manual for SAP2000*; Computer and Structures Inc.: Berkley, CA, USA, 2010.

84. Bondonet, G.; Filiatrault, A. Frictional Response of PTFE Sliding Bearings at High Frequencies. *J. Bridge Eng.* **1997**, *2*, 139–148. [CrossRef]

85. Mosqueda, G.; Whittaker Andrew, S.; Fenves Gregory, L. Characterization and Modeling of Friction Pendulum Bearings Subjected to Multiple Components of Excitation. *J. Struct. Eng.* **2004**, *130*, 433–442. [CrossRef]

86. Constantinou, M.C.; Whittaker, A.S.; Kalpakidis, Y.; Fenz, D.M.; Warn, G.P. *Performance of Seismic Isolation Hardware under Service and Seismic Loading*; MCEER-07-0012; State University of New York at Buffalo (NY): Buffalo, NY, USA, 2007.

87. Dolce, M.; Cardone, D.; Croatto, F. Frictional Behavior of Steel-PTFE Interfaces for Seismic Isolation. *Bull. Earthq. Eng.* **2005**, *3*, 75–99. [CrossRef]

88. Quaglini, V.; Bocciarelli, M.; Gandelli, E.; Dubini, P. Numerical Assessment of Frictional Heating in Sliding Bearings for Seismic Isolation. *J. Earthq. Eng.* **2014**, *18*, 1198–1216. [CrossRef]

89. Gandelli, E. Advanced Tools for the Design of Sliding Isolation Systems for Seismic-Retrofitting of Hospitals. Ph.D. Thesis, Politecnico di Milano, Milan, Italy, 2017.

90. De Domenico, D.; Ricciardi, G.; Infanti, S.; Benzoni, G. Frictional Heating in Double Curved Surface Sliders and Its Effects on the Hysteretic Behavior: An Experimental Study. *Front. Built Environ.* **2019**, *5*, 74. [CrossRef]

91. Kumar, M.; Whittaker, A.S.; Constantinou, M.C. Characterizing Friction in Sliding Isolation Bearings. *Earthq. Eng. Struct. Dyn.* **2015**, *44*, 1409–1425. [CrossRef]

92. Barone, S.; Calvi, G.M.; Pavese, A. Experimental Dynamic Response of Spherical Friction-Based Isolation Devices. *J. Earthq. Eng.* **2019**, *23*, 1465–1484. [CrossRef]

93. Buckle, I.G.; Constantinou, M.; Dicleli, M.; Ghasemi, H. *Seismic Isolation of Highway Bridges*; MCEER-06-SP07; State University of New York at Buffalo (NY): Buffalo, NY, USA, 2006.

94. Constantinou, M.C.; Kalpakidis, Y.; Filiatrault, A.; Ecker Lay, R.A. *LRFD-Based Analysis and Design Procedures for Bridge Bearings and Seismic Isolators*; MCEER-11-0004; State University of New York at Buffalo (NY): Buffalo, NY, USA, 2011.

95. Buckle, I.; Al-Ani, M.; Monzon, E. *Seismic Isolation Design Examples of Highway Bridges*; NCHRP 20-7/Task 262 (M2); University of Nevada Reno: Reno, NV, USA, 2011.

96. American Association of State Highway and Transportation Officials. *Guide Specifications for Seismic Isolation Design*; American Association of State Highway and Transportation Officials: Washington, DC, USA, 1999.

97. American Society of Civil Engineers. *Minimum Design Loads and Associated Criteria for Buildings and other Structures*; ASCE/SEI 7-2016; American Society of Civil Engineers: Reston, VA, USA, 2016.

98. *CEN 1337-1*; Structural Bearings—Part 1: General Design Rules. European Committee for Standardization: Bruxelles, Belgium, 2000.

99. *CEN 15129*; Anti-Seismic Devices. European Committee for Standardization: Bruxelles, Belgium, 2009.

100. BSLEO. *Building Standard Law*; [2000/2016]; Building Center of Japan, Chiyoda-ku: Tokyo, Japan, 2016.

101. *Guideline for the Design of Seismic Isolation Systems for Buildings—Draft for Trial Use*; New Zealand Society for Earthquake Engineering Inc.: Wellington, New Zealand, 2019.

102. Sarlis, A.A.; Constantinou, M.C. *Model of Triple Friction Pendulum Bearing for General Geometric and Frictional Parameters and for Uplift Conditions*; MCEER-13-0010; State University of New York at Buffalo (NY): Buffalo, NY, USA, 2013.

103. Bao, Y.; Becker, T.C.; Hamaguchi, H. Failure of Double Friction Pendulum Bearings under Pulse-Type Motions. *Earthq. Eng. Struct. Dyn.* **2017**, *46*, 715–732. [CrossRef]

104. Nagarajaiah, S.; Reinhorn, A.M.; Constantinou, M.C. Experimental Study of Sliding Isolated Structures with Uplift Restraint. *J. Struct. Eng.* **1992**, *118*, 1666–1682. [CrossRef]

105. Xiong, W.; Zhang, S.-J.; Jiang, L.-Z.; Li, Y.-Z. Introduction of the Convex Friction System (CFS) for Seismic Isolation. *Struct. Control Health Monit.* **2017**, *24*, e1861. [CrossRef]

106. Xiong, W.; Zhang, S.-J.; Jiang, L.-Z.; Li, Y.-Z. The Multangular-Pyramid Concave Friction System (MPCFS) for Seismic Isolation: A Preliminary Numerical Study. *Eng. Struct.* **2018**, *160*, 383–394. [CrossRef]

107. Roussis, P.C.; Constantinou, M.C. Uplift-Restraining Friction Pendulum Seismic Isolation System. *Earthq. Eng. Struct. Dyn.* **2006**, *35*, 577–593. [CrossRef]

108. Guéraud, R.; Noël-Leroux, J.-P.; Livolant, M.; Michalopoulos, A.P. Seismic Isolation Using Sliding-Elastomer Bearing Pads. *Nucl. Eng. Des.* **1985**, *84*, 363–377. [CrossRef]

109. Mostaghel, N.; Khodaverdian, M. Dynamics of Resilient-Friction Base Isolator (R-FBI). *Earthq. Eng. Struct. Dyn.* **1987**, *15*, 379–390. [CrossRef]

110. Kelly, J.M.; Chalhoub, M.S. *Earthquake Simulator Testing of a Combined Sliding Bearing and Rubber Bearing Isolation System*; Earthquake Engineering Research Center, College of Engineering, University of California: Berkeley, CA, USA, 1990.

111. Constantinou, M.C.; Mokha, A.S.; Reinhorn, A.M. Study of Sliding Bearing and Helical-steel-spring Isolation System. *J. Struct. Eng.* **1991**, *117*, 1257–1275. [CrossRef]

112. Wei, X.; Li-Zhong, J.; Zhi-Hui, Z.; Yao-Zhuang, L. Introduction of Flat-Spring Friction System for Seismic Isolation. *Soil Dyn. Earthq. Eng.* **2021**, *145*, 106649. [CrossRef]

113. Chakraborty, S.; Roy, K.; Ray-Chaudhuri, S. Design of Re-Centering Spring for Flat Sliding Base Isolation System: Theory and a Numerical Study. *Eng. Struct.* **2016**, *126*, 66–77. [CrossRef]

114. Tsopelas, P.; Constantinou, M.C.; Okamoto, S.; Fujii, S.; Ozaki, D. Experimental Study of Bridge Seismic Sliding Isolation Systems. *Eng. Struct.* **1996**, *18*, 301–310. [CrossRef]

115. Makris, N.; Chang, S.-P. Effect of Viscous, Viscoplastic and Friction Damping on the Response of Seismic Isolated Structures. *Earthq. Eng. Struct. Dyn.* **2000**, *29*, 85–107. [CrossRef]

116. Chang, S.-P.; Makris, N.; Whittaker, A.S.; Thompson, A.C.T. Experimental and Analytical Studies on the Performance of Hybrid Isolation Systems. *Earthq. Eng. Struct. Dyn.* **2002**, *31*, 421–443. [CrossRef]
117. Soneji, B.B.; Jangid, R.S. Passive Hybrid Systems for Earthquake Protection of Cable-Stayed Bridge. *Eng. Struct.* **2007**, *29*, 57–70. [CrossRef]
118. Providakis, C.P. Effect of Supplemental Damping on LRB and FPS Seismic Isolators under Near-Fault Ground Motions. *Soil Dyn. Earthq. Eng.* **2009**, *29*, 80–90. [CrossRef]
119. Krishnamoorthy, A. Variable Curvature Pendulum Isolator and Viscous Fluid Damper for Seismic Isolation of Structures. *J. Vib. Control* **2011**, *17*, 1779–1790. [CrossRef]
120. Krishnamoorthy, A. Seismic Control of Continuous Bridges Using Variable Radius Friction Pendulum Systems and Viscous Fluid Dampers. *Int. J. Acoust. Vib.* **2015**, *20*, 24–35. [CrossRef]
121. Wolff, E.D.; Ipek, C.; Constantinou, M.C.; Tapan, M. Effect of Viscous Damping Devices on the Response of Seismically Isolated Structures. *Earthq. Eng. Struct. Dyn.* **2014**, *44*, 185–198. [CrossRef]
122. Zhou, Y.; Chen, P. Shaking Table Tests and Numerical Studies on the Effect of Viscous Dampers on an Isolated RC Building by Friction Pendulum Bearings. *Soil Dyn. Earthq. Eng.* **2017**, *100*, 330–344. [CrossRef]
123. Chen, X.; Xiong, J. Seismic Resilient Design with Base Isolation Device Using Friction Pendulum Bearing and Viscous Damper. *Soil Dyn. Earthq. Eng.* **2022**, *153*, 107073. [CrossRef]
124. Hartog, J.P.D. Forced Vibrations with Combined Viscous and Coulomb Damping. *Lond. Edinb. Dublin Philos. Mag. J. Sci.* **1930**, *9*, 801–817. [CrossRef]
125. Yeh, G.C.K. Forced Vibrations of a Two-degree-of-freedom System with Combined Coulomb and Viscous Damping. *J. Acoust. Soc. Am.* **1966**, *39*, 14–24. [CrossRef]
126. Soong, T.T.; Manolis, G.D. Active Structures. *J. Struct. Eng.* **1987**, *113*, 2290–2302. [CrossRef]
127. Cha, J.Z.; Pitarresi, J.M.; Soong, T.T. Optimal Design Procedures for Active Structures. *J. Struct. Eng.* **1988**, *114*, 2710–2723. [CrossRef]
128. Inaudi, J.; López-Almansa, F.; Kelly, J.M.; Rodellar, J. Predictive Control of Base-Isolated Structures. *Earthq. Eng. Struct. Dyn.* **1992**, *21*, 471–482. [CrossRef]
129. Chang, C.-M.; Spencer, B.F., Jr. Active Base Isolation of Buildings Subjected to Seismic Excitations. *Earthq. Eng. Struct. Dyn.* **2010**, *39*, 1493–1512. [CrossRef]
130. Soong, T.T.; Reinhorn, A.M.; Aizawa, S.; Higashino, M. Recent Structural Applications of Active Control Technology. *J. Struct. Control* **1994**, *1*, 1–21. [CrossRef]
131. Soong, T.T.; Masri, S.F.; Housner, G.W. An Overview of Active Structural Control under Seismic Loads. *Earthq. Spectra* **1991**, *7*, 483–505. [CrossRef]
132. Datta, T.K. A State-of-the-Art Review on Active Control of Structures. *ISET J. Earthq. Technol.* **2003**, *40*, 1–17.
133. Soong, T.T.; Spencer, B.F. Active, Semi-Active and Hybrid Control of Structures. *Bull. N. Z. Soc. Earthq. Eng.* **2000**, *33*, 387–402. [CrossRef]
134. Feng, M.Q.; Shinozuka, M.; Fujii, S. Friction-controllable Sliding Isolation System. *J. Eng. Mech.* **1993**, *119*, 1845–1864. [CrossRef]
135. Nagarajaiah, S.; Feng, M.Q.; Shinozuka, M. Control of Structures with Friction Controllable Sliding Isolation Bearings. *Soil Dyn. Earthq. Eng.* **1993**, *12*, 103–112. [CrossRef]
136. Nagarajaiah, S.; Riley, M.A.; Reinhorn, A. Control of Sliding-isolated Bridge with Absolute Acceleration Feedback. *J. Eng. Mech.* **1993**, *119*, 2317–2332. [CrossRef]
137. Yang, J.N.; Wu, J.C.; Kawashima, K.; Unjoh, S. Hybrid Control of Seismic-Excited Bridge Structures. *Earthq. Eng. Struct. Dyn.* **1995**, *24*, 1437–1451. [CrossRef]
138. Yang, J.N.; Wu, J.C.; Reinhorn, A.M.; Riley, M. Control of Sliding-Isolated Buildings Using Sliding-Mode Control. *J. Struct. Eng.* **1996**, *122*, 179–186. [CrossRef]
139. Dyke, S.J.; Spencer, B.F.; Sain, M.K.; Carlson, J.D. Seismic Response Reduction Using Magnetorheological Dampers. *IFAC Proc. Vol.* **1996**, *29*, 5530–5535. [CrossRef]
140. Dyke, S.J.; Spencer, B.F.; Sain, M.K.; Carlson, J.D. Modeling and Control of Magnetorheological Dampers for Seismic Response Reduction. *Smart Mater. Struct.* **1996**, *5*, 565–575. [CrossRef]
141. Spencer, B.F.; Dyke, S.J.; Sain, M.K.; Carlson, J.D. Phenomenological Model for Magnetorheological Dampers. *J. Eng. Mech.* **1997**, *123*, 230–238. [CrossRef]
142. Nagarajaiah, S.; Sahasrabudhe, S.; Iyer, R. Earthquake Protection of Bridges Using Sliding Isolation System and MR Dampers. In *Structural Control for Civil and Infrastructure Engineering*; World Scientific: Singapore, 2001; pp. 375–383.
143. Sahasrabudhe, S.S.; Nagarajaiah, S. Semi-Active Control of Sliding Isolated Bridges Using MR Dampers: An Experimental and Numerical Study. *Earthq. Eng. Struct. Dyn.* **2005**, *34*, 965–983. [CrossRef]
144. Kim, H.-S.; Roschke, P.N. Design of Fuzzy Logic Controller for Smart Base Isolation System Using Genetic Algorithm. *Eng. Struct.* **2006**, *28*, 84–96. [CrossRef]
145. Wongprasert, N.; Symans, M.D. Numerical Evaluation of Adaptive Base-Isolated Structures Subjected to Earthquake Ground Motions. *J. Eng. Mech.* **2005**, *131*, 109–119. [CrossRef]
146. Krishnamoorthy, A.; Bhat, S.; Bhasari, D. Radial Basis Function Neural Network Algorithm for Semi-active Control of Base-isolated Structures. *Struct Control Health Monit.* **2017**, *24*, e1984. [CrossRef]

147. Nagarajaiah, S.; Sahasrabudhe, S. Seismic Response Control of Smart Sliding Isolated Buildings Using Variable Stiffness Systems: An Experimental and Numerical Study. *Earthq. Eng. Struct. Dyn.* **2006**, *35*, 177–197. [CrossRef]
148. Peng, P.; Dongbin, Z.; Yi, Z.; Yachun, T.; Xin, N. Development of a Tunable Friction Pendulum System for Semi-Active Control of Building Structures under Earthquake Ground Motions. *Earthq. Eng. Struct. Dyn.* **2018**, *47*, 1706–1721. [CrossRef]
149. Zhang, D.; Pan, P.; Zeng, Y.; Guo, Y. A Novel Robust Optimum Control Algorithm and Its Application to Semi-Active Controlled Base-Isolated Structures. *Bull. Earthq. Eng.* **2020**, *18*, 2431–2460. [CrossRef]
150. Zeng, Y.; Pan, P.; Guo, Y. Development of Distributed Tunable Friction Pendulum System (DTFPS) for Semi-Active Control of Base-Isolated Buildings. *Bull. Earthq. Eng.* **2021**, *19*, 6243–6268. [CrossRef]
151. Rai, D.C.; Prasad, A.M.; Jain, S.K.; Rao, G.; Patel, P. Hospitals and Schools. *Earthq. Spectra* **2002**, *18*, 265–277. [CrossRef]
152. Mayes, R.L.; Brown, A.G.; Pietra, D. Using Seismic Isolation and Energy Dissipation to Create Earthquake-Resilient Buildings. *BNZSEE* **2012**, *45*, 117–122. [CrossRef]
153. Taflanidis, A.A.; Beck, J.L. Life-Cycle Cost Optimal Design of Passive Dissipative Devices. *Struct. Saf.* **2009**, *31*, 508–522. [CrossRef]
154. Goda, K.; Lee, C.S.; Hong, H.P. Lifecycle Cost–Benefit Analysis of Isolated Buildings. *Struct. Saf.* **2010**, *32*, 52–63. [CrossRef]
155. Kilar, V.; Petrovčič, S.; Koren, D.; Šilih, S. Cost Viability of a Base Isolation System for the Seismic Protection of a Steel High-Rack Structure. *Int. J. Steel Struct.* **2013**, *13*, 253–263. [CrossRef]
156. Jiang, S.; Yao, S.; Liu, D. Economic Performance Analysis of Seismic Isolation, Energy Dissipation, and Traditional Seismic Structures. *E3S Web Conf.* **2021**, *248*, 01032. [CrossRef]
157. Mayes, R.L.; Jones, L.R.; Kelly, T.E. The Economics of Seismic Isolation in Buildings. *Earthq. Spectra* **1990**, *6*, 245–263. [CrossRef]
158. Egbelakin, T.; Ogunmakinde, O.E.; Omotayo, T.; Sojobi, A. Demystifying the Barriers and Motivators for the Adoption of Base Isolation Systems in New Zealand. *Buildings* **2022**, *12*, 522. [CrossRef]
159. Clemente, P.; Buffarini, G. Base Isolation: Design and Optimization Criteria. *J. Anti-Seism. Syst. Int. Soc.* **2010**, *1*, 17–40. [CrossRef]
160. Castaldo, P.; Palazzo, B.; Della Vecchia, P. Life-Cycle Cost and Seismic Reliability Analysis of 3D Systems Equipped with FPS for Different Isolation Degrees. *Eng. Struct.* **2016**, *125*, 349–363. [CrossRef]
161. Saitta, F.; Clemente, P.; Buffarini, G.; Bongiovanni, G.; Salvatori, A.; Grossi, C. Base Isolation of Buildings with Curved Surface Sliders: Basic Design Criteria and Critical Issues. *Adv. Civ. Eng.* **2018**, *2018*, 1569683. [CrossRef]

MDPI

St. Alban-Anlage 66

4052 Basel

Switzerland

www.mdpi.com

Buildings Editorial Office

E-mail: buildings@mdpi.com

www.mdpi.com/journal/buildings